Apoptosis, Cell Signaling, and Human Diseases

Apoptosis, Cell Signaling, and Human Diseases

Molecular Mechanisms, Volume 2

Edited by

Rakesh Srivastava

Department of Biochemistry
University of Texas Health Center at Tyler
Tyler, TX

HUMANA PRESS  TOTOWA, NEW JERSEY

Library of Congress Cataloging in Publication Data
Apoptosis, cell signaling, and human diseases : molecular mechanisms / edited by Rakesh Srivastava.
            p. ; cm.
      Includes bibliographical references and index.
      ISBN 1-58829-677-6 (v. 1 : alk. paper) — ISBN 1-58829-882-5 (v. 2 : alk. paper)
      1. Apoptosis. 2. Pathology, Molecular. 3. Cellular signal transduction. I. Srivastava, Rakesh, 1956-
      [DNLM: 1. Apoptosis—physiology. 2. Cell Transformation, Neoplastic—genetics. 3. Gene Therapy. 4. Signal Transduction. QU 375 A64473 2007]
      QH671.A6547 2007
      571.9'36—dc22   2006030613

Preface

The aim of *Apoptosis, Cell Signaling, and Human Diseases: Molecular Mechanisms* is to present recent developments in cell survival and apoptotic pathways and their involvement in human diseases, such as cancers and neurodegenerative disorders. This requires an integration of knowledge from several fields of research, including pathology, genetics, virology, cell biology, medicine, immunology, and molecular biology. This edition of the book examines the impact of molecular biology on disease mechanisms. With recent advances in technology such as microarray and proteomics, new biomarkers and molecular targets have been identified. These potential targets will be very useful for the development of novel and more effective drugs for the treatment of human diseases. The challenge now is not only to understand disease mechanisms but also to apply this knowledge to find therapies that are more effective.

Cellular processes play major roles in cell survival and apoptosis. These events are essential for tissue homeostasis and the maintenance of proper growth and development of multicellular organisms. Imbalance in survival and apoptotic pathways may lead to several diseases. Therefore, understanding the molecular mechanisms of cell survival and apoptotic pathways is essential for the treatment and prevention of human diseases. The main focus of *Apoptosis, Cell Signaling, and Human Diseases: Molecular Mechanisms* is to discuss the recent development in cell signaling events, growth, metastasis, and angiogenesis, mechanisms of drug resistance, and targeted therapy for human diseases. Volume 1 contains 15 chapters divided into two sections: "Malignant Transformation and Metastasis" and "Molecular Basis of Disease Therapy"; Volume 2 contains 18 chapters, also divided into two sections: "Kinases and Phosphatases" and "Molecular Basis of Cell Death." Scientists well known in their fields have contributed to this book.

In part I, the pathophysiological processes including the mechanisms by which normal cells are transformed to malignant cells, regulation of cell growth, differentiation and apoptosis by oncogenes and tumor suppressor genes, consequences of DNA damage and the ability of cells to repair damaged DNA in response to stress stimuli, molecular events involved in metastasis and angiogenesis, and roles of transcription factors and cytokines in cell survival and apoptosis, are discussed. The recent development in technology has allowed us to identify new diseases before the appearance of the symptoms. The delay in identification of the disease may be fatal to human life. The incorporation of concepts of engineering to the principles of biology has further revolutionized the field of medicine. Nanotechnology, bioinformatics, microarray and proteomics are powerful tools that are being used in drug discovery and development, and treatment of human diseases.

In part II, biological significance of the kinases, and the cell signaling events that control cell survival and apoptosis are discussed. The identification of over 500 protein kinases encoded by the human genome sequence offers one measure of the importance of protein kinase networks in cell biology. Phosphorylation and dephosphorylation of protein kinases such as protein kinase A (PKA), protein kinase C (PKC), cyclin-dependent kinase (CDK), phosphatidylinositol 3-kinase (PI3K), Akt, and MAP kinase (MAPK) are important for regulating cell cycle, survival, and apoptosis. High-throughput technologies for inactivating genes are producing an inspiring amount of data on the cellular and organismal effects of reducing the levels of individual protein kinases. Despite these technical advances, our understanding of kinase networks remains imprecise. Major challenges include correctly assigning kinases to particular networks, understanding how they are regulated, and identifying the relevant in vivo substrates. Genetic methods provide a way of addressing these questions, but their application requires understanding the mutations and how they affect protein-protein interacts.

Apoptosis is a genetically controlled process that plays important roles in embryogenesis, metamorphosis, cellular homeostasis, and as a defensive mechanism to remove infected, damaged, or mutated cells. Molecules involved in cell death pathways are potential therapeutic targets in immunologic, neurologic, cancer, infectious, and inflammatory diseases. Although a number of stimuli triggers apoptosis, it is mainly mediated through at least three major pathways that are regulated by (i) the death receptors, (ii) the endoplasmic reticulum (ER), and (iii) the mitochondria. Under certain conditions, these pathways may cross talk to enhance apoptosis. Death receptor pathways are involved in immune-mediated neutralization of activated or autoreactive lymphocytes, virus-infected cells, and tumor cells. Consequently, dysregulation of the death receptor pathway has been implicated in the development of autoimmune diseases, immunodeficiency, and cancer. Increasing evidence indicates that the mitochondrial and ER pathways of apoptosis play a critical role in death receptor-mediated apoptosis. Dysregulation of these pathways may contribute to drug resistance.

A lot of progress has been made in understanding the mechanisms of apoptosis. Mitochondria are critical death regulators of the intrinsic apoptotic pathway in response to DNA damage, growth factor withdrawal, hypoxia, or oncogene deregulation. Activation of the mitochondrial pathway results in disruption of mitochondrial homeostasis, and release of mitochondrial proteins. The release of mitochondrial apoptogenic factors is regulated by the pro- and anti-apoptotic Bcl-2 family proteins, which either induce or prevent the permeabilization of the outer mitochondrial membrane. Activation of the death receptor pathway also links the cell-intrinsic pathway through Bid. Mitochondrial membrane permeabilization induces the release of mitochondrial proteins (e.g., cytochrome c, Smac/DIABLO, AIF, Omi/HtrA2, and endonuclease G), which are regulated by proapoptotic and antiapoptotic proteins of Bcl-2 family, and in caspase-dependent and -independent apoptotic pathways. The antiapoptotic members (e.g. Bcl-2 or Bcl-X_L) inhibit the release of mitochondrial apoptogenic factors whereas the proapoptotic members (e.g. Bax, and Bak) trigger the release.

Recent studies suggest that, in addition to mitochondria and death receptors, other organelles, including the endoplasmic reticulum (ER), Golgi bodies, and lysosomes, are also major points of integration of proapoptotic signaling and damage sensing. Each organelle possesses sensors that detect specific alterations, locally activate signal transduction pathways, and emit signals that ensure inter-organellar cross-talk.

The genomic responses in intracellular organelles, after DNA damage, are controlled and amplified in the cross-signaling via mitochondria; such signals induce apoptosis, autophagy, and other cell death pathways.

Chromatin remodeling agents modulate gene expression in tumor cells. Acetylation and deacetylation are catalyzed by specific enzyme families, histone acetyltransferases (HATs) and deacetylases (HDACs), respectively. Since aberrant acetylation of histone and nonhistone proteins has been linked to malignant diseases, HDAC inhibitors bear great potential as new drugs due to their ability to modulate transcription, induce differentiation and apoptosis, and inhibit angiogenesis. The preclinical data on HDAC inhibitors are very promising, and several HDAC inhibitors are currently under clinical trials for the treatment of cancers.

Apoptosis, Cell Signaling, and Human Diseases: Molecular Mechanisms will be valuable to graduate students, postdoctoral and medical fellows, and scientists with a working knowledge of biology and pathology who desire to learn about the molecular mechanisms of human diseases and therapy. I hope that individuals of diverse backgrounds will find these volumes very useful.

Rakesh Srivastava, PhD

Contents

Contributors

BALTAZAR D. AGUDA, PhD • *Bioinformatics Institute, Singapore*

DARIO C. ALTIERI, MD • *Department of Cancer Biology , University of Massachusetts Medical School, Worcester, MA*

SHRIKANT ANANT, PhD • *Division of Gastroenterology, Department of Internal Medicine, Washington University School of Medicine, St. Louis, MO*

CHRISTIAN BELTINGER, MD • *University Children's Hospital, Ulm, Germany*

FRED E. BERTRAND, PhD • *Department of Microbiology & Immunology, Brody School of Medicine at East Carolina University, Greenville, NC*

JOHN BLENIS, PhD • *Department of Cell Biology, Harvard Medical School, Boston, MA*

AMINA BOUADIS, BS • *Women's Cancers Section, Laboratory of Pathology, Center for Cancer Research, National Cancer Institute, Bethesda, MD*

JULIE L. BRONDER, PhD • *Women's Cancers Section, Laboratory of Pathology, Center for Cancer Research, National Cancer Institute, Bethesda, MD*

ZHEQING CAI, PhD • *Vascular Program, Institute for Cell Engineering; Departments of Pediatrics, Medicine, Oncology, Radiation Oncology; and Institute of Genetic Medicine, The Johns Hopkins University School of Medicine, Baltimore, MD*

FUMIN CHANG, PhD • *Department of Microbiology & Immunology, Brody School of Medicine at East Carolina University, Greenville, NC*

YOON S. CHO-CHUNG, MD, PhD • *Cellular Biochemistry Section, Basic Research Laboratory, Center for Cancer Research, National Cancer Institute, National Institutes of Health, Bethesda, MD*

SAMANTHA COORAY, PhD • *Department of Virology, Wright Flemming Institute, Imperial College Faculty of Medicine, St. Mary's Campus, Norfolk Place, London, UK*

TASMAN JAMES DAISH, PhD • *Hanson Institute, Adelaide, Australia*

KLAUS-MICHAEL DEBATIN, MD • *University Children's Hospital, Ulm, Germany*

STEIN OVE DØSKELAND, PhD • *Department of Biomedicine, Section of Anatomy and Cell Biology, University of Bergen, Bergen, Norway*

PAUL G. EKERT, MDDS, PhD, FRACP • *The Walter and Eliza Hall Institute of Medical Research, Victoria, Australia*

XIANJUN FANG, PhD • *Department of Biochemistry, Medical College of Virginia , Virginia Commonwealth School of Medicine, Richmond, VA*

EVA L. FELDMAN, MD, PhD • *Department of Neurology, University of Michigan, Ann Arbor, MI*

RICHARD A. FRANKLIN, PhD • *Department of Microbiology & Immunology and Leo Jenkins Cancer Center, Brody School of Medicine at East Carolina University, Greenville, NC*

GRO GAUSDAL, CSci • *Department of Biomedicine, Section of Anatomy and Cell Biology, University of Bergen, Bergen, Norway*

MARIA SAVERIA GILARDINI MONTANI, PhD • *Department of Environmental Science, La Tuscia University, Viterbo, Italy*

FRANK GILLARDON, PhD • *Boehringer INGELHEIM Pharma GmbH & Co. KG, CNS Research, Biberach an der Riss, Germany*

LARS HERFINDAL, PhD • *Department of Biomedicine, Section of Anatomy and Cell Biology, University of Bergen, Bergen, Norway*

CHRISTINE E. HORAK, PhD • *Women's Cancers Section, Laboratory of Pathology, Center for Cancer Research, National Cancer Institute, Bethesda, MD*

STEVEN IDELL, PhD • *The Texas Lung Injury Institute, The University of Texas Health Center at Tyler, Tyler, TX*

GERGELY JARMY, PhD • *University Children's Hospital, Ulm, Germany*

PATRICK G. JOHNSTON, PhD • *Drug Resistance Laboratory, Centre for Cancer Research and Cell Biology, Queen's University Belfast, Belfast, N. Ireland*

JAMES P. KEHRER, PhD • *College of Pharmacy, Washington State University, Pullman, WA*

MICHAEL KHAN, PhD, FRCP • *Molecular Medicine , Biomedical Research Institute , University of Warwick, Coventry, UK*

CAMILLA KRAKSTAD, PhD • *Department of Biomedicine, Section of Anatomy and Cell Biology, University of Bergen, Bergen, Norway*

SHARAD KUMAR, PhD • *Hanson Institute, Adelaide, Australia*

MARIKKI LAIHO, MD, PhD • *Hartman Institute and Molecular Cancer Biology Research Program, Biomedicum Helsinki, University of Helsinki, Helsinki, Finland*

CHRISTIAN M. LASTOSKIE • *Department of Civil and Mechanical Engineering, University of Michigan, Ann Arbor, MI*

DANIEL B. LONGLEY, PhD • *Drug Resistance Laboratory, Centre for Cancer Research and Cell Biology, Queen's University Belfast, Belfast, N. Ireland*

PETER LOW, PhD • *Department of Anatomy, Cell and Developmental Biology, Lorand Eotvos University, Budapest, Hungary*

YOHEI MAESHIMA, MD, PhD • *Assistant Professor of Medicine, Okayama University Graduate School of Medicine and Dentistry, Department of Medicine and Clinical Science, Okayama, Japan*

LIVIO MALLUCCI, MD • *King's College London, Biomedical and Health Sciences, Pharmaceutical Science Research Division, London, UK*

JAMES A. MCCUBREY, PhD • *Department of Microbiology and Immunology, East Carolina University, Brody School of Medicine, Greenville, NC*

MAURIZIO MEMO, PhD • *Department of Biomedical Sciences and Biotechnologies, University of Brescia Medical School, Brescia, Italy*

MARIA V. NESTEROVA • *Cellular Biochemistry Section, Basic Research Laboratory, CCR, National Cancer Institute, Bethesda, MD*

FELICIA NG • *Bioinformatics Institute, Singapore*

STELLA PELENGARIS, Bsc, PhD • *Molecular Medicine , Biomedical Research Institute , University of Warwick, Coventry, UK*

MARÍA PÉREZ-CARO, PhD • *Instituto Biologia Molecular y Celular del Cancer (IBMCC), CSIC/University of Salamanca, Salamanca, Spain*

MAURICE REIMANN, PhD • *Hematology, Oncology, and Tumor Immunology, Charité - Universitätsmedizin Berlin (CVK), Berlin, Germany*

MARY E. REYLAND, PhD • *Department of Craniofacial Biology, University of Colorado Health Sciences Center, Denver, CO*

PHILIPPE P. ROUX, PhD • *Department of Cell Biology, Harvard Medical School, Boston, MA*

MAHDIEH SADIDI • *Department of Neurology, University of Michigan, Ann Arbor, MI*

ISIDRO SÁNCHEZ-GARCÍA, MD, PhD • *Instituto Biologia Molecular y Celular del Cancer (IBMCC), CSIC/University of Salamanca, Salamanca, Spain*

ANN MARIE SASTRY • *Department of Mechanical Engineering, University of Michigan, Ann Arbor, MI*

CLEMENS A. SCHMITT, MD • *Hematology, Oncology, and Tumor Immunology, Charité - Universitätsmedizin Berlin (CVK), and Max-Delbrück-Center for Molecular Medicine, Berlin, Germany*

GREGG L. SEMENZA, MD, PhD • *Vascular Program, Institute for Cell Engineering; Departments of Pediatrics, Medicine, Oncology, Radiation Oncology; and Institute of Genetic Medicine, The Johns Hopkins University School of Medicine, Baltimore, MD*

SHARMILA SHANKAR, PhD • *Department of Biochemistry, University of Texas Health Center at Tyler, Tyler, TX*

SREERAMA SHETTY, PhD • *The Texas Lung Injury Institute, The University of Texas Health Center at Tyler, Tyler, TX*

JOHN SILKE, PhD • *The Walter and Eliza Hall Institute of Medical Research, Parkville, Australia*

TOMASZ SKORSKI, MD, PhD • *Department of Microbiology and Immunology, School of Medicine, Temple University, Philadelphia, PA*

DANIEL SLIVA, PhD • *Department of Medicine, Indiana School of Medicine, and Cancer Research Laboratory, Methodist Research Institute, Indianapolis, IN*

SARAH SPIEGEL, PhD • *Department of Biochemistry, Medical College of Virginia , Virginia Commonwealth School of Medicine, Richmond, VA*

RAKESH SRIVASTAVA, PhD • *Department of Biochemistry, University of Texas Health Center at Tyler, Tyler, TX*

PATRICIA S. STEEG, PhD • *Molecular Therapeutics Program, Molecular Targets Faculty, and Women's Cancers Section, Laboratory of Pathology, Center for Cancer Research, National Cancer Institute, Bethesda, MD*

LINDA S. STEELMAN, BS, MS • *Department of Microbiology & Immunology, Brody School of Medicine at East Carolina University, Greenville, NC*

KELLI A. SULLIVAN • *Department of Neurology, University of Michigan, Ann Arbor, MI*

SRIPATHI M. SUREBAN • *Department of Internal Medicine, Washington University School of Medicine, St. Louis, MO*

DEAN G. TANG, PhD • *The University of Texas M. D. Anderson Cancer Center, Department of Carcinogenesis, Science Park-Research Division, Smithville, TX*

DAVID M. TERRIAN, PhD • *Leo Jenkins Cancer Center and Department of Anatomy & Cell Biology, Brody School of Medicine at East Carolina University, Greenville, NC*

ANDREA M. VINCENT • *Department of Neurology, University of Michigan, Ann Arbor, MI*

JIWU WEI, MD • *University Children's Hospital, Ulm, Germany*

VALERIE WELLS, PhD • *King's College London, Biomedical and Health Sciences, Pharmaceutical Science Research Division, London, UK*

WEE KHENG YIO • *Bioinformatics Institute, Singapore*

JAI Y. YU, BSc • *The Walter and Eliza Hall Institute of Medical Research, Parkville, Australia*

Contents of Volume 1

I

KINASES AND PHOSPHATASES

1

Significance of Protein Kinase A in Cancer

Maria V. Nesterova and Yoon S. Cho-Chung

Summary

The system of cyclic nucleotides is one of the regulatory cascades whose initial step begins at the level of the external signal perception and whose final step is completed by phosphorylation of a number of substrate proteins mediating the diverse functional responses of the cell. The study of cyclic adenosine monophosphate (cAMP) has indicated a dual effect for this nucleotide on proliferation and differentiation processes. Elevation of intracellular cAMP in normal and transformed cells can lead to the initiation of cell proliferation, or is accompanied by corresponding morphological changes and the induction of differentiation. The main target of cAMP action in the cell is cAMP-dependent protein kinase, which may exist as two different isozymes, designated as type I (PKA-I) and type II (PKA-II). These two isoforms are essentially distinct in their physicochemical properties. Such evident differences of the protein kinases, both under the control of cAMP, indicate that they act at different stages of cell life. The relative ratio of PKA-I and PKA-II varies not only throughout the cell cycle in cells of the same type, but also among tissues and during development. The data reported to date suggest that PKA-I acts as a positive growth effector, whereas PKA-II inhibits cell division. Some exceptions were described, particularly the Carney complex, which is completely attributed to mutations in the regulatory subunit of PKA-I. Thus, by modulating PKA isozymes, one can regulate the process of cell division by controlling the balance in the proliferation–differentiation system. This chapter describes a strategy for using cAMP analogs, antisense oligonucleotides, and cAMP response element (CRE) transcription factor decoy oligonucleotides to modulate the cAMP-mediated system to encounter and combat diseases, such as cancer. In connection with a newly discovered excreted form of PKA, termed extracellular PKA (ECPKA), we also discuss the possibility of using PKA as a tool in the diagnosis of cancer.

Key Words: Protein kinase A; 8-Cl-cAMP; antisense PKA RI; cancer; growth inhibition; tumor reversion; CRE-transcription factor decoy.

1. Introduction

Cells can respond to extracellular signals through the activation of cell-surface receptors and second-messenger pathways. The first-discovered and best-characterized second-messenger system in eukaryotic cells is the cyclic adenosine monophosphate (cAMP)-mediated system *(1–3)*. Today, it is generally accepted that, in keeping with Greengard's hypothesis, most cAMP effects are related to the activating action on cAMP-dependent protein kinase A (PKA) and the phosphorylation of its protein substrates *(4)*. In the absence of cAMP, PKA is an enzymatically inactive tetrameric holoenzyme consisting of two catalytic subunits (C) bound to a regulatory subunit (R) dimer. Cyclic AMP binds in cooperative manner to two sites on each regulatory subunits. Upon the binding of four molecules of cAMP, the enzyme dissociates into an R subunit dimer with four molecules

From: *Apoptosis, Cell Signaling, and Human Diseases: Molecular Mechanisms, Volume 2*
Edited by R. Srivastava © Humana Press Inc., Totowa, NJ

of cAMP bound and two free and active C subunits that phosphorylate serine and threonine residues on specific substrate proteins *(5,6)*.

Cyclic AMP regulates a number of different cellular processes, such as cell growth and differentiation, metabolic processes, lipolysis, ion channel conductivity, synaptic release of neurotransmitters, and gene expression. The question rises of how one enzyme with such broad specificity like PKA can mediate so precisely the enormous variety of diverse physiological events. The substrate specificity of PKA can be achieved on the different levels of biological regulation. The first and most basic level of such regulation is the fact that PKA has several isoforms. Initially, two izosymes of PKA, termed type I and type II (PKA-I and PKA-II, respectively), were found based on their pattern of elution from diethylaminoethyl (DEAE)-cellulose columns *(7,8)*.

However, molecular cloning has identified a significant heterogeneity in both R and C subunits, which revealed multiplicity of PKA isozymes. Cloning of cDNAs for regulatory subunits have identified two RI subunits, named RIα *(9,10)* and RIβ *(11,12)*, and two RII subunits, termed RIIα *(13,14)* and RIIβ *(15,16)*, as separate gene products. Furthermore, three distinct C subunits were identified and designated as Cα *(17)*, Cβ *(18,19)*, and Cγ *(20)*, respectively.

The PKA-I and PKA-II holoenzymes have distinct biochemical properties *(8,21)*. It has been shown that association and dissociation rates of PKA isozymes are affected differently by factors such as salt and MgATP concentrations *(8,22,23)*. In addition, autophosphorylation of PKA-II favors its dissociation *(24)*. Furthermore, PKA-I holoenzymes are more easily dissociated by cAMP in vitro than PKA-II holoenzymes *(5,25)*. Importantly, the preferential association of RII with C occurs in intact cells despite the observation that the purified preparations of RI and RII show similar affinities for the C subunit *(26)*. If RII is overexpressed in cells, the C subunit preferably binds to RII, increasing its holoenzyme PKA-II and depleting the RI holoenzyme, PKA-I *(27)*. In contrast, when RI is overexpressed, it is present at free dimers, cannot compete with RII for C-binding, and does not downregulate PKA-II or upregulate PKA-I *(27)*. More recently it was shown, however, that depletion of the RIIβ or RIβ subunit by gene targeting did not result in quantitative compensation by RIIα in the RIIβ knockout mice or by RIIα/RIIβ in the RIβ knockouts as would be expected. Instead, the authors found the induction of RIα and PKA-I assembly in both the RIβ and RIIβ knockouts as a result of a four- to fivefold increase in the half-life of RIα protein when binding to the C subunit forming PKA-I holoenzyme *(28)*.

These data from the embryonic cells *(28)* are totally opposite to that observed in the adult cells *(27)*. Thus, the PKA-I and PKA-II assembly in vivo is clearly distinctive between cells in the developmental stage, such as knockout mice, and the mature-adult cells *(28)*. It indicates that the PKA-I is essential for developmental embryonic cells, whereas PKA-I is no longer needed in the fully developed adult cells. Therefore, these data obtained from the knockout mice cannot be extrapolated directly to that of the adult cells.

The presence of an isozyme consisting of an RIα-RIβ heterodimer with associated phosphotransferase activity has been reported *(29)*. This isozyme was eluted at the position of PKA-II by DEAE-cellulose chromatography, indicating different biochemical properties of holoenzymes containing R-subunit homodimers vs heterodimers. The existence of multiple R and C subunits make possible the formation of a number of PKA holoenzymes with different biological characteristics and activities, which can be a part of a regulatory system affecting the specificity of cAMP/PKA signaling pathway in cell development, differentiation, and malignant transformation.

2. Protein Kinase A Isozymes: Normal Cells vs Cancer Cells

2.1. Tissue Specificity and Intracellular Distribution

In earlier studies, it was shown that the relative content of PKA-I and PKA-II varies among tissues in adult animals and also in the same tissue among the different species of animals *(8,30)*, although the total R subunit/C subunit molar ratio in all normal tissues examined was found to be 1:1 *(30)*. It has become clear that considerable heterogeneity in R isoforms exists not only among species and tissues, but also within a single cell type under different physiological/pathological conditions *(31)*. The distinct localization of PKA isoforms within the cell presents another level for regulation of the ability of a single second messenger, cAMP, to produce a variety of effects and provides the possibility for the phosphorylation of locally accessible substrates, ultimately resulting in specific regulatory effects. The RI isoforms are thought to be primarily soluble and cytoplasmic, whereas between 30 and 75% (depending on the tissue source) of the cellular RII pool is in particular compartments and in association with certain subcellular structures, such as with specific A-kinase anchor proteins (AKAPs) *(32,33)*. It has been shown that RIα and RIIα may be associated with microtubules *(34)*.

RIIα and RIIβ subunits were found associated with pericentriolar matrix of the centrosome during interphase *(35)*, and the catalytic subunits bind to microtubules or mitotic spindels *(36)*. Localization on these structures, which are major components of the cytoskeleton and mitotic apparatus, suggests that PKA can play an important role in different phases of the cell cycle. It was found that RIIα is tightly bound to centrosomal structures during interphase through interaction with the AKAP450, but dissociates and redistributes from centrosomes at mitosis. Furthermore, RIIα is solubilized from particular cell fractions and changes affinity toward AKAP450 as a result of phosphorylation by CDK1 *(37,38)*. It was found in colon cancer cells LS-174T that plasma membranes, Golgi apparatus, and mitochondria contained mainly RIIα, and in the fractions of lysosomes and microsomes RIα and RIIα were found in nearly equal amounts *(39)*. The analysis of purified plasma membranes and the surface of intact cells revealed the presence of PKA on the external surface of LS-174T human colon carcinoma cells *(40)*. Immunoprecipitation identified the cAMP-binding protein as the RIIα regulatory subunit of PKA. Moreover, during the logarithmic stage of growth, both the RIα and RIIα subunits of PKA were localized on the cell surface.

The cAMP signaling into the nucleus deserves special attention. It has been proposed that the cellular localization of catalytic subunit may be responsible for determining its physiological substrates. The most dominating theory is that in the basal state, PKA resides in the cytoplasm as an inactive holoenzyme and induction of cAMP liberates the catalytic subunit, which passively diffuses into the nucleus and induces cellular gene expression *(41–43)*. However, in earlier studies, PKA regulatory subunits were found in the nucleus *(44–46)*. It was observed that the localization of R-subunits in the nucleus was dependent on the cAMP level especially in cancer cells *(44)*. Recently, it was shown that the two types of PKA in NG108-15 cells (neuroblastoma-glyoma hybrid) differentially mediate the forskolin- and ethanol-induced cyclic AMP response element-binding protein (CREB) phosphorylation and cAMP response element (CRE)–mediated gene transcription *(47)*. It was discovered that activated type II PKA is translocated into the nucleus where it phosphorylates CREB. By contrast, activated type I PKA does not translocate to the nucleus but is required for CRE-mediated gene transcription for the activation of coactivator, CBP. In the recent study of PC3M prostate cancer cells with the aid of confocal microscopy, the total colocalization of Cα and RIIβ was found in both the cytoplasm and the nucleus of RIIβ-transfected, growth-arrested cancer cells *(48)*. By contrast, such total colocalization of RIIβ and Cα was not found in the parental-nontransfectant, RIα-α and Cα-overexpressing cells *(48)*. It was shown that *v-Ki-Ras* oncogene alters cAMP nuclear signaling by regulating the location and the expression of PKA-RIIβ *(49)*. They demonstrated that localization of PKA-II influences cAMP signaling to the nucleus. *Ras* alters the localization and the expression of PKA-II Translocation of PKA-II to the cytoplasm reduces the expression of cAMP-dependent gene products (including RIIβ, thyroid stimulating hormone [TSH] receptor, and thyroglobulin), and the loss of RIIβ permanently downregulates thyroid-specific expression pattern *(49)*. Taken together, these data suggest the special role for RIIβ in cAMP-mediated nuclear signaling.

2.2. Ontogeny and Cell Differentiation

The data described previously show us the importance of PKA in cell physiology. The changes in the amount or activity of PKA isozymes have been correlated with highly regulated processes, such as ontogeny and cell differentiation. Examples of such studies are described here.

Two studies *(50,51)* of the development of PKA in mouse heart showed that the type I/type II isozyme activity ratio (or RI/RII ratio) decreased from 3.0 in neonates (7- or 14-d-old) to 1.0 in adult hearts. It was found that in rodents, after the third postnatal wk, that no more cardiac cell division occurs, such that only cellular hypertrophy is responsible for further cardiac growth *(52)*. This implies that high levels of type I kinase (or RI) are associated with the developmental phase of cell growth. A study of the differentiation of ovarian follicle granulosa cells from immature hypophysectomized rats treated with injections of estradiol and follicle-stimulating hormone showed a 10- to 20-fold increase in the RIIβ content of granulosa cells, without a corresponding increase in the catalytic activity of PKA *(53)*.

An inverse relationship between estrogen receptor and RII cAMP receptor has been shown *(54)* in the growth and regression of hormone-dependent mammary tumors for which estrogen has a trophic role. Differential expression of type I and type II protein kinase is also cell-cycle specific. In Chinese hamster ovary (CHO) cells, PKA-I activity is high during mitosis and relatively low during G1 and S phases, whereas type II activity is low during mitosis and increases at the G1-to-S phase border, again decreasing by mid- to late-S phase *(55)*. Experiments on peripheral blood lymphocytes suggested that the mitogen concanavalin A selectively activated PKA-I, whereas N^6,O^2-dibutyryl-cAMP activated PKA-II as well *(56)*. This raised the question of whether the two isozymes might mediate different effects on cell replication in these cells. However, some reports have concluded that both isozymes convey inhibitory signals in lymphocytes *(57)*. The most recent paper confirmed the significance of PKA-I in receptor-mediated T-lymphocyte activation *(58)*. Another report *(59)* showed that

PKA-I mediates the inhibitory effects of cAMP on cell replication in T-lymphocytes. It was concluded that activation of PKA-I is sufficient to inhibit T-lymphocyte proliferation. The membrane-bound PKA-II may mediate cAMP actions not related to cell replication. Interesting results were obtained by transfection of a human neoplastic B precursor cell line (Reh) with plasmid-containing $C\alpha$ gene *(60)*. The growth rate of clones transfected with $C\alpha$ was retarded. In contrast, overexpression of the RIα had no effect on cAMP-dependent inhibition of cell proliferation. Furthermore, expression of mutant regulatory subunit, which renders cAMP-dependent protein kinase unresponsive to cAMP, clearly protected against the inhibitory effect of cAMP. These data provide evidence that activation of the C subunit of PKA mediated the inhibitory action of cAMP on cell proliferation in Reh cells *(60)*. Thus, lymphoid cells provide us with the important possibility of exploring how mitogenic signals regulate the levels of PKA subunits.

Several examples of changes in PKA those that are associated with cell differentiation induced in malignant cells by means other than cAMP stimulus, have been described. Friend erythroleukemia cells—virus-transformed murine erythroleukemia cells—can be stimulated by a wide variety of agents to express several characteristics of erythroid cell differentiation. Induction of differentiation by dimethyl sulfoxide was shown to result in a threefold increase in RII, a threefold decrease in RI, and an RII/RI ratio of 11, compared with 1.2 in control cells *(61)*. Murine embryonal sarcoma cells, the pluripotent stem cells of malignant teratocarcinoma, responded to retinoic acid followed by cAMP (but not the reverse order) by differentiating and becoming parietal endoderm cells *(62)*. Because cAMP alone cannot produce this effect, it was suggested that the retinoic acid induced changes, which altered the response to cAMP *(63)*. The treatment of F-9 cells with retinoic acid resulted in an increase in RI, RII, and histone kinase activity in cytosol and a preferential increase in the amount of RII associated with plasma membrane. A selective activation of type II isozyme was also correlated with the growth arrest induced by calcitonin in human breast cancer cell line *(64)*. These data indicate the distinct roles of cAMP receptor isoforms in the regulation of ontogeny and cell differentiation and support a hypothesis that protein kinase type II is primarily involved in differentiation processes, whereas protein kinase type I relates cell proliferation *(65,66)*.

2.3. RIα/PKA-I: Cell Transformation and Human Tumors

Because the changing ratio of PKA-I and PKA-II appears to be involved in the ontogenic and differentiation processes as described above, it is logical to expect changes or abnormalities in the PKA of neoplastic cells. RI is the major subunit of PKA detected in various types of human cancer cell lines *(67,68)*. These cell lines include hormone-dependent (MCF-7, T47D, and ZR-75-D) and hormone-independent (MCF-7$_{ras}$, MDA-MB-231, and BT-20) breast cancers; WiDr, LS-174, and HT-29 colon carcinomas; A549 lung carcinoma; HT-1080 fibrosarcoma; A4573 Ewing's sarcoma; FOG and U251 gliomas; and HL-60, K-562, K-562$_{myc}$, and Molt-4 leukemias.

The comparison of primary breast tumors with normal breast tissues showed significantly higher levels of the RI/RII ratio and an increase in RI in tumors *(69,70)*. Extensive studies of PKA composition in mammary tumors in cancer patients revealed *(71)* that overexpression of the R subunits of PKA (in particular, RI) is associated with high proliferation in normal breast, malignant transformation in the breast, poor prognosis in established breast cancer, and resistance to antiestrogens. These data, together with the observation that successful antiestrogen therapy is associated with reduced expression of RI mRNA, suggest that targeting R subunits is an appropriate therapeutic strategy for breast cancer.

The ratio of type I to type II protein kinase in renal cell carcinomas was about twice that in renal cortex, although the total soluble cAMP-dependent protein kinase activity was similar in both normal and malignant tissues *(72)*. In the surgical specimens of Wilms' tumor, the type I/type II PKA ratio was two times greater than that in normal kidney, and the RI/RII ratio was 0.79 for normal kidney and 2.95 for kidney with tumor *(73)*. Specimens of colorectal cancer contained R subunits, with no detectable levels of RII, whereas both distant and adjacent mucosa contained detectable amounts of RII with decreased levels of RI *(74)*. Changes in cAMP-dependent protein kinase activity have been reported in human thyroid carcinomas *(75)*, in human cerebral tumors *(76)*, and in leukemia cells from patients *(77)*. The comparison between RI and RII patterns in normal peripheral blood lymphocytes and those in lymphocytes from patients with chronic lymphocytic leukemia is characterized by low absolute amounts of RI and RII and an altered molecular species of RII *(78)*.

2.4. RIα/PKA-I: Carcinogenesis

Alterations in PKA isozymes have been found during carcinogenesis. During dimethyl-benz(a)anthracene-induced mammary carcinogenesis in rats, the transient increase in type I PKA and RI subunit coincided with the initial action of carcinogen in the mammary gland *(79)*. The incidence of gastric adenocarcinoma by *N*-methyl-*N*'-nitro-*N*-nitroso-guanidine and the trophic action of gastrin on gastric carcinoma production was correlated with an increase in type I PKA activity *(80)*. Changes in the cAMP-binding affinity to RII but not to RI have been shown to correlate with the progression of urethan-induced mouse lung tumors *(81)*. Thus, RII of normal adult lung contained both high- and low-affinity cAMP-binding sites, whereas RII in tumor exhibited only the low-affinity site, and these changes in the cAMP-binding property correlated in degree with tumor size and extent of anaplasticity. In contrast, a decreased expression of PKA-I and RI subunit was found in mouse tumor cell lines of lung epithelial origin as compared with the immortalized non-tumorigenic cell lines *(82)*.

Changes in PKA isozyme patterns have been commonly observed in in vivo cell transformation. SV40 viral transformation of BALB 3T3 fibroblasts was accompanied by an increase in PKA-I activity and RI subunit level; normal BALB 3T3 cells contain only type II kinase isozyme *(83)*. Transformation of rat 3Y1 cells by the highly oncogenic human adenovirus type 12 correlated with a three- to sixfold increase in PKA-I and RI subunit levels *(84)*. A marked increase in RI with a decrease in RII subunits was detected in Harvey murine sarcoma virus-transformed NIH/3T3 clone 13-3B-4 cells *(85)* and in NRK cells transformed with transforming growth factor (TGF)α or *v-Ki-ras* oncogene *(86)*. TGFα-induced transformation of mouse mammary epithelial cells brought about a marked increase in the RI mRNA level along with a decrease in RIIα and RIIβ mRNA levels *(87)*.

The dual signals, positive and negative, transduced by cAMP may depend on the availability of RI and RII subunits, respectively. We have previously shown *(88–91)* that downregulation of RIα receptor and the compensatory increase in RIIβ receptor, obtained either by treatment with site-selective cAMP analogs, such as 8-Chloro-adenosine 3′:5′-phosphate (8-Cl-cAMP) or with an antisense oligonucleotide targeted against the RIα mRNA (see following sections), lead to arrest of cancer cell growth and induction of differentiation. Moreover, the role of RII/PKA-II has been implicated in the control of gene transcription. In the rat cancer cell line A126-1B2, a mutant of PC-12 pheochromocytoma that contains a deficient PKA-II but contains the wild-type level of PKA-I, reconstitution of functional levels of RIIβ/PKAII rescued the cAMP-dependent cell-differentiation pathway and the inducibility of the CRE-containing somatostatin gene transcription *(92)*. We have hypothesized that in cancer cells, a reduced availability of RIIβ receptor, or its functional inactivation, may play a key role in sustaining the high levels of RIα expression and transformed phenotypes. However, the direct link between overexpression of RI subunit and subsequent transformation of normal cell to malignant cell is still unknown.

2.5. RIα: A Partner for Cell-Cycle Regulator

Recently, RFC40 (RFC2), a novel partner for RIα the second subunit of replication factor C (RFC) complexwas identified. It was shown, that this interaction may be associated with cell survival *(93)*. The RFC complex, which consists of five subunits (RFC1-5), functions as a clamp loader to load the sliding clamp, proliferating cell nuclear antigen (PCNA) onto the DNA, in an ATP-dependent manner, for the processive DNA synthesis catalyzed by DNA polymerases δ and ε. The small subunits of the RFC complex have been shown to form different complexes involved in cell-cycle checkpoint, sister chromatid cohesion and chromosome transmission, and genomic stability *(94,95)*. In the course of this study, a nonconventional nuclear localization sequence (NLS) was identified for RIα and the investigators suggested, that nuclear localization of RFC40 may be dependent on RIα. Furthermore, when this process was disrupted, the progression of cells to S phase became impaired leading to GI arrest. Thus, this novel protein–protein interaction may also provide a basis for the regulation of cell-cycle progression by RIα.

2.6. RIα/RIIβ: Point Mutation for Distinct Tumor Phenotype

One important method to study the significance of the protein is the introduction of mutations into its structure. It has been shown that replacement of the pseudophosphorylation site Ala-97 in bovine RIα with Ser significantly reduces MgATP binding affinity for PKA-I holoenzyme *(96)*. Conversely, replacement of the autophosphorylation site Ser-145 in the R subunit of yeast PKA with Ala significantly enhances the affinity of R for C subunit *(97)*. In our study, we have shown *(98)* that overexpression of RIIβ in human colon carcinoma cells induces growth arrest and phenotypic changes that can be abolished by site-directed mutation (autophosphorylation site) in RIIβ. Our results suggest a role for RIIβ in the suppression of tumor cell growth, and thus abnormal expression of R subunit isoforms of PKA may be involved in neoplastic transformation. Introduction of Ala[99] to Ser mutation in human RIα RIα-P) causes reduced kinase activation by cAMP and arrest of hormone-dependent breast cancer cell growth *(99)*.

2.7. Microarray Validation: RIIβ—Induction of Tumor Reversion

The use of DNA microarray technology made possible genome-wide analysis of gene expression patterns in cancer cells transfected with the isoforms of PKA regulatory subunits. We used PC3M prostate carcinoma cells as a model to overexpress wild-type and mutant R and C subunit genes and examined the effects of differential expression of these genes on tumor growth *(48)*. RIIβ- and RIα-P–overexpressing cells markedly suppressed genes that define the proliferation signature; RIα-P, Ala^{99} to Ser mutation of human RIα is a functional mimic of RIIβ. This cluster of genes was highly expressed in parental-, Cα-, and RIα-overexpressing cells. In this proliferation signature, genes encoding cell-cycle control proteins, transcription factors, growth factor receptors, protein kinases and phosphatases, and BRCA-1-associated protein-1 were predominant. RIIβ- and RIα-P–overexpressing cells also downregulated the transformation signature, including genes encoding the *ras* homolog gene family, v-Akt murine thymoma, viral oncogene homolog 2, TGF-β1 and receptor III, and matrix metalloproteinase 1 (MMP-1). Conversely, RIIβ-overexpressing cells upregulated the differentiation signature. This cluster was dominated by genes encoding the PKA RIIβ subunit, thioredoxin peroxidase, COX 11 homolog, E-cadherin, ornithine decarboxylase antizyme, thyroid hormone receptor interactor 3, guanine nucleotide binding protein, Notch homolog 4, retinoblastoma-binding protein 2, plasminogen activator inhibitor type II, protein phosphatase 4 and 2, catalytic subunits, p53-binding protein 1, general transcription factor IIA1, and caspase 4. Strikingly, this cluster of differentiation genes induced in RIIβ-overexpressing cells was unaltered in RIα-and Cα-overexpressing cells. Further characterization of these differentiation genes will shed light on the mechanism of PKA signaling in apoptosis/ differentiation of tumor cells.

2.8. RIα Mutation in Carney Complex

Cancer is associated with extensive genetic alterations and chromosomal disturbances, resulting in mutations of many genes. A series of recently published articles has shown the first human disease mapping to mutations in RIα–Carney complex *(100)*. Carney complex describes the association of spotty skin pigmentation, myxomas, and endocrine overactivity; the disease is in essence the latest form of multiple endocrine neoplasia to be described. It affects the pituitary, thyroid, adrenal, and gonadal glands. Primary pigmented nodular adrenocortical disease (PPNAD), a micronodular form of bilateral adrenal hyperplasia that causes a unique, inherited Cushing syndrome, is also the most common endocrine manifestation of Carney complex. PPNAD and Carney complex are genetically heterogeneous, but one of the responsible genes is *PRKAR1A*, at least for those families that map to 17q22-24 (the chromosome region that harbors PRKAR1A). The presence of inactivating germline mutations and the loss of wild-type allele in Carney complex lesions indicated that *PRKAR1A* could function as a tumor-suppressor gene in these tissues. Moreover, the experiments with transgenic mice bearing an antisense construct of RIα showed the possibility of the development of endocrine and other tumors with increased levels of RIIβ *(101–103)*. Another group of investigators *(104)* evaluated the expression of RIα, RIIβ, and RIIβ in a series of 30 pituitary adenomas and the effects of subunit activation on cell proliferation. In these tumors, neither mutation of *PRKAR1A* nor loss of heterozygosity was identified. By real-time PCR , the mRNA of the three R subunits was detected in all of the tumors. RIα mRNA was most predominantly found in the majority of samples. By contrast, immunohistochemistry documented low or absent RIα protein levels in all tumors, whereas RIIα and RIIβ proteins were highly expressed, thus resulting in an unbalanced

RI/RII ratio. The low levels of RIα resulted, at least in part, from proteosome-mediated degradation. The effect of the RI/RII ratio on proliferation was assessed in GH3 cells (growth hormone–secreting pituitary adenoma), which showed a similar unbalanced pattern of R subunits expression, and in growth hormone-secreting adenomas. 8-Cl–cAMP and RIα RNA silencing (siRNA) stimulated cell proliferation and increased cyclin D1 expression, respectively, in human and rat adenomatous somatotrophs. These data show that a low RI/RII ratio resulted in proliferation of transformed somatotrophs and are consistent with the Carney complex model in which RIα inactivating mutations further unbalance this ratio in favor of RII subunits.

Taken together, the balance between two isoforms of PKA appears to be critical for maintenance of the certain state of the cell. Therefore, by modulating the activity of these isozymes, one can regulate cell processes by controlling the balance in the proliferation/differentiation system. Thus, PKA can be a target for antitumor drugs and can be used in the diagnosis of cancer. The following sections of this review will describe the new developments in the use of PKA as a target in cancer therapy.

3. PKA—Target for Anticancer Drugs

3.1. 8-Cl-cAMP

3.1.1. Site-Selective cAMP Analogs

In earlier studies, it was shown that growth and morphology of a number of cultured cell lines can be influenced by cAMP or its derivatives *(105,106)*. Exposure of various lines of transformed cells to cyclic AMP or its dibutyryl derivative resulted in inhibition of growth rate without affecting cell viability. The possibility of using the analogs of cAMP for inhibition of tumor growth in vivo was shown for the first time in experiments with hormone-dependent mammary tumors *(107)*. It was shown that estrogen concentration did not change, but acid ribonuclease activity and synthesis increased during treatment with the $N^6,O^{2'}$-dibutyryl cyclic AMP and during tumor regression as a result of hormonal deprivation. Growth arrest, thus, appears to derive from enhanced tissue catabolism *(107)*.

The study of mechanism of action and biology of PKA brought up the idea to synthesize the analogs of cAMP that preferentially bind and activate PKA-I vs PKA-II. cAMP binds to the regulatory subunits of PKA, at two different binding sites, termed Site A (Site 2) and Site B (Site 1) *(108,109)*. These binding sites have been identified by differences in their rate of cyclic nucleotide dissociation and specificity for cyclic nucleotide binding. Unlike parental cAMP, site-selective cAMP analogs demonstrate selective binding for either one of the two known cAMP binding sites in the R subunit, resulting in preferential binding and activation of either protein kinase isozymes *(108,109)*. cAMP, at high millimolar concentrations, saturates both PKA-I and PKA-II maximally and equally without discrimination; therefore, selective modulation of cAMP receptor isoforms has not been possible in previous studies where high concentrations of cAMP/cAMP analogs have been used *(109)*.

With the use of site-selective cAMP analogs, it became possible to correlate the specific effect of cellular protein kinase isozymes with cAMP-mediated responses in intact cells *(110)*. It was found that site-selective cAMP analogs demonstrate a major regulatory effect on growth in a broad spectrum of human cancer cell lines, including breast, colon, lung and gastric carcinomas, fibrosarcomas, gliomas, and leukemias, as well as on growth in athymic mice of human cancer xenografts of various cell types, including breast, colon, and lung carcinomas *(67,68,87,110)*. The effect of the analogs on growth inhibition appeared to be selective toward transformed cancer cells as opposed to nontransformed cells. The analogs produced little or no growth inhibition of NIH/3T3 cells, normal mammary epithelial cells, and normal peripheral blood lymphocytes *(110)*. In contrast with the previously studied cAMP analogs, the site-selective analogs demonstrated their growth-inhibitory effect in micromolecular concentrations and were able to inhibit the growth of cancer cell lines that are resistant to the cAMP analogs previously studied, such as dibutyryl-cAMP. Of these site-selective cAMP analogs, the most potent in growth inhibition was 8-Cl-cAMP.

3.1.2. 8-Cl-cAMP: A Clinically Relevant cAMP Analog

8-Cl-cAMP, which belongs to the isozyme site discriminator class of site-selective cAMP analogs, activates and downregulates PKA-I, but not PKA-II, by binding to both site A and site B of RI, and to site B of PKA-II *(111,112)*. The important property of site-selective analogs was discovered by the effect of 8-Cl-cAMP on in vivo growing tumors *(68)*. The growth of transplantable hormone-independent DMBA- and metastatic NMU-rat mammary carcinomas, which are completely resistant to dibutyryl cAMP, was markedly inhibited by 8-Cl-cAMP treatment, and in the treated NMU–tumor-bearing animal, no metastatic lesions were detected at the time of sacrifice, whereas the animals with untreated control tumor displayed metastatic lesions *(68)*.

3.1.3. 8-Cl-cAMP vs 8-Cl-adenosine

In the cell culture, in vitro, and in rodent serum, in vivo, 8-Cl cAMP can be metabolized by serum phosphodiesterase and 5′-nucleotidase to 8-Cl-adenosine, which is strongly cytotoxic *(113)*. It should be noted, however, that in human serum, 8-Cl-cAMP was found not to be metabolized (NCI Decision Network Meeting, 1989). The universal potent growth inhibitory effect of 8-Cl-cAMP, demonstrated in a broad spectrum of human cancer cell lines *(67,68,87,110)*, attracted attention of several investigators to perform 8-Cl-cAMP study. However, their seemingly wrong experimental approaches, such as short time course (2 d instead of more than 3 d) and high concentrations of 8-Cl-cAMP (50–100 μM instead of 5–10 μM), which produced the metabolites of 8-Cl-cAMP and resulted in growth inhibition of both cancer cells as well as normal cells, led to the conclusion that the effect of 8-Cl-cAMP was actually that of its metabolite, 8-Cl-adenosine *(114)*. These investigators completely ignored the newly discovered fact that the growth inhibitory effect of site-selective cAMP analogs is different from that in previous reports that have shown a strong cytotoxicity using some of the amino-substituted C-8 analogs and cyclic nucleotides of purine analogs; the cytotoxicity was mainly caused by the adenosine metabolites of their nucleotides *(115,116)*. Our laboratory performed an extensive analysis of the effects of bovine (commercial) serum on the growth inhibitory effect of 8-Cl-cAMP vs 8-Cl-adenosine *(117)*. We used 11 heat-inactivated sera of different sources and one non-heat-inactivated serum. 8-Cl-cAMP (5 μM for 3 d) exhibited varying degrees (0–51%) of growth inhibition on the same cell line cultured in the medium containing heat-inactivated serum from a different source. 8-Cl-adenosine (5 μM for 3 d) produced a constant growth inhibition (60%) regardless of serum source. The phosphodiesterase activities for neither the cAMP nor 8-Cl-cAMP correlated with the varying degrees of growth inhibition exerted by 8-Cl-cAMP in the presence of different serum supplements. The expression of RIα subunit of PKA in the same cell line varied widely with the different sources of serum supplement. 8-Cl-cAMP–induced growth inhibition was dependent on the basal levels of RIα and correlated with specific down-regulation of the RIα 8-Cl-cAMP, but not 8-CPT-cAMP (8-Cl-phenyl-thio-cAMP or N^6-benzyl-cAMP), inhibited cell growth in serum-free medium. These results show that the growth inhibitory effect of 8-Cl-cAMP can vary with serum factors that modulate the cellular RIα expression but not the hydrolysis of 8-Cl-cAMP.

3.1.4. 8-Cl-cAMP: Antiproliferative and Proapoptotic Agent

It was shown that 8-Cl-cAMP-induced growth inhibition of LS-174T human colon carcinoma cells was preceded by an increase in the transcription of *RIIβ* gene and a decreased transcription of *RIα* gene and with corresponding changing levels of RIIβ and RIα proteins *(88)*. 8-Cl-cAMP also suppresses *c-myc* and *c-ras* oncogene expression and induces changes in cell morphology *(86)*. The inhibition of growth of LX-1 human lung carcinoma in athymic mice by 8-Cl-cAMP correlated with an increase in RIIβ mRNA and protein levels and with a decrease in RIα mRNA and protein levels *(89)*. A marked increase in RIIβ mRNA along with a decrease in RIα mRNA also preceded the megacaryocytic differentiation induced in K-562 human leukemia cells by site-selective cAMP analogs *(118)*. The antagonistic effect of 8-Cl-cAMP toward TGF-α also brought about changing levels of cAMP receptor mRNA *(87)*. Recently, it was shown that treatment of MCF-7 breast cancer cells with 8-Cl-cAMP not only led to reduction of RIα but also decreased the amount of RFC40-RIα complex, a cell-cycle regulator complex, together with a decrease in cell survival *(93)*.

Two independent, experimental approaches have been used to determine the mechanism of the antitumor activity exhibited by 8-Cl-cAMP: overexpression of Bcl-2 or treatment with ZVAD (a broad-range caspase inhibitor) to specifically block apoptotic cell death without affecting the cell-proliferation pathway; and assessment of the effect of 8-Cl-cAMP in the cell overexpressing RIIβ, which exhibits retarded cell growth and a reverted phenotype but does not undergo spontaneous apoptosis *(119,120)*. At approx 5 d of 8-Cl-cAMP treatment, Bcl-2 is transiently downregulated, and Bad expression continuously increased. Overexpression of Bcl-2 blocks 8-Cl-cAMP-induced apoptosis but has no effect on the accompanying inhibition of cell proliferation. Suppression of apoptosis by ZVAD does not abrogate 8-Cl-cAMP–induced inhibition of cell proliferation, and 8-Cl-cAMP exhibits no additive effect on the inhibition of cell proliferation in cells overexpressing RIIβ. These results indicate that 8-Cl-cAMP inhibits cancer cell growth through both an antiproliferation mechanism and a proapototic mechanism *(120)*. Most likely, 8-Cl-cAMP, being a selective activator of PKA-I but not PKA-II *(68,88,121)* will promote the phosphorylation of Bcl-2, but not Bad, leading to Bcl-2 inactivation and apoptosis. In fact, Bad phosphorylation by PKA-II in mitochondria is shown to activate Bcl-2 *(122)*. Further studies are required to refine the mechanism of action of 8-Cl-cAMP in tumor growing inhibition.

3.1.5. 8-Cl-cAMP: Inhibitor of Clonogenic Growth

Important preclinical studies were conducted for evaluation the effects of 8-Cl-cAMP on clonogenic growth of blast progenitors from 15 patients with acute myeloblastic leukemia and 3 patients affected by advanced myelodysplastic syndrome *(123)*. It was shown that 8-Cl-cAMP was more effective in inhibiting the self-renewing clonogenic cells than the terminally dividing blast cells. In addition, in four out of six cases studied, 8-Cl-cAMP was able to induce a morphologic and/or immunophenotypic maturation of leukemic blasts. An evident reduction of RI levels in fresh leukemic cells after exposure to 8-Cl-cAMP was also detected. The results showing that 8-Cl-cAMP is a powerful inhibitor of clonogenic growth of leukemic blast progenitors by primarily suppressing their self-renewal capacity indicate that this site-selective cAMP analog represents a potent biological agent for acute myeloblastic leukemia therapy in humans *(123)*.

3.1.6. Phase I and Ex Vivo Clinical Studies

A phase I study of 8-Cl-cAMP provided promising results for use of this compound in a clinical setting *(124)*. Thirty-six courses of 8-Cl-cAMP were administered to 17 patients by continuous intravenous infusion of the drug for 5 d/wk for 2 wk followed by a 1-wk rest period. Six increasing dose levels, from 0.01 to 0.25 mg/kg/h, were explored. A grade 4 and a grade 3 increase in serum creatinine and a grade 2 increase in blood urea nitrogen observed at dose level VI (0.25 mg/kg/h) in two patients were the dose-limiting toxicity. Level V (0.2 mg/kg/h) was the maximum tolerated dose in which a grade 1 increase in serum creatinine was observed. An increase in calcium levels was observed in several patients. Pharmacokinetic analysis demonstrated that 8-Cl-cAMP at level IV (0.125 mg/kg/h), a dose devoid of toxicity, achieved plasma concentrations in the potential therapeutic range. Interleukin (IL)-2 receptor α expression, natural killer (NK)cell number, and cytolytic activity of peripheral blood lymphocytes were markedly increased with 8-Cl-cAMP administration at all doses *(124)*.

Recently, it was found that activation of PKA by cAMP agonists, such as 8-Cl-cAMP, selectively causes rapid apoptosis in v-abl transformed fibroblasts by inhibiting the Raf-1 kinase *(125)*. It was investigated whether 8-Cl-cAMP is useful for the treatment of chronic myelogenous leukemia (CML), which is marked by the expression of the $p210^{bcr/abl}$ *$p210^{bcr/abl}$* oncogene. The study of the effects of 8-Cl-cAMP on primary leukemia cells, bone marrow cells (BMCs) from eight CML patients (one at diagnosis, three in chronic, and four in accelerated phase), were treated. Ex vivo treatment of BMCs obtained in chronic phase of CML with 100 μM 8-Cl-cAMP for 24 to 48 h led to the selective purging of Philadelphia Chromosome (Ph I chromosome) without toxic side effects on BMCs of healthy donors as measured by colony-forming unit (CFU) assays. BMCs from patients in accelerated phase showed selective, but incomplete elimination of Ph I chromosome positive colony forming cells. The mechanism of 8-Cl-cAMP was investigated in cells transformed by $p210^{bcr/abl}$, a cell culture model for CML. The results showed that 8-Cl-cAMP reduced DNA synthesis, and viability independent of Raf inhibition as Raf inhibitors had no effect. MEK inhibitors interfered with DNA synthesis, but not with viability. These data indicate that 8-Cl-cAMP could be useful to purge malignant cells from the bone marrow of patients with CML and certain other forms of leukemias.

Further clinical applications of 8-Cl-cAMP may include its combination with other anticancer drugs, differentiation agents, and conventional chemotherapeutic agents.

3.2. Antisense Oligonucleotides

3.2.1. Antisense PKA RIα

In the past years, selective nucleic acid therapeutics, those that can be used to modulate PKA isozyme expression, have been developed. These made a major contribution to the understanding of several PKA-I functions and led to the development of potential anticancer drugs. Results from this approach provided the first direct evidence that the RIα is a positive effector of cancer cell growth. It was found that downregulation of RIα by 21-mer antisense oligonucleotide directed to codons 1–7 of human RIα (15–30 μ*M*) led to growth arrest and differentiation in HL-60 leukemia cells *(126)* and the inhibition of growth in human cancer cells of epithelial origin, including breast (MCF-7), colon (LS-174T), and gastric (TMK-1) carcinoma and neuroblastoma (SK-N-SH) cells, with no sign of cytotoxicity *(91)*. The effect of RIα antisense oligonucleotide correlated with a decrease in RIα protein and a concomitant increase in RIIβ protein levels *(127)*. The increase in RIIβ may, therefore, be responsible for the differentiation in these cells exposed to RIα antisense. In fact, exposure of HL-60 cells to 21-mer RIIβ antisense resulted in a blockade of cAMP-induced growth inhibition and differentiation without apparent effect on the differentiation induced by phorbol esters *(127)*. Thus, RIIβ cAMP receptor, but not RIα, is the mediator of cAMP-induced differentiation in HL-60 cells. The increase in RIIβ at mRNA and protein level has also been correlated with differentiation of K562 chronic myelocytic leukemia *(118)* and Friend erythrocytic leukemia cells *(128)*.

It has been shown that the sequence-specific inhibition of RIα gene expression inhibits in vivo tumor growth *(129)*. A single subcutaneous injection into nude mice bearing LS-174T human colon carcinoma with RIα antisense phosphorothiate oligonucleotide (directed to 8–13 codons of human RIα) resulted in an almost complete suppression of tumor growth for 7 d. There was no apparent sign of systemic toxicity. Even after 14 d, tumor growth was significantly inhibited in the antisense-treated animals. In contrast, tumors in untreated, saline-treated, or control antisense-treated animals showed continued growth. The RIα levels in tumors from the antisense-treated animals were markedly decreased within 24 h and remained at low levels for up to 2 to 3 d. Specific targeting of RIα by the antisense is evident, as RIIα levels remained unchanged. At day 3 after antisense treatment, tumors that contained unreduced amounts of RIα contained a new species of R, RIIβ along with a reduced amount of RIIα. The increase in RIIβ expression was also found in tumors that contained decreased levels of RIα without reduction in RIIα content. RIIβ appeared 24 h to 3 d after antisense treatment but was not detected in control tumors. These data show that the antisense-targeted suppression of RIα brought about a compensatory increase in RIIβ levels.

3.2.2. Antisense RIα Stabilizes the Competitor Molecule of RIIβ

Examination of RIIβ mRNA levels and the rate of RIIβ protein synthesis in the control and antisense-treated LS-174T colon cancer cells revealed that the mechanism of RIIβ compensation demonstrated in the RIα antisense-treated cells does not involve transcriptional or translational control *(130)*. This implies that the increased RIIβ protein observed in the antisense-treated tumor cells must be due to stabilization of the protein. Pulse-chase experiments were performed to determine the half-life of RIIβ protein in control and antisense-treated LS-174T cells. The half-life of RIIα in control cells was approx 2.0 h as measured by immunoprecipitation of [^{35}S]-labeled RIIβ proteinfrom cell extracts after a cold chase with unlabeled methionine. In contrast, the half-life of RIIβ protein in antisense-treated cells was 11 h. This represents a 5.5-fold increase in the half-life of the RIIβ protein on treatment with RIα antisense. The compensatory stabilization of RIIβ protein may represent an important biochemical mechanism of RIα antisense that ensures depletion of PKA-I, leading to sustained inhibition of tumor cell growth *(130)*.

3.2.3. Second Generation Antisense RIα

The "second generation" of structurally modified oligonucleotides has improved the effectiveness of antisense oligos through a wide variety of sugar modifications. The most important modifications involve the 2′ position, such as 2′-*O*-methyl, 2′-*O*-methoxy-ethyl, 2′-*O*-alkyl, or other groups. These analogs generally have increased affinity for RNA and are more resistant to nucleases. Nevertheless, these oligos do not support RNase H activity, and for this reason, their antisense effect is limited to a physical block of translation. Other representatives of "second generation" oligonucleotides have modified phosphate linkages or ribosyl moieties, as well as oligonucleotides with an altered backbone *(131)*.

One of the most interesting examples of such second generation ODNs is the RNA-DNA mixed backbone ODN of RIα antisense, HYB165 (GEM 231) *(132)*. The polyanionic nature of the antisense RIα PS-ODN is minimized, and the immunostimulatory effect (GCGT motif) is blocked in RNA-DNA mixed-backbone RIα antisense ODN *(133,134)*. Such second-generation ODNs have been shown to improve antisense activity, be more resistant to nucleases, form more stable duplexes with RNA and retain the capability to induce RNase H *(135–137)*. Studies conducted in both in vitro and animal models have demonstrated that, following treatment with GEM231, downregulation of PKA-I is balanced by a rapid compensatory increase of PKA-IIβ isoform and is associated with early inhibition of expression of growth factors and their receptors (TGF-α, endothelial growth factor receptor [EGFR], and erbB-2), oncogenes (*myc* and *ras*), and angiogenic factors (vascular endothelial growth factor [VEGF] and basic fibroblast growth factor [bFGF]), as well as the induction of apoptosis and, finally, growth arrest *(138–142)*.

3.2.4. Microarray—Genomic View of Antisense RIα

Our laboratory has conducted an analysis by DNA microarray of normal and cancer cells treated in vitro and in vivo with antisense RIα. We have demonstrated that this agent is able to modulate a wide set of genes related to cell proliferation and transformation *(143)*. It was shown that in a sequence-specific manner, antisense targeted to protein kinase A RIα alters expression of the clusters of coordinately expressed genes at a specific stage of cell growth, differentiation, and activation. The genes that define the proliferation-transformation signature are downregulated, whereas those that define differentiation-reverse transformation signature are upregulated in antisense-treated cancer cells and tumors, but not in host livers. In this differentiation signature, the genes showing the highest induction include genes for the G proteins Rap 1 and Cdc42. The expression signature induced by the endogenously supplied antisense oligodeoxy-nucleotide overlaps strikingly with that induced by endogenous antisense gene over-expression. Defining antisense DNAs on the basis of their effects on global gene expression can lead to identification of clinically relevant antisense therapeutics and can identify which molecular and cellular events might be important in complex biological processes, such as cell growth and differentiation *(143)*.

3.2.5. Antisense RIα: Combinatorial Therapy With Cytotoxic Drugs and Monoclonal Antibodies

Several studies have shown that the antisense RIα is able to cooperate with a variety of anticancer drugs of different classes, following intraperitonial as well as oral admini-stration of the antisense. In particular, synergistic antitumor activity associated with increased apoptosis can be obtained with taxanes, topoisomerase I and II inhibitors, and platinum derivatives, both in vitro and in nude mice bearing a wide variety of human cancer types *(144–147)*. The biochemical and molecular basis of the cooperative effect observed include the sensitization of cells to certain anticancer agents following down-regulation of PKA-I and compensatory increase of PKA-II, the pharmacokinetic inter-actions of antisense with certain drugs, and finally, the involvement of PKA-I in signaling pathways hit by cytotoxic drugs.

A group of investigators proposed that the functional interactions of EGFR and PKA-I may provide the basis for the development of a therapeutic strategy based on the com-bination of their selective inhibitors *(148)*. The combination of MabC225 (chimeric monoclonal antibody [MAb] against EGFR-erbitux) with either the site-selective cAMP analog 8-Cl-cAMP or, later, with GEM231, has been the first demonstration of the fea-sibility and the antitumor activity of the combined blockade of pathways that are criti-cal for cancer cell proliferation, survival, and progression and the regression of tumor xenografs in vivo in nude mice *(149)*. The combination of low doses of these agents causes a marked cooperative antitumor effect in vitro and in vivo in nude mice accom-panied by a significant prolongation of survival with no sign of toxicity.

Several studies established a link between PKA, bcl-2, and apoptosis. PKA seems to be involved in bcl-2 phosphorylation following treatment with paclitaxel and microtubule-damaging agents, whereas the PKA-I subunit RIα is directly bound to cytochrome *c* oxidase, and PKA-I inhibition causes cytochrome *c* release and apop-tosis *(150)*. Antisense PKA-I is able to inhibit bcl-2 expression and function, as well as induce cleavage of PARP, activation of caspase 3, and finally, apoptosis *(151)*. More recently, it has been demonstrated that the PKA RIα antisense can induce phosphorylation of bcl-2 and hypohposphorylation of BAD, thus causing bcl-2 inactivation and induction of apoptosis in androgen-independent human prostate cancer cells *(152)*. These data also provide an explanation for the enhanced apoptotic activity observed when antisense RIα is associated with cytotoxic drugs. For these reasons, it was investigated whether the combined blockade of PKA and bcl-2 by antisense strategy may represent a potential therapeutic approach *(153)*. It was demonstrated that oral administration of GEM231 in combination with intraperi-toneal injection of antisense bcl-2, oblimersen, has a marked antitumor effect and causes a significant prolongation of survival in nude mice bearing human colon cancer xenografs *(153)*.

3.2.6. Antisense RIα: Phase II Clinical Study

Antisense RIα has completed phase I studies *(131)* and is now entering phase II evaluation in combination with cytotoxic drugs. It could have therapeutic applications in combination with other agents through enhancement of their antitumor activity and the triggering of apoptosis, as well as after conventional therapy, by turning off mitogenic signals and inducing a state of tumor dormancy. This therapeutic strategy would allow the use of lower doses of cytotoxic drugs or radiation and a more selective and long-term control of cancer with moderate toxicity.

3.2.7. Antisense RIα: Chemoprevention

Because PKA-I seems to participate in the signals triggered by proteins implicated in the process of neoplastic transformation, antisense RIα may have a role in the field of cancer chemoprevention. It was shown that, in DMBA-induced mammary carcinogenesis, RIα antisense inhibited the tumor production in a sequence-specific manner *(154)*. The results demonstrated that RIα antisense produces dual anticarcinogenic effects: (1) increasing DMBA detoxification in the liver by increasing phase II enzyme activities, via increasing CRE-binding-protein phosphorylation and enhancing CRE-and AP-1 directed transcription; and (2) activating DNA repair processes in the mammary gland by downregulating PKA-I.

3.3. CRE-Transcription Factor Decoy

The CRE consensus sequence is intimately involved in the transcription of a wide range of genes *(155)*. The promoter regions of several of these genes have been studied, and a common CRE sequence has been found upstream of the transcription start site *(156)*. All of the cAMP responsive gene promoter regions have the same eight-base enhancer sequence, the CRE, which is the palindromic sequence 5′-TGACGTCA-3′ *(157)*. Protein that binds to the CREs has been identified as 43-kDa in size, contains a basic leucine zipper DNA-binding motif, and is activated after phosphorylation by cAMP-dependent protein kinase *(158)*. Functional studies have shown that this transcription factor, termed the CRE-binding protein (CREB), couples gene activation to a wide variety of cellular signals, and thus coordinates a multitude of genes that regulate numerous cellular processes, including cell growth and differentiation *(159–161)*.

The ubiquitous nature of the CRE consensus site makes it a good target for chemotherapy. Synthetic double-stranded phosphorothioate oligonucleotoides with high affinity for a target transcription factor can be introduced into cells as decoy *cis*-elements to bind the factors and alter gene expression. Because the CRE *cis*-element is palindromic, a synthetic single-stranded oligonucleotide composed of the CRE sequence self-hybrydizes to form a duplex/ hairpin. It has been shown that a palindromic trioctamer of this sequence can interfere with CREB binding, and specifically inhibits PKA subunit expression, interfering with the CRE-PKA pathway *(162)*. This oligonucleotide restrained tumor cell proliferation, without affecting the growth of noncancerous cells. Furthermore, in animal studies, CRE-decoy oligo inhibited tumor growth in nude mice without obvious toxicity *(162)*.

CREB is known to heterodimerize with a variety of other transcription factors, including members of the Jun/Fos family *(163)*. *c-fos* is also induced by cAMP, suggesting crosstalk between the CRE and AP-1 pathways. The CRE decoy brings about a marked decrease in AP-1 binding resulting from decreases in c-fos expression *(162)*. This result demonstrates CRE decoy inhibition of transcription factor binding at two different *cis*-elements: CRE and AP-1. Moreover, it was shown, that CRE decoy upregulates p53 and inhibits the cyclin D1/Cdk4/pRB signaling pathway *(164)*.

In a recent report, the effect of CRE-decoy oligo alone in a panel of three colorectal cancer cell lines and the effect of combining this oligonucleotide with etoposide, 5-fluorouracil on cell growth and viability have been investigated *(165)*. Simple drug–drug interaction studies showed that combining CRE-decoy oligo with chemotherapy resulted in an enhancement of the antiproliferative effects. Furthermore, this cytostatic effect was protracted and associated with an increase in senescence-associated β-galactosidase activity at pH 6.0. There is a possible role for p21^{waf1} in mediating this effect, as the enhancement of cell growth inhibition was not observed in cells lacking the ability to correctly upregulate this protein. Additionally, significant decreases in cyclin-dependent kinase (CDK)-1 and CDK-4 function were seen in the responsive cells. These data provide a possible model of drug interaction in colorectal cell lines, which involves the complex interplay of the molecules regulating the cell cycle. Clinically, the cytostatic ability of CRE-decoy oligo could improve and enhance the antiproliferative effects of conventional cytotoxic agents.

These results support the ability of the CRE decoy oligonucleotide to regulate the expression of cAMP-responsive genes underlying tumorigenesis and tumor progression.

4. PKA—A Cancer Diagnostic Tool

Recently, the presence of active PKA in the form of free C subunit was found in the sera of patients with cancer, as well as in conditioned medium of cancer cells in tissue culture *(166–168)*. This ECPKA phosphorylates Kemptide, a PKA-specific synthetic peptide substrate containing the consensus phosphorylation site of PKA *(166)*. The ECPKA activity is specifically inhibited by PKI, a PKA peptide inhibitor, but not by PKC-specific peptide inhibitor, and is not activated by cAMP *(166–168)*. Biochemical and immunological characterization have shown that ECPKA is identical to the free Cα subunit of intracellular PKA *(166)*. A striking correlation was found between PKA-I overexpression in the cell and ECPKA secretion. Cα and RIα transfectants, which upregulate PKA-I, increased ECPKA expression *(166)*. Conversely, downregulation of PKA-I in RIIβ transfectants correlated with the sharp downregulation of ECPKA *(166)*. Overexpression of RIα-P, which functionally mimics RIIβ, downregulated ECPKA expression as compared with that by RIα, and mutant RIIβ-P, a functional mimic of RIα, upregulated ECPKA as compared with RIIβ cells.

Overexpression of mutant Cα, which lacks the N-terminal myristate, upregulated total cellular PKA activity and PKA-I holoenzyme, but barely increased ECPKA, indicating a structural requirement of ECPKA secretion *(166)*. Increase of functional PKA-I via RIα or Cα overexpression in the cell, which may promote cell proliferation and neoplastic transformation, enhances ECPKA secretion; conversely, ECPKA secretion is decreased in tumor cells that are growth arrested via RIIβ overexpression, which down-regulates PKA-I and upregulates PKA-IIβ (RIIβ containing PKA-II). Recently, the clinical significance of ECPKA in melanoma patients has been demonstrated *(169)*. The results showed the presence of ECPKA activity in the serum of melanoma patients, which correlated with the appearance and size of the tumor. Most importantly, surgical removal of melanoma caused a precipitous decrease in ECPKA activity in the sera of patients, suggesting that ECPKA may be a novel predictive marker in melanoma.

5. Conclusion

Cyclic AMP-dependent protein phosphorylation has been implicated in the regulation of a wide variety of cellular processes. Activation occurs by binding of cAMP to R subunits of PKA resulting in a release of free C subunits and subsequent phosphorylation of a variety of protein substrates. How can one enzyme with broad substrate specificity be in charge of regulation of so many cellular processes? The control of cAMP-dependent phosphorylation occurs on many different levels. First, this enzyme exists in the form of two isozymes, which have several isoforms of R and C subunits. The differences in promoter organization allow the differential regulation of the expression of the enzyme in response for extracellular stimuli. The difference in physicochemical properties allows for activation of the enzyme at various intracellular conditions. The localization of PKA in the cell is another extremely important point necessary to achieve the specificity of biological function of PKA. There are many experimental evidences for distinct function of PKA-I and PKA-II, providing molecular proof for the importance of intracellular balanced expression of the two isoforms, which can play a critical role in controlling cell growth and differentiation. Any disturbance of the balance between subunits can lead to dramatic changes in the cell.

The key role in biological regulation makes PKA an important target for antitumor drugs. One possibility for modulating the balance between PKA isozymes is the use of site-selective analogs of cAMP. It was discovered that site-selective cAMP analogs can act as novel biological agents capable of inducing growth inhibition and differentiation in a broad spectrum of human cancer cell lines, including carcinomas, sarcomas, and leukemias, without causing cytotoxicity. These studies resulted in the selection of 8-Cl-cAMP, the most potent site-selective cAMP analog for preclinical and clinical phase I studies.

RIα, which has completed phase I studies and is now entering phase II evaluation in combination with cytotoxic drugs, could have therapeutic applications in combination with other agents by enhancing their antitumor activity and triggering apoptosis, as well as after conventional therapy, by turning off mitogenic signals and inducing a state of tumor dormancy.

The CRE transcription factor complex is a pleiotropic activator that participates in the induction of a wide variety of cell proliferation genes. Decoy technology is useful in studying the role of CRE-directed transcription in tumorigenesis, tumor progression, and cancer therapy. The CRE decoy is harmless to normal cells, but it is a potent inhibitor of cancer cell growth both in vitro and in vivo.

Various cancer cells excrete PKA into conditioned media. Compared with serum taken from healthy persons, serum taken from cancer patients exhibit marked upregulation of ECPKA expression. Thus, it may be possible to use serum PKA as a tool for diagnosing cancer.

Thus, the diversity and complexity of the cAMP signaling are highly dependent on different stages of normal cellular development and differentiation, and such signaling is disrupted in an abnormal physiology such as cancer. We believe that the large amount of experimental evidence now accumulated strongly supports the idea that PKA is a valuable target for cancer treatment and diagnosis.

References

1. Sutherland EW, Rall TW. Fractionation and characterization of a cyclic adenine ribonucleotide formed by tissue particles. J Biol Chem 1958;232:1077–1092.
2. Walsh DA, Perkins JP, Krebs EG. An adenosine 3′,5′-monophosphate-dependant protein kinase from rabbit skeletal muscle. J Biol Chem 1968;243:3763–3765.
3. Gill GN, Garren LD. Role of the receptor in the mechanism of action of adenosine 3′:5′-cyclic monophosphate. Proc Natl Acad Sci USA 1971;68:786–790.
4. Kuo JF, Greengard P. Cyclic nucleotide-dependent protein kinases IV. Widespread occurrence of adenosine 3′:5′-monophosphate-dependent protein kinase in various tissues and phyla of the animal kingdom. Proc Natl Acad Sci USA 1969;64:1349–1355.
5. Beebe SJ, Corbin JD. Cyclic nucleotide-dependent protein kinases. In: Krebs EG, Boyer PD, eds. The Enzymes: Control by Phosphorylation. Orlando and London. Academic Press, New York, 1986:43–111.
6. Døskeland SO, Maronde E, Gjertsen BT. The genetic subtypes of cAMP-dependent protein kinase—functionally different or redundant? Biochim Biophys Acta 1993;1178(3): 249–258.
7. Reimann EM, Walsh DA, Krebs EG. Purification and properties of rabbit skeletal muscle adenosine 3′:5′-monophosphate-dependent protein kinases. J Biol Chem 1971;246:1986–1995.
8. Corbin JD, Keely SL, Park CR. The distribution and dissociation of cyclic adenosine 3′,5′-monophosphate-dependent protein kinases in adipose, cardiac, and other tissues. J Biol Chem 1975;250:218–225.
9. Lee DC, Carmichael DF, Krebs EG, McKnight GS. Isolation of a cDNA clone for the type I regulatory subunit of bovine cAMP-dependent protein kinase. Proc Natl Acad Sci USA 1983;80:3608–3612.
10. Sandberg M, Tasken K, Oyen O, Hansson V, Jahnsen T. Molecular cloning, cDNA structure and deduced amino acid sequence for a type I regulatory subunit of cAMP-dependent protein kinase from human testis. Biochem Biophys Res Commun 1987;149:939–945.
11. Clegg CH, Cadd GG, McKnight GS. Genetic characterization of a brain-specific form of the type I regulatory subunit of cAMP-dependent protein kinase. Proc Natl Acad Sci USA 1988;85:3703–3707.
12. Solberg R, Tasken K, Keiserud A, Jahnsen T. Molecular cloning, cDNA structure and tissue-specific expression of the human regulatory subunit RI beta of cAMP-dependent protein kinases. Biochem Biophys Res Commun 1991;176:166–172.
13. Scott JD, Glaccum MB, Zoller MJ, Uhler MD, Helfman DM, McKnight GS, et al. The molecular cloning of a type II regulatory subunit of the cAMP-dependent protein kinase from rat skeletal muscle and mouse brain. Proc Natl Acad Sci USA 1987;84:5192–5196.
14. Øyen O, Myklebust F, Scott JD, Hansson V, Jahnsen T. Human testis cDNA for the regulatory subunit RII alpha of cAMP-dependent protein kinase encodes an alternate amino-terminal region. FEBS Lett 1989;246:57–64.
15. Jahnsen T, Hedin L, Kidd VJ, Beattie WG, Lohmann SM, Walter U, et al. Molecular cloning, cDNA structure, and regulation of the regulatory subunit of type II cAMP-dependent protein kinase from rat ovarian granulosa cells. J Biol Chem 1986;261:12,352–12,361.
16. Levy FO, Oyen O, Sandberg M, Tasken K, Eskild W, Hansson V, et al. Molecular cloning, complementary deoxyribonucleic acid structure and predicted full-length amino acid sequence of the hormone-inducible regulatory subunit of 3′,5′-cyclic adenosine monophosphate-dependent protein kinase from human testis. Mol Endocrinol 1988;2: 1364–1373.

17. Uhler MD, Carmichael DF, Lee DC, Chrivia JC, Krebs EG, McKnight GS. Isolation of cDNA clones coding for the catalytic subunit of mouse cAMP-dependent protein kinase. Proc Natl Acad Sci USA 1986;83:1300–1304.

18. Uhler MD, Chrivia JC, McKnight GS. Evidence for a second isoform of the catalytic subunit of cAMP-dependent protein kinase. J Biol Chem 1986;261:15,360–15,363.

19. Showers MO, Maurer RA. A cloned bovine cDNA encodes an alternate form of the catalytic subunit of cAMP-dependent protein kinase. J Biol Chem 1986;261:16,288–16,291.

20. Beebe SJ, Oyen O, Sandberg M, Froysa A, Hansson V, Jahnsen T. Molecular cloning of a unique tissue-specific protein kinase (C-gamma) from human testis—representing a third isoform for the catalytic subunit of the cAMP-dependent protein kinase. Mol Endocrinol 1990;4:465–475.

21. Ulmasov KhA, Nesterova MV, Severin ES. cAMP-dependent protein kinase from pig brain: subunit structure, mechanism of autophosphorylation and dissociation into subunits under the action of cAMP. Biochemistry (N.Y.) 1980;45:639–647.

22. Corbin JD, Keely SL, Soderling TR, Park CR. Hormonal regulation of adenosine 3′,5′-monophosphate-dependent protein kinase. Adv Cyclic Nucleotide Res 1975;5:265–279.

23. Hofman F, Beavo JA, Bechtel PJ, Krebs EG. Comparison of adenosine 3′,5′-monophosphate-dependent protein kinase from rabbit skeletal and bovine heart muscle. J Biol Chem 1975;250:7795–7801.

24. Rangel-Aldao R, Rosen OM. Mechanism of self-phosphorylation of adenosine 3′:5′-monophosphate-depedent protein kinase from bovine cardiac muscle. J Biol Chem 1976;251:3375–3380.

25. Dostmann WR, Taylor SS, Genieser HG, Jastorff B, Doskeland SO, Ogreid D. Probing the cyclic nucleotide binding sites of cAMP-dependent protein kinases I and II with analogs of adenosine 3′,5′-cyclic phosphorothioates. J Biol Chem 1990;265:10,484–10,491.

26. Hofman F. Apparent constants for the interaction of regulatory and catalytic subunit of cAMP-dependent protein kinase I and II. J Biol Chem 1980;255:1559–1564.

27. Otten AD, McKnight GS. Overexpression of the type II regulatory subunit of the cAMP-dependent protein kinase eliminates the type I holoenzyme in mouse cells. J Biol Chem 1989;264:20,255–20,260.

28. Amieux PS, Cummings DE, Motamed K, Brandon EP, Wailes LA, Le K, et al. Compensatory regulation of RIalpha protein levels in protein kinase A mutant mice. J Biol Chem 1997;272:3993–3998.

29. Tasken K, Skalhegg BS, Solberg R, Andersson KB, Taylor SS, Lea T, et al. Novel isozymes of cAMP-dependent protein kinase exist in human cells due to formation of RI alpha-RI beta heterodimeric complexes. J Biol Chem 1993;268:21,276–21,283.

30. Sugden PH, Corbin JD. Adenosine 3′:5′-cyclic monophosphate-binding proteins in bovine and rat tissues. Biochem J 1976;159:423–427.

31. Hofmann F, Bechtel PJ, Krebs EG. Concentrations of cyclic AMP-dependent protein kinase subunits in various tissues. J Biol Chem 1977;252(4):1441–1447.

32. Scott JD, McCartney S. Localization of A-kinase through anchoring proteins. Mol Endocrinol 1994;8:5–11.

33. Meinkoth JL, Ji Y, Taylor SS, Feramisco JR. Dynamics of the distribution of cyclic AMP-dependent protein kinase in living cells. Proc Natl Acad Sci USA 1990;87:9595–9599.

34. Vallee RB, DiBartolomeis MJ, Theurkauf WE. A protein kinase bound to the projection portion of MAP 2 (microtubule-associated protein 2). J Cell Biol 1981;90(3):568–576.

35. Imaizumi-Scherrer T, Faust DM, Barradeau S, Hellio R, Weiss MC. Type I protein kinase a is localized to interphase microtubules and strongly associated with the mitotic spindle. Exp Cell Res 2001;264:250–265.

36. Keryer G, Skalhegg BS, Landmark BF, Hansson V, Jahnsen T, Tasken K. Differential local-ization of protein kinase A type II isozymes in the Golgi-centrosomal area. Exp Cell Res 1999;249:131–146.

37. Keryer G, Yassenko M, Labbe JC, Castro A, Lohmann SM, Evain-Brion D, et al. Mitosis-specific phosphorylation and subcellular redistribution of the RIIalpha regulatory subunit of cAMP-dependent protein kinase. J Biol Chem 1998;273:34,594–34,602.

38. Carlson CR, Witczak O, Vossebein L, et al. CDK1-mediated phosphorylation of the RIIalpha regulatory subunit of PKA works as a molecular switch that promotes dissociation of RIIalpha from centrosomes at mitosis. J Cell Sci 2001;114:3243–3254.

39. Kondrashin AA, Nesterova MV, Cho-Chung YS. Subcellular distribution of the R-subunits of cAMP-dependent protein kinase in LS-174T human colon carcinoma cells. Biochem Mol Biol Int 1998;45:237–244.

40. Kondrashin A, Nesterova M, Cho-Chung YS. Cyclic adenosine 3′:5′-monophosphate-dependent protein kinase on the external surface of LS-174T human colon carcinoma cells. Biochemistry 1999;38:172–179.

41. Riabowol KT, Fink JS, Gilman MZ, Walsh DA, Goodman RH, Feramisco JR. The catalytic subunit of cAMP-dependent protein kinase induces expression of genes containing cAMP-responsive enhancer elements. Nature 1988;336:83–86.

42. Solberg R, Tasken K, Wen W, et al. Human regulatory subunit RI beta of cAMP-dependent protein kinases: expression, holoenzyme formation and microinjection into living cells. Exp Cell Res 1994;214:595–605.

43. Neary CL, Cho-Chung YS. Nuclear translocation of the catalytic subunit of protein kinase A induced by an antisense oligonucleotide directed against the RIalpha regulatory subunit. Oncogene 2001;20:8019–8024.

44. Nesterova MV, Ulmasov KHA, Abdukarimov A, Aripdzhanov AA, Severin ES. Nuclear translocation of cAMP-dependent protein kinase. Expl Cell Res 1981;132:367–373.

45. Aprikian AG, Nesterova MV, Glukhov AI, Severin ES. Binding of a holoenzyme and cAMP-dependent protein kinase subunits with cell nuclei. Biochemistry (NY)1987;52: 1118–1124.

46. Kapoor CL, Grantham F, Cho-Chung YS. Nucleolar accumulation of cyclic adenosine 3′:5′-monophosphate receptor proteins during regression of MCF-7 human breast tumor. Cancer Res 1984;44:3554–3560.

47. Constantinescu A, Wu M, Asher O, Diamond I. cAMP-dependent protein kinase type I regu-lates ethanol-induced cAMP response element-mediated gene expression via activation of CREB-binding protein and inhibition of MAPK. J Biol Chem 2004;279: 43,321–43,329.

48. Neary CL, Nesterova M, Cho YS, Cheadle C, Becker KG, Cho-Chung YS. Protein kinase A isozyme switching: eliciting differential cAMP signaling and tumor reversion. Oncogene 2004;23:8847–8856.

49. Feliciello A, Giuliano P, Porcellini A, et al. The v-Ki-Ras oncogene alters cAMP nuclear signaling by regulating the location and the expression of cAMP-dependent protein kinase IIβ. J Biol Chem 1996; 271:25,350–25,359.

50. Haddox MK, Roeske WR, Russell DH. Independent expression of cardiac type I and II cyclic AMP-dependent protein kinase during murine embryogenesis and postnatal deve-lopment. Biochim Biophys Acta 1979;585:527–534.

51. Malkinson AM, Hogy L, Gharrett AJ, Gunderson TJ. Ontogenetic studies of cyclic AMP-dependent protein kinase enzymes from mouse heart and other tissues. J Exp Zool 1978; 205:423–431.

52. Claycomb WC. Biochemical aspects of cardiac muscle differentiation. Deoxyribonucleic acid synthesis and nuclear and cytoplasmic deoxyribonucleic acid polymerase activity. J Biol Chem 1975;250:3229–3235.

53. Jonassen JA, Bose K, Richards JS. Enhancement and desensitization of hormone-responsive adenylate cyclase in granulosa cells of preantral and antral ovarian follicles: effects of estradiol and follicle-stimulating hormone. Endocrinology 1982;111:74–79.

54. Fuller DJ, Byus CV, Russell DH. Specific regulation by steroid hormones of the amount of type I cyclic AMP-dependent protein kinase holoenzyme. Proc Natl Acad Sci USA 1978; 75:223–227.

55. Costa M, Gerner EW, Russell DH. Cell cycle-specific activity of type I and type II cyclic adenosine 3':5'-monophosphate-dependent protein kinases in Chinese hamster ovary cells. J Biol Chem 1976;251(11):3313–3319.

56. Byus CV, Klimpel GR, Lucas DO, Russell DH. Type I and type II cyclic AMP-dependent protein kinase as opposite effectors of lymphocyte mitogenesis. Nature 1977;268:63–64.

57. Kammer GM. The adenylate cyclase-cAMP-protein kinase A pathway and regulation of the immune response. Immunol Today 1988;9:222–229.

58. Laxminarayana D, Kammer GM. Activation of type I protein kinase A during receptor-mediated human T lymphocyte activation. J Immunol 1996;156:497–506.

59. Skalhegg BS, Landmark BF, Doskeland SO, Hansson V, Lea T, Jahnsen T. Cyclic AMP-dependent protein kinase type I mediates the inhibitory effects of 3',5'-cyclic adenosine monophosphate on cell replication in human T lymphocytes. J Biol Chem 1992;267: 15,707–15,714.

60. Tasken K, Andersson KB, Erikstein BK, Hansson V, Jahnsen T, Blomhoff HK. Regulation of growth in a neoplastic B cell line by transfected subunits of 3',5'-cyclic adenosine monophosphate-dependent protein kinase. Endocrinology 1994;135:2109–2119.

61. Schwartz DA, Rubin CS. Regulation of cAMP-dependent protein kinase subunit levels in Friend erythroleukemic cells. J Biol Chem 1983;258:777–784.

62. Strickland S, Smith KK, Marotti KR. Hormonal induction of differentiation in teratocarcinoma stem cells: generation of parietal endoderm by retinoic acid and dibutyryl cAMP. Cell 1980;21:347–355.

63. Plet A, Evain D, Anderson WB. Effect of retinoic acid treatment of F9 embryonal carcinoma cells on the activity and distribution of cyclic AMP-dependent protein kinase. J Biol Chem 1982;257:889–893.

64. Ng KW, Livesey SA, Larkins RG, Martin TJ. Calcitonin effects on growth and on selective activation of type II isoenzyme of cyclic adenosine 3':5'-monophosphate-dependent protein kinase in T 47D human breast cancer cells. Cancer Res 1983;43:794–800.

65. Cho-Chung YS. Hypothesis: cyclic AMP and its receptor protein in tumor growth regulation in vivo. J Cyclic Nucl Res 1980;6:163–177.

66. Russell DH. Type I cyclic AMP-dependent protein kinase as a positive effector of growth. Adv Cycl Nucl Res 1978;9:493–506.

67. Katsaros D, Tortora G, Tagliaferri P, et al. Site-selective cyclic AMP analogs provide a new approach in the control of cancer cell growth. FEBS Lett 1987;223:97–103

68. Cho-Chung YS, Clair T, Tagliaferri P, et al. Site-selective cyclic AMP analogs as new biological tools in growth control, differentiation and proto-oncogene regulation. Cancer Inv 1989;7:161–177.

69. Eppenberger U, Briedermann K, Handshin JC, et al. Cyclic AMP-dependent protein kinase type I and type II and cyclic AMP binding in human mammary tumours. Adv Cyclic Nucleotide Res 1980;12:123–128.

70. Katsaros D, Ally S, Cho-Chung YS. Site-selective cyclic AMP analogues are antagonistic to estrogen stimulation of growth and proto-oncogene expression in human breast-cancer cells. Int J Cancer 1988;41:863–867.

71. Miller WR. Regulatory subunits of PKA and breast cancer. Ann NY Acad Sci 2002;968: 37–48.

72. Fossberg TM, Doskeland SO, Ueland PM. Protein kinases in human renal cell carcinoma and renal cortex. A comparison of isozyme distribution and of responsiveness to adenosine 3′:5′-cyclic monophosphate. Arch Biochem Biophys 1978;189:372–381.

73. Nakajima F, Imashuku S, Wilimas J, Champion JE, Green AA. Distribution and properties of type I and type II binding proteins in the Wilms' tumors. Cancer Res 1984;44: 5182–5187.

74. Bradbury AW, Miller WR, Clair T, Yokozaki H, Cho-Chung YS. Overexpressed type I regulatory subunit (RI) of cAMP-dependent protein kinase (PKA) as tumor marker in colorectal cancer. Proc Am Assoc Cancer Res 1990;31:172.

75. Sand G, Jortay A, Pochet R, Dumont JE. Adenylate cyclase and protein phosphokinase activities in human thyroid. Comparison of normal glands, hyperfunctional nodules and carcinomas 1976;12:447–453.

76. Trabucchi M, Canal N, Frattola L. Cyclic nucleotides in human cerebral tumors: role of the protein kinase system. Adv Cyclic Nucleotide Protein Phosphorylation Res 1984;17: 671–676.

77. Pena JM, Itarte E, Domingo A, Cusso R. Cyclic adenosine 3′:5′-monophosphate-dependent and-independent protein kinases in human leukemic cells. Cancer Res 1983;43:1172–1175.

78. Weber W, Schwoch G, Wielckens K, Gartemann A, Hilz H. cAMP receptor proteins and protein kinases in human lymphocytes: fundamental alterations in chronic lymphocytic leukemia cells. Eur J Biochem 1981;120:585–592.

79. Cho-Chung YS, Clair T, Shepheard C. Anticarcinogenic effect of N6,O2-dibutyryl cyclic adenosine 3′,5′-monophosphate on 7,12-dimethylbenz(α) anthracene mammary tumor induction in the rat and its relationship to cyclic adenosine 3′,5′-monophosphate metabolism and protein kinase. Cancer Res 1983;43:2736–2740.

80. Yasui W, Tahara E. Effect of gastrin on gastric mucosal cyclic adenosine 3′,5′-monophosphate-dependent protein kinase activity in rat stomach carcinogenesis induced by N-methyl-N-nitro-N-nitrosoguanidine. Cancer Res 1985;45:4763–4767.

81. Butley MS, Stoner GD, Beer DG, Beer DS, Mason RJ, Malkinson AM. Changes in cyclic adenosine 3′:5′-monophosphate-dependent protein kinases during the progression of urethan-induced mouse lung tumors. Cancer Res 1985;45:3677–3685.

82. Lange-Carter CA, Fossli T, Jahnsen T, Malkinson AM. Decreased expression of the type I isozyme of cAMP-dependent protein kinase in tumor cell lines of lung epithelial origin. J Biol Chem 1990;265:7814–7818.

83. Wehner JM, Malkinson AM, Wiser MF, Sheppard JR. Cyclic AMP-dependent protein kinases from Balb 3T3 cells and other 3T3 derived lines. J Cell Physiol 1981;108:175–184.

84. Ledinko N, Chan IJ. Increase in type I cyclic adenosine 3′:5′-monophosphate-dependent protein kinase activity and specific accumulation of type I regulatory subunits in adenovirus type 12-transformed cells. Cancer Res 1984;44:2622–2627.

85. Tagliaferri P, Katsaros D, Clair T, Neckers L, Robins RK, Cho-Chung YS. Reverse transformation of Harvey murine sarcoma virus-transformed NIH/3T3 cells by site-selective cyclic AMP analogs. J Biol Chem 1988;263:409–416.

86. Tortora G, Ciardiello F, Ally S, Clair T, Salomon DS, Cho-Chung YS. Site-selective 8-chloroadenosine 3′,5′-cyclic monophosphate inhibits transformation and transforming growth factor alpha production in Ki-ras-transformed rat fibroblasts. FEBS Lett 1989;242: 363–367.

87. Ciardiello F, Tortora G, Kim N, et al. 8-chloro-cAMP inhibits transforming growth factor α transformation of mammary epithelial cells by restoration of the normal mRNA patterns for cAMP-dependent protein kinase regulatory subunit isoforms which show disruption upon transformation. J Biol Chem 1990;265:1016–1020.

88. Ally S, Tortora G, Clair T, et al. Selective modulation of protein kinase isozymes by the site-selective analog 8-chloroadenosine 3′,5′-cyclic monophosphate provides a biological means for control of human colon cancer cell growth. Proc Natl Acad Sci USA 1988;85: 6319–6322.

89. Ally S, Clair T, Katsaros D, et al. Inhibition of growth and modulation of gene expression in human lung carcinoma in athymic mice by site-selective 8-Cl-cyclic adenosine monophosphate. Cancer Res 1989;49:5650–5655.

90. Tortora G, Yokozaki H, Pepe S, Clair T, Cho-Chung YS. Differentiation of HL-60 leukemia cells by type I regulatory subunit antisense oligodeoxynucleotide of cAMP-dependent protein kinase. Proc Natl Acad Sci USA 1991;88:2011–2015.

91. Yokozaki H, Budillon A, Tortora G, et al. An antisense oligodeoxynucleotide that depletes RIα subunit of cyclic AMP-dependent protein kinase induces growth inhibition in human cancer cells. Cancer Res 1993;53:868–872.

92. Tortora G, Cho-Chung YS. Type II regulatory subunit of protein kinase restores cAMP-dependent transcription in a cAMP-unresponsive cell line. J Biol Chem 1990;265: 18,067–18,070.

93. Gupte RS, Weng Y, Liu L, Lee MY. The Second Subunit of the Replication Factor C complex (RFC40) and the Regulatory Subunit (RIalpha) of Protein Kinase A Form a Protein Complex Promoting Cell Survival. Cell Cycle 2005;4:323–329.

94. Levin DS, Vijayakumar S, Liu X, Bermudez VP, Hurwitz J, Tomkinson AE. A conserved interaction between the replicative clamp loader and DNA ligase in eukaryotes: implications for Okazaki fragment joining. J Biol Chem 2004;279:55,196–55,201.

95. Parrilla-Castellar ER, Arlander SJ, Karnitz L. Dial 9-1-1 for DNA damage: the Rad9-Hus1-Rad1 (9-1-1) clamp complex. DNA Repair (Amst) 2004;3:1009–1014.

96. Durgerian S, Taylor SS. The consequences of introducing an autophosphorylation site into the type I regulatory subunit of cAMP-dependent protein kinase. J Biol Chem 1989;264: 9807–9813.

97. Kuret J, Johnson KE, Nicolette C, Zoller MJ. Mutagenesis of the regulatory subunit of yeast cAMP-dependent protein kinase: Isolation of site-directed mutants with altered binding affinity for catalytic subunit. J Biol Chem 1988;263:9149–9154.

98. Nesterova MV, Yokozaki H, McDuffie L, Cho-Chung YS. Overexpression of RIIβ regulatory subunit of protein kinase A in human colon carcinoma cell induces growth arrest and phenotypic changes that are abolished by site-directed mutation of RIIβ. Eur J Biochem 1996;235:486–494.

99. Lee GR, Kim SN, Noguchi K, Park SD, Hong SH, Cho-Chung YS. Ala99ser mutation in RI alpha regulatory subunit of protein kinase A causes reduced kinase activation by cAMP and arrest of hormone-dependent breast cancer cell growth. Mol Cell Biochem 1999;195:77–86.

100. Kirschner LS, Carney JA, Pack SD, et al. Mutations of the gene encoding the protein kinase A type I-alpha regulatory subunit in patients with the Carney complex. Nat Genet 2000; 26: 89–92.

101. Bossis I, Voutetakis A, Bei T, Sandrini F, Griffin KJ, Stratakis CA. Protein kinase A and its role in human neoplasia: the Carney complex paradigm. Endocr Relat Cancer 2004;11: 265–280.

102. Griffin KJ, Kirschner LS, Matyakhina L, et al. Down-regulation of regulatory subunit type 1A of protein kinase A leads to endocrine and other tumors. Cancer Res 2004;64: 8811–8815.

103. Griffin KJ, Kirschner LS, Matyakhina L, et al. A transgenic mouse bearing an antisense construct of regulatory subunit type 1A of protein kinase A develops endocrine and other tumours: comparison with Carney complex and other PRKAR1A induced lesions. J Med Genet 2004;41:923–931.

104. Lania AG, Mantovani G, Ferrero S, et al. Proliferation of transformed somatotroph cells related to low or absent expression of protein kinase a regulatory subunit 1A protein. Cancer Res 2004;64:9193–9198.

105. Burk RR. Reduced adenyl cyclase activity in a polyomavirus transformed cell line. Nature 1968;219:1272–1275.
106. Johnson GS, Friedman RM, Pastan I. Restoration of several morphological characteristics of normal fibroblasts in sarcoma cells treated with adenosine-3′,5′-cycle monophosphate and its derivatives. Proc Natl Acad Sci USA 1971;68:425–429.
107. Cho-Chung YS, Gullino PM. *In vivo* inhibition of growth of two hormone-dependent mammary tumors by dibutyryl cyclic AMP. Science 1974;183:87–88.
108. Døskeland SO. Evidence that rabbit muscle protein kinase has two kinetically distinct binding sites for adenosine 3′,5′-cyclic monophosphate. Biochem Biophys Res Commun 1978;83:542–549.
109. Rannels SR, Corbin JD. Two different intrachain cAMP binding sites of cAMP-dependent protein kinases. J Biol Chem 1980;255:7085–7088.
110. Cho-Chung YS. Commentary: site-selective 8-chloro-cyclic adenosine 3′,5′-monophosphate as a biologic modulator of cancer: restoration of normal control mechanisms. J Natl Cancer Inst 1989;81:982–987.
111. Ogreid D, Ekanger R, Suva RH, et al. Activation of protein kinase isozymes by cyclic nucleotide analogs used singly or in combination. Principles for optimizing the isozyme specificity of analog combinations. Eur J Biochem 1985;150:219–227.
112. Cho-Chung YS. Role of cyclic AMP receptor proteins in growth, differentiation, and suppression of malignancy: new approaches to therapy. Cancer Res 1990;50:7093–7100.
113. Halgren RG, Traynor AE, Pillay S, et al. 8Cl-cAMP cytotoxicity in both steroid sensitive and insensitive multiple myeloma cell lines is mediated by 8Cl-adenosine. Blood 1998;92:2893–2898.
114. Tagliaferri P, Katsaros D, Clair T, et al. Synergistic inhibition of growth of breast and colon human cancer cell lines by site-selective cyclic AMP analogues. Cancer Res 1988;48:1642–1650.
115. Koontz JW, Wicks WD. Cytotoxic effects of two novel 8-substituted cyclic nucleotide derivaties in cultured rat hepatoma cells. Mol Pharmacol 1980;18:65–71.
116. Koontz JW, Wicks WD. Comparison of the effects of 6- thio- and 6-methylthiopurine ribonucleoside cyclic monophosphates with their corresponding nucleosides on the growth of rat hepatoma cells. Cancer Res 1977;37:651–657.
117. Cho-Chung YS, Budillon A, Nesterova M, Tortora G, Kondrashin A, Lee GR. Development of 8-Cl-cAMP as differentiation agent. In:Challenges of Modern Medicine, Serono Symposium Publication series/Diffrentiation Therapy 1995;10:183–198.
118. Tortora G, Clair T, Katsaros D, et al. Induction of megakaryocytic differentiation and modulation of protein kinase gene expression by site-selective cAMP analogs in K-562 human leukemic cells. Proc Natl Acad Sci USA 1989;86:2849–2852.
119. Srivastava RK, Srivastava AR, Cho-Chung YS, Longo DL. Synergistic effects of retinoic acid and 8-chloro-adenosine 3′,5′-cyclic monophosphate on the regulation of retinoic acid receptor beta and apoptosis: involvement of mitochondria. Clin Cancer Res 1999;5:1892–1904.
120. Kim SN, Kim SG, Park JH, et al. Dual anticancer activity of 8-Cl-cAMP: inhibition of cell proliferation and induction of apoptotic cell death. Biochem Biophys Res Commun 2000;273:404–410.
121. Rohlff CT, Clair T, Cho-Chung YS. 8-Cl-cAMP induces trancation and down-regulation of the RIα subunit and up-regulation of the RIIβ subunit of cAMP-dependent protein kinase leading to type II holoenzyme-dependent growth inhibition and differentiation of HL-60 leukemia cells. J Biol Chem 1993;268:5774–5782.
122. Harada H, Becknell B, Wilm M, et al. Phosphorylation and inactivation of BAD by mitochondria-anchored protein kinase A. Mol Cell 1999;3:413–422.

123. Pinto A, Aldinucci D, Gattel V, et al. Inhibition of the self-renewal capacity of blast progenitors from acute myeloblastic leukemia patients by site-selective 8-chloroadenosine 3′,5′-cyclic monophosphate. Proc Natl Acad Sci USA 1992;89:8884–8888.

124. Tortora G, Ciardiello F, Pepe S, et al. Phase I clinical study with 8-chloro-cAMP and evaluation of immunological effects in cancer patients. Clin Cancer Res 1995;4:377–384.

125. Weissinger EM, Oettrich K, Evans C, et al. Activation of protein kinase A (PKA by 8-Cl-cAMP as a novel approach for antileukaemic therapy. Br J Cancer 2004;91:186–192.

126. Tortora G, Pepe S, Yokozaki H, Clair T, Rohlff C, Cho-Chung YS. A RIa subunit antisense oligonucleotide of cAMP-dependent protein kinase inhibits proliferation of human HL-60 promyelocytic lekemia. Proc Am Acssoc Cancer Res 1990;31:38.

127. Tortora G, Clair T, Cho-Chung YS. An antisense oligodeoxynucleotide targeted against the type RII beta regulatory subunit mRNA of protein kinase inhibits cAMP-induced differentiation in HL-60 leukemia cells without affecting phorbol ester effects. Proc Natl Acad Sci USA 1990;87:705–708.

128. Schwartz DA, Rubin CS. Identification and differential expression of two forms of regulatory subunits (RII) of cAMP-dependent protein kinase II in Friend urythroleukemic cells. J Biol Chem 1985;260:6063–6296.

129. Nesterova M, Cho-Chung YS. A single-injection protein kinase A-directed antisense treatment to inhibit tumour growth. Nat Med 1995;1:528–633.

130. Nesterova M, Noguchi K, Park YG, Lee YN, Cho-Chung YS. Compensatory stabilization of RIIβ protein, cell cycle deregulation, and growth arrest in colon and prostate carcinoma cells by antisense-directed down-regulation of protein kinase A RIα protein. Clin Cancer Res 2000;6:3434–3441.

131. Agrawal S, Jiang Z, Zhao Q, et al. Mixed-backbone oligonucleotides as second generation antisense oligonucleotides: in vitro and in vivo studies. Proc Natl Acad Sci USA 1997;94: 2620–2625.

132. Chen HX, Marshall JL, Ness E, Martin RR, Dvorchik B, Rizvi N, et al. A safety and pharmacokinetic study of a mixed-backbone oligonucleotide (GEM 231) targeting the type I protein kinase A by two-hour infusions in patients with refractory solid tumors. Clin Cancer Res 2000;6:1259–1266.

133. Agrawal S, Zhao Q. Mixed backbone oligonucleotides: improvement in oligonucleotide-induced toxicity in vivo. Antisense Nucleic Acid Drug Dev 1998;8:135–139.

134. Krieg AM, Yi AK, Matson S, Waldschmidt TJ, Bishop GA, Teasdale R, et al. CpG motifs in bacterial DNA trigger direct B-cell activation. Nature 1995;374(6522):546–549.

135. Metelev V, Liszlewicz J, Agrawal S. Study of antisense oligonucleotide phosphorothioates containing segments of oligodeoxynucleotides and 2′-O-methyloligoribonucleotides. Bioorg Med Chem Lett 1994;4:2929–2934.

136. Monia BP, Lesnik EA, Gonzalez C, Lima WF, McGee D, Guinosso CJ, et al. Evaluation of 2′-modified oligonucleotides containing 2′-deoxygaps as antisense inhibitors of gene expression. J Biol Chem 1993;268:14,514–14,522.

137. Shibahara S, Mukai S, Morisawa H, Nakashima H, Kobayashi S, Yamamoto N. Inhibition of human immunodeficiency virus (HIV-1) replication by synthetic oligo-RNA derivatives. Nucleic Acids Res 1989;17:239–252.

138. Tortora G, Ciardiello F. Protein kinase A as target for novel integrated strategies of cancer therapy. Ann NY Acad Sci 2002;968:139–147.

139. Ciardiello F, Pepe S, Bianco C, et al. Down-regulation of RIα subunit of cAMP-dependent protein kinase induces growth inhibition of human mammary epithelial cells transformed by c-Ha-ras and c-erbB-2 proto-oncogenes. Int J Cancer 1993;53:438–443.

140. Turini ME, DuBois RN. Cyclooxygenase-2: a therapeutic target. Annu Rev Med 2002; 53: 35–57.

141. Srivastava RK, Srivastava AR, Korsmeyer SJ, Nesterova M, Cho-Chung YS, Longo DL. Involvement of microtubules in the regulation of Bcl2 phosphorylation and apoptosis through cyclic AMP-dependent protein kinase. Mol Cell Biol 1998;18:3509–3517.

142. Alper O, Hacker NF, Cho-Chung YS. Protein kinase A-Ialpha subunit-directed antisense inhibition of ovarian cancer cell growth: crosstalk with tyrosine kinase signaling pathway. Oncogene 1999;18:4999–5004.

143. Cho YS, Kim M-K, Cheadle C, Neary C, Becker KG, Cho-Chung YS. Antisense DNAs as multisite genomic modulators identified by DNA microarray. Proc Natl Acad Sci USA 2001;98:9819–9823.

144. Tortora G, Caputo R, Damiano V, et al. Synergistic inhibition of human cancer cell growth by cytotoxic drugs and mixed backbone antisense oligonucleotide targeting protein kinase A. Proc Natl Acad Sci USA 1997;94:12,586–12,591.

145. Wang H, Cai Q, Zeng X, Yu D, Agrawal S, Zhang R. Antitumor activity and pharmacokinetics of a mixed-backbone antisense oligonucleotide targeted to the RIalpha subunit of protein kinase A after oral administration. Proc Natl Acad Sci USA 1999;96:13,989–13,994.

146. Tortora G, Bianco R, Damiano V, et al. Oral antisense that targets protein kinase A cooperates with taxol and inhibits tumor growth, angiogenesis, and growth factor production. Clin Cancer Res 2000;6:2506–2512.

147. Wang H, Hang J, Shi Z, et al. Antisense oligonucleotide targeted to RIalpha subunit of cAMP-dependent protein kinase (GEM231) enhances therapeutic effectiveness of cancer chemotherapeutic agent irinotecan in nude mice bearing human cancer xenografts: in vivo synergistic activity, pharmacokinetics and host toxicity. Int J Oncol 2002;21:73–80.

148. Tortora G, Damiano V, Bianco C, et al. The RIα subunit of protein kinase A (PKA) binds to Grb2 and allows PKA interaction with the activated EGF-receptor. Oncogene 1997;14:923–928.

149. Tortora G, Ciardiello F. Targeting of epidermal growth factor receptor and protein kinase A: molecular basis and therapeutic applications. Ann Oncol 2000;11:777–783.

150. Yang WL, Iacono L, Tang WM, Chin KV. Novel function of the regulatory subunit of protein kinase A: regulation of cytochrome c oxidase activity and cytochrome *c* release. Biochemistry 1998;37:14,175–14,180

151. Srivastava RK, Srivastava AR, Park YG, Agrawal S, Cho-Chung YS. Antisense depletion of RIalpha subunit of protein kinase A induces apoptosis and growth arrest in human breast cancer cells. Breast Cancer Res Treat 1998;49:97–107.

152. Cho YS, Kim MK, Tan L, Srivastava R, Agrawal S, Cho-Chung YS. Protein kinase A RIα antisense inhibition of PC3M prostate cancer cell growth: *Bcl-2* hyperphosphorylation, Bax up-regulation, and Bad-hypophosphorylation. Clin Cancer Res 2002;8:607–614.

153. Tortora G, Caputo R, Damiano V, et al. Combined blockade of protein kinase A and bcl-2 by antisense strategy induces apoptosis and inhibits tumor growth and angiogenesis. Clin Cancer Res 2001;7:2537–2544.

154. Nesterova MV, Cho-Chung YS. Antisense protein kinase A RIalpha inhibits 7,12-dimethyl-benz(a)anthracene-induction of mammary cancer: blockade at the initial phase of carcinogenesis. Clin Cancer Res 2004;10:4568–4577.

155. Mayr B, Montminy M. Transcriptional regulation by the phosphorylation-dependent factor CREB. Nat Rev Mol Cell Biol 2001;2:599–609.

156. Conkright MD, Guzman E, Flechner L, et al. Genome-wide analysis of CREB target genes reveals a core promoter requirement for cAMP responsiveness. Mol Cell 2003;11:1101–1108.

157. Montminy MR, Bilezikjian LM. Binding of a nuclear protein to the cyclic-AMP response element of the somatostatin gene. Nature 1987;328:175–178.

158. Montminy MR, Sevarino KA, Wagner JA, Mandel G, Goodman HM. Identificaiton of a cyclic-AMP-responsive element with the rat somatostatin gene. Proc Natl Acad Sci USA 1986;83:6682–6686.

159. Moriuchi A, Ido A, Nagata Y, et al. A CRE and the region occupied by a protein induced by growth factors contribute to up-regulation of cyclin D1 expression in hepatocytes. Biochem Biophys Res Commun 2003;300:415–421.

160. Droogmans L, Cludts I, Cleuter Y, Kettmann R, Burny A. Nucleotide sequence of the bovine interleukin-6 gene promoter. DNA Seq 1992;3:115–117.

161. Russell DL, Doyle KM, Gonzales-Robayna I, Pipaon C, Richards JS. Egr-1 induction in rat granulosa cells by follicle-stimulating hormone and luteinizing hormone: combinatorial regulation by transcription factors cyclic adenosine 3′,5′-monophosphate regulatory element binding protein, serum response factor, sp1, and early growth response factor-1. Mol Endocrinol 2003;17:520–533.

162. Park YG, Nesterova M, Agrawal S, Cho-Chung YS. Dual blockade of cyclic AMP response element (CRE)- and AP-1-directed transcription by CRE transcription factor decoy oligonucleotide: Gene-specific inhibition of tumor growth. J Biol Chem 1999;274:1573–1580.

163. Habener JF. Cyclic AMP response element binding proteins: a cornucopia of transcription factors. Mol Endocrinol 1990;4:1087–1094.

164. Park YG, Park S, Lim SO, et al. Reduction in cyclin D1/Cdk4/retinoblastoma protein signaling by CRE-decoy oligonucleotide. Biochem Biophys Res Commun 2001;281:1213–1219.

165. Liu WM, Scott KA, Shahin S, Propper DJ. The in vitro effects of CRE-decoy oligonucleotides in combination with conventional chemotherapy in colorectal cancer cell lines. Eur J Biochem 2004;271(13):2773–2781.

166. Cho YS, Park YG, Lee YN, et al. Extracellular protein kinase A as a cancer biomarker: its expression by tumor cells and reversal by a myristate-lacking Calpha and RIIbeta subunit overexpression. Proc Natl Acad Sci USA 2000;97:835–840.

167. Cho YS, Lee YN, Cho-Chung YS. Biochemical characterization of extracellular cAMP-dependent protein kinase as a tumor marker. Biochem Biophys Res Commun 2000;278:679–684.

168. Cvijic ME, Kita T, Shih W, DiPaola RS, Chin KV. Extracellular catalytic subunit activity of the cAMP-dependent protein kinase in prostate cancer. Clin Cancer Res 2000;6(6):2309–2317.

169. Kita T, Goydos J, Reitman E, et al. Extracellular cAMP-dependent protein kinase (ECPKA) in melanoma. Cancer Lett 2004;208:187–191.

2

Protein Kinase C and Apoptosis

Mary E. Reyland

Summary

The protein kinase C (PKC) family consists of ten structurally related serine/threonine protein kinases. PKC isoforms are critical regulators of cell proliferation and survival and their expression or activity is altered in some human diseases, particularly cancer. The development and utilization of PKC isoform specific tools, including dominant inhibitory kinases, mouse models in which specific PKC isoforms have been disrupted, and PKC isoform specific antisense/siRNA, has allowed studies to define isoform-specific functions of PKC in the apoptotic pathway. From these approaches a pattern is emerging in which the conventional isoforms, particularly PKCα and PKCβ, and the atypical PKCs, PKCι/λ and PKCζ, appear to be anti-apoptotic/pro-survival. The novel isoform, PKCδ, is primarily pro-apoptotic, whereas PKCε in most studies appears to suppress apoptosis. The identification of both pro- and anti-apoptotic isoforms suggests that PKC isoforms may function as molecular sensors, promoting cell survival under favorable conditions, and executing the death of abnormal or damaged cells when needed. This chapter discusses what is currently known about the contribution of specific isoforms to apoptosis, and how signal transduction by PKC integrates with other molecular regulators to promote or inhibit apoptosis.

1. Introduction

Apoptosis was originally described by Kerr, Wyllie, and Currie as a series of morphologic changes to the cell which include membrane blebbing, nuclear condensation and DNA digestion *(1)*. It is now appreciated that this program of cell death is initiated by physiological stimuli, during development, and by a wide range of cellular toxins. Apoptosis is essential for the maintenance of tissue homeostasis in complex organisms, and alterations in this pathway underlie a variety of disease processes. During development, apoptosis mediates cell turnover and tissue remodeling and is important for the elimination of self-reactive cells in the immune system. Apoptosis is also important for the clearance of altered or damaged cells, and notably does so without eliciting an inflammatory response. However, inappropriate activation or inhibition of apoptosis is associated with a wide range of human diseases including cancer, autoimmune disease, and neurodegenerative disorders. In cancer and autoimmune disease, failure to eliminate defective or unwanted cells may contribute to disease *(2–8)*, whereas in neurodegenerative disorders there is an inappropriate loss of cells. Included in this later group are diseases such as heart failure and other types of acute and chronic tissue injury where apoptosis may contribute to excessive cell loss.

Genetic disruption of the apoptotic pathway is an extremely common feature of tumor cells and the ability to evade apoptosis is considered an essential "hallmark of cancer" *(9,10)*. Correlations between the expression of specific apoptotic markers and

From: *Apoptosis, Cell Signaling, and Human Diseases: Molecular Mechanisms, Volume 2*
Edited by R. Srivastava © Humana Press Inc., Totowa, NJ

clinical outcome underscore the relevance of this pathway to cancer biology *(11,12)*. For instance, increased expression of the anti-apoptotic protein, Bcl-2, or reduced expression of the pro-apoptotic protein, Bax, correlates with poor prognosis and increased metastasis in breast and other tumor types *(11,12)*. Bax expression is lost in a subset of human colon cancers and its loss is associated with increased cancer cell growth in vivo and in vitro *(13)*. Likewise, inhibition of apoptosis may underlie the resistance of many tumors to chemotherapeutic drugs. Loss of Bax in glioblastoma multiforme tumors results in resistance to apoptotic stimuli in vitro *(14)*. Mouse models also suggest that targeted suppression of the apoptotic pathway promotes tumor progression in mice expressing activated oncogenes. For instance, overexpression of Bcl-2 increases c-myc induced tumorigenesis in the mammary gland *(15)*, whereas loss of the pro-apoptotic protein, Bax, accelerates mammary tumor development in C3(1)/SV40-Tag transgenic mice *(16)*. Taken together, these studies suggest that normal apoptosis is critical for tumor suppression and that inactivation, or aberrant regulation of this pathway, may have important consequences for tumorigenesis or tumor progression.

PKC family members have been implicated in a wide range of cellular responses including cell permeability, contraction, migration, hypertrophy, proliferation, apoptosis, and secretion. In many cases these functions appear to be cell or tissue specific, implying that the specification of these responses relies on the interaction of PKC isoforms with other regulatory pathways in the cell. In particular, protein kinase cascades are emerging as important modulators of the apoptotic response *(17)*. These include the phosphoinositide 3-kinase/AKT (PI3-kinase/AKT) pathway, the c-Jun-N-terminal-Kinase (JNK) and p38 pathways, the Janus-Kinase-Signal Transducer and Activator of Transcription (JAK-STAT) pathway *(18–20)*, and many isoforms of PKC. Activation of these cascades can result in direct phosphorylation of apoptotic proteins, or regulate apoptosis by activating or inhibiting the transcription of pro- or anti-apoptotic genes. Recently, tools to decipher the function of specific PKC isoforms have been developed, including "knock-out" mouse models, enabling investigators to probe the roles played by specific members of this family. This chapter will focus on the evidence that specific members of this family play distinct roles in regulating apoptosis.

2. Apoptosis

Many laboratories have been involved in deciphering the molecular players that execute apoptosis as well as molecular regulators of the pathway. From these studies it has become clear that despite the disparity in signals that induce apoptosis, execution of the pathway relies on a common set of biochemical mediators. Critical genes in the apoptotic pathway were identified first in *Caenorhabditis elegans,* and homologs of these genes have since been cloned in mammalian cells, revealing a highly conserved pathway from nematodes to mammals *(21,22)*. Essential players in this pathway include the Bcl-2 family of pro- and anti-apoptotic proteins, and the cysteine-dependent aspartate-directed (caspase) proteases *(23,24)*.

The Bcl-2 family consists of pro- and anti-apoptotic proteins that regulate the release of pro-apoptogenic factors from the mitochondria, such as cytochrome *c*, which is essential for caspase activation *(23,25)*. Apoptosis is suppressed through heterodimerization of anti-apoptotic Bcl-2 proteins, such as Bcl-2 and Bcl-xL, with pro-apoptotic proteins such as Bak and Bax, thus the ratio of pro- to anti-apoptotic Bcl-2 proteins is an important determinant of cell fate. In nonapoptotic cells, anti-apoptotic proteins bind to and neutralize pro-apoptotic proteins. Apoptotic stimuli alleviate the Bcl-2 mediated suppression of pro-apoptotic Bax and Bak, allowing these proteins to oligomerize into transmembrane pores in the mitochondria, induce cytochrome *c* release and activate caspases. Key to this pathway are the "BH3" only proteins, pro-apoptotic members of the Bcl2 family such as Bim, Bid, Bik, PUMA, Noxa, and Bad, which act as apical damage sensors *(26)*. BH3-only proteins are thought to function by antagonizing the action of pro-survival Bcl-2 proteins *(26)*. The diversity of this subfamily of Bcl-2 proteins suggests that they may have evolved in to response to diverse types of cell stress.

Caspases are expressed as inactive zymogens and are processed to an active form in response to apoptotic stimuli *(24,27,28)*. Activated initiator caspases cleave and activate other caspases, resulting in a cascade of caspase activation. The job of activated caspases is to dismantle the cell through cleavage of cell proteins, thus, caspase activation is central to the process of apoptosis and activation of caspases is generally viewed as an irreversible commitment to cell death. Whereas activation of the apoptotic pathway is critical for removal of unwanted cells, studies from mice lacking specific components of the apoptotic cascade suggest that this pathway is also absolutely required for development, because loss of caspase-3, -7, -8, or -9, or Bcl-2 results in either embryonic or perinatal death *(29–31)*.

Two pathways for the activation of caspases have been described (*see* Fig. 1) *(32)*. These pathways differ in the mechanism by which initiator caspases are activated, whereas the activation of downstream, or effector caspases, such as caspase 3, 6, and 7 is common to both pathways. The receptor-mediated, or extrinsic, pathway is initiated by ligand binding to death receptors such as tumor necrosis factor (TNF), Fas, and TNF-related apoptosis-inducing ligand (TRAIL) receptors *(33)*. Ligand binding leads to the formation of signaling complexes which activate caspases and lead to cell death. Key to this pathway is formation of the death inducing signaling complex (DISC). Death receptors contain a cytoplasmic domain, known as the "death domain" which, upon ligand binding, interacts with the death domain of the adaptor protein, Fas-associated death domain protein (FADD) or TRAIL-associated death domain protein (TRADD). Pro-caspase-8 is then recruited to the complex to form the DISC, resulting in auto-cleavage and activation of caspase-8 *(34)*. Activated caspase 8 in turn cleaves and activates downstream "effector" caspases, such as caspase-3, leading to cleavage of cellular proteins and cell death. Although this pathway was originally described as mitochondrial independent, it is now clear that active caspase-8 can cleave Bid, a member of the Bcl-2 family, and that cleaved Bid can amplify the death signal by promoting the release of apoptogenic proteins from the mitochondria *(35)*.

Drugs, chemicals, irradiation, and cell stress activate caspases via the intrinsic or mitochondria-dependent pathway. Although the specific cell signals delivered by these agents differ, all appear to converge at the mitochondria resulting in the release of cytochrome *c* and loss of mitochondrial membrane potential. Cytochrome *c*, together with Apaf1, ATP and pro-caspase-9, forms the "apoptosome" and leads to activation of caspase-9, and the subsequent activation of effector caspases. In addition to the Bcl-2 proteins discuss above, caspase activation is also regulated by the release of mitochondrial proteins such as the inhibitor of apoptosis proteins (IAP) that inhibit activated caspases, and SMAC/DIABLO which binds and inhibits IAPs *(36,37)*. Finally, a group of

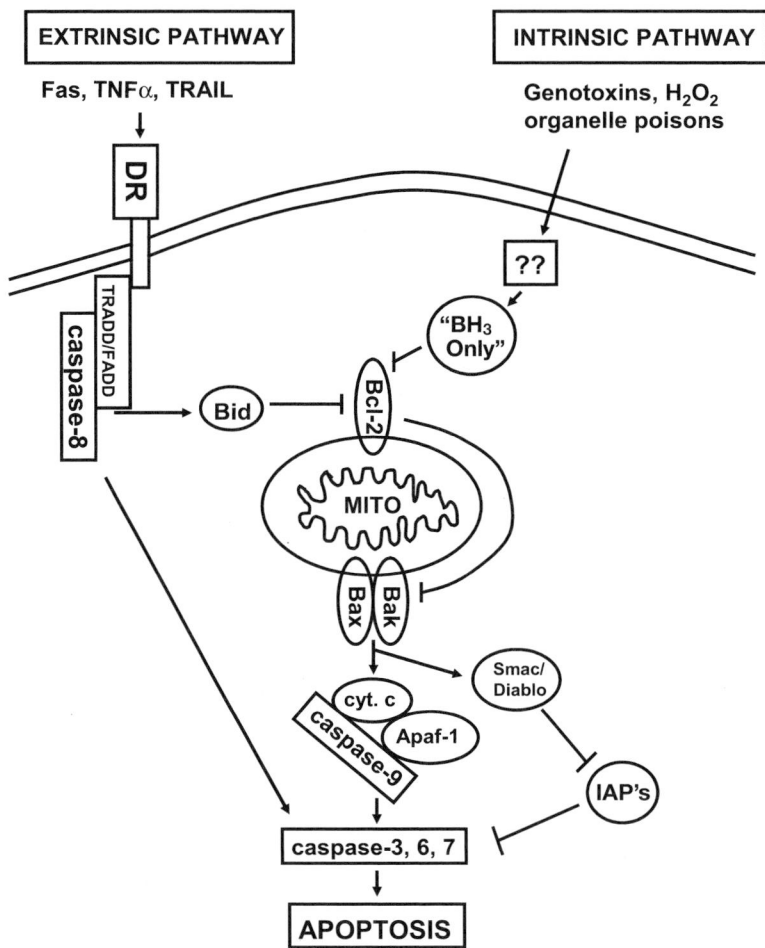

Fig. 1. Intrinsic and extrinsic apoptotic pathways. Apoptosis can be activated through the extrinsic/death receptor dependent pathways, or the intrinsic/mitochondrial dependent pathway. Both pathways converge to activate a common set of effector caspases. See text for details.

mitochondrial proteins have been identified that induce apoptosis independently of caspase activation *(38,39)*. These include Apoptosis Inducing Factor *(38,39)* and endonuclease G *(40)*, which are released from the mitochondria in response to an apoptotic signal and translocate to the nucleus to trigger nuclear condensation and DNA fragmentation *(39)*.

3. PKC Structure/Activation

The PKC family contains 10 structurally related serine/threonine protein kinases that were originally characterized by their dependency upon lipids for activity *(see* Fig. 2) *(41,42)*. The lipid dependence of these enzymes has facilitated the identification of upstream activators. Physiologic regulators of PKC, including growth factors and hormones, activate PKC via receptor stimulated activation of phosphatidylinositol-specific phospholipase C (PI-PLC). Activation of PI-PLC results in the generation of diacylglycerol (DAG), an increase in intracellular Ca++ via generation of Ins(1,4,5)P3, and the

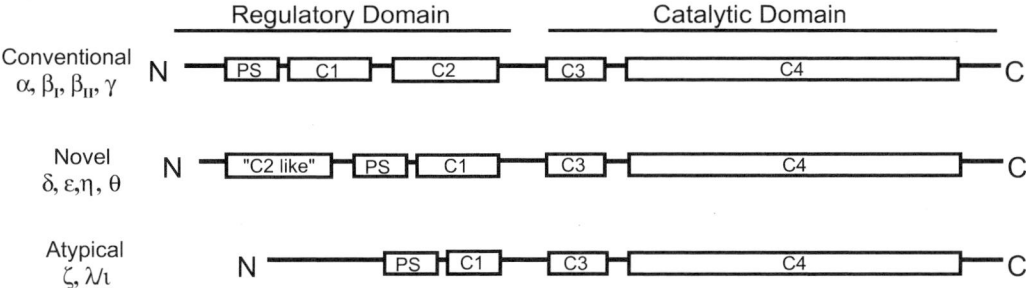

Fig. 2. Structural representation of PKC isozymes and their subfamilies. PS, pseudosubstrate; C1, binds DAG/phorbol esters; C2, binds phosphatidyl serine and Ca2+; C3 and C4, kinase domain.

subsequent activation of PKC. Sub-families of PKC are defined by their requirement for these activators, with the classical isoforms (PKCα, β, and γ) requiring DAG and calcium, the novel isoforms (PKCδ, ε, η, and θ) requiring DAG, but not calcium, and the atypical isoforms (PKCζ and λ/ι) requiring neither. The domains critical for binding these activators have been defined and reside in the N-terminal regulatory portion of the protein. These include the C1 domain which binds DAG and the C2 domain which binds Ca++. These domains also function to target PKC to membranes via DAG and anionic phospholipids *(43)*. The N-terminal regulatory domain also contains binding sites for anchoring proteins which are thought to target the activated kinase to specific subcellular sites. These include the Receptors for Activated C Kinase (RACK's) as well as other PKC-interacting proteins *(44)*.

The C-terminal kinase domain of PKC is highly conserved between isoforms and phosphorylation at three sites in this domain is required to generate a mature form of the kinase that can be recruited to membranes *(45–51)*. The first of these phosphorylation events occurs at a conserved threonine in the activation loop; phosphorylation at this site appears to be essential for activity of most isoforms *(48,51)*. Several laboratories have identified the PIP3 regulated kinase, PDK-1, as the kinase responsible for PKC activation loop phosphorylation *(46,52)*. Phosphorylation at two additional C-terminal sites contributes to the stability of the kinase. These include an autophosphorylation site and a C-terminal hydrophobic site *(48)*. Phosphorylation at these sites renders PKC protease and phosphatase resistant, and catalytically competent. However PKC is still in an inactive conformation in which the substrate binding pocket is occupied by the pseudosubstrate domain. Generation of the second messengers, DAG and Ca^{++}, increases the affinity of "primed" PKC for the membrane resulting in release of the pseudosubstrate from the substrate binding pocket and activation of the kinase. Tyrosine phosphorylation of some PKC isoforms, particularly PKCδ, is seen in response to many stimuli including apoptotic stimuli such as UV, H_2O_2 and etoposide *(53–63)*. Tyrosine residues important for apoptosis have been identified by mutagenesis and will be addressed below.

4. PKC and Apoptosis

PKC plays a fundamental role in the regulation of cell proliferation and differentiation and recent studies suggest that it is also involved in the regulation of cell survival. Early approaches to defining the role of PKC in apoptosis relied upon activation of PKC by phorbol-12-myristate-13-acetate (PMA), which targets the conventional and novel isoforms, and inhibition by pharmacological agents. These agents are problematic both because of their broad specificity within the PKC family, and in the case of inhibitors, their potential for inhibition of other enzymes. Nonetheless, using these types of approaches, investigators have clearly demonstrated a role for PKC in regulating apoptosis induced by both death receptors (extrinsic pathway), and by DNA damaging agents and cell toxins (intrinsic pathway). Whether alterations in PKC activity enhance or suppress apoptosis appears to depend on the initiating signal as well as the specific cell type. Most studies indicate that activation of PKC with PMA blocks Fas, TRAIL, and TNF-α induced apoptosis *(64–69)*. In some cases this appears to result from disruption of DISC formation *(66,67)*. The protective effect of PMA on Fas-induced apoptosis has also been attributed to activation of the extracellular regulated kinases (ERK) and NF-κB pathways *(70)*. However, in some studies PMA induces apoptosis, or sensitizes cells to death receptor induced apoptosis *(71–73)*. This supports the notion that the functional outcome of PKC activation reflects the specific PKC isoform expression profile of a given cell type.

The particular cellular mix of PKC's maybe even more critical in the context of the intrinsic apoptosis pathway. Activation of PKC with PMA blocks irradiation induced apoptosis in Jurkat cells *(74)* and singlet oxygen induced apoptosis in HL-60 cells *(75)*. However, in some cell types, including salivary epithelial cells *(71)* and prostate cancer cells *(72,76)*, PMA induces apoptosis. Likewise, the PKCδ inhibitor, rottlerin, suppresses genotoxin induced apoptosis in most cells, however in some cell types pretreatment with rottlerin enhances apoptosis *(77,78)*. The complexity and potential redundancy of the PKC signaling network has prompted the development of PKC isoform specific tools including dominant inhibitory kinases, mouse models in which specific PKC isoforms have been disrupted, and PKC isoform-specific antisense/siRNA to define isoform-specific functions of PKC in the apoptotic pathway. From these approaches a pattern is emerging in which the conventional isoforms, particularly PKCα and PKCβ, and the atypical PKC's, PKCι/λ, and PKCζ, appear to be anti-apoptotic/pro-survival. The novel isoform, PKCδ, is primarily pro-apoptotic, whereas PKCε in most studies appears to suppress apoptosis. What is currently known about the contribution of specific isoforms to apoptosis, and how signal transduction by specific PKC isoforms integrates with other molecular regulators to promote or inhibit apoptosis is discussed in the next section.

5. Pro-survival PKC Isoforms

Pro-survival PKC isoforms have been defined chiefly based on the ability of these protein kinases to suppress apoptotic signaling and/or promote cell survival.

5.1. PKCα

Most evidence suggests that PKCα promotes cell survival and that loss of PKCα activity either induces death outright, or sensitizes cells to death signals. A notable exception is LNCaP prostate cancer cells *(79,80)*. In these cells PKCα activation by a synthetic DAG analog that activated PKCα, but not PKCδ, was shown to induce apoptosis, and this could be blocked by expression of a dominant-negative PKCα mutant *(79)*. However, in melanoma cell lines *(81)*, COS1 cells, bladder carcinoma cell lines

(82), glioma cells *(83)* and salivary gland epithelial cells *(84)*, depletion of PKCα activity by expression of dominant negative of PKCα, or by PKCα depletion with antisense or siRNA, induces apoptosis. In the case of salivary epithelial cells and glioma cells, PKCα and PKCδ have been shown to be reciprocal regulators of apoptosis, with PKCα promoting cell survival and PKCδ promoting cell death *(83,84)*. Finally, ceramide induced apoptosis is thought to function at least in part by inhibiting phosphorylation, and thereby activation, of PKCα *(85)*.

An important question remaining is whether apoptosis induced by loss of PKCα is secondary to loss of a proliferative signal, or occurs through a direct effect on the apoptotic machinery. Proteins involved in the execution of apoptosis have been identified as potential targets of PKCα. Overexpression of PKCα increases Bcl-2 phosphorylation at serine 70 and suppresses apoptosis in human pre-B REH cells *(86)*. Interestingly, phosphorylation at serine 70 has been shown to stabilize and increase the anti-apoptotic function of Bcl-2 *(87)*. In a similar vein, depletion of PKCα in COS cells induces apoptosis and this correlates with down regulation of Bcl-2 expression *(82)*. PKCα has also been implicated in transduction of the Akt/PKB survival signal via direct phosphorylation of Akt at serine 473, and by of activation of the serine/threonine protein kinase Raf-1 *(88,89)*. Overexpression of PKCα in 32D myeloid progenitor cells activates endogenous Akt, consistent with its pro-survival function *(90)*.

Overexpression of PKCα is seen in a variety of human tumors, arguing that it may be a pro-proliferative signal as well as an anti-apoptotic signal in these cells *(91,92)*. Overexpression of PKCα increases the proliferative capacity of thymocytes, MCF-7 breast cancer cells and glioma cells *(91,93,94)*. PKCα has been shown to regulate a number of pathways involved in cell proliferation including the MAPK, AP-1 and NF-kB pathways, and promotes cell-cycle progression in some cells. The increased expression of PKCα in tumor cells has prompted the development of therapies directed at reducing its expression or activity *(91,93,94)*. Studies in vitro using a PKCα antisense oligonucleotide to decrease PKCα expression in tumor cell lines showed decreased proliferation and increased expression of p53, suggesting that PKCα depletion therapy may sensitize tumors cells to apoptosis *(95–97)*. Likewise, PKCα antisense oligonucleotides significantly reduced tumor growth in a xenograph model *(97)*. However, clinical trials to assess the efficacy of this strategy for the treatment of human tumors have been disappointing.

5.2. PKCβ

Two forms of PKCβ, which differ in their C-terminus, are generated by alternative splicing, PKCβI and PKCβII. Studies from mice deficient for both isoforms indicate that PKCβ is essential for signaling via the B-cell antigen receptor and that loss of PKCβ results in decreased NF-κB activation and B cell survival *(100)*. Most in vitro data indicates a pro-survival function for PKCβII. In vivo studies show that overexpression of PKC $β_{II}$ protects small cell lung cancer cells against c-myc induced apoptosis *(101)* and Whitman et al. have shown that activation of PKCβII suppresses Ara-C induced apoptosis in HL-60 cells and increases the level of the anti-apoptotic protein, Bcl-2 *(102)*. Nuclear translocation and activation of PKCβII is also associated with v-Abl mediated suppression of apoptosis in IL-3 dependent hematopoietic cells *(103)*. Reported substrates for PKCβII include the pro-survival kinase, Akt, and lamin B1. In antigen stimulated mast cells PKCβII phosphorylates Akt at serine 473, consistent with a pro-survival function *(104)*.

PKCβII phosphorylation on lamin B1 in etoposide treated rat fibroblasts appears to precede caspase-6 cleavage of lamin B1 and dissolution of the nuclear membrane.

In line with a pro-survival/anti-apoptotic function, animal cancer models as well as human tumor studies suggest that PKCβII expression may contribute to tumor promotion or progression. For instance, PKCβII expression is specifically increased in patients with diffuse large B-cell lymphomas and inhibition of PKCβ in cultured cells from these patients induces apoptosis *(106)*. In mice, expression of a PKCβII transgene results in hyperplasia of intestinal epithelial cells and increased sensitivity to chemical carcinogens *(107,108)*. In a mouse xenograph model, growth of hepatocellular carcinomas induced by overexpression of vascular endothelial growth factor (VEGF) could be blocked, and apoptosis induced, by oral administration of an inhibitor of PKCβ *(109)*. Recently, the PKC-β selective inhibitor, Enzastaurin, has been shown to suppress growth and induce apoptosis in xenographs of human colon cancer and glioblastoma *(110)*.

Whereas PKCβII functions primarily to promote cell survival, the role of PKCβI in survival/apoptosis is less clear. Inhibition of PKCβI in W10 B cells increases apoptosis, presumably by suppressing activation of the ERK pathway *(111)*. Likewise, overexpression of PKCβI suppresses the apoptotic response of gastric cancer cells to the COX-2 inhibitor, SC-236, whereas antisense depletion of PKCβI sensitizes these cells to chemotherapeutic drugs *(95,112)*. In contrast, a pro-apoptotic function for PKCβI has been demonstrated using a HL-60 variant, HL-525 cells, which are deficient in PKCβ *(113)*. HL-525 cells are suppressed in death receptor induced apoptosis, but this response can be recovered by transfection of PKCβI *(113,114)*. Likewise, in U-937 myeloid leukemia cells PMA treatment induces apoptosis in a PKCβ dependent manner *(73)*.

5.3. PKCε

PKCε expression/activation is often associated with cell transformation and tumorigenesis and the ability of PKCε to promote tumorigenesis is in many cases related to the suppression of apoptosis *(115–120)*. Early in vitro studies showed that PKCε is required for PMA mediated protection of U937 cells from TNF-α or calphostin C induced apoptosis *(121)*. Caspase cleavage of PKCε occurs in some cells undergoing apoptosis, and Basu et al. have shown that in MCF-7 cells treated with TNF-α, caspase cleavage generates an active, anti-apoptotic form of PKCε *(122,123)*. In glioma cells, expression of PKCε suppresses TRAIL induced apoptosis. This protection can be enhanced by expression of a caspase-resistant form of PKCε, suggesting the caspase cleavage of PKCε contributes to its pro-survival function *(124)*. Likewise, expression of PKCε promotes survival of lung cancer cells and increases their resistance to chemotherapeutic drugs *(120)*. In contrast, PKCε has been shown to be required for UV induced apoptosis through regulation of the JNK and ERK signaling pathways via activation of Ras/Raf *(125)*. This may be a common effector of PKCε because studies from other labs suggest that the oncogenic function of PKCε is through activation of the Ras/Raf pathway *(126–128)*. PKCε has also been shown to enhance survival through activation of the NF-κB pathway *(129)*.

Other studies have suggested that changes in PKCε expression may be associated with tumor progression in humans. The PKCε gene is amplified in 28% of thyroid cancers and a chimeric/truncated version of PKCε has been cloned from human thyroid cancer cells *(116)*. Expression of this chimeric/truncated PKCε protein in PCCL3 cells made them resistant to apoptosis, suggesting that this may contribute to tumor proliferation or progression *(116)*. In contrast, a later study from the same group showed no mutations in PKCε in a study of 31 thyroid cancers; however some tumors had decreased expression of PKCε *(130)*. Analysis of a panel of melanoma cell lines showed that PMA sensitizes cells to TRAIL induced apoptosis and that sensitization correlates with low expression of PKCε*(131)*.

Overexpression of PKCε is a common feature of human prostate tumors *(132)*. Studies in human prostate carcinoma cells show that overexpression of PKCε is associated with conversion from an androgen dependent to androgen independent state, and that in a CWR22 xenograft model, PKCε is upregulated in recurrent prostate tumors *(117)*. This study further shows that endogenous PKCε is required for resistance to apoptosis in CWR-R1 cells, a cell line selected from the recurrent CWR22 tumors *(117)*. In other studies from the same lab, the resistance of prostate cancer cells to apoptosis was shown to result from the interaction of PKCε with the pro-apoptotic protein, Bax *(133)*. Moreover, the association of PKCε with Bax correlated with the progression of prostate cancer cells to an apoptosis resistant state *(133)*.

5.4. PKCλ and PKCζ

The atypical PKC isoforms, PKCλ/ι, and PKCζ are associated with survival in many cells. These isoforms are downstream effectors of PI-3 kinase, are required for mitogenic activation in oocytes and fibroblasts, and for NF-κB activation in NIH3T3 cells and U937 cells *(134,135)*. Most studies indicate a correlation between the expression or activation of PKCλ/ι and/or PKCζ and sensitivity to apoptosis. PKCζ suppresses Fas-FasL induced apoptosis in Jurkat cells by inhibiting DISC formation *(136)*. Murray and Fields have shown that PKCλ/ι protects human K562 leukemia cells from apoptosis induced by taxol or okadaic acid *(137)*. Protection against these agents appears to be specific for PKCλ/ι, as expression of PKCζ had no effect *(137)*. This same lab has also shown that Bcr-Abl mediated resistance to apoptosis requires PKCι*(138)*. Plo et al. have reported that overexpression of PKCζ protects U937 cells from etoposide and mitoxantrone-induced apoptosis via inhibition of topoisomersase II activity *(139)*. In line with these studies, inhibition of PKCζ sensitizes U937 cells to etoposide and TNF-α induced apoptosis *(140)*. Berra et al. have shown that exposure of cells to apoptotic stimuli such as UV radiation, leads to a dramatic decrease in the activity of the atypical PKC isoforms, PKCζ and/or PKCλ *(141)*. PKCζ is cleaved by caspases and down regulation of cleaved PKCζ has been shown to occur through the ubiquitin-proteasome pathway *(142,143)*.

Studies investigating the molecular mechanisms by which atypical PKC's inhibit apoptosis have revealed two potential pathways. PKCζ activity in apoptotic cells is negatively regulated by it's interaction with the pro-apoptotic protein, Prostate Apoptosis Response-4 (PAR-4). Expression of PAR-4 induces apoptosis by coordinately activating the Fas-FasL death receptor pathway and suppressing the pro-survival NF-κB pathway *(144)*. Data from several laboratories suggests that PAR-4 functions in part by regulating pro-survival signaling mediated by the atypical PKCs. PAR-4 has been shown to interact with, and suppress, the enzymatic activity of PKCζ and PKCλ in apoptotic cells *(145)*. Further studies revealed that the atypical PKC's are critical for activation of the NF-κB pathway, and that sequestration of PKCζ and PKCλ by PAR-4 is one mechanism by which PAR-4 suppresses NF-κB activation *(146–149)*. PKCζ has also been shown to interact with p38 in apoptotic chondrocytes, resulting in inhibition of PKCζ*(150)*. Recently it has been shown that p38 interacts with the regulatory domain of PKCζ and suppresses PKCζ activity by blocking autophosphorylation of the kinase *(151)*. Interestingly, PKCζ and PKCλ have also been shown to bind to, and inactivate, the pro-survival kinase, Akt, in response to ceramide and growth factors *(152–154)*. This finding suggests that through regulation of pro-survival pathways, the atypical PKC's may act as a switch between cell death and cell proliferation.

6. Pro-apoptotic PKC Isoforms

The novel isoforms, PKCδ and PKCθ, are often grouped together as pro-apoptotic. This is based primarily on the finding that both isoforms are cleaved by caspase-3 to generate a constitutively activated form of the kinase, which, when introduced into cells can induce apoptosis *(155,156)*. However, the contribution of PKCθ to apoptosis has been investigated to only a limited extent, with some studies indicating that it is required for T-cell survival *(157,158)*, whereas others indicate a pro-apoptotic function *(155,156,159,160)*. In contrast, the function of PKCδ in apoptotic cells has been studied extensively and it is clear that in most cellular contexts PKCδ promotes, and in many instances is required for, execution of the apoptotic program.

6.1. PKCδ

PKCδ is a ubiquitously expressed isoform that has been implicated in both regulation of cell proliferation and cell death. In addition, recent studies on PKCδ$^{-/-}$ mice have identified diverse roles for this signaling molecule in control of immunity *(161,162)*, apoptosis *(163)*, and cell migration *(164)*. In most cells overexpression of PKCδ results in inhibition of proliferation, and/or apoptosis, and loss of PKCδ is associated with cell transformation *(165–169)*. However, some studies have attributed an anti-apoptotic or pro-survival function to PKCδ, particularly in transformed or tumor cells *(170–176)*. PKCδ is required for activation of ERK downstream of the epidermal growth factor receptor (EGF) *(177–179)* and for signal transduction downstream of the insulin growth factor-1 (IGF-1) receptor in some tumor cells *(180–182)*. These studies suggest that PKCδ has the potential to both positively and negatively regulate cell proliferation and cell death, and hence may act as a "switch" to direct a cell into an appropriate pathway depending on the cellular milieu *(183)*.

PKCδ is activated by numerous apoptotic stimuli and PKCδ activity is required for apoptosis induced by ultraviolet (UV) irradiation, genotoxins, taxol and brefeldin A *(184,185)*, phorbol ester *(186)*, oxidative stress *(187)*, and death receptors *(188)*. Suppression of PKCδ function through treatment with the chemical inhibitor rottlerin *(185)*, expression of kinase dead PKCδ (PKCδKD) *(184)* or the PKCδ regulatory domain *(186)*, or by introduction of a competitive PKCδ specific RACK RBS peptide *(189)*, all inhibit apoptosis in different cell types.

Our laboratory and other groups have utilized inhibitors of PKCδ to probe where in the apoptotic pathway PKCδ functions. In parotid C5 cells exposed to etoposide, expression of PKCδKD inhibits distal apoptotic events such as DNA fragmentation and the morphological features of apoptosis, as well as proximal apoptotic events including loss of mitochondrial membrane potential *(184)*. Indeed, other reports have shown inhibition of mitochondrial apoptotic features upon suppression of PKCδ function

(186,187). This indicates that PKCδs role in apoptosis is at, or prior to, signaling events at the mitochondria. Consistent with these findings, mice in which the PKCδ gene has been disrupted are defective in mitochondria-dependent apoptosis *(163)*.

6.2. Caspase-Cleavage of PKCδ

Caspase cleavage of PKCδ is emerging as an important mechanism for amplification of the apoptotic pathway. Early studies in U937 cells showed that irradiation activated a 40-kD myelin basic protein kinase that was subsequently identified as a stable, proteolytically cleaved, yet catalytically competent fragment of PKCδ *(190)*. Cleavage of PKCδ was shown to occur in the hinge domain of the protein at a consensus caspase-3 cleavage sequence, thus causing the release of the N-terminal regulatory region of the enzyme and generating a C-terminal, constitutively active kinase fragment (*see* Fig. 3). Production of this PKCδ fragment could be inhibited by overexpression of Bcl-2 or treatment of cells with a pharmacological caspase inhibitor. In an accompanying study, expression of the PKCδ catalytic fragment (PKCδCF) was shown to be sufficient to rapidly induce apoptotic cell death independent of a killing stimulus, thus revealing a role in apoptosis for PKCδ *(191)*.

The current evidence suggests that full length PKCδ, as well as PKCδCF, contribute to apoptosis. It is known that expression of the caspase generated PKCδCF is sufficient to induce cell death, and Sitailo et al have shown that expression of PKCδCF is associated with activation of the pro-apoptotic protein, Bax, and cytochrome *c* release *(192,193)*. Furthermore, a caspase resistant mutant of PKCδ protects keratinocytes from UV-induced apoptosis *(194)*. However, evidence supports a role for full length PKCδ (PKCδFL) in apoptosis as well, because a caspase uncleavable form of PKCδ can induce apoptosis *(192)* and PKCδ is important for cell death in response to the toxins ceramide and phorbol esters which do not induce PKCδ cleavage *(76,195–197)*. Thus PKCδ may function at two or more points in the apoptotic pathway; a likely scenario being that PKCδFL contributes to activation of caspase-3 and generation of PKCδCF which then feeds back to amplify the apoptotic pathway. How PKCδ regulates specific apoptotic events is not clear, although insight can be gathered from the large number of studies that have analyzed PKCδ activation, subcellular localization and phosphorylation targets in apoptotic cells.

6.3. Tyrosine Phosphorylation of PKCδ

Phosphorylation of PKCδ on tyrosine residues is an early response to many stimuli including apoptotic stimuli such as UV, H_2O_2 and etoposide*(53–56)*. Functionally important tyrosine residues in PKCδ have been identified by mutagenesis and include Y64 and Y187 in glioma cells treated with etoposide *(53)*, Y311, Y332 and Y512 in response to H_2O_2 *(56)*, and Y52, Y64 and Y155 in response to Sindbis virus infection *(198)*. Recently, Okhrimenko et al. have identified PKCδ Y155 as being important for the anti-apoptotic effects of PKCδ in glioma cells treated with TRAIL *(176)*. These studies suggest that phosphorylation of PKCδ on specific tyrosine residues may regulate the cell or stimulus specific functions of PKCδ.

Although in many cases the tyrosine kinases upstream of PKCδ have not been defined, studies point to nonreceptor tyrosine kinases, specifically c-Abl and the Src-like kinase, Lyn, as being important in response to apoptotic agents *(54,56,58–60,199,200)*.

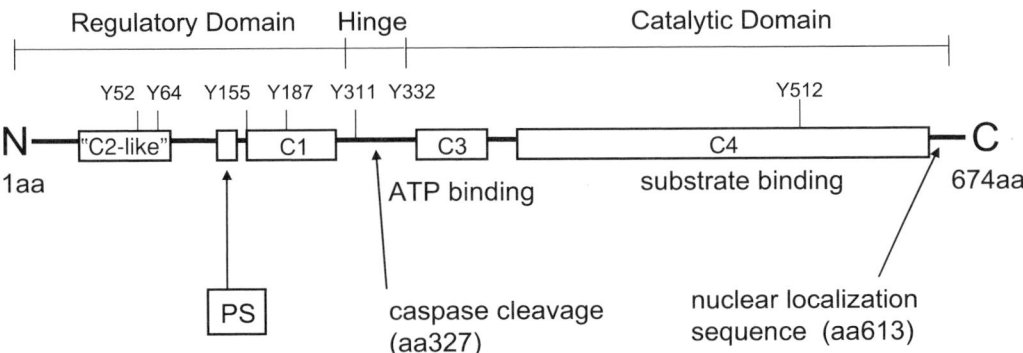

Fig. 3. PKCδ and apoptosis. Sites in the mouse PKCδ protein important for regulation and activation in apoptotic cells are shown. These include Y52, Y64, Y155, Y187, Y311, Y332, and Y512; the caspase cleavage site at aa327; and the C-terminal nuclear localization sequence. *See* text for more details.

Apoptotic agents activate c-Abl and Lyn and induce their association with PKCδ *(55,201,202)*. Yuan et al. report that in MCF-7 cells c-Abl constitutively associates with PKCδ, and that PKCδ becomes phosphorylated on tyrosine in cells induced to undergo apoptosis with ionizing radiation *(55)*. Studies from this same lab show that in H_2O_2 treated cells, c-Abl phosphorylates PKCδ and PKCδ can also phosphorylate and activate c-Abl *(201)*. Yoshida et al. have reported that expression of dominant negative Lyn can suppress tyrosine phosphorylation of PKCδ in Ara-C treated cells *(202)*. Whereas Lyn and c-Abl are found in both the nucleus and cytosol, it is the nuclear form of these kinases that is activated by DNA damaging agents *(203,204)*. Tyrosine phosphorylation may link PKCδ to downstream events in the apoptotic pathway by regulating its catalytic activity, subcellular localization, caspase cleavage or interaction with other proteins such as substrates.

6.4. Subcellular Localization of PKCδ

PKCδ localizes to a variety of subcellular compartments in apoptotic cells in a cell and stimulus dependent fashion. Translocation of PKCδ to the mitochondria has been demonstrated in response to PMA, whereas oxidative stress induces mitochondrial accumulation of PKCδ in U937 cells *(186,187,196,197)*. In these studies translocation of PKCδ to the mitochondria results in cytochrome *c* release and loss of mitochondrial membrane potential. Mitochondrial association of PKCδ in prostate cancer cells amplifies ceramide formation and promotes apoptosis *(205)*. However, treatment of HeLa cells with ceramide induces PKCδ localization to the Golgi compartment, and translocation to the Golgi was shown to be indispensable for apoptosis *(195)*. In contrast, treatment of glioma cells with TRAIL results in the accumulation of PKCδ in the endoplasmic reticulum and suppression of apoptosis *(176)*.

PKCδ translocates from the cytosol to the nucleus in response to etoposide, γ-irradiation, Fas ligand and IL-2 deprivation *(53,55,192,206)*. We have defined a nuclear localization sequence (NLS) in the carboxy-terminus of PKCδ that is required for nuclear import of both full-length PKCδ and the PKCδCF and have shown that nuclear localization of PKCδCF is required for its ability to induce apoptosis *(192)*. Strikingly, parotid C5 cells transfected with GFP-tagged PKCδ catalytic fragments exhibit rapid nuclear accumulation and induction of apoptosis, independent of an apoptotic stimulus. This suggests that caspase cleavage facilitates nuclear translocation of PKCδ in parotid C5 cells, while in glioma cells, nuclear localization may precede caspase cleavage *(53,192)*. Taken together, these studies suggest a nuclear function for PKCδ in some apoptotic cells. This hypothesis is supported by the observation that the many of PKCδ substrates in apoptotic cells are nuclear proteins.

6.5. Targets of PKCδ in Apoptotic Cells

Substrates of PKCδ in apoptotic cells include transcription factors, protein kinases, structural proteins, DNA repair and checkpoint molecules, membrane lipid modification enzymes, and Bcl-2 family members. Intriguingly, these substrates are localized to mitochondria, plasma membrane, and nuclear compartments, suggesting that PKCδ can regulate apoptosis from various organelles within the cell. In UV exposed cells, PKCδ can phosphorylate and activate phospholipid scramblase 3 (PLS3) at the mitochondria *(207)*. In Fas treated cells, PKCδ phosphorylates and activates another member of this family, PSL1, at the plasma membrane *(208)*. In apoptotic monocytes PKCδ associates with, phosphorylates, and increases the activity of caspase-3 *(209)*.

The majority of PKCδ's substrates in apoptotic cells are nuclear proteins. For instance, lamin B, a nuclear structural protein, is phosphorylated by PKCδ in Ara-c exposed cells and contributes to this protein's degradation and the subsequent destruction of the nuclear infrastructure *(210)*. Also, within the nucleus, PKCδ interacts with and phosphorylates DNA-PK in cells exposed to genotoxins *(211,212)*. Phosphorylation of DNA-PK inhibits DNA binding, suggesting that PKCδ mediated phosphorylation inactivates the DNA double strand break repair function of this protein *(212)*. PKCδ also phosphorylates the multifunctional molecule hRad9, which regulates DNA repair and can act as a BH3 only death molecule by binding Bcl-2 *(211)*. PKCδ may also regulate the transcription of death genes through activation of the transcription factors p53, p73β, and STAT1. One study found that downregulation of PKCδ inhibits the basal transcription of p53 whereas other studies report PKCδ dependent accumulation of the p53 protein in apoptotic cells *(213–215)*. p73β, a transcription factor that mediates genotoxin induced cell killing, is phosphorylated by PKCδCF, inducing p73β dependent reporter transcription *(216)*. Our laboratory has shown that PKCδ and STAT1 interact in etoposide treated cells and that STAT1 is required for PKCδCF mediated DNA fragmentation and apoptosis *(217)*.

Although possibly not direct targets, activated PKCδ has been shown to interface with downstream signaling cascades to regulate the apoptotic machinery. Indeed, in apoptotic cells, the PI3-kinase/AKT pathway, the ERK, JNK and p38 pathways and the JAK-STAT pathway all appear to be regulated at least in part by PKCδ. PKCδ activates the JNK pathway in irradiation and Ara-c induced apoptosis, possibly through phosphorylation and activation of MEKK1 *(202)*. PKCδ also enhances radiation-induced apoptosis via ERK1/2 activation and suppression of radiation-induced G2-M arrest, and *(218)*. In keratinocytes p38δ is a downstream effector of PKCδ *(219)*. In cardiomyocytes exposed to ischemia, Akt is dephosphorylated and inactivated in a PKCδ dependent manner and this is accompanied by dephosphorylation and loss of Bad sequestration, allowing this potent apoptotic protein to induce cell death *(220)*. Finally, the binding of HSP25 to PKCδ has been shown to inhibit cell death, suggesting that the pro-apoptotic function of PKCδ may be negatively regulated through specific protein-protein interactions *(221)*.

7. Conclusions

The identification of both pro- and anti-apoptotic isoforms suggests that PKC may function as a molecular sensor, promoting cell survival under favorable conditions and executing the death of abnormal or damaged cells when needed. Although it is convenient for the purpose of discussion to classify PKC isoforms as pro- or anti-apoptotic, most PKC isoforms classified as primarily pro-survival/anti-apoptotic can also function to promote apoptosis and vis a versa. This observation underscores the importance of cellular context and begs the question of how the function of these isoforms is specified. Although this is not well understood, it is clear that specification of function is likely to involve post-translational modifications as well as function-specific changes in subcellular localization. Post-translational modifications, such as phosphorylation, may allow for protein–protein interactions with other molecular regulators of apoptosis, and/or target the active kinase to a specific subcellular location where these interactions may occur. Understanding the molecular basis for regulation of apoptosis by PKC isoforms may contribute to the development of therapeutic strategies to treat diseases such as cancer and neurodegenerative disorders.

References

1. Kerr JF WA, Currie AR. Apoptosis: A basic biological phenomenon with wide-ranging implications in tissue kinetics. Br J Cancer 1972;26:239–257.
2. Hayashi T, Faustman DL. Role of defective apoptosis in Type 1 diabetes and other auto-immune diseases. Recent Prog Horm Res 2003;58:131–153.
3. Ohsako S, Elkon K. Apoptosis in the effector phase of autoimmune diabetes, multiple sclerosis and thyroiditis. Cell Death Diff 1999;6:13–21.
4. Eguchi K. Apoptosis in autoimmune diseases. Intern Med 2001;40:275–284.
5. Evan G, Vousden K. Proliferation, cell cycle and apoptosis in cancer. Nature 2001;411: 342–348.
6. Rathmell JC, Thompson CB. Pathways of apoptosis in lymphocyte development, homeostasis and disease. Cell 2002;109:S97–S107.
7. O'Reilly LA, Strasser A. Apoptosis and autoimmune disease. Inflamm Res 1999;48:5–21.
8. Humphreys-Beher MG, Peck AB, Dang H, Talal N. The role of apoptosis in the initiation of the autoimmune response in Sjogren's syndrome. Clin Exp Immunol 1999;116: 383–387.
9. Johnstone R, Ruefli A, Lowe S. Apoptosis: A link between cancer genetics and chemotherapy. Cell 2002;108:135–144.
10. Hanahan D, Weinberg R. The hallmarks of cancer. Cell 2000;100:57-70.
11. Rogers C, Loveday R, Drew P, Greenman J. Molecular prognostic indicators in breast cancer. Eur J Surg Oncol 2002;28:467–478.
12. Konstantinidou A, Korkolopoulou P, Patsouris E. Apoptotic markers for tumor recurrence: A minireview. Apoptosis 2002;7:461–470.
13. Ionov Y, Yamamoto H, Krajewski S, Reed JC, Perucho M. Mutational inactivation of the pro-apoptotic gene BAX confers selective advantage during tumor clonal evolution. Proc Natl Acad Sci USA 2000;97:10,872–10,877.
14. Cartron P-F, Juin P, Oliver L, Martin S, Meflah K, Vallette FM. Nonredundant role of Bax and Bak in Bid-mediated apoptosis. Mol Cell Biol 2003;23:4701–4712.

15. Jager R, Herzer U, Schenkel J, Weiher H. Overexpression of Bcl-2 inhibits alveolar apoptosis during involution and accelerates c-myc-induced tumorigenesis of the mammary gland in transgenic mice. Oncogene 1997;15:1787–1795.
16. Shibata M-A, Liu M-L, Knudson MC, et al. Haploid loss of Bax leads to accelerated mammary tumor development in C3(1)/SV40-TAg transgenic mice: reduction in protective apoptotic response at the preneoplastic stage. EMBO J 1999;18:2692–2701.
17. Utz P, Anderson P. Life and death decisions: regulation of apoptosis by proteolysis of signaling molecules. Cell Death Differ 2000;7:589–602.
18. Stephanou A, Scarabelli TM, Townsend PA, et al. The carboxy-terminal activation domain of the STAT-1 transcription enhances ischemia/reperfusion-induced apoptosis in cardiac myocytes. FASEB J 2002;16:1841–1843.
19. Stephanou A, Brar BK, Scarabelli TM, et al. Ischemia-induced STAT-1 expression and activation play a critical role in cardiomyocyte apoptosis. J Biol Chem 2000;275: 10,002–10,008.
20. Kumar A, Commane M, Flickinger TW, Horvath CM, Stark GR. Defective TNF-alpha-induced apoptosis in STAT1-null cells due to low constitutive levels of caspases. Science 1997;278: 1630–1632.
21. Kroemer G. The proto-oncogene Bcl-2 and its role in regulating apoptosis. Nat Med 1997;3:614–620.
22. Ellis H, Horvitz HR. Genetic control of programmed cell death in the nematode C. elegans. Cell 1986;44:817–829.
23. Kuwana T, Newmeyer DD. Bcl-2-family proteins and the role of mitochondria in apoptosis. Curr Opin Cell Biol 2003;15:691–699.
24. Riedl SJ, Shi Y. Molecular mechanisms of caspase regulation during apoptosis. Nat Rev Mol Cell Biol 2004;5:897–907.
25. Cory S, Huang DC, Adams JM. The Bcl-2 family: roles in cell survival and oncogenesis. Oncogene 2003;22:8590–8607.
26. Gelinas C, White E. BH3-only proteins in control: specificity regulates MCL-1 and BAK-mediated apoptosis. Genes Dev 2005;19:1263–1268.
27. Antonsson B, Martinou JC. The Bcl-2 protein family. Exp Cell Res 2000;256:50–57.
28. Wolf B, Green D. Suicidal tendencies: Apoptotic cell death by caspase family proteinases. J Biol Chem 1999;274:20,049–20,052.
29. Ranger AM, Malynn BA, Korsmeyer SJ. Mouse models of cell death. Nat Genet 2001;28: 113–118.
30. Varfolomeev EE, Schuchmann M, Luria V, et al. Targeted disruption of the mouse caspase 8 gene ablates cell death induction by the TNF receptors, Fas/Apo1, and DR3 and is lethal prenatally. Immunity 1998;9:267–276.
31. Zheng TS, Flavell RA. Divinations and surprises: genetic analysis of caspase function in mice. Exp Cell Res 2000;256:67–73.
32. Adams JM. Ways of dying: multiple pathways to apoptosis. Genes Dev 2003;17:2481–2495.
33. Thorburn A. Death receptor-induced cell killing. Cell Signal 2004;16:139–144.
34. Wang S, WS. E-D. TRAIL and apoptosis induction by TNF-family death receptors. Oncogene 2003;22:8628–8633.
35. Luo X, Budihardjo I, Zou H, Slaughter C, Wang X. Bid, a Bcl2 interacting protein, mediates cytochrome c release from mitochondria in response to activation of cell surface death receptors. Cell 1998;21:481–490.
36. Salvesen G, Duckett C. IAP proteins:blocking the road to death's door. Nat Rev Mol Cell Biol 2002;3:401–410.
37. Shi Y. Mechanisms of caspase activation and inhibition during apoptosis. Mol Cell 2002;9:459–470.

38. Bras M QB, Susin SA. Programmed cell death via mitochondria: different modes of dying. Biochemistry 2005;70:231–239.
39. Lorenzo HK, Susin SA. Mitochondrial effectors in caspase-independent cell death. FEBS Letters 2004;557:14–20.
40. Li LY, Luo X, Wang X. Endonuclease G is an apoptotic DNase when released from mitochondria. Nature 2001;412:95–99.
41. Dempsey EC, Newton AC, Mochly-Rosen D, et al. Protein kinase C isozymes and the regulation of diverse cell responses. Am J Physiol 2000;279:L429–438.
42. Newton AC. Protein kinase C: Structural and spatial regulation by phosphorylation, cofactors, and macromolecular interactions. Chem Rev 2001;101:2352–2364.
43. Newton A. Regulation of the ABC kinases by phosphorylation: Protein kinase C as a paradigm. Biochemical J 2003;370:361–371.
44. Poole AW, Pula G, Hers I, Crosby D, Jones ML. PKC-interacting proteins: from function to pharmacology. Trends Pharm Sci 2004;25:528–535.
45. Bornancin F, Parker PJ. Phosphorylation of protein kinase C-alpha on serine 657 controls the accumulation of active enzyme and contributes to its phosphatase- resistant state [published erratum appears in J Biol Chem 1997 May 16;272:13458]. J Biol Chem 1997;272:3544–3549.
46. Le Good JA, Ziegler WH, Parekh DB, Alessi DR, Cohen P, Parker PJ. Protein kinase C isotypes controlled by phosphoinositide 3-Kinase through the protein kinase PDK1. Science 1998;281:2042–2045.
47. Parekh D, Ziegler W, Yonezawa K, Hara K, Parker PJ. Mammalian TOR controls one of two kinase pathways acting upon nPKCdelta and nPKCepsilon. J Biol Chem 1999;274: 34,758–34,764.
48. Parekh DB, Ziegler W, Parker PJ. Multiple pathways control protein kinase C phosphorylation. EMBO J 2000;19:496–503.
49. Li W, Zhang J, Bottaro DP, Li W, Pierce JH. Identification of serine 643 of protein kinase C-delta as an important autophosphorylation site for its enzymatic activity. J Biol Chem 1997;272:24,550–24,555.
50. Stempka L, Schnolzer M, Radke S, Rincke G, Marks F, Gschwendt M. Requirements of protein kinase C-delta for catalytic function. Role of glutamic acid 500 and autophosphorylation on serine 643. J Biol Chem 1999;274:8886–8892.
51. Stempka L, Girod A, Muller H-J, et al. Phosphorylation of protein kinase C delta at threonine 505 is not a prerequisite for enzymatic activity. Expression of rat PKC delta and an alanine 505 mutant in bacterial in a functional form. J Biol Chem 1997;272:6805–6811.
52. Dutil EM, Toker A, Newton A. Regulation of conventional protein kinase C isozymes by phophoinositide-dependent kinase 1(PDK-1). Curr Biol 1998;8:1366–1375.
53. Blass M, Kronfeld I, Kazimirsky G, Blumberg P, Brodie C. Tyrosine phosphorylation of protein kinase C delta is essential for its apoptotic effect in response to etoposide. Mol Cell Biol 2002;22:182–195.
54. Song JS, Swann PG, Szallasi Z, Blank U, Blumberg PM, Rivera J. Tyrosine phosphorylation-dependent and -independent associations of protein kinase C delta with src family kinases in the RBL-2H3 mast cell line: Regulation of src family kinase activity by protein kinase C. Oncogene 1998;16:3357–3368.
55. Yuan Z-M, Utsugisawa T, Ishiko T, et al. Activation of protein kinase C delta by the c-Abl tyrosine kinase in response to ionizing radiation. Oncogene 1998;16:1643–1648.
56. Konishi H, Yamauchi E, Taniguchi H, et al. Phosphorylation sites of protein kinase C delta in H2O2-treated cells and its activation by tyrosine kinase in vitro. Proc Natl Acad Sci USA 2001;98:6587–6592.
57. Denning MF, Dlugosz AA, Threadgill DW, Magnuson T, Yuspa SH. Activation of the epidermal growth factor receptor signal transduction pathway stimulates tyrosine phosphorylation of protein kinase C delta. J Biol Chem 1996;271:5325–5331.

58. Brodie C, Bogi K, Acs P, Lorenzo P, Baskin L, Blumberg P. Protein kinase C delta inhibits the expression of glutamine synthase in glial cells via PKCdelta regulatory domain and its tyrosine phosphorylation. J Biol Chem 1998;273:30,713–30,718.

59. Kronfeld I, Kazimirsky G, Lorenzo PS, Garfield SH, Blumberg PM, Brodie C. Phosphorylation of PKCdelta on distinct tyrosine residues regulates specific cellular functions. J Biol Chem 2000:35,491–35,498.

60. Blake R, Garcia-Paramio P, Parker P, Courtneidge S. Src promotes PKCdelta degradation. Cell Growth Differ 1999;10:231–241.

61. Benes C, Soltoff S. Modulation of PKCdelta tyrosine phosphorylation and activity in salivary and PC-12 cells by Src kinases. Am J Physiol Cell Physiol 2001;280:1498–1510.

62. Szallasi Z, Denning MF, Chang E, et al. Development of a rapid approach to identification of tyrosine phosphorylation sites: application to PKC delta phosphorylated upon activation of the high affinity receptor for IgE in rat basophilic leukemia cells. Biochem Biophys Res Commun 1995;214:888–894.

63. Li W, Mischak H, Yu JC, et al. Tyrosine phosphorylation of protein kinase C-delta in response to its activation. J Biol Chem 1994;269:2349–2352.

64. Herrant M, Liuciano F, Loubat A, Auberger P. The protective effect of phorbol esters on Fas-mediated apoptosis in T cells. Transcriptional and postranscriptional regulation. Oncogene 2002;21:4957–4968.

65. Gomez-Angelats M, Bortner CD, Cidlowski JA. Protein kinase C (PKC) inhibits Fas receptor-induced apoptosis through modulation of the loss of K+ and cell shrinkage. J Biol Chem 2000;275:19,609–19,619.

66. Gomez-Angelats M, Cidlowski JA. Protein kinase C regulates FADD recruitment and death-inducing signaling complex formation in Fas/CD95-induced apoptosis. J Biol Chem 2001;276:44,944–44,952.

67. Harper N, Hughes MA, Farrow SN, Cohen GM, MacFarlane M. Protein kinase C modulates tumor necrosis factor-related apoptosis-inducing by targeting the apical events of death receptor signaling. J Biol Chem 2003;278:44,338–44,347.

68. Sarker M, Ruiz-Ruiz C, López-Rivas A. Activation of protein kinase C inhibits TRAIL-induced caspases activation, mitochondrial events and apoptosis in a human leukemic T cell line. Cell Death Differ 2001;2:172–181.

69. Lee JY, Hannun YA, Obeid LM. Functional dichotomy of protein kinase C (PKC) in tumor necrosis factor-alpha (TNF-alpha) signal transduction in L929 cells. Translocation and inactivation of PKC by TNF-alpha. J Biol Chem 2000;275:29,290–29,298.

70. Engedal N, Blomhoff HK. Combined action of ERK and NFkappa B mediates the protective effect of phorbol ester on Fas-induced apoptosis in Jurkat Cells. J Biol Chem 2003;278:10,934–10,941.

71. Reyland M, Barzen K, Anderson S, Quissell D, Matassa A. Activation of PKC is sufficient to induce an apoptotic program in salivary gland acinar cells. Cell Death Differ 2000;7:1200–1209.

72. Yin L, Bennani-Baiti N, Powell CT. Phorbol ester-induced apoptosis of C4-2 cells requires both a unique and a redundant protein kinase C signaling pathway. J Biol Chem 2005;280:5533–5541.

73. Ito Y, Mishra N, Yoshida K, Kharbanda S, Saxena S, Kufe D. Mitochondrial targeting of JNK/SAPK in the phorbol ester response of myeloid leukemia cells. Oncogene 2001;8:794–800.

74. Haimovitz-Friedman A BN, McLoughlin M, Ehleiter D, Michaeli J, Vlodavsky I, Fuks Z. Protein kinase C mediates basic fibroblast growth factor protection of endothelial cells against radiation-induced apoptosis. Cancer Res 1994;54:2591–2597.

75. Zhuang S DJ, Kochevar IE. Protein kinase C inhibits singlet oxygen-induced apoptosis by decreasing caspase-8 activation. Oncogene 2001;20:6764–6776.

76. Fujii T, Garcia-Bermejo ML, Bernabo JL, et al. Involvement of protein kinase C delta in phorbol ester-induced apoptosis in LNCaP prostate cancer cells. Lack of proteolytic cleavage of PKCδ. J Biol Chem 2000;275:7574–7582.

77. Ni H, Ergin M, Tibudan SS, Denning MF, Izban KF, S. A. Protein kinase C-delta is commonly expressed in multiple myeloma cells and its downregulation by rottlerin causes apoptosis. Br J Cancer 2003;121:849–856.

78. Zhang J, Liu N, Zhang J, Liu S, Liu Y, Zheng D. PKCdelta protects human breast tumor MCF-7 cells against tumor necrosis factor-related apoptosis-inducing ligand-mediated apoptosis. J Cell Biochem 2005;96:522–532.

79. Garcia-Bermejo ML, Leskow FC, Fujii T, et al. Diacylglycerol (DAG)-lactones, a new class of protein kinase C (PKC) agonists, induce apoptosis in LNCaP prostate cancer cells by selective activation of PKCalpha. J Biol Chem 2002;277:645–655.

80. Gavrielides M, Frijhoff A, Conti C, MG K. Protein kinase C and prostate carcinogenesis: targeting the cell cycle and apoptotic mechanisms. Curr Drug Targets 2004;5:431–443.

81. Jorgensen K, Skrede M, Cruciani V, Mikalsen S-O, Slipicevic A, Florenes VA. Phorbol ester phorbol-12-myristate-13-acetate promotes anchorage-independent growth and survival of melanomas through MEK-independent activation of ERK1/2. Biochem Biophys Res Commun 2005;329:266–274.

82. Whelan DHR, Parker PJ. Loss of protein kinase C function induces an apoptotic response. Oncogene 1998;16:1939–1944.

83. Mandil R, Ashkenazi E, Blass M, et al. Protein kinase C alpha and protein kinase C delta play opposite roles in the proliferation and apoptosis of glioma cells. Cancer Res 2001;61:4612–4619.

84. Matassa A, Kalkofen R, Carpenter L, Biden T, Reyland M. Inhibition of PKCalpha induces a PKCdelta-dependent apoptotic program in salivary epithelial cells. Cell Death Differ 2003;10:269–277.

85. Lee JY, Hannun YA, Obeid LM. Ceramide inactivates cellular protein kinase Calpha. J Biol Chem 1996;271:13,169–13,174.

86. Ruvolo PP, Deng X, Carr BK, May WS. A functional role for mitochondrial protein kinase C alpha Bcl-2 phosphorylation and suppression of apoptosis. J Biol Chem 1998;273: 25,436–25,442.

87. Deng X, Ito T, Carr B, Mumby M, May WS, Jr. Reversible phosphorylation of Bcl2 following Interleukin 3 or Bryostatin 1 is mediated by direct interaction with protein phosphatase 2A. J Biol Chem 1998;273:34,157–34,163.

88. Majewski M, Nieborowska-Skorska M, Salomoni P, et al. Activation of mitochondrial Raf-1 Is involved in the anti-apoptotic effects of Akt. Cancer Res 1999;59:2815–2819.

89. Partovian C, Simons M. Regulation of protein kinase B/Akt activity and Ser473 phosphorylation by protein kinase C alpha in endothelial cells. Cell Signal 2004;16:951–957.

90. Li W, Zhang J, Flechner L, et al. Protein kinase C-alpha overexpression stimulates Akt activity and suppresses apoptosis induced by interleukin 3 withdrawal. Oncogene 1999;18:6564–6572.

91. Michie AM, Nakagawa R. The link between PKCalpha regulation and cellular transformation. Immunology Letters 2005;96:155–162.

92. Lahna M, Köhlerd G, Sundella K, et al. Protein kinase C alpha expression in breast and ovarian cancer. Oncology 2004;67:1–10.

93. Iwamoto T, Hagiwara M, Hidaka H, Isomura T, Kioussis D, Nakashima I. Accelerated proliferation and interleukin-2 production of thymocytes by stimulation of soluble anti-CD3 monoclonal antibody in transgenic mice carrying a rabbit protein kinase C alpha. J Biol Chem 1992;267:18,644–18,648.

94. Ways DK, Kukoly CA, deVente J, et al. MCF-7 breast cancer cells transfected with protein kinase C-alpha exhibit altered expression of other protein kinase C isoforms and display a more aggressive neoplastic phenotype. J Clin Invest 1995;95:1906–1915.

95. Jiang X-H, Tu S-P, Cui J-T, et al. Antisense targeting protein kinase Calpha and beta1 inhibits gastric carcinogenesis. Cancer Res 2004;64:5787–5794.

96. Shen L, Dean N, Glazer R. Induction of p53-dependent, insulin-like growth factor-binding protein-3-mediated apoptosis in glioblastoma multiforme cells by a protein kinase C alpha antisense oligonucleotide. Mol Pharmacol 1999;55:396–402.

97. Lahn M, Sundell K, Moore S. Targeting protein kinase C-alpha in cancer with the phosphorothioate antisense oligonucleotide Aprinocarsen. Ann NY Acad Sci 2003;1002: 263–270.

98. Grossman S, Alavi J, Supko J, et al. Efficacy and toxicity of the antisense oligonucleotide aprinocarsen directed against protein kinase C- delivered as a 21-day continuous intravenous infusion in patients with recurrent high-grade astrocytomas. Neuro-Oncology 2005;1:32–40.

99. Vansteenkiste J, Canon J-L, Riska H, et al. Randomized phase II evaluation of aprinocarsen in combination with gemcitabine and cisplatin for patients with advanced/metastatic non-small cell lung cancer. Invest New Drugs 2005;23:263–269.

100. Guo B, Su TT, Rawlings DJ. Protein kinase C family functions in B-cell activation. Curr Opin Immunol 2004;16:367–373.

101. Barr L, Campbell S, Baylin S. Protein kinase C-beta 2 inhibits cycling and decreases c-myc-induced apoptosis in small cell lung cancer cells. Cell Growth Differ 1997;8:381–392.

102. Whitman SP, Civoli F, Daniel LW. Protein Kinase C beta II activation by 1-beta -D-Arabinofuranosylcytosine is antagonistic to stimulation of apoptosis and Bcl-2alpha downregulation. J Biol Chem 1997;272:23,481–23,484.

103. Evans C, Lord J, Owen-Lynch P, Johnson G, Dive C, Whetton A. Suppression of apoptosis by v-ABL protein tyrosine kinase is associated with nuclear translocation and activation of protein kinase C in an interleukin-3-dependent haemopoietic cell line. J Cell Sci 1995;108:2591–2598.

104. Kawakami Y, Nishimoto H, Kitaura J, et al. Protein Kinase C betaII regulates Akt phosphorylation on ser-473 in a cell type- and stimulus-specific fashion. J Biol Chem 2004;279:47,720–47,725.

105. Chiarini A, Whitfield JF, Armato U, Dal Pra I. Protein kinase C-beta II is an apoptotic lamin kinase in polyomavirus-transformed, etoposide-treated pyF111 rat fibroblasts. J Biol Chem 2002;277:18,827–18,839.

106. Su T, Guo B, Kawakami Y, et al. PKC- beta controls IB kinase lipid raft recruitment and activation in response to BCR signaling. Nat Immunol 2002;3:780–786.

107. Murray NR, Davidson LA, Chapkin RS, Clay Gustafson W, Schattenberg DG, Fields AP. Overexpression of protein kinase C beta II induces colonic hyperproliferation and increased sensitivity to colon carcinogenesis. J Cell Biol 1999;145:699–711.

108. Gokmen-Polar Y, Murray NR, Velasco MA, Gatalica Z, Fields AP. Elevated protein kinase C beta II is an early promotive event in colon carcinogenesis. Cancer Res 2001;61:1375–1381.

109. Yoshiji H, Kuriyama S, Ways DK, et al. Protein kinase C lies on the signaling pathway for vascular endothelial growth factor-mediated tumor development and angiogenesis. Cancer Res 1999;59:4413–4418.

110. Graff JR, McNulty AM, Hanna KR, et al. The protein kinase Cbeta-selective inhibitor, Enzastaurin (LY317615.HCl), suppresses signaling through the AKT pathway, induces apoptosis, and suppresses growth of human colon cancer and glioblastoma xenografts. Cancer Res 2005;65:7462–7469.

111. Cao M-Y, Shinjo F, Heinrichs S, Soh J-W, Jongstra-Bilen J, Jongstra J. Inhibition of anti-IgM-induced translocation of protein kinase C beta I inhibits ERK2 activation and increases apoptosis. J Biol Chem 2001;276:24,506–24,510.

112. Jiang X-H, Lam S-K, Lin M, et al. Novel target for induction of apoptosis by cyclooxygenase-2 inhibitor SC-236 through a protein kinase C-1-dependent pathway. Oncogene 2002;21:6113–6122.

113. Laouar A, Glesne D, Huberman E. Involvement of protein kinase C-beta and ceramide in tumor necrosis factor-alpha -induced but not Fas-induced apoptosis of human myeloid leukemia cells. J Biol Chem 1999;274:23,526–23,534.

114. Macfarlane D, Manzel L. Activation of beta-isozyme of protein kinase C (PKC beta) is necessary and sufficient for phorbol ester-induced differentiation of HL-60 promyelocytes. Studies with PKC beta-defective PET mutant. J Biol Chem 1994;269:4327–23484.

115. Wheeler DL, Martin KE, Ness KJ, et al. Protein Kinase C epsilon is an endogenous photosensitizer that enhances ultraviolet radiation-induced cutaneous damage and development of squamous cell carcinomas1. Cancer Res 2004;64:7756–7765.

116. Knauf JA, Elisei R, Mochly-Rosen D, et al. Involvement of protein kinase Cepsilon (PKCe) in thyroid cell death. A truncated chimeric PKC epsilon cloned from a thyroid cancer cell line protects thyroid cells from apoptosis. J Biol Chem 1999;274:23,414–23,425.

117. Wu D, Foreman TL, Gregory CW, et al. Protein Kinase Cepsilon has the potential to advance the recurrence of human prostate cancer. Cancer Res 2002;62:2423–2429.

118. Tachado SD MM, Wescott GG, Foreman TL, Goodwin CD, McJilton MA, Terrian DM. Regulation of tumor invasion and metastasis in protein kinase C epsilon-transformed NIH3T3 fibroblasts. J Cell Biochem 2002;85:785–797.

119. Perletti GP, Folini M, Lin HC, Mischak H, Piccinini F, Tashjian AHJ. Overexpression of protein kinase C epsilon is oncogenic in rat colonic epithelial cells. Oncogene 1996;12: 847–854.

120. Ding L, Wang H, Lang W, Xiao L. Protein kinase C-epsilon promotes survival of lung cancer cells by suppressing apoptosis through dysregulation of the mitochondrial caspase pathway. J Biol Chem 2002;277:35,305–35,313.

121. Mayne GC, Murray AW. Evidence that protein kinase Cepsilon mediates phorbol ester inhibition of calphostin C- and tumor necrosis factor-alpha-induced apoptosis in U937 histiocytic lymphoma cells. J Biol Chem 1998;273:24,115–24,121.

122. Koriyama H, Kouchi Z, Umeda T, et al. Proteolytic activation of protein kinase C delta and epsilon by caspase-3 in U937 cells during chemotherapeutic agent-induced apoptosis. Cell Signal 1999;11:831–838.

123. Basu A, Lu D, Sun B, Moor AN, Akkaraju GR, Huang J. Proteolytic activation of protein kinase C-epsilon by caspase-mediated processing and transduction of anti-apoptotic signals. J Biol Chem 2002;277:41,850–41,856.

124. Okhrimenko H, Lu W, Xiang C, Hamburger N, Kazimirsky G, Brodie C. Protein kinase C-epsilon regulates the apoptosis and survival of glioma cells. Cancer Res 2005;65: 7301–7309.

125. Chen N, Ma W, Huange C, Dong Z. Translocation of Protein Kinase C-epsilon and Protein kinase C-delta to membrane is required for ultraviolet b-induced activation of mitogen-activated protein kinases and apoptosis. J Biol Chem 1999;274:15,389–15,394.

126. Cacace AM UM, Philipp A, Han EK, Kolch W, Weinstein IB. PKC epsilon functions as an oncogene by enhancing activation of the Raf kinase. Oncogene 1996;13:2517–2526.

127. Perletti GP CP, Brusaferri S, Marras E, Piccinini F, Tashjian AH Jr. Protein kinase C epsilon is oncogenic in colon epithelial cells by interaction with the Ras signal transduction pathway. Oncogene 1998;16:3345–3348.

128. Piiper A, Elez R, You S-J, et al. Cholecystokinin stimulates extracellular signal-regulated kinase through activation of the epidermal growth factor receptor, Yes, and protein kinase C. Signal amplification at the level of Raf by activation of protein kinase Cε. J Biol Chem 2003;278:7065–7072.

129. Catley MC, Cambridge LM, Nasuhara Y, et al. Inhibitors of protein kinase C (PKC) prevent activated transcription: role of events downstream of NF-KappaB DNA binding. J Biol Chem 2004;279:18,457–18,466.

130. Knauf JA, Ward LS, Nikiforov YE, et al. Isozyme-Specific Abnormalities of PKC in thyroid cancer: Evidence for post-transcriptional changes in PKC epsilon. J Clin Endocrinol Metab 2002;87:2150–2159.

131. Gillespie S, Zhang XD, Hersey P. Variable expression of protein kinase Cepsilon in human melanoma cells regulates sensitivity to TRAIL-induced apoptosis. Mol Cancer Ther 2005;4:668–676.

132. Cornford P, Evans J, Dodson A, et al. Protein Kinase C Isoenzyme Patterns characteristically modulated in early prostate cancer. Am J Pathol 1999;154:137–144.

133. McJilton MA VSC, Wescott GG, Wu D, Foreman TL, Gregory CW, Weidner DA, Harris Ford O, Morgan Lasater A, Mohler JL, Terrian DM. Protein kinase Cepsilon interacts with Bax and promotes survival of human prostate cancer cells. Oncogene 2003;22:7958–7968.

134. Wooten M, Seibenhener M, Zhou G, Vandenplas M, Tan T. Overexpression of atypical PKC in PC12 cells enhances NGF-responsiveness and survival through an NF-kappaB dependent pathway. Cell Death Differ 1999;6:753–764.

135. Lu Y JL, Brasier AR, Fields AP. NF-kappaB/RelA transactivation is required for atypical protein kinase C iota-mediated cell survival. Oncogene 2001;20:4777–4792.

136. Leroy I, de Thonel A, Laurent G, Quillet-Mary A. Protein kinase C zeta associates with death inducing signaling complex and regulates Fas ligand-induced apoptosis. Cell Signal 2005;17:1149–1157.

137. Murray NR, Fields AP. Atypical protein kinase C iota protects human leukemia cells against drug-induced apoptosis. J Biol Chem 1997;272:27,521–27,524.

138. Jamieson L, Carpenter L, Biden TJ, Fields AP. Protein Kinase Ciota activity is necessary for Bcr-Abl-mediated resistance to drug-induced apoptosis. J Biol Chem 1999;274: 3927–3930.

139. Plo I, Hernandez H, Kohlhagen G, Lautier D, Pommier Y, Laurent G. Overexpression of the atypical protein kinase C zeta reduces topoisomerase II catalytic activity, cleavable complexes formation, and drug-induced cytotoxicity in monocytic U937 leukemia cells. J Biol Chem 2002;277:31,407–31,415.

140. Filomenko R, Poirson-Bichat F, Billerey C, et al. Atypical protein kinase C zeta as a target for chemosensitization of tumor cells. Cancer Res 2002;62:1815–1821.

141. Berra E, Municio M, Sanz L, Frutos S, Diaz-Meco M, Moscat J. Positioning atypical protein kinase C isoforms in the UV-induced apoptotic signaling cascade. Mol Cell Biol 1997;17:4346–4354.

142. Frutos S, Moscat J, Diaz-Meco MT. Cleavage of zeta PKC but not lambda /iota PKC by caspase-3 during UV-induced apoptosis. J Biol Chem 1999;274:10,765–10,770.

143. Smith L, Chen L, Reyland ME, et al. Activation of atypical protein kinase C zeta by caspase processing and degradation by the ubiquitin-proteasome system. J Biol Chem 2000;275:40,620–40,627.

144. Gurumurthy S, Goswami A, Vasudevan KM, Rangnekar VM. Phosphorylation of Par-4 by protein kinase A is critical for apoptosis. Mol Cell Biol 2005;25:1146–1161.

145. Diaz-Meco M, Municio M, Frutos S, et al. The product of *par*-4, a gene induced during apoptosis, interacts selectively with the atypical isoforms of protein kinase C. Cell 1996; 86:777–786.

146. Diaz-Meco MT, Lallena M-J, Monjas A, Frutos S, Moscat J. Inactivation of the inhibitory kappa B protein kinase/Nuclear Factor kappa B pathway by Par-4 expression potentiates Tumor Necrosis Factor alpha -induced apoptosis. J Biol Chem 1999;274:19,606–19,612.

147. Duran A, Diaz-Meco M-T, Moscat J. Essential role of RelA Ser311 phosphorylation by PKC in NF-B transcriptional activation. EMBO J 2003;22:3910–3918.

148. Leitges M, Sanz L, Martin P, et al. Targeted Disruption of the zetaPKC Gene Results in the Impairment of the NF-kappaB Pathway. Mol Cell 2001;8:771–780.

149. Wang G, Silva J, Krishnamurthy K, Tran E, Condie BG, Bieberich E. Direct Binding to ceramide activates protein kinase C zeta before the formation of a pro-apoptotic complex with PAR-4 in differentiating stem cells. J Biol Chem 2005;280:26,415–26,424.

150. Kim S-J, Kim H-G, Oh C-D, et al. p38 Kinase-dependent and -independent Inhibition of Protein Kinase C zeta and -alpha regulates nitric oxide-induced apoptosis and dedifferentiation of articular chondrocytes. J Biol Chem 2002;277:30,375–30,381.

151. Kim J-S, Park Z-Y, Yoo Y-J, Yu S-S, Chun J-S. p38 kinase mediates nitric oxide-induced apoptosis of chondrocytes through the inhibition of protein kinase C zeta by blocking autophosphorylation. 2005;12:201–212.

152. Mao M, Fang X, Lu Y, Lapushin R, Bast R, Mills G. Inhibition of growth-factor-induced phosphorylation and activation of protein kinase B/Akt by atypical protein kinase C in breast cancer cells. Biochem J 2000;352:475–482.

153. Doornbos RP, Theelen M, van der Hoeven PCJ, van Blitterswijk WJ, Verkleij AJ, van Bergen en Henegouwen PMP. Protein kinase C zeta is a negative regulator of protein kinase B activity. J Biol Chem 1999;274:8589–8596.

154. Powell DJ, Hajduch E, Kular G, Hundal HS. Ceramide disables 3-phosphoinositide binding to the pleckstrin homology domain of protein kinase B (PKB)/Akt by a PKCzeta-dependent mechanism. Mol Cell Biol 2003;23:7794–7808.

155. Datta R, Kojima H, Yoshida K, Kufe D. Caspase-3-mediated cleavage of protein kinase C theta in induction of apoptosis. J Biol Chem 1997;272:20,317–20,320.

156. Schultz A, Jönsson J-I, Larsson C. The regulatory domain of protein kinase C localises to the Golgi complex and induces apoptosis in neuroblastoma and Jurkat cells. Cell Death Diff 2003;10:662–675.

157. Villalba M, Bushway P, Altman A. Protein Kinase C-theta Mediates a selective T cell survival signal via phosphorylation of BAD. J Immunol 2001;166:5955–5563.

158. Barouch-Bentov R, Lemmens EE, Hu J, et al. Protein Kinase C-theta is an early survival factor required for differentiation of effector CD8+ T Cells. J Immunol 2005;175:5126–5134.

159. Ahmed S, Shibazaki M, Takeuchi T, Kikuchi H. Protein kinase C theta activity is involved in the 2,3,7,8-tetrachlorodibenzo-p-dioxin-induced signal transduction pathway leading to apoptosis in L-MAT, a human lymphoblastic T-cell line. FEBS Journal 2005;272:903–915.

160. Schaack J, Langer S, Guo X. Efficient selection of recombinant adenoviruses using vectors that express beta-galactosidase. J Virol 1995;69:3920–3923.

161. Mecklenbrauker I, Kalled SL, Leitges M, Mackay F, Tarakhovsky A. Regulation of B-cell survival by BAFF-dependent PKCdelta-mediated nuclear signaling. Nature 2004;431:456–461.

162. Miyamoto A, Nakayama K, Imaki H, et al. Increased proliferation of B cells and autoimmunity in mice lacking protein kinase Cdelta. Nature 2002;416:865–869.

163. Leitges M, Mayr M, Braun U, et al. Exacerbated vein graft arteriosclerosis in protein kinase C delta-null mice. J Clin Invest 2001;108:1505–1512.

164. Jackson D, Zheng Y, Lyo D, et al. Suppression of cell migration by protein kinase Cdelta. Oncogene 2005;24:3067–3072.

165. Acs P, Beheshti M, Szallasi Z, Li L, Yuspa SH, Blumberg PM. Effect of a tyrosine 155 to phenylalanine mutation of protein kinase Cdelta on the proliferative and tumorigenic properties of NIH 3T3 fibroblasts. Carcinogenesis 2000;21:887–891.

166. Toyoda M, Gotoh N, Handa H, Shibuya M. Involvement of MAP kinase-independent protein kinase C signaling pathway in the EGF-induced p21(WAF1/Cip1) expression and growth inhibition of A431 cells. Biochem Biophys Res Commun 1998;250:430–435.

167. Mischak H, Goodnight JA, Kolch W, et al. Overexpression of protein kinase C-delta and epsilon in NIH 3T3 cells induces opposite effects on growth, morphology, anchorage dependence, and tumorigenicity. J Biol Chem 1993;268:6090–6096.

168. Watanabe T, Ono Y, Taniyama Y, et al. Cell division arrest induced by phorbol ester in CHO cells overexpressing protein kinase-delta subspecies. Proc Natl Acad Sci USA 1992;89: 10,159–10,163.

169. Lu Z, Hornia A, Jiang Y-W, Zang Q, Ohno S, Foster D. Tumor promotion by depleting cells of protein kinase Cdelta. Mol Cell Biol 1997;17:3418–3428.

170. Kiley S, Clark K, Duddy S, Welch D, Jaken S. Increased protein kinase Cdelta in mammary tumor cells:Relationship to transformation and metastatic progression. Oncogene 1999;18:6748–6757.

171. Kiley SC, Clark KJ, Goodnough M, Welch DR, Jaken S. Protein Kinase C delta involvement in mammary tumor cell metastasis. Cancer Res 1999;59:3230–3238.

172. Li W, Jiang Y-X, Zhang J, et al. Protein kinase C delta is an important signaling molecule in insulin-like growth factor I receptor-mediated cell transformation. Mol Cell Biol 1998;18:5888–5898.

173. Kilpatrick LE, Lee JY, Haines KM, Campbell DE, Sullivan KE, Korchak HM. A role for PKC-delta and PI 3-kinase in TNF-alpha -mediated anti-apoptotic signaling in the human neutrophil. Am J Physiol Cell Physiol 2002;283:C48–C57.

174. Peluso JJ, Pappalardo A, Fernandez G. Basic fibroblast growth factor maintains calcium homeostasis and granulosa cell viability by stimulating calcium efflux via a PKCdelta-dependent pathway. Endocrinology 2001;142:4203–4211.

175. Wert M, Palfrey C. Divergence in the anti-apoptotic signaling pathways used by nerve growth factor and basic fibroblast growth factor (bFGF) in PC12 cells: rescue by bFGF involves protein kinase Cdelta. Biochem J 2000;352:175–182.

176. Okhrimenko H, Lu W, Xiang C, et al. Roles of tyrosine phosphorylation and cleavage of protein Kinase Cdelta in its protective effect against tumor necrosis factor-related apoptosis inducing ligand-induced apoptosis. J Biol Chem 2005;280:23,643–23,652.

177. Fan C, Katsuyama M, Yabe-Nishimura C. Protein kinase C-delta mediates up-regulation of NOX1, a catalytic subunit of NADPH oxidase, via transactivation of the EGF receptor: possible involvement of PKCdelta in vascular hypertrophy. Biochem J 2005;390: 761–767.

178. Ueda Y, Hirai S-i, Osada S-i, Suzuki A, Mizuno K, Ohno S. Protein Kinase C delta Activates the MEK-ERK Pathway in a manner independent of Ras and dependent on Raf. J Biol Chem 1996;271:23,512–23,519.

179. Keshamouni VG, Mattingly RR, Reddy KB. Mechanism of 17-beta -estradiol-induced Erk1/2 activation in breast cancer cells. J Biol Chem 2002;277:22,558–22,565.

180. Mingo-Sion AM, Ferguson HA, Koller E, Reyland ME, Berg CLVD. PKCdelta and mTOR interact to regulate stress and IGF-I induced IRS-1 Ser312 phosphorylation in breast cancer cells. Breast Can Res Treat 2005;91:259–269.

181. Datta K, Nambudripad R, Pal S, Zhou M, Cohen HT, Mukhopadhyay D. Inhibition of insulin-like growth factor-I-mediated cell signaling by the von Hippel-Lindau gene product in renal cancer. J Biol Chem 2000;275:20,700–20,706.

182. Li W, Jiang Y-X, Zhang J, et al. Protein kinase C-delta is an important signaling molecule in insulin-like growth factor I receptor-mediated cell transformation. Mol Cell Biol 1998;18:5888–5898.

183. Jackson DN, Foster DA. The enigmatic protein kinase Cdelta: complex roles in cell proliferation and survival. FASEB J 2004;18:627–636.

184. Matassa A, Carpenter L, Biden T, Humphries M, Reyland M. PKCdelta is required for mitochondrial-dependent apoptosis in salivary epithelial cells. J Biol Chem 2001;276: 29,719–29,728.

185. Reyland M, Anderson S, Matassa A, Barzen K, Quissell D. Protein kinase C delta is essential for etoposide-induced apoptosis in salivary acinar cells. J Biol Chem 1999;274: 11,915–11,923.

186. Majumder PK, Pandey P, Sun X, et al. Mitochondrial translocation of protein kinase Cdelta in phorbol ester-induced cytochrome c release and apoptosis. J Biol Chem 2000;275: 21,793–21,796.

187. Majumder PK, Mishra NC, Sun X, et al. Targeting of protein kinase Cdelta to mitochondria in the oxidative stress response. Cell Growth Differ 2001;12:465–470.

188. Khwaja A, Tatton L. Caspase-mediated proteolysis and activation of protein kinase Cdelta plays a central role in neutrophil apoptosis. Blood 1999;94:291–301.

189. Chen L, H. , Hahn G, Wu C-H, et al. Opposing cardioprotective actions and parallel hypertrophic effects of delta PKC and epsilon PKC. Proc Natl Acad Sci USA 2001;98: 11,114–11,119.

190. Emoto Y, Manome Y, Meinhardt G, et al. Proteolytic activation of protein kinase C δ by an Ice-like protease in apoptotic cells. EMBO J 1995;14:6148–6156.

191. Ghayur T, Hugunin M, Talanian RV, et al. Proteolytic activation of protein kinase C delta by an ICE/CED 3-like protease induces characteristics of apoptosis. J Exp Med 1996;184:2399–2404.

192. DeVries TA, Neville MC, Reyland ME. Nuclear import of PKCdelta is required for apoptosis: identification of a novel nuclear import sequence. EMBO J 2002;21:6050–6060.

193. Sitailo LA, Tibudan SS, Denning MF. Bax activation and induction of apoptosis in human keratinocytes by the protein kinase C delta catalytic domain. J Invest Dermatol 2004;123:434–443.

194. D'Costa AM, Denning MF. A caspase-resistant mutant of PKC-delta protects keratinocytes from UV-induced apoptosis. Cell Death Differ 2005;12:224–232.

195. Kajimoto T, Shirai Y, Sakai N, et al. Ceramide-induced apoptosis by translocation, phosphorylation, and activation of protein kinase Cdelta in the Golgi complex. J Biol Chem 2004;279:12,668–12,676.

196. Li B, Murphy K, Laucirica R, Kittrell F, Medina D, Rosen J. A transgenic mouse model for mammary carcinogenesis. Oncogene 1998;16:997–1007.

197. Li L, Lorenzo P, Bogi K, Blumberg P, Yuspa S. Protein kinase C delta targets mitochondria, alters mitochondrial membrane potential, and induces apoptosis in normal and neoplastic keratinocytes when overexpressed by an adenoviral vector. Mol Cell Biol 1999;19: 8547–8558.

198. Zrachia A, Dobroslav M, Blass M, et al. Infection of glioma cells with Sindbis virus induces selective activation and tyrosine phosphorylation of protein kinase C delta. J Biol Chem 2002;277:23,693–23,701.

199. Zang Q, Lu Z, Curto M, Barile N, Shalloway D, Foster DA. Association between v-Src and protein kinase C delta in v-Src-transformed fibroblasts. J Biol Chem 1997;272:13,275–13,280.

200. Shanmugam M, Krett NL, Peters CA, et al. Association of PKC delta and active src in PMA-treated MCF-7 human breast cancer cells. Oncogene 1998;16:1649-1654.

201. Sun X, Wu F, Datta R, Kharbanda S, Kufe D. Interaction between protein kinase C delta and the c-Abl tyrosine kinase in the cellular response to oxidative stress. J Biol Chem 2000;275:7470–7473.

202. Yoshida K, Miki Y, Kufe D. Activation of SAPK/JNK signaling by protein kinase Cdelta in response to DNA damage. J Biol Chem 2002;277:48,372–48,378.

203. Kharbanda S, Pandey P, Jin S, et al. Functional interaction between DNA-PK and c-Abl in response to DNA damage. Nature 1997;386:732–735.

204. Kharbanda S, Saleem A, Yuan Z, et al. Nuclear signaling induced by ionizing radiation involves colocalization of the activated p56/p53lyn tyrosine kinase with p34cdc2. Cancer Res 1996;56:3617–3621.

205. Sumitomo M, Ohba M, Asakuma J, et al. Protein kinase Cdelta amplifies ceramide formation via mitochondrial signaling in prostate cancer cells. J Clin Invest 2002;109:827–836.

206. Scheel-Toellner D, Pilling D, Akbar AN, et al. Inhibition of T cell apoptosis by IFN-β rapidly reverses nuclear translocation of protein kinase C-delta. Eur J Immunol 1999;29:2603–2612.
207. Liu J, Chen J, Dai Q, Lee RM. Phospholipid Scramblase 3 is the mitochondrial target of protein kinase C delta-induced apoptosis. Cancer Res 2003;63:1153–1156.
208. Frasch S, Henson P, Kailey J, et al. Regulation of phospholipid scramblase activity during apoptosis and cell activation by protein kinase C delta. J Biol Chem 2000;275: 23,065–23,073.
209. Voss OH, Kim S, Wewers MD, Doseff AI. Regulation of monocyte apoptosis by the protein kinase Cdelta-dependent phosphorylation of caspase-3. J Biol Chem 2005;280:17,371–17,379.
210. Cross T, Griffiths G, Deacon E, et al. PKC-delta is an apoptotic lamin kinase. Oncogene 2000;4:2331–2337.
211. Yoshida K, Wang H-G, Miki Y, Kufe D. Protein kinase Cdelta is responsible for constitutive and DNA damage-induced phosphorylation of Rad9. EMBO J 2003;22:1431–1441.
212. Bharti A, Kraeft S-K, Gounder M, et al. Inactivation of DNA-dependent protein kinase by protein kinase C delta:Implications for apoptosis. Mol Cell Biol 1998;18:6719–6728.
213. Ryer EJ, Sakakibara K, Wang C, et al. Protein Kinase C delta Induces apoptosis of vascular smooth muscle cells through induction of the tumor suppressor p53 by both p38-dependent and p38-independent mechanisms. J Biol Chem 2005;280:35,310–35,317.
214. Abbas T, White D, Hui L, Yoshida K, Foster DA, Bargonetti J. Inhibition of human p53 basal transcription by down-regulation of protein kinase Cdelta. J Biol Chem 2004;279: 9970–9977.
215. Niwa K, Inanami O, Yamamori T, Ohta T, Hamasu T, Kuwabara M. Redox Regulation of PI3K/Akt and p53 in bovine aortic endothelial cells exposed to hydrogen peroxide. Antioxidants & Redox Signaling 2003;5:713–722.
216. Ren J, Datta R, Shioya H, et al. p73beta is regulated by protein kinase Cdelta catalytic fragment generated in the apoptotic response to DNA damage. J Biol Chem 2002;277: 33,758–33,765.
217. DeVries TA, Kalkofen RL, Matassa AA, Reyland ME. Protein kinase Cdelta regulates apoptosis via activation of STAT1. J Biol Chem 2004;279:45,603–45,612.
218. Lee Y-J, Soh J-W, Dean NM, et al. Protein Kinase Cdelta overexpression enhances radiation sensitivity via extracellular regulated protein kinase 1/2 activation, abolishing the radiation-induced G2-M Arrest. Cell Growth Differ 2002;13:237–246.
219. Efimova T, Broome A-M, Eckert RL. Protein kinase Cdelta regulates keratinocyte death and survival by regulating activity and subcellular localization of a p38delta-extracellular signal-regulated kinase 1/2 complex. Mol Cell Biol 2004;24:8167–8183.
220. Murriel CL, Churchill E, Inagaki K, Szweda LI, Mochly-Rosen D. Protein kinase Cdelta activation induces apoptosis in response to cardiac ischemia and reperfusion damage: a mechanism involving Bad and the mitochondria. J Biol Chem 2004;279:47,985–47,991.
221. Lee Y-J, Lee D-H, Cho C-K, et al. HSP25 Inhibits protein kinase Cdelta-mediated cell death through direct interaction. J Biol Chem 2005;280:18,108–18,119.

3

The Role of Phosphoinositide 3-Kinase-Akt Signaling in Virus Infection

Samantha Cooray

Summary

Successful virus infection of host cells requires efficient viral replication, production of virus progeny and spread of newly synthesized virus particles. This success, however also depends on the evasion of a multitude of antiviral signaling mechanisms. Many viruses are capable of averting antiviral signals through modulation of host cell signaling pathways. Apoptotic inhibition, for example, is a universal intracellular antiviral response, which prolongs cellular survival and allows viruses to complete their life cycle. Ongoing apoptotic inhibition contributes to the establishment of latent and chronic infections, and has been implicated in viral oncogenesis. The phosphoinositide 3-kinase (PI3K)-Akt pathway has become recognized as being pivotal to the inhibition of apoptosis and cellular survival. Thus, modulation of this pathway provides viruses with a mechanism whereby they can increase their survival, in addition to other established mechanisms such as expression of viral oncogenes and direct inhibition of proapoptotic proteins. Recent research has revealed that this pathway is up-regulated by a number of viruses during both short-term acute infections and long-term latent or chronic infections. During acute infections PI3K-Akt signaling helps to create an environment favorable for virus replication and virion assembly. In the case of long-term infections, modulation of PI3K-Akt signaling by specific viral products is believed to help create a favorable environment for virus persistence, and contribute to virus-mediated cellular transformation.

Key Words: Phosphoinositide 3-kinase; Akt; virus; transformation; signaling; survival.

1. Introduction

Efficient virus replication and production of virus progeny is dependent on the ability of viruses to survive in a hostile host environment. In order to survive inside cells, viruses have evolved mechanisms by which they can modulate cellular events, particularly those governing apoptosis and cellular survival. Virus-mediated apoptotic inhibition is a well-established survival mechanism. During acute infections, such as those caused by respiratory viruses like influenza A and respiratory syncitial virus (RSV), apoptotic inhibition plays a role in maintenance of cell viability during virus replication and growth. During long-term infections such as latent herpesviruses infections, or chronic hepatitis B and C virus infections, apoptotic inhibition plays a role in prolonged survival of infected cells. In latently infected cells apoptotic mechanisms are often held in check by specific viral proteins and the cell cycle can also be modulated, creating an environment favorable for cellular transformation and tumor development. During chronic infection, it is believed that multiple biochemical changes occur within the host cell, resulting from both virus-dependent and -independent mechanisms, which can lead to cellular transformation. The molecular mechanisms that result in cellular transformation

From: *Apoptosis, Cell Signaling, and Human Diseases: Molecular Mechanisms, Volume 2*
Edited by R. Srivastava © Humana Press Inc., Totowa, NJ

in vivo as a result of long-term virus infections are not well defined, and are likely to depend also on the virus species, the cell type infected, and the genetics of the host.

Viruses block apoptosis through inhibition of classical apoptotic pathway proteins such as death receptors, caspases, and p53, and expression of viral homologues to anti-apoptotic proteins such as Bcl-2 *(1)*. However, numerous mitogenic signaling pathways within the eukaryotic cell also regulate the balance between apoptosis and cell survival. Phosphoinositide 3-kinases (PI3Ks) are pivotal to several signal transduction pathways and act on a number of downstream signaling molecules to regulate cellular events such as cellular survival, differentiation and proliferation *(2)*. Akt kinase is one such molecule, and PI3K-Akt signaling has been demonstrated to be extremely important in cell survival *(3)*. Constitutive up-regulation of PI3K-Akt cell survival signaling has also been implicated in oncogenesis, as it averts apoptotic cell death during uncontrolled cellular proliferation *(4)*.

Activation of the PI3K-Akt signaling pathway during virus infection is emerging as a common mechanism for virus survival during early replication, the establishment of latent and chronic infections, and virus-mediated cellular transformation. This chapter will describe the PI3K-Akt pathway in detail and discuss the viral modulation of this pathway as a means for survival in different types of virus infection.

2. PI3K-Akt Signaling

2.1. PI3Ks

PI3Ks are a family of enzymes that phosphorylate the 3′ hydroxyl group of the inositol ring of phosphatidylinositol (PtdIns) and related inositol phospholipids, generating 3′-phosphoinositide products (*see* Fig. 1) *(5)*. These phosphoinositide products act as second messengers, which aid the recruitment of numerous proteins into signaling complexes at the plasma membrane, and in this way can activate numerous downstream signaling events.

There are three classes of PI3Ks (I, II, and II), which differ in their substrate specificity and regulation. Class I PI3Ks, however, are by far the best studied, as they function to regulate downstream signaling events in response to external mitogenic stimuli. In addition, it is only the effect of signaling downstream class I PI3Ks that has been studied in the context of virus infections, therefore only this class will be discussed. Further information on class II and III PI3Ks can be found in a number of recent reviews *(2,6)*.

Class I PI3Ks are heterodimeric proteins consisting of a catalytic subunit (110 kDa, p110) and a regulatory (or adaptor) subunit. This class of PI3Ks has been further subdivided into subgroups I_A and I_B which are activated downstream of tyrosine kinases and G-protein-coupled receptors (GPCRs) respectively *(6)*. In mammals the p110 catalytic subunit of class I_A PI3Ks has 3 isoforms (α, β, δ), each encoded by a separate gene. They also have seven adaptor subunits, generated by expression and alternative splicing of three different genes (p85α, p85β, and p55γ). The prototype p85α subunit has an Src-homology 3 (SH3) domain, a proline-rich domain and two Src-homology 2 (SH2) domains that mediate protein-protein interactions (Table 1). The p110 catalytic subunits form functional complexes with the adaptor subunits by binding to a region between their SH2 domains (the inter-SH2 region) (Table 1) *(7)*. Forms of class I_A PI3Ks have also been identified in *Drosophila melanogaster*, *Caenohabiditis elegans*, and the slime mold *Dictyostelum discoideum* *(6)*.

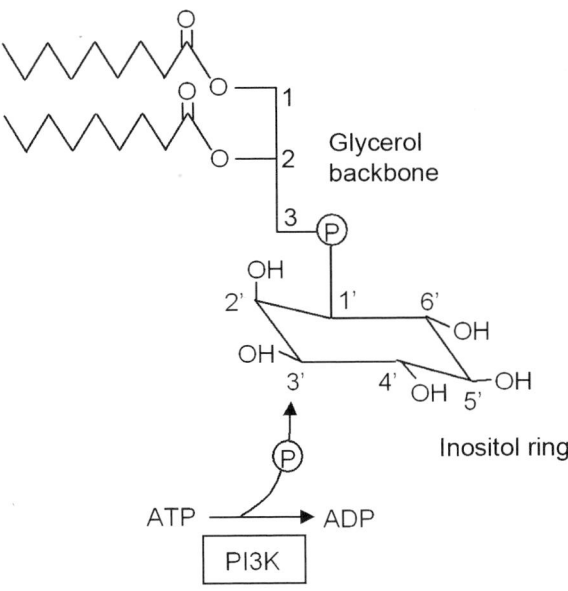

Fig. 1. The site of action of PI3Ks. Phosphatidylinositol (PtdIns) is composed of a glycerol backbone with fatty acid side chains attached at positions 1 and 2 and an inositol ring attached at position 3. The fatty acid side chains lie within the inner leaflet of the plasma membrane, with the inositol ring in the cytoplasm. PI3Ks transfer a phosphate group (P) from adenosine triphosphate (ATP) to the 3′ hydroxyl of the inositol ring of phosphatidylinositol. This results in the production of adenosine diphosphate (ADP) and phosphoinositide-3-phosphate (PtdIns-3-P). PI3Ks can also transfer a phosphate group to phosphoinositides already phosphorylated at other inositol hydroxyl positions (1′, 2′, 4′, and 5′) to produce 3′ phosphoinositides.

Activation of class I_A PI3Ks is mediated by the adaptor subunits. In quiescent cells the p85α regulatory subunit has been shown to inhibit the catalytic activity of the p110 (8). Upon mitogenic stimulation p85α mediates the translocation of p110 to the plasma membrane through binding of its SH2 domains to autophosphorylated receptor tyrosine kinases (RTKs) (*see* Fig. 2) (9). This binding event is also believed to result in a conformational change which releases the p110 subunit from the inhibitory binding of p85α, allowing it to be free to phosphorylate its lipid substrates (10–12). Class I_A PI3Ks can also be activated by tyrosine kinases in the cytoplasm downstream of other types of receptor. Src-family kinases, for example, have been shown to bind constitutively to class I_A PI3Ks and increase activity through phosphorylation of the p85 subunit at Tyr-688 (13,14).

Class I_B PI3Ks appear only to exist in mammals and consist of a single p110γ catalytic subunit, associated with a single 101 kDa (p101) adaptor subunit (Table 1) (2,5,15). Unlike the class I_A adaptor subunits, p101 does not contain any Src homology domains and interacts with p110γ instead via its proline-rich domain (Table 1) (5). Class I_B PI3Ks are activated by binding of the catalytic subunit to the $G_{βγ}$ subunits of heterotrimeric GTP-binding proteins (16). It has been observed that class I_A PI3Ks can also be activated by G-proteins, and both class I_A and I_B PI3Ks can bind Ras. In addition,

Table 1
The Properties of Class I_A and I_B PI3Ks

PI3K Class	Regulated by	*In vitro* substrates
IA 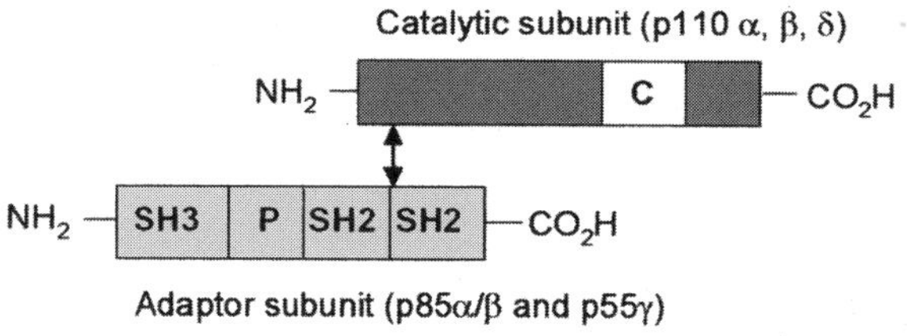	Tyrosine Kinases Ras GPCRs?	PtdIns PtdIns-4-P PtdIns-4,5-P_2
IB 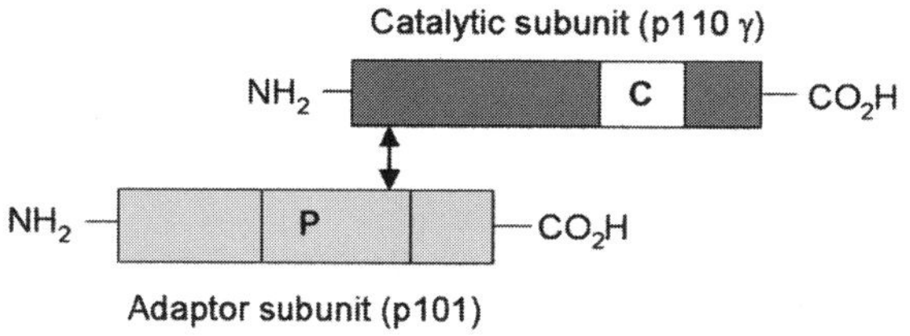	GPCRs Ras	PtdIns PtdIns-4-P PtdIns-4,5-P_2

Key
C – catalytic domain
SH2 – Src homology 2 domain
SH3 – Src homology 3 domain
P – proline-rich domain(s)

the HA-Ras isoform is associated with activation of PI3K and Akt rather than the classical mitogen activated Raf-MEK-ERK signaling cascade (*see* Fig. 2) *(17)*. However, the regulation and activation of PI3K-Akt signaling events by Ras has not been well studied, and the influence this has on downstream signaling events remains to be determined *(16,18,19)*.

The primary target for class I PI3Ks, both in vitro and in vivo, is phosphatidylinositol-4,5-diphosphate (PtdIns-4,5-P_2), and hence the primary product is phosphatidylinositol-3,4,5-triphosphate (PtdIns-3,4,5-P_3) (*see* Fig. 2) *(20)*. The production of PtdIns-3,4,5-P_3 is regulated by the phosphatase, PTEN, which catalyses the dephosphorylation of PtdIns-3,4,5-P_3 to PtdIns-4,5-P_2 (*see* Fig. 2) *(21,22)*.

2.2. Activation and Antiapoptotic Function of Akt

The production of PtdIns-3,4,5-P_3 by class I PI3Ks results in the recruitment of a wide variety of signal transduction proteins to the plasma membrane, which is facilitated by their lipid-binding pleckstrin homology (PH) domains (*see* Fig. 2) *(23)*. Once at the plasma membrane these proteins are most commonly activated by secondary phosphorylation events. One such protein, considered largely responsible for the antiapoptotic and cell survival mechanisms downstream of PI3K, is Akt (*see* Fig. 2).

Akt was discovered as a cellular homologue (c-*Akt*) of the viral oncogene (v-*Akt*) from the acutely transforming retrovirus AKT8, isolated from a murine T-cell lymphoma

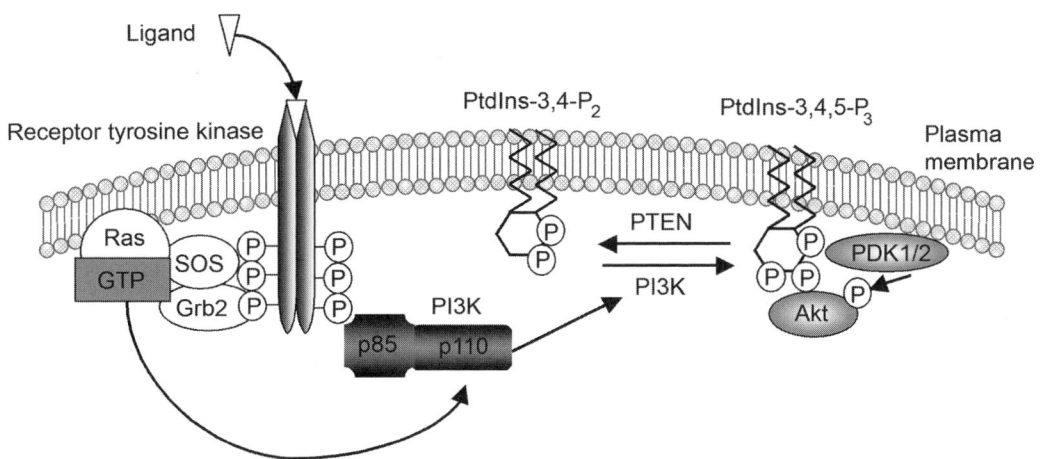

Fig. 2. Activation of class I_A PI3Ks. RTKs bind to extracellular mitogens, which trigger their activation through dimerization and autophosphorylation of their cytoplasmic domains. The p85 (or p55) adaptor subunits of class I_A PI3Ks recruit the p110 catalytic subunits to the plasma membrane through binding of their SH2 domains to the phosphorylated tyrosines on the RTK. This binding also causes a conformational change resulting in activation of p110. p110 is also known to be activated by G-proteins such as Ras. Activated PI3Ks then phosphorylate plasma membrane PtdIns such as PtdIns-4,5-P_2 to produce Ptd-3,4,5-P_3. Various signaling proteins such as Akt are recruited to the plasma membrane via binding of their PH domains to phosphorylated PtdIns. Once there, they are subsequently phosphorylated and activated by kinases such as PDK1/2.

(24–26). It was simultaneously identified as a novel kinase with many similarities to protein kinase A (PKA) and protein kinase C (PKC), and therefore was also named protein kinase B (PKB) *(27)*. In mammals, there are three isoforms of Akt (Akt 1, 2, and 3 or PKB α, β, and γ) that have a broad tissue distribution. All three isoforms are composed of an N-terminal PH domain, a central catalytic domain, and a C-terminal hydrophobic domain. As mentioned above, binding of the PH domain of Akt to the phosphoinositide products of PI3K results in its recruitment to the plasma membrane. Once there, Akt is activated by phosphorylation at Thr-308 of the catalytic domain by phosphoinositide-dependent kinase 1 (PDK-1), and at Ser-473 of the C-terminal hydrophobic region (Akt1/PKBα) by another kinase, termed phosphoinositide-dependent kinase 2 (PDK-2), which is yet to be identified *(28)*.

Upon activation, Akt phosphorylates a wide variety of targets at Ser/Thr residues, which are involved in the regulation of cell differentiation, proliferation and survival /apoptotic inhibition (*see* Fig. 3). A number of Bad proapoptotic proteins are inactivated by Akt phosphorylation. These include the Bcl-2 family member Bad (Ser-136), the cell death effector protease caspase-9 (Ser-196) and glycogen synthase kinase-3 beta (GSK-3β) (Ser-9) (*see* Fig. 3) *(29)*. Akt-phosphorylated BAD binds to 14-3-3 proteins and is sequestered in the cytosol. This interaction prevents it from heterodimerizing with and inactivating anti-apoptotic Bcl_2-family members, such as Bcl-2 and Bcl-xL, at the mitochondrial membrane *(15,30)*. Akt-phosphorylation and inactivation of human caspase-9, and several other caspases, blocks activation of the proteolytic apoptotic caspase cascade *(31)*. GSK-3β has been shown to induce apoptosis and this is blocked by Akt phosphorylation *(32)*. GSK-3β

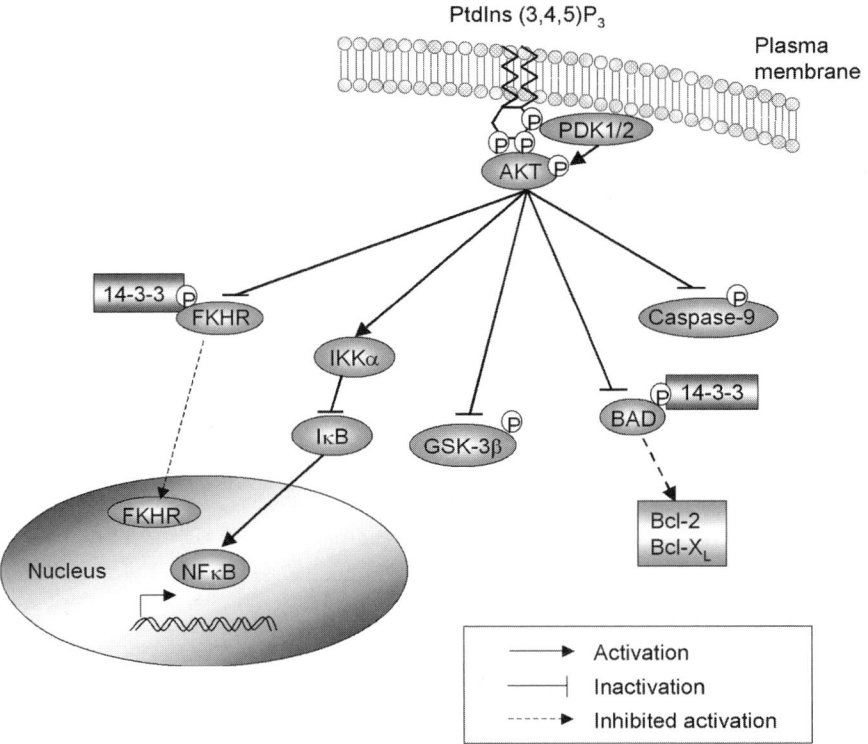

Fig. 3. Akt-mediated survival. Activated Akt mediates cell survival by phosphorylating and inhibiting a number of Bad proapoptotic proteins, including, caspase-9, Bad and GSK-3β. Akt phosphorylates and inhibits forkhead transcription factors like FKHR, preventing them from migrating to the nucleus and up-regulating the expression of Bad proapoptotic genes. Akt can also activate transcription of Bad prosurvival genes by activation IKKα, which degrades IκB and releases the transcription factor NF-κB. The inhibition of Bad and FKHR is aided by 14-3-3 proteins, which bind and sequester their phosphorylated forms in the cytoplasm.

normally phosphorylates and inhibits glycogen synthase, therefore it has been suggested that Akt inactivation of GSK-3β may also increase glycogen synthesis *(33)*.

In addition, Akt inhibits the transcription of proapoptotic genes by phosphorylating members of the forkhead family of transcription factors such as FKHR. FKHR predominantly resides in the nucleus, where it regulates the transcription of a number of genes crucial to apoptosis, for example FasL and Bim *(34)*. Akt-phosphorylation promotes the export of FKHR from the nucleus into the cytosol, where it is bound and inhibited by 14-3-3 proteins *(34,35)*. Akt can also activate the transcription of antiapoptotic genes (e.g., inhibitors of apoptosis [AIFs]) through phosphorylation of IκB kinase α (IKKα). IKKα phosphorylates the NF-κB inhibitor IκB and facilitates its ubiquitin-mediated degradation. This permits NF-κB to translocate to the nucleus where it can up-regulate gene expression *(36–40)*.

It is considered that the ability of Akt to simultaneously block apoptotic factors and increase survival factors correlates with protection against apoptotic stimuli in a variety of cell types in which activated Akt has been constitutively over-expressed.

3. PI3K-Akt Signaling in Virus Infection

3.1. PI3K-Akt Mediated Cell Survival During the Early Stages of Acute Virus Infections

Activation of Akt downstream of PI3K has been shown to be important in cell survival and apoptotic inhibition during different types of virus infection. Recent research on viruses that cause acute infections, for example, suggests that activation of PI3K-Akt signaling may contribute to survival during the early stages of infection, when virus replication and protein synthesis are taking place. However, activation of such survival responses may also be induced by the infected cells themselves to permit sufficient time for the activation of antiviral cellular defense mechanisms and viral clearance by the host immune system.

Early activation of PI3K-Akt survival signaling has been most commonly observed in vitro with RNA viruses that cause acute infections such as human RSV, severe acute respiratory syndrome (SARS) coronavirus, and rubella virus (RV). Infections with these viruses usually results in the induction of apoptosis, which is considered, for nonlytic viruses, to facilitate spread of progeny *(1)*. As these viruses cannot evade immune system detection, they have to replicate and spread rapidly in order to survive, which may be aided by initial activation of survival responses followed by induction of apoptosis *(1)*.

RSV is an important cause of serious respiratory tract illness in children and immunocompromised adults *(41)*. This virus preferentially infects airway epithelial cells, leading to a severe inflammatory response characterized by the up-regulation of inflammatory cytokines and chemokines, as well as signal transducers and activators of transcription (STATS) *(42–44)*. This response has been shown in vitro to be dependent on an increase in the transcriptional activity of NF-κB *(42,43)*. The induction of NF-κB activity during RSV infection in A549 airway epithelial cells has been shown to result from activation of PI3K and Akt *(45)*. Interestingly, although RSV infection ultimately leads to cell death, a large proportion of RSV infected cells remain viable well into the infection. This maintenance of cell viability is also dependent on the activation of PI3K, Akt, and NF-κB, as inhibition of PI3K with the drug LY294002 has been shown to cause a rapid increase in the speed and magnitude of RSV-induced apoptosis *(45)*. Further studies have demonstrated that RSV infection of A549 cells and primary tracheobronchial epithelial cells leads to the activation of ceramidase and sphingosine kinase resulting in an up-regulation of the production of the prosurvival molecule sphingosine 1-phosphate (S1P) *(46)*. S1P subsequently mediates the downstream activation of PI3K-Akt (as well as extracellular regulated kinase [ERK]) leading to cell survival and apoptotic inhibition in the initial stages of RSV infection *(46)*. This data suggests that PI3K-Akt signaling contributes to cell survival to preserve host cells until the life cycle of RSV is complete.

Similar signaling events have been observed during RV infection in vitro. RV infection has been shown to increase the phosphorylation of Akt and GSK-3β, and like RSV, inhibition of PI3K with LY294002 increases the speed and magnitude of RV-induced caspase-dependent apoptosis *(47)*. However, in contrast to RSV, these survival signals occur concomitantly with, and no prior to, apoptotic signals that occur early in RV infection *(48–52)*. However extensive apoptosis does occur at later stages of the virus life cycle, implying that the cell survival signals are eventually overridden following the production and release of large amounts of virus progeny. This suggests that in RV infection, the signals regulating downstream Akt survival events as well as apoptosis, be they viral or cellular, probably differ from those of RSV. The significance of cell survival and apoptosis during RV associated disease such as acute lymphadenopathy, rash and congenital rubella syndrome, is unknown.

Like RV, infection with SARS-associated coronavirus (CoV) in Vero E6 cells has been shown to increase phosphorylation of Akt and GSK-3β as well as a PKCζ, another downstream mediator of PI3K survival signaling *(53,54)*. However, in contrast, the survival response resulting from PI3K-Akt signaling was deemed to be weak, as LY294002 treatment did not result in an increase in apoptotic DNA laddering *(54)*. The authors conclude that the weak activation of Akt and GSK-3β in SARS-CoV infected cells is not sufficient to prevent virus-induced apoptosis at any stage of infection. The inability of cells to mount an adequate survival response may contribute to the pathology of SARS in vivo, however further studies need to be done to investigate this.

Coxsackievirus B3 (CVB3) is the causative agent of acute myocarditis, although infection with CVB3 can also lead to chronic cardiomyopathy *(55,56)*. CVB3 infection of cardiac myocytes results in caspase-dependent apoptotic cell death, the extent of which is considered to influence not only the fate of infected cells, but also the severity of the disease. Esfandiarei and colleagues *(57)* have demonstrated that CVB3 infection of human lung epithelial (HeLa) cells results in a gradual increase in both Akt and GSK-3β phosphorylation, and akin to RV and RSV, inhibition of PI3K with LY294002 increases CVB3-induced apoptosis. An increase in apoptosis was also detected when a dominant negative mutant of Akt1 was transfected into cells prior to CVB3 infection *(57)*. LY29002 and dominant negative Akt were also used to show that PI3K-Akt signaling was necessary for viral RNA synthesis and viral protein expression. Interestingly activation of Akt was not dependent on the caspase cascade, suggesting that PI3K-Akt signaling and caspase-dependent apoptosis work independently of each other. However, unlike studies with the viruses mentioned above, the time course over which the apoptotic and survival signals were activated were not analyzed. Therefore it is difficult to say whether or not PI3K-Akt signaling is activated for cell survival early in infection, although the dependence of such signals for RNA synthesis is certainly suggestive of this. The activation and involvement of PI3K-Akt signaling during CVB3-induced chronic cardiomyopathy has not been investigated.

It is tempting to speculate that infection of cells with viruses such as RV, RSV, SARS-CoV, and CVB3 all result in the initial activation of PI3K-Akt and other survival signals to support their replication, followed thereafter by induction apoptosis to facilitate virus spread. However, as the viral products that could potentially mediate such effects have not been identified, one cannot conclude that the effects are virus- rather than cell-mediated.

3.2. PI3K-Akt Mediated Survival in Lytic/Latent Viral Infections

PI3K-Akt mediated survival has also been found to be important during both the lytic and latent stages of lytic/latent virus infections. The lytic stage of a lytic/latent infection is similar to an acute infection in that virus replication and production of virus progeny results in death of infected cells, which can also correlate with the appearance of disease symptoms. However such viruses can avoid host immune system detection and clearance by establishing a latent infection. During viral latency, a limited set of viral proteins is expressed and infectious viral progeny are not produced *(58)*. Viruses that are able to establish latent infections such as human papillomavirus (HPV), and herpesviruses such as human cytomegalovirus (HCMV), Epstein-Barr virus (EBV), and Kaposi's sarcoma-associated herpesvirus (KSHV) maintain their genomes as extrachromosomal episomes during latency *(59)*. Lytic replication can be re-initiated at any time and this is often accompanied by the reappearance of disease symptoms.

Activation of PI3K-Akt signaling is believed to contribute to the maintenance of the latent state by suppressing apoptosis, and hence the elimination of virus-infected cells. Long periods of latency, or reactivation from latent to lytic state, can also lead to transformation of infected cells. Signaling downstream of PI3K and Akt has been shown to contribute to both reactivation from latency, and virus-mediated transformation. Viral proteins expressed during the lytic or latent stages of infection have been shown to activate PI3K either through direct interaction with the catalytic or adaptor subunits, or by facilitating the association of PI3K with receptor or non-receptor tyrosine kinases.

Although a leading cause of congenital defects worldwide *(60)*, HCMV infection in healthy individuals in usually asymptomatic. However like other herpesviruses, HCMV is capable of establishing life-long latent infections *(61)*. Establishment of HCMV latent infection in vitro has been demonstrated to result in cellular transformation. However, the molecular mechanisms and viral proteins involved in the establishment of HCMV latency and transformation have not been well characterized. Studies looking at the activation of signaling pathways such as PI3K-Akt during HCMV infection have focused on virus entry and replication during the initial stages of primary lytic infections in vitro.

Entry of HCMV into host cells is facilitated by binding of its envelope glycoproteins gB (UL55) and gH (UL75) to receptors on the surface of the host cell. This binding has been demonstrated to result in the activation of several downstream intracellular signaling molecules which are important for viral DNA replication *(62–64)*. However, exactly how these molecules are activated downstream of the host cell receptors is not well understood *(64–66)*. Recently it was demonstrated that following HCMV infection of human embryonic lung fibroblasts (HELs), PI3K was strongly activated via phosphorylation of its p85 adaptor subunit. This resulted in the subsequent activation of Akt, p70 S6 kinase, and NF-κB *(67)*. p70 S6 kinase is another downstream target of Akt, which is associated with cellular proliferation rather than survival, as it phosphorylates ribosomal protein S6 to elevate protein production *(68,69)*. Activation of PI3K-Akt signaling did not produce a cell survival response in HCMV-infected cells, as inhibition of PI3K with LY294002 did not induce apoptosis. This is in contrast to RSV, where inhibition PI3K, Akt, and NF-κB results in increased apoptosis during the initial stages of infection. Inhibition of PI3K did, however, block DNA replication, and inhibited expression of viral proteins IE1-72, IE2-86, UL44, and UL84 *(67)*. Up-regulation of transcription factor NF-κB and p70 S6 kinase downstream of PI3K and Akt may result in cellular proliferation rather than survival, and such signaling appears to be important for transcription and translation of immediate early genes and completion of the lytic cyle *(67)*. The HCMV proteins involved in the up-regulation of PI3K-Akt signaling remain to be determined, and the involvement of PI3K-Akt signaling during HCMV latency and cell mediated transformation in vitro has not yet been investigated.

EBV (human herpesvirus 4) is the causative agent of infectious mononucleosis *(59)* and is implicated in the development of a variety of B-cell and epithelial-cell malignancies including Burkitt's lymphoma, Hodgkin's disease, and nasopharyngeal carcinoma *(70,71)*. Like HCMV, EBV can establish latent infections leading to the transformation of cells in vitro. EBV infection of primary human B-cells results in their transformation to lymphoblastoid cell lines (LCLs). During latent infection, EBV constitutively expresses a restricted set of proteins, which are also detected both in vitro in EBV-transformed B-cells and in a number of the EBV-associated malignancies *(72)*. Two of these latently encoded proteins, the integral membrane proteins LMP1 and LMP2A, interfere with PI3K-Akt signaling and upregulate PI3K-Akt mediated cell survival.

LMP1 behaves like a constitutively active tumor necrosis factor receptor (TNFR), and facilitates the recruitment of TNFR-associated death domain proteins (TRADD and RIP), and TNFR-associated factors (TRAFs) to the plasma membrane *(73–75)*. In this way LMP1 is able to regulate a number of mitogenic signaling cascades and has been shown to be essential for in vitro B-cell transformation *(76–78)*. The C-terminal cytoplasmic domain of LMP1 has been shown to bind to the p85 adaptor subunit of PI3K, leading to the activation of Akt *(see* Fig. 4). LMP1 activation of the PI3K-Akt pathway is thought to significantly contribute to cell survival and the morphological changes observed in B-cell transformation, as inhibition of PI3K with LY2940092 reverses the transformed phenotype *(79)*.

LMP2A, like LMP1, has also been shown to activate PI3K and Akt in B-cells, although a direct interaction has not been demonstrated *(80)*. However, it is possible that direct binding occurs via the C-terminal cytoplasmic tail of LMP2A, which is phosphorylated, providing binding sites for the SH2 domains of Src protein tyrosine kinases (PTKs) like Syk, as well as Lyn, both important mediators of B-cell signal transduction. Phosphorylation and subsequent activation of such tyrosine kinases in B-cells, is required for activation of PI3K by LMP2A. This is presumably because although LMP2A can recruit molecules via is phosphorylated cytoplasmic tail, it probably lacks the tyrosine kinase activity of autocatalytic RTKs *(see* Fig. 4) *(80)*. LMP2A activation of PI3K-Akt signaling does not appear to contribute to B-cell survival in vitro *(80)*. However, in primary B-cells from LMP2A transgenic mice, Ras is constitutively activated, as is PI3K and Akt *(81)*. These cells also show constitutive and up-regulation of antiapoptotic Bcl-2 family protein Bcl-xL, but no activation of mitogen activated kinases Raf, MEK1/2 and ERK1/2, which are traditionally associated with Ras *(81)*. Thus the Ras isoform involved is likely to be HA-Ras, which is associated with activation of PI3K rather than Raf. In addition, inhibitors of Ras, PI3K, and Akt but not Raf resulted in an increase in apoptosis in these cells. These findings suggest that during B-cell development LMP2A mimics a B-cell receptor (BCR) through constitutive activation of Ras, PI3K, Akt, and Bcl-xL proteins. This allows excess BCR negative B-cells, which would normally be eliminated by apoptosis, to survive in peripheral lymphoid organs and this may be involved in the development of EBV-associated B-cell lymphomas such as Hodgkin's disease *(81)*.

Expression of LMP2A in the human epithelial keratinocytes, also activates PI3K-Akt cell survival signals, leading to cellular transformation *(82)*. In Ramos Burkitt's lymphoma and HSC-39 epithelial gastric carcinoma cell lines, LMP2A expression protects against transforming growth factor (TGF)-β induced caspase-dependent

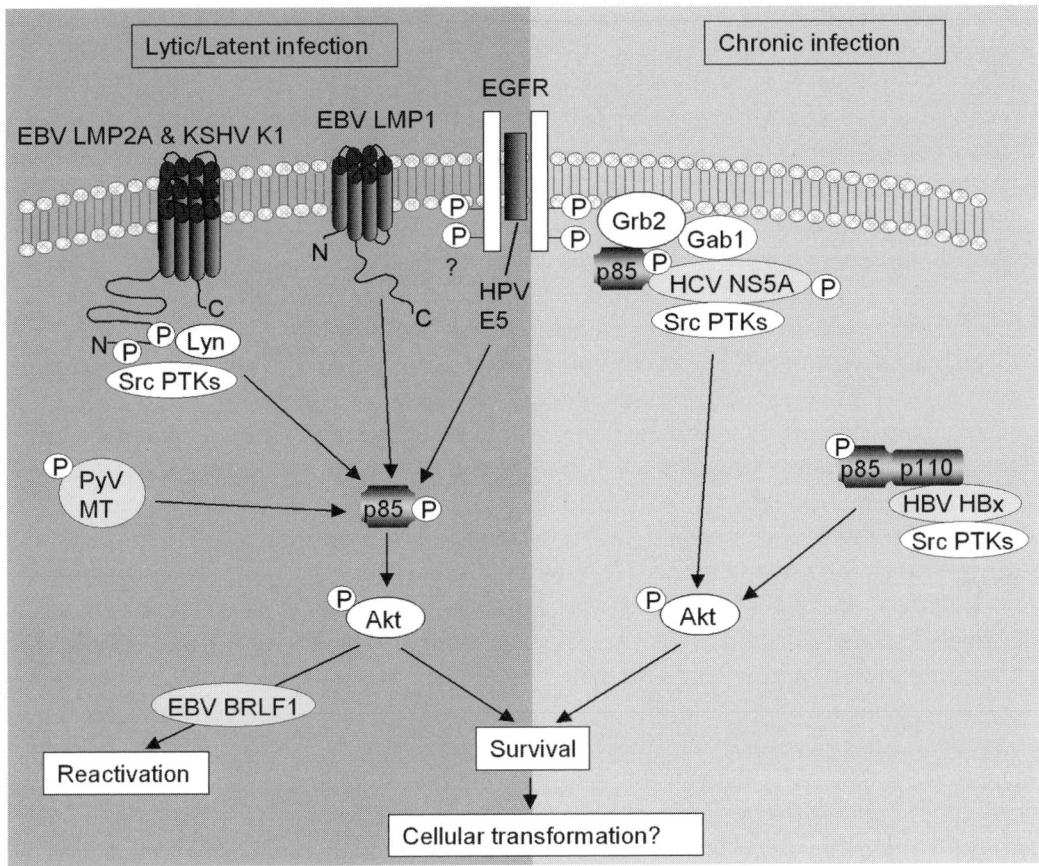

Fig. 4. Viral proteins that mediate PI3K-Akt survival during lytic/latent infections, chronic infections and cellular transformation. Viruses that cause lytic/latent infections express proteins that activate PI3K-Akt mediated survival to maintain the latent state such as EBV LMP1 and LMP2A, and KSHV K1. This is believed to contribute to cellular transformation following prolonged periods of latency, or as in the case of HPV E5, upon reactivation from latency. PI3K-Akt has also been shown to be required for the reactivation to the lytic state mediated by EBV protein BRLF1. PI3K-Akt signaling also facilitates PyV-mediated transformation in non-permissive cells. Viruses that cause chronic infections also express proteins that activate PI3K-Akt survival signals such as HCV NS5A and HBV HBx protein. However, whether constitutive up-regulation of PI3K and Akt contributes to cellular transformation after long periods of chronic infection is unknown. Activation of PI3K by viral proteins requires viral-protein mediated translocation to plasma membrane and binding to activated RTKs and/or the recruitment of cytoplasmic Src PTKs.

apoptosis via activation of PI3K and Akt *(83)*. Inhibition of PI3K with LY294002 was shown to inhibit Akt phosphorylation and block the antiapoptotic effect of LMP2A. These findings demonstrate that LMP2A can protect both B-cells and epithelial cells from apoptosis, perhaps providing a clonal selective advantage to such cells resulting in their immortilization. The role of LMP1 and LMP2A mediated apoptotic protection via PI3K and Akt has yet to be demonstrated for epithelial cells in vivo, and the involvement of PI3K-Akt signaling in EBV-epithelial cell malignancies is less well understood.

The latent form of EBV is periodically converted to the lytic form by expression of two proteins which work in conjunction to activate transcription, BRLF1 and BZLF1 *(84,85)*. PI3K-Akt signaling has been shown to be necessary for EBV reactivation from latency as PI3K and Akt are activated by BRLF1, and inhibition of PI3K abrogates BRLF1 transcriptional activity and ability to disrupt viral latency *(86)*. BZLF1, however, does not activate PI3K, but PI3K-Akt signaling may be required for the synergistic action of BRLF1 and BZLF1.

KSHV or human herpesvirus 8 has been identified as the etiologic agent of Kaposi's sarcoma (KS), of which there are various types including transplant-KS, endemic-KS, classical-KS, and acquired immunodeficiency syndrome (AIDS)-associated KS *(87–89)*. All forms of KS are histologically identical, and are angiogenic multicellular tumours *(88)*. However AIDS-associated KS is the most aggressive, and is the most common tumor to arise in human immunodeficiency virus (HIV)-infected individuals *(90)*. KSHV encodes an array of "pirated" regulatory proteins, which control cell growth, and are thought to contribute to KSHV latency and development of KS, although the molecular mechanisms involved are not well understood *(91–93)*. One such protein is K1 a transforming BCR-like transmembrane protein, which is similar to EBV LMP2A, and can also recruit tyrosine kinases like Syk to the plasma membrane via its cytoplasmic domain *(94,95)*. The cytoplasmic domain of K1 can also induce phosphorylation of a number of signaling molecules, perhaps via Src PTKs, including the p85 subunit of PI3K *(see* Fig. 4) *(94)*. This phosphorylation has been shown to correspond to the up-regulation of PI3K activity in B-cells over-expressing K1 *(96)*. Increased PI3K activity leads to the phosphorylation and activation of Akt, and inhibition of PTEN phosphatase and forkhead transcription factors *(see* Fig. 4) *(96)*. In addition K1 expression can protect cells from both FKHR- and Fas-mediated apoptosis *(96)*. This suggests that K1 may protect KHSV-infected cells early in the virus life cycle and contribute to the survival of tumorigenic cells during the development of KS.

HPV can also result in immortilization of infected cells, however, unlike the herpesviruses, HPV-mediated transformation occurs upon reactivation to the lytic state rather than after long periods of latency. HPV causes benign epithelial warts and is associated with the development of cervical and urogenital cancers *(97)*. The high-risk HPV type 16 (HPV 16), which is regularly detected in cervical cancers, encodes a putative transmembrane membrane protein E5, that can activate the PI3K-Akt pathway *(see* Fig. 4) *(98,99)*. HPV 16 E5 has been shown to interact with the epidermal growth factor receptor (EGFR) in human epithelial keratinocytes, stimulating activation through facilitating dimerization and autophosphorylation *(100,101)*. EGFR activation by HPV16 E5 up-regulates PI3K-Akt survival signaling, which can protect cells against ultraviolet (UV) induced apoptosis *(99)*. Whether PI3K and Akt activation also contributes to HPV reactivation from latency is unknown. However, E5 is necessary for full activation of the HPV transforming protein E7, therefore induction of PI3K-Akt dependent apoptotic inhibition by E5 may contribute to E7-mediated oncogenesis *(99,102–104)*. E5-mediated activation of PI3K, like EBV LMP1 and LMP2A, probably occurs at the plasma membrane through association of the phosphorylated cytoplasmic domain of the EGFR with p85 adaptor subunit of PI3K *(see* Fig. 4).

The *Polyomaviridae* differ from the herpesviruses and HPV, in that they only persist and stimulate cellular proliferation and transformation in non-permissive host cells that do not support their replication *(105)*. During the early stages of polyomavirus infection, the "tumor," or T-antigens, are produced which are able to stimulate resting cells to re-enter the cell cycle, and have transforming capability. Primate polyomaviruses encode two T-antigens, large T (LT) and small T (ST), whose transforming capability result, in part, from inhibition of apoptosis by blocking the activity of the p53 tumor suppressor. The LT antigen from mouse polyomavirus (PyV) does not have a binding site for p53. However PyV does encode a novel middle T (MT) antigen, a cytosolic phosphoprotein that interacts with a number of SH2 containing proteins, including PI3K, phospholipase C γ (PLCγ) and Shc *(106–108)*. The SH2 domain of the PI3K p85 subunit associates with the phosphorylated Tyr-315 of MT, which leads to its activation subsequent activation of Akt *(108–110)*. Recent studies suggest that MT may utilize the PI3K-Akt pathway to block apoptosis during viral transformation, independently of p53 *(109)*.

3.3. PI3K-Akt Mediated Survival During Chronic Viral Infections

Another strategy by which viruses can persist in the infected host is through establishment of a chronic infection. In contrast to viruses which persist in a latent state, viruses that cause chronic infections continuously replicate and produce infectious progeny *(58)*. Chronic infections result from failure of the host immune system to clear the initial infection, and thus disease symptoms are ongoing. In some circumstances malignant transformation can result from chronic infection of a specific cell type, although unlike latent infections, expression of particular viral proteins are usually not involved. Instead chronic infection is believed to lead to a series of biochemical events that are thought to bring about a cellular environment favorable for tumor development. PI3K-Akt signaling has been proposed to be involved in the survival of the host cell during chronic infection and contribute to cellular transformation.

Hepatitis B virus (HBV) and hepatitis C virus (HCV) infect hepatocytes and cause acute liver disease. A small percentage of HBV and a large percentage of HCV infections become chronic, and after many years can lead to hepatocellular carcinoma (HCC). During chronic HBV infection and HBV-associated HCC, the viral DNA becomes integrated at random into the host cell genome, which causes gene duplications, deletions, and chromosomal translocations. The core and polymerase regions of the genome are often destroyed but interestingly the gene encoding the hepatitis B X protein (HBx) remains intact. The HBx protein transcriptionally transactivates a variety of viral and cellular promoter and enhancer elements *(111)*. HBx also indirectly activates transcription factors through up-regulation of several mitogenic signaling pathways, including the Ras-Raf-MEK-ERK and JNK pathways *(112,113)*. In hepatoma cells, transcriptionally active HBx has been shown to associate with the catalytic subunit of PI3K (*see* Fig. 4), leading to increased phosphorylation of the p85 adaptor subunit and activation of PI3K *(114)*. This is in contrast to the other viral proteins discussed herein, which interact with the p85 adaptor subunit, suggesting that HBx may be novel in its mechanism of PI3K activation.

HBx-induced PI3K activation was further demonstrated to block TGF-β-induced apoptosis through downstream activation of Akt, phosphorylation Bad, and subsequent inhibition of caspase-3 *(114,115)*. This inhibition of TGF-β-induced apoptosis via PI3K and Akt, is similar to that mediated by EBV LMP2A, and also requires Src PTKs.

Src PTK activity is elevated following HBx expression and Src kinase inhibitors block PI3K protection against TGF-β-induced apoptosis *(116)*. This suggests that like EBV LMP2A and KSHV K1, HBx probably facilitates the activation of PI3K by bringing it into close proximity with the Src PTKs, although this event would not require recruitment to the plasma membrane.

HBx-induced apoptotic inhibition via PI3K-Akt signaling may provide HBV-infected hepatocytes with a selective growth advantage. This could be important during the initial stages of HCC development, however the situation in vivo is likely to be more complex and further studies are required to unravel the complexities of tumor development following chronic HBV infection *(114,115)*.

The molecular effects of HCV infection in hepatocytes that contribute to chronic infection and HCC are less well defined. However, many studies have focused on the HCV nonstructural protein NS5A following the discovery that mutations in this protein correlate to resistance to interferon treatment *(117)*. NS5A is an HCV nonstructural protein thought to play a role in virus replication, although its exact function is unknown. Like polyomavirus MT, NS5A is a cytosolic phosphoprotein that can interact with and regulate a number of signaling molecules *(118–121)*. The C-terminus of NS5A contains a highly conserved polyproline motif which can interact with the Src homology 3 (SH3) domains of the adaptor protein Grb2 and the PI3K p85 subunit and form a complex with the EGFR substrate Grb2-associated binder 1 (Gab1) *(see* Fig. 4) *(122–124)*. In human lung fibroblasts and hepatoma cells stably expressing NS5A, or harboring subgenomic HCV replicons the NS5A-PI3K p85 complex has been found to increase p85 phosphorylation, PI3K kinase activity and downstream phosphorylation of Akt and Bad *(122,124)*. This results in an increased protection against apoptotic stimuli *(124)*. Thus in a mechanism similar to that of HPV E5, the complex that NS5A forms with the PI3K p85 subunit as well as Grb2 and Gab1 may facilitate binding of p85 to the phosphorylated cytoplasmic tail of EGFR and subsequent activation of PI3K. These findings suggest that NS5A is important for survival during HCV infection. However whether NS5A up-regulates survival signals during chronic liver disease in vivo and the development of HCC requires further investigation.

3.4. HIV: A Law Unto Itself

In terms of virus infection, human immunodeficiency virus type 1 (HIV-1) is characteristically unique. HIV-1 is able to cause acute cytopathic infection but at the same time can evade the host immune system by establishing a latent infection in target CD4$^+$ cells through integration of proviral DNA into the host genome. However unlike retroviruses such as human T-cell leukemia virus (HTLV), and viruses with lytic/latent infectious cycles, persistence of HIV is not known to directly lead to malignant transformation. Therefore, studies on PI3K-Akt signaling during HIV-1 infection have focused on viral replication and acute pathogenicity, and HIV involvement in the development of KS. Several HIV-1 proteins have been shown to interact with PI3K, either directly or indirectly in association with other proteins.

HIV-1 entry is facilitated by binding of the HIV-1 glycoprotein gp120 to the CD4 surface molecule on T-cells and macrophages, and also requires the presence of chemokine coreceptors *(125–127)*. The interaction of gp120 with CD4 in primary CD4$^+$ T-cells and macrophages has been shown to result in rapid phosphorylation of the p85 adaptor subunit.

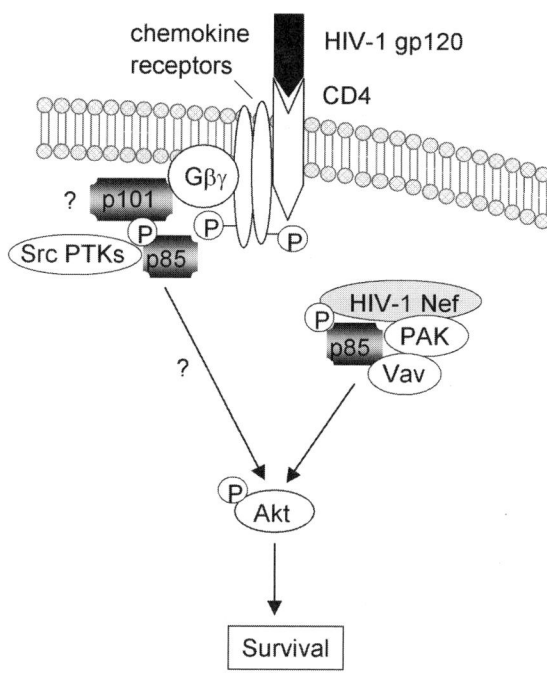

Fig. 5. HIV proteins regulating PI3K-Akt signaling. Binding of HIV-1 gp120 to CD4+ T-cells and macrophages, and subsequent virus entry causes the recruitment and activation of PI3K, which is required for viral replication and reverse transcription. At a later stage of HIV-1 infection the proline-rich protein Nef recruits PI3K, PAK and Vav into a signaling complex, which causes the activation of PI3K and Akt, and inhibition of caspase-dependent apoptosis. This probably protects cells from premature apoptosis and allows HIV-1 to complete its life cycle.

Full PI3K activation, however, requires the chemokine receptor, and like EBV LPM2A and HBV HBx, also requires Src PTKs (*see* Fig. 5) *(128,129)*. The chemokine receptors are G-protein linked serpentine receptors, and the activation of PI3K by gp120 binding is impaired by pertussis toxin, which is a G-protein inhibitor. This suggests that gp120 binding to CD4 and its coreceptors stimulates the activation of class I_B rather than class I_A PI3Ks. However, it was the class I_A PI3K p85 adaptor subunit that was shown to be phosphorylated, an event that possibly be mediated by Src PTKs, which may also be recruited to the plasma membrane. This suggests that perhaps both class I_A and I_B PI3Ks are activated downstream of HIV-1 binding and entry; however, whether this leads to activation of Akt and downstream survival events, remains unknown. Like PI3K signals triggered downstream of HCMV binding and entry, activation of PI3K is important during the HIV-1 life cycle. Inhibition of PI3K with LY294002 was shown to affect viral replication and reverse transcription, but was not required for viral DNA integration or gene expression *(129)*.

The HIV-1 Nef protein is proline-rich and enhances virus infectivity through its interaction with the SH3 domains of a variety of signaling proteins including Src PTKs, T-cell receptors, G-proteins and p21-activated kinase (PAK) *(130–133)*. Nef is able to directly bind to the C-terminal end of the PI3K p85 subunit and recruit PAK and guanosine 5′ triphosphate (GTP) exchange factor Vav into a signaling complex (*see* Fig. 5)

(134). Inhibition of PI3K with LY294002 in HIV-1 infected Jurkat and Cos-1 cells prevented activation of PAK and decreased the production of viral progeny *(134)*. Nef expression at the plasma membrane in NIH3T3 cells blocks apoptosis, which requires both PAK and PI3K *(135)*. Nef appears to play an important role in apoptotic inhibition and stimulation of cell survival, via molecules such as PI3K and PAK, during acute HIV infection. These data suggest that the regulation of PI3K signaling by different HIV-1 proteins is important during various stages of the virus life cycle. Activation of PI3K during acute HIV-1 infection, in common with other viruses that cause acute infection, is likely to help premature host cell death prior to production of new virus progeny *(1,135,136)*. However, further studies are required to show whether whether Akt-mediated survival is up-regulated downstream of PI3K during HIV-1 infection.

HIV-1 infection leads to a progressive decline in the $CD4^+$ T-cells and macrophages it infects. This results in immunodeficiency and permits infection and development of disease by other opportunistic agents. HIV-1 proteins have been shown to modulate the host cell environment and contribute to the disease symptoms caused by other infectious agents. The HIV-1 Tat protein, for example, is thought to contribute to the aggressiveness of AIDS-associated KS *(88,89,137)*. Tat is able to stimulate a variety of signaling mechanisms in KS cells, including activation of PI3K *(137–139)*. Tat inhibits apoptosis and increases cell viability via phosphorylation of Akt and Bad downstream of PI3K, which is down-regulated by chemotherapeutic agent vincristine, used to treat KS *(140,141)*. Inhibition of PI3K was shown to block Tat-induced Akt activation, Bad phosphorylation, and downstream apoptotic inhibition *(141)*. Therefore, Tat-induced PI3K-Akt survival during KSHV transformed cells, may contribute to tumor cell survival and the aggressive nature of AIDS-associated KS.

4. Conclusion

In the past 20 yr extensive research in the field of cell biology has lead to the discovery and characterization of many molecules and signaling cascades, which regulate cell proliferation, apoptosis, and survival. The PI3K-Akt signaling pathway has received considerable attention, because its importance in cell survival and apoptotic inhibition was realized. As a result, a vast amount of research is emerging into the involvement of this pathway in virus infection. Many of the major breakthroughs in cell biology, leading to the characterization of various signaling molecules and their involvement in disease states, have been made through the identification of unregulated viral counterparts. Thus an understanding of cell signaling in the context of virus infection not only contributes to our understanding of the effects of various signaling proteins, but also to our understanding of virus-host dynamics and virus disease states.

PI3K-Akt signaling appears to be important during the early stages of acute infections, with viruses such as CVB3, RV, RSV, and SARS-CoV. Inhibition of PI3K early in the virus life cycles induces premature apoptotic cell death and has a negative affect on virus replication and production. The induction of survival signals may only be required for virus replication and protein production as virus particle budding and release is often facilitated by apoptosis. However, the host cell itself may initiate induction the of PI3K-Akt survival to allow antiviral mechanisms to get under way.

HIV-1 can also cause acute infection, but at the same time is able to persist for long periods in the host. Activation of PI3K during acute HIV-1 infection in vitro has been shown to be important for many stages in the virus life cycle. However, unlike the other viruses that cause acute infections it is not know whether activation of PI3K leads to Akt-mediated survival signaling and the effect of such signaling on virus persistence in the host.

A number of viruses including EBV, KSHV, and HPV, have the ability to establish long-term latent infections in the host. After long periods of latency or upon reactivation from latency, such infections can ultimately lead to virus-mediated cellular transformation. It appears that the gene products of latently infecting viruses can constitutively up-regulate PI3K-Akt cell survival signals and therefore continuously block apoptotic signals. This contributes to both virus survival in the latent state and allows proliferative signals to go unchecked resulting in oncogenic transformation. However, activation of this pathway is not only required for viral transformation but also for other stages of the virus life cycle. EBV BZLF1-mediated reactivation from latency, for example, requires the activation of PI3K and Akt. Productive polyomavirus infection requires the up-regulation of PI3K-Akt cell survival and cellular proliferation.

Long-term infections can also be established by chronically infecting viruses such as HBV and HCV, which, after prolonged periods, can also lead to cellular transformation. However, both HBV and HCV viral products have been shown to induce PI3K-Akt survival signals, which blocks apoptosis. Whether this situation occurs in vivo in chronic infection and the cellular transformation that ensues has not been studied, and this is partly due to the lack of efficient cell culture systems for such viruses.

Another limitation to studies examining the effect of virus infection on host cell signaling is the use of continuous tumorigenic cell lines with altered biological properties. The cross-regulation between multiple signaling pathways, which may differ in cell systems in vivo and in vitro, also makes it difficult to obtain results which are conclusive. However, the use of transgenic animals, as in the case of EBV LMP2A and the ongoing development of new techniques such RNA interference (RNAi), will allow for better understanding of the modulation of cell signaling cascades. In future, this may help to identify new cellular and viral proteins, and lead to a more in depth understanding of cellular transformation and other viral and cellular diseases.

References

1. Roulston A, Marcellus RC, Branton PE. Viruses and apoptosis. Annu Rev Microbiol 1999; 53:577–628.
2. Cantrell DA. Phosphoinositide 3-kinase signalling pathways. J Cell Sci 2001;114:1439–1445.
3. Chan TO, Rittenhouse SE, Tsichlis PN. AKT/PKB and other D3 phosphoinositide-regulated kinases: kinase activation by phosphoinositide-dependent phosphorylation. Annu Rev Biochem 1999;68:965–1014.
4. Chang F, Lee JT, Navalonic PM, et al. Involvement of PI3K/Akt pathway in cell cycle progression, apoptosis, and neoplastic transformation: a target for cancer chemotherapy. Leukemia 2003;17:590–603.
5. Vanhaesebroeck B, Leevers SJ, Ahmadi K, et al. Synthesis and function of 3-phosphorylated inositol lipids. Annu Rev Biochem 2001;70:535–602.
6. Vanhaesebroeck B, Waterfield MD. Signaling by distinct classes of phosphoinositide 3-kinases. Exp Cell Res 1999;253:239–254.
7. Vanhaesebroeck B, Leevers SJ, Panayotou G, Waterfield MD. Phosphoinositide 3-kinases: a conserved family of signal transducers. Trends Biochem Sci 1997;22:267–272.

8. Yu J, Zhang Y, McIlroy J, Rordorf-Nikolic T, Orr GA, Backer JM. Regulation of the p85/p110 phosphatidylinositol 3′-kinase: stabilization and inhibition of the p110alpha catalytic subunit by the p85 regulatory subunit. Mol Cell Biol 1998;18:1379–1387.

9. Skolnik EY, Margolis B, Mohammadi M, et al. Cloning of PI3 kinase-associated p85 utilizing a novel method for expression/cloning of target proteins for receptor tyrosine kinases. Cell 1991;65:83–90.

10. Carpenter CL, Auger KR, Chanudhuri M, et al. Phosphoinositide 3-kinase is activated by phosphopeptides that bind to the SH2 domains of the 85-kDa subunit. J Biol Chem 1993; 268:9478–9483.

11. Nolte RT, Eck MJ, Schlessinger J, Shoelson SE, Harrison SC. Crystal structure of the PI 3-kinase p85 amino-terminal SH2 domain and its phosphopeptide complexes. Nat Struct Biol 1996;3:364–374.

12. Shoelson SE, Sivaraja M, Williams KP, Hu P, Schlessinger J, Weiss MA. Specific phospho-peptide binding regulates a conformational change in the PI 3-kinase SH2 domain associated with enzyme activation. EMBO J 1993;12:795–802.

13. Cipres A, Carrasco S, Merida I. Deletion of the acidic-rich domain of the IL-2Rbeta chain increases receptor-associated PI3K activity. FEBS Lett 2001;500:99–104.

14. Gonzalez-Garcia A, Merida I, Martinez AC, Carrera AC. Intermediate affinity interleukin-2 receptor mediates survival via a phosphatidylinositol 3-kinase-dependent pathway. J Biol Chem 1997;272:10,220–10,226.

15. Vanhaesebroeck B, Alessi DR. The PI3K-PDK1 connection: more than just a road to PKB. Biochem J 2000;346:561–576.

16. Stephens L, Smrcha A, Cooke F, Jackson T, Sternweis P, Hawkins P. A novel phospho-inositide 3 kinase activity in myeloid-derived cells is activated by G protein bg subunits. Cell 1994;77:83–93.

17. Yan J, Roy S, Apolloni A, Lane A, Hancock JF. Ras isoforms vary in their ability to activate Raf-1 and phosphoinositide 3-kinase. J Biol Chem 1998;273:24,052–24,056.

18. Rodriguez-Viciana P, Warne PH, Dhand R, et al. Phosphatidylinositol-3-OH kinase as a direct target of Ras. Nature 1994;370:527–532.

19. Rodriguez-Viciana P, Warne PH, Vanhaesebroeck B, Waterfield MD, Downward J. Activation of phosphoinositide 3-kinase by interaction with Ras and by point mutation. EMBO J. 1996;15:2442–2451.

20. Hawkins PT, Jackson TR, Stephens LR. Platelet-derived growth factor stimulates synthesis of PtdIns(3,4,5)P3 by activating a PtdIns(4,5)P2 3-OH. Nature 1992;358:157–159.

21. Maehama T, Dixon JE. PTEN: a tumour suppressor that functions as a phospholipid phos-phatase. Trends Cell Biol 1999;9:125–128.

22. Leslie NR, Downes CP, Maehama T, Dixon JE. PTEN: The down side of PI3-kinase signaling. Cell Signal 2002;14:285–295.

23. Bottomley MJ, Salim K, Panayotou G. Phospholipid-binding protein domains. Biochim Biophys Acta 1998;1436:165–183.

24. Staal SP, Hartley JW, Rowe WP. Isolation of transforming murine leukemia viruses from mice with a high incidence of spontaneous lymphoma. Proc Natl Acad Sci USA 1977;74:3065–3067.

25. Bellacosa A, Testa JR, Staal SP, Tsichlis PN. A retroviral oncogene, akt, encoding a serine-threonine kinase containing an SH2-like region. Science 1991;254:274–277.

26. Jones PF, Jakubowicz T, Pitossi FJ, Maurer F, Hemmings BA. Molecular cloning and identification of a serine/threonine protein kinase of the second-messenger subfamily. Proc Natl Acad Sci USA 1991;88:4171–4175.

27. Coffer PJ, Woodgett JR. Molecular cloning and characterisation of a novel putative protein-serine kinase related to the cAMP-dependent and protein kinase C families. Eur J Biochem 1991;201:475–481.

28. Scheid MP, Woodgett JR. Unraveling the activation mechanism of protein kinase B/Akt. FEBS letters 2003;546:108–112.
29. Lawlor MA, Alessi DR. PKB/Akt: a key mediator of cell proliferation, survival and insulin responses? J Cell Sci 2001;114:2903–2910.
30. Datta SR, Dudek H, Tao X, et al. Akt phosphorylation of BAD couples survival signals to the cell-intrinsic death machinery. Cell 1997;91:231–241.
31. Cardone MH, Roy N, Stennicke HR, et al. Regulation of cell death protease caspase-9 by phosphorylation. Science 1998;282:1318–1321.
32. Pap M, Cooper GM. Role of glycogen synthase kinase-3 in the phosphatidylinositol 3-kinase/Akt cell survival pathway. J Biol Chem 1998;273:19,929–19,932.
33. Cross DA, Alessi DR, Cohen P, Andjelkovich M, Hemmings BA. Inhibition of glycogen synthase kinase-3 by insulin mediated by protein kinase B. Nature 1995;378:785–789.
34. Burgering BM, Medema RH. Decisions on life and death: FOXO Forkhead transcription factors are in command when PKB/Akt is off duty. J Leukoc Biol 2003;73:689–701.
35. Rena G, Guo S, Cichy SC, Unterman TG, Cohen P. Phosphorylation of the transcription factor forkhead family member FKHR by protein kinase B. J Biol Chem 1999;274:17,179–17,183.
36. Khwaja A. Akt is more than just a Bad kinase. Nature 1999;401:33–34.
37. Ozes ON, Mayo LD, Gustin JA, Pfeffer SR, Pfeffer LM, Donner DB. NF-kappaB activation by tumour necrosis factor requires the Akt serine-threonine kinase. Nature 1999;401:82–85.
38. Yang CH, Murti A, Pfeffer SR, Kim JG, Donner DB, Pfeffer LM. Interferon alpha/beta promotes cell survival by activating nuclear factor kappa B through phosphatidylinositol 3-kinase and Akt. J Biol Chem 2001;276:13,756–13,761.
39. Romashkova JA, Makarov SS. NF-kappaB is a target of AKT in anti-apoptotic PDGF signalling. Nature 1999;401:86–90.
40. Hatano E, Brenner DA. Akt protects mouse hepatocytes from TNF-alpha- and Fas-mediated apoptosis through NF-kappa B activation. Am J Physiol Gastrointest Liver Physiol 2001;281:G1357–G1368.
41. Collins PL, Chanock RM, Murphy BR. Respiratory Syncytial Virus. In: Knipe DM, Howley PM, Griffin DE, et al., eds. Fields Virology., Lippincott Williams & Wilkins, Philadelphia, 2001:1443–1486.
42. Bitko V, Velazquez A, Yang L, Yang YC, Barik S. Transcriptional induction of multiple cytokines by human respiratory syncytial virus requires activation of NF-kappa B and is inhibited by sodium salicylate and aspirin. Virology 1997;232:369–378.
43. Haeberle HA, Takizawa R, Casola A, et al. Respiratory syncytial virus-induced activation of nuclear factor-kappa B in the lung involves alveolar macrophages and toll-like receptor 4-dependent pathways. J Infect Dis 2002;186:1199–1206.
44. Kong X, San Juan H, Kumar M, et al. Respiratory syncytial virus infection activates STAT signaling in human epithelial cells. Biochem Biophys Res Commun 2003;306:616–622.
45. Thomas KW, Monick MM, Staber JM, Yarovinsky T, Carter AB, Hunninghake GW. Respiratory syncytial virus inhibits apoptosis and induces NF-kappa B activity through a phosphatidylinositol 3-kinase-dependent pathway. J Biol Chem 2002;277:492–501.
46. Monick MM, Cameron K, Powers LS, et al. Spingosine kinase mediates activation of extracellular signal related kinase and Akt by respiratory syncytial virus. Am J Respir Cell Mol Biol 2004;30:844–852.
47. Cooray S, Jin L, Best JM. The involvement of survival signaling pathways in rubella-virus induced apoptosis. Virol J 2005;2:1.
48. Cooray S, Best JM, Jin L. Time-course induction of apoptosis by wild-type and attenuated strains of rubella virus. J Gen Virol 2003;84:1275–1279.

49. Domegan LM, Atkins GJ. Apoptosis induction by the Therien and vaccine RA27/3 strains of rubella virus causes depletion of oligodendrocytes from rat neural cell cultures. J Gen Virol 2002;83:2135–2143.

50. Duncan R, Muller J, Lee N, Esmaili A, Nakhasi HL. Rubella virus-induced apoptosis varies among cell lines and is modulated by Bcl-XL and caspase inhibitors. Virology 1999; 255:117–128.

51. Hofmann J, Pletz MW, Liebert UG. Rubella virus-induced cytopathic effect in vitro is caused by apoptosis. J Gen Virol 1999;80:1657–1664.

52. Pugachev KV, Frey TK. Rubella virus induces apoptosis in culture cells. Virology 1998; 250:359–370.

53. Kronfield I, Kazimirsky G, Gelfand EW, Brodie C. NGF rescues human B lymphocytes from anti-IgM induced apoptosis by activation of PKC. Eur J Immunol 2002;32: 136–143.

54. Mizutani T, Fukushi S, Masayuki S, Kurane I, Morikawa S. Importance of Akt signaling pathway for apoptosis in SARS-CoV-infected Vero E6 cells. Virology 2004;327:169–174.

55. Carthy CM, Granville DJ, Watson KA, et al. Caspase activation and specific cleavage of substrates after coxsackievirus B3-induced cytopathic effect in HeLa cells. J Virol 1998; 72:7669–7675.

56. Joo CH, Hong HN, Kim EO, et al. Coxsackievirus B3 induces apoptosis in the early phase of murine myocarditis: a comparative analysis of cardiovirulent and noncardiovirulent strains. Intervirology 2003;46:135–140.

57. Esfandiarei M, Luo H, Yanagawa B, et al. Protein kinase B/Akt regulates coxsackievirus B3 replication through a mechanism which is not caspase dependent. J Virol 2004;78: 4289–4298.

58. Tyler KL, Nathanson N. Pathogenesis of Viral Infections. In: Knipe DM, Howley PM, Griffin DE, et al., eds. Fields Virology. Lippincott Williams & Wilkins, Philadelphia, 2001:199–244.

59. Kieff E, Rickinson AB. Epstein-Barr Virus and Its Replication. In: Knipe DM, Howley PM, Griffin DE, et al., eds. Fields Virology. Lippincott Williams & Wilkins, Philadelphia, 2001:2511–2574.

60. Bale JF, Blackman JA, Sato Y. Outcome in children with symptomatic congenital cytomeglovirus infection. J Child Neurol 1990;5:131–136.

61. Pass RF. Cytomeglovirus. In: Knipe DM, Howley PM, Griffin DE, et al., eds. Feilds Virology. Lippincott Williams & Wilkins, Philadelphia, 2001:2675–2706.

62. Yurochko AD, Hwang ES, Rasmussen L, Keay S, Pereira L, Huang ES. The human cytomegalovirus UL55 (gB) and UL75 (gH) glycoprotein ligands initiate the rapid activation of Sp1 and NF-kappaB during infection. J Virol 1997;71:5051–5059.

63. Yurochko AD, Mayo MW, Poma EE, Baldwin AS, Jr., Huang ES. Induction of the transcription factor Sp1 during human cytomegalovirus infection mediates upregulation of the p65 and p105/p50 NF-kB promoters. J Virol 1997;71:4638–4648.

64. Johnson RA, Huong SM, Huang ES. Activation of the mitogen-activated protein kinase p38 by human cytomegalovirus infection through two distinct pathways: a novel mechanism for activation of p38. J Virol 2000;74:1158–1167.

65. Bresnahan WA, Thompson EA, Albrecht T. Human cytomegalovirus infection results in altered Cdk2 subcellular localization. J Gen Virol 1997;78:1993–1997.

66. Johnson RA, Ma XL, Yurochko AD, Huang ES. The role of MKK1/2 kinase activity in human cytomegalovirus infection. J Gen Virol 2001;82:493–497.

67. Johnson RA, Wang X, Ma XL, Huong SM, Huang ES. Human cytomegalovirus up-regulates the phosphatidylinositol 3-kinase (PI3-K) pathway: inhibition of PI3-K activity inhibits viral replication and virus-induced signaling. J Virol 2001;75:6022–6032.

68. Johnson MD, Okedli E, Woodard A, Toms SA, Allen GS. Evidence for phosphatidylinositol 3-kinase-Akt-p70S6K pathway activation and transduction of mitogenic signals by platelet-derived growth factor in meningioma cells. J Neurosurg 2002;97:668–675.

69. Weng QP, Andrabi K, Kozlowski MT, Grove JR, Avruch J. Multiple independent inputs are required for activation of the p70 S6 kinase. Mol Cell Biol 1995;15:2333–2340.

70. Katz BZ, Raab-Traub N, Miller G. Latent and replicating forms of Epstein-Barr virus DNA in lymphomas and lymphoproliferative diseases. J Infect Dis 1989;160:589–598.

71. Raab-Traub N, Flynn K, Pearson G, et al. The differentiated form of nasopharyngeal carcinoma contains Epstein-Barr virus DNA. Int J Cancer 1987;39:25–29.

72. Rickinson AB, Kieff E. Epstein-Barr Virus. In: Knipe DM, Howley PM, Griffin DE, et al., eds. Fields Virology. Lippincott Williams & Wilkins, Philadelphia, 2001:2575–2628.

73. Gires O, Zimber-Strobl U, Gonnella R, et al. Latent membrane protein 1 of Epstein-Barr virus mimics a constitutively active receptor molecule. EMBO J 1997;16:6131–6140.

74. Izumi KM, Cahir McFarland ED, Ting AT, Riley EA, Seed B, Kieff ED. The Epstein-Barr virus oncoprotein latent membrane protein 1 engages the tumor necrosis factor receptor-associated proteins TRADD and receptor-interacting protein (RIP) but does not induce apoptosis or require RIP for NF-kappaB activation. Mol Cell Biol 1999;19:5759–5767.

75. Mosialos G, Birkenbach M, Yalamanchili R, VanArsdale T, Ware C, Kieff E. The Epstein-Barr virus transforming protein LMP1 engages signaling proteins for the tumor necrosis factor receptor family. Cell 1995;80:389–399.

76. Eliopoulos AG, Young LS. Activation of the cJun N-terminal kinase (JNK) pathway by the Epstein-Barr virus-encoded latent membrane protein 1 (LMP1). Oncogene 1998;16:1731–1742.

77. Gires O, Kohlhuber F, Kilger E, et al. Latent membrane protein 1 of Epstein-Barr virus interacts with JAK3 and activates STAT proteins. EMBO J 1999;18:3064–3073.

78. Roberts ML, Cooper NR. Activation of a ras-MAPK-dependent pathway by Epstein-Barr virus latent membrane protein 1 is essential for cellular transformation. Virology 1998;240:93–99.

79. Dawson CW, Tramountanis G, Eliopoulos AG, Young LS. Epstein-Barr virus latent membrane protein 1 (LMP1) activates the phosphatidylinositol 3-kinase/Akt pathway to promote cell survival and induce actin filament remodelling. J Biol Chem 2003;278:3694–3704.

80. Swart R, Ruf IK, Sample J, Longnecker R. Latent membrane protein 2A-mediated effects on the phosphatidylinositol 3-Kinase/Akt pathway. J Virol 2000;74:10,838–10,845.

81. Portis T, Longnecker R. Epstein-Barr virus (EBV) LMP2A mediates B-lymphocyte survival through constitutive activation of the Ras/PI3K/Akt pathway. Oncogene 2004;2004:8619–8628.

82. Scholle F, Bendt KM, Raab-Traub N. Epstein-Barr virus LMP2A transforms epithelial cells, inhibits cell differentiation, and activates Akt. J Virol 2000;74:10,681–10,689.

83. Fukuda M, Longnecker R. Latent membrane protein 2A inhibits transforming growth factor-b1-indiced apoptosis through the phosphatidylinositol 3-kinase/Akt pathway. J Virol 2004;78:1697–1705.

84. Flemington E, Speck SH. Epstein-Barr virus BZLF1 trans activator induces the promoter of a cellular cognate gene, c-fos. J Virol 1990;64:4549–4552.

85. Zalani S, Holley-Guthrie E, Kenney S. Epstein-Barr viral latency is disrupted by the immediate-early BRLF1 protein through a cell-specific mechanism. Proc Natl Acad Sci USA 1996;93:9194–9199.

86. Darr DC, Mauser A, Kenney S. Epstein-Barr virus immediate-early protein BRLF1 induces the lytic form of viral replication through a mechanism involving phosphatidylinositol-3 kinase activation. J Virol 2001;75:6135–6142.

87. Chang Y, Moore PS. Kaposi's Sarcoma (KS)-associated herpesvirus and its role in KS. Infect Agents Dis 1996;5:215–222.

88. Antman K, Chang Y. Kaposi's sarcoma. N Engl J Med 2000;342:1027–1038.

89. Gallo RC. The enigmas of Kaposi's sarcoma. Science 1998;282:1837–1839.

90. Chang Y, Cesarman E, Pessin MS, et al. Identification of herpesvirus-like DNA sequences in AIDS-associated Kaposi's sarcoma. Science 1994;265:1865–1869.

91. Bais C, Santomasso B, Coso O, et al. G-protein-coupled receptor of Kaposi's sarcoma-associated herpesvirus is a viral oncogene and angiogenesis activator. Nature 1998;391: 86–89.

92. Cheng EH, Nicholas J, Bellows DS, et al. A Bcl-2 homolog encoded by Kaposi's sarcoma-associated virus, human herpesvirus 8, inhibits apoptosis but does not heterodimerize with Bax or Bak. Proc Natl Acad Sci USA 1997;94:690–694.

93. Moore PS, Boshoff C, Weiss RA, Chang Y. Molecular mimicry of human cytokine and cytokine response pathway genes by KSHV. Science 1996;274:1739–1744.

94. Lee H, Guo J, Li M, et al. Indentification of an immunoreceptor tyrosine-based activation motif of K1 transforming protein of Kaposi's sarcoma-associated herpesvirus. Nat Med 1998;404:782–787.

95. Lagunoff M, Majeti A, Weiss A, Ganem D. Deregulated signal transduction by the K1 gene product of Kaposi's sarcoma-associated herpesvirus. Proc Natl Acad Sci USA 1999;96: 5704–5709.

96. Tomlinson CC, Damania B. The K1 protein of Kaposi's sarcoma-associated herpesvirus activates the Akt signaling pathway. J Virol 2004;78:1918–1927.

97. zur Hausen H, de Villiers EM. Human papillomaviruses. Annu Rev Microbiol 1994;48: 427–447.

98. zur Hausen H. Papillomaviruses and cancer: from basic studies to clincal application. Nat Rev Cancer 2002;2:342–350.

99. Zhang B, Spandau DF, Roman A. E5 protein of human papillomavirus type 16 protects human foreskin keratinocyes from UV B-irradiation-induced apoptosis. J Virol 2002;76:220–231.

100. Crusius K, Auvinen E, Steuer B, Gaissert H, Alonso A. The human papillomavirus type 16 E5-protein modulates ligand-dependent activation of the EGF receptor family in the human epithelial cell line HaCaT. Exp Cell Res 1998;241:76–83.

101. Hwang ES, Nottoli T, Dimaio D. The HPV-16 E5 protein: expression, detection, and stable complex formation with transmembrane proteins in COS cells. Virology 1995;211:227–233.

102. Hawley-Nelson P, Vousden KH, Hubbert NL, Lowy DR, Schiller JT. HPV16 E6 and E7 proteins cooperate to immortalize human foreskin keratinocytes. EMBO J 1989;8: 3905–3910.

103. Kaur P, McDougall JK, Cone R. Immortalization of primary human epithelial cells by cloned cervical carcinoma DNA containing human papillomavirus type 16 E6/E7 open reading frames. J Gen Virol 1989;70:1261–1266.

104. Munger K, Phelps WC, Bubb V, Howley PM, Schlegl R. The E6 and E7 genes of the human papillomavirus type 16 together are necessary and sufficient for transformation of primary human keratinocytes. J Virol 1989;63:4417–4421.

105. Gottlieb KA, Villarreal LP. Natural biology of polyomavirus middle T antigen. Microbiol Mol Biol Rev 2001;65:288–318.

106. Campbell KS, Ogris E, Burke B, et al. Polyoma middle tumor antigen interacts with SHC via the NPTY (Asn-Pro-Thr-Tyr) motif in middle tumor antigen. Proc Natl Acad Sci USA 1994;91:6344–6348.

107. Su W, Liu W, Schaffhausen BS, Roberts TM. Association of Polyomavirus middle tumor antigen with phospholipase C-gamma 1. J Biol Chem 1995;270:12,331–12,334.

108. Whitman M, Kaplan DR, Schaffhausen B, Cantley L, Roberts TM. Association of phosphatidylinositol kinase activity with polyoma middle-T competent for transformation. Nature 1985;315:239–242.

109. Dahl J, Jurczak A, Cheng LA, Baker DC, Benjamin TL. Evidence of a role for phosphatidylinositol 3-kinase activation in the blocking of apoptosis by polyomavirus middle T antigen. J Virol 1998;72:3221–3226.

110. Summers SA, Lipfert L, Birnbaum MJ. Polyoma middle T antigen activates the Ser/Thr kinase Akt in a PI3-kinase-dependent manner. Biochem Biophys Res Commun 1998;246:76–81.

111. Murakami S. Hepatitis B virus X protein: structure, function and biology. Intervirology 1999;42:81–99.

112. Benn J, Schneider RJ. Hepatitis B virus HBx protein activates Ras-GTP complex formation and establishes a Ras, Raf, MAP kinase signaling cascade. Proc Natl Acad Sci USA 1994;91:10,350–10,354.

113. Benn J, Su F, Doria M, Schneider RJ. Hepatitis B virus HBx protein induces transcription factor AP-1 by activation of extracellular signal-regulated and c-Jun N-terminal mitogen-activated protein kinases. J Virol 1996;70:4978–4985.

114. Lee YI, Kang-Park S, Do SI. The hepatitis B virus-X protein activates a phosphatidylinositol 3-kinase-dependent survival signaling cascade. J Biol Chem 2001;276:16,969–16,977.

115. Shih WL, Kuo ML, Chuang SE, Cheng AL, Doong SL. Hepatitis B virus X protein inhibits transforming growth factor-beta-induced apoptosis through the activation of phosphatidylinositol 3-kinase pathway. J Biol Chem 2000;275:25,858–25,864.

116. Shih WL, Kuo ML, Chuang SE, Cheng AL, Doong SL. Hepatitis B virus X protein activates a survival signaling by linking SRC to phosphatidylinositol 3-kinase. J Biol Chem 2003;278:31,807–31,813.

117. Enamoto N, Sakuma I, Asahina Y, et al. Mutations in the nonstructural protein 5A gene and response to interferon in patients with chronic hepatitis C virus 1b infection. N Engl J Med 1996;334:77–81.

118. Lan KH, Sheu ML, Hwang SJ, et al. HCV NS5A interacts with p53 and inhibits p53-mediated apoptosis. Oncogene 2002;21:4801–4811.

119. Qadri I, Iwahashi M, Simon F. Hepatitis C virus NS5A protein binds TBP and p53, inhibiting their DNA binding and p53 interactions with TBP and ERCC3. Biochim Biophys Acta 2002;1592:193–204.

120. Tan SL, Nakao H, He Y, et al. NS5A, a nonstructural protein of hepatitis C virus, binds growth factor receptor-bound protein 2 adaptor protein in a Src homology 3 domain/ligand-dependent manner and perturbs mitogenic signaling. Proc Natl Acad Sci USA 1999;96:5533–5538.

121. Georgopoulou U, Caravokiri K, Mavromara P. Suppression of the ERK1/2 signaling pathway from HCV NS5A protein expressed by herpes simplex recombinant viruses. Arch Virol 2003;148:237–251.

122. He Y, Nakao H, Tan SL, et al. Subversion of cell signaling pathways by hepatitis C virus nonstructural 5A protein via interaction with Grb2 and P85 phosphatidylinositol 3-kinase. J Virol 2002;76:9207–9217.

123. Macdonald A, Crowder K, Street A, McCormick C, Harris M. The hepatitis C virus NS5A protein binds to members of the Src family of tyrosine kinases and regulates kinase activity. J Gen Virol 2004;85:721–729.

124. Street A, Macdonald A, Crowder K, Harris M. The hepatitis C virus NS5A protein activates a phosphoinositide 3-kinase-dependent survival signaling cascade. J Biol Chem 2004;279:12,232–12,241.

125. Choe H, Farzan M, Sun Y, et al. The beta-chemokine receptors CCR3 and CCR5 facilitate infection by primary HIV-1 isolates. Cell 1996;85:1135–1148.

126. Feng Y, Broder CC, Kennedy PE, Berger EA. HIV-1 entry cofactor: functional cDNA cloning of a seven-transmembrane, G protein-coupled receptor. Science 1996;272:872–877.

127. Dalgleish AG, Beverley PC, Clapham PR, Crawford DH, Greaves MF, Weiss RA. The CD4 (T4) antigen is an essential component of the receptor for the AIDS retrovirus. Nature 1984;312:763–767.

128. Briand G, Barbeau B, Tremblay M. Binding of HIV-1 to its receptor induces tyrosine phosphorylation of several CD4-associated proteins, including the phosphatidylinositol 3-kinase. Virology 1997;228:171–179.

129. Francois F, Klotman ME. Phosphatidylinositol 3-kinase regulates human immunodeficiency virus type 1 replication following viral entry in primary CD4+ T lymphocytes and macrophages. J Virol 2003;77:2539–2549.

130. Arora VK, Fredericksen BL, Garcia JV. Nef: agent of cell subversion. Microbes Infect 2002;4:189–199.

131. Baur AS, Sass G, Laffert B, Willbold D, Cheng-Mayer C, Peterlin BM. The N-terminus of Nef from HIV-1/SIV associates with a protein complex containing Lck and a serine kinase. Immunity 1997;6:283–291.

132. Fackler OT, Luo W, Geyer M, Alberts AS, Peterlin BM. Activation of Vav by Nef induces cytoskeletal rearrangements and downstream effector functions. Mol Cell 1999;3:729–739.

133. Simmons A, Aluvihare V, McMichael A. Nef triggers a transcriptional program in T cells imitating single-signal T cell activation and inducing HIV virulence mediators. Immunity 2001;14:763–777.

134. Linnemann T, Zheng YH, Mandic R, Peterlin BM. Interaction between Nef and phosphatidylinositol-3-kinase leads to activation of p21-activated kinase and increased production of HIV. Virology 2002;294:246–255.

135. Wolf D, Witte V, Laffert B, et al. HIV-1 Nef associated PAK and PI3-kinases stimulate Akt-independent Bad phosphorylation to induce anti-apoptotic signals. Nat Med 2001;11: 1217–1224.

136. Geleziunas R, Xu W, Takeda K, Ichijo H, Greene WC. HIV-1 Nef inhibits ASK1-dependent death signalling providing a potential mechanism for protecting the infected host cell. Nature 2001;410:834–838.

137. Ensoli B, Barillari G, Salahuddin SZ, Gallo RC, Wong-Staal F. Tat protein of HIV-1 stimulates growth of cells derived from Kaposi's sarcoma lesions of AIDS patients. Nature 1990; 345:84–86.

138. Albini A, Benelli R, Presta M, et al. HIV-tat protein is a heparin-binding angiogenic growth factor. Oncogene 1996;12:289–297.

139. Ensoli B, Buonaguro L, Barillari G, et al. Release, uptake, and effects of extracellular human immunodeficiency virus type 1 Tat protein on cell growth and viral transactivation. J Virol 1993;67:277–287.

140. Cantaluppi V, Biancone L, Boccellino M, et al. HIV type 1 Tat protein is a survival factor for Kaposi's sarcoma and endothelial cells. AIDS Res Hum Retroviruses 2001;17:965–976.

141. Deregibus MC, Cantaluppi V, Doublier S, et al. HIV-1-Tat protein activates phosphatidylinositol 3-kinase/ AKT-dependent survival pathways in Kaposi's sarcoma cells. J Biol Chem 2002;277:25,195–25,202.

4

Cyclin-Dependent Kinase 5

A Target for Neuroprotection?

Frank Gillardon

Summary

Cyclin-dependent kinase 5 (CDK5) activity is mainly restricted to the nervous system where it plays a central role in neuronal development and neurotransmission. There is increasing evidence that overactivation of CDK5 contributes to the pathogenesis of both chronic and acute neurodegenerative diseases suggesting that deregulated CDK5 may represent a therapeutic target for neuroprotection.

By high-throughput screening we identified small molecule CDK5 inhibitors that prevented neuronal cell death in various paradigms. The compounds blocked the cell death program upstream of mitochondrial depolarization and cytochrome c release. Phosphoproteome analysis revealed potential mechanisms underlying neuroprotection by CDK5 inhibitors, but also indicated interference with the physiological function of CDK5. Microarray analysis demonstrated rapid changes in gene expression following administration of CDK5 inhibitors to cultured neurons. Although compound-related effects cannot be excluded, our data advise caution when considering CDK5 inhibitors as therapeutic agents in chronic neurodegenerative diseases.

Key Words: CDK5; neurodegeneration; kinase inhibitors; proteomics; gene expression profiling.

1. Introduction

Cyclin-dependent kinases (CDKs) are serine/threonine kinases that are involved in the regulation of cell-cycle progression. Although CDK5 shows 60% homology to CDK1/2, CDK5 is not involved in cell proliferation and its activity is mainly restricted to postmitotic neurons. During neuronal differentiation expression of cell-cycle CDKs is switched off, whereas expression of CDK5 and its neuron-specific activator p35 becomes activated (reviewed in refs. *1,2*). *CDK5* knockout mice die in utero and embryos exhibit severe structural disorganization of the brain indicating a role for CDK5 in neuronal development. Blockade of CDK5 activity in cultured neurons points to a physiological function in neurite outgrowth and synaptic signaling (reviewed in ref. *3*).

Overactivation of CDK5 has been detected in the central nervous system of both Alzheimer's disease patients and transgenic mouse models of amyotrophic lateral sclerosis and Niemann-Pick disease, respectively (reviewed in refs. *1,3,4*). Re-expression and activation of mitotic CDKs has also been shown in Alzheimer's disease brains and various animal models for neurodegeneration. Moreover, nonselective CDK inhibitors and dominant-negative kinase mutants prevent neuronal cell loss in these models demonstrating that CDKs contribute to neurodegeneration *(1,2)*. Because the pathophysiological relevance of CDK5 has been extensively reviewed, this chapter will focus

From: *Apoptosis, Cell Signaling, and Human Diseases: Molecular Mechanisms, Volume 2*
Edited by R. Srivastava © Humana Press Inc., Totowa, NJ

on preclinical development of small molecule CDK5 inhibitors and their characterization by phosphoproteome and transcriptome analysis.

2. Small Molecule CDK Inhibitors and Neuroprotection

To determine the relevance of various CDKs for neuronal cell death, we used small molecule inhibitors for mitotic CDKs that were originally developed for anti-proliferative treatment of tumors. All compounds completely blocked proliferation of tumor cell lines with an EC_{50} in the low-nanomolar range, thus demonstrating that they are biologically active and cross the cell membrane. The compounds were named Indolinones, which refers to their basic chemical structure *(8)*. In cell-free kinase assays, some Indolinones also inhibited recombinant CDK5 (Table 1), whereas other kinases that may contribute to neuronal cell death (e.g., c-Jun N-terminal kinases, p38 mitogen-activated protein kinases [MAPK]) were not significantly affected. An increase in adenosine triphosphate (ATP) concentration shifted the dose-response curve to the right demonstrating competition with ATP for binding to CDKs *(see* Fig. 1A). Although the ATP binding pocket is highly conserved between members of the CDK family *(see* Fig. 1B), some of the compounds exhibit selectivity towards individual CDKs (e.g., Indolinone B) (Table 1). Similar findings have been reported using other ATP competitive protein kinase inhibitors, and structure analysis indicates that specificity is conferred by side-chain interaction with nonconserved residues adjacent to the ATP binding site *(5,6)*. However, we were unable to identify a highly selective (selectivity factor >50) CDK5 inhibitor by high-throughput screening of more than 800,000 compounds.

Oxidative stress plays a central role in both acute and chronic neurodegenerative diseases. Therefore, we tested the compounds in cultures of cerebellar granule neurons following administration of buthionine sulfoximine (BSO), an irreversible inhibitor of glutathione synthase, which induces delayed neuronal cell death by cellular glutathione depletion and subsequent free radical stress *(7)*. The CDK5 inhibitors Indolinone A (0.3 µM) and roscovitine (30 µM) completely prevented the decline of Alamar Blue fluorescence as a measure for mitochondrial activity after BSO exposure (Table 1). The neuroprotective effect of indolinone A was not mediated via scavenging of reactive oxygen species (ROS) following glutathione depletion, because indolinone A did not prevent the BSO-induced increase in oxidant-sensitive dichlorodihydrofluorescein fluorescence. Indolinone A also promoted survival of cerebellar granule neurons in other paradigms of cell death (e.g., staurosporine treatment, potassium/serum deprivation, colchicine exposure) with an EC50 in the sub-micromolar range *(8)*. In cultured rat cortical neurons, administration of Indolinone A reduced cell death following DNA damage induced by the topoisomerase I inhibitor camptothecin.

It should be mentioned however, that Indolinones were ineffective in other in vitro models for neurodegeneration. Excitotoxic cell death in organotypic hippocampal slice cultures following glutamate exposure was not reduced by Indolinone A treatment *(see* Fig. 2A). This is somewhat surprising, since CDK5 phosphorylates NMDA receptors, thereby enhancing ion channel activity *(9)*. Furthermore, viral overexpression of dominant-negative CDK5 mutants protected hippocampal neurons from cell death following forebrain ischemia, where glutamate excitotoxicity plays a major role.

Uptake of the parkinsonism-inducing toxin, 1-methyl-4-phenylpyridinium, into cultured dopaminergic neurons impairs mitochondrial complex 1 activity and causes oxygen radical generation *(10)*. Treatment of cultured neurons from rat midbrain with

Table 1
Inhibition of CDKs and Neuroprotection by Indolinones

	Indo-A	Indo-B	Indo-C
CDK1/IC50 (μM)	0.025	2.5	0.009
CDK2/IC50 (μM)	0.036	2.0	0.093
CDK4/IC50 (μM)	0.0001	0.0008	>10.0
CDK5/IC50 (μM)	0.005	1.4	1.9
Glutathione depletion/EC50 (μM)	0.13	>1.0	>1.0
Serum deprivation/EC50 (μM)	0.08	>1.0	>1.0
Focal cerebral ischemia/infarct reduction (%)	58.8	n.d.	nd
Optic nerve transection/neuron survival (%)	62.7	−5.6	nd

Indolinone A did not protect tyrosine hydroxylase-immunoreactive dopaminergic neurons against 1-methyl-4-phenylpyridinium toxicity. Moreover, Indolinone A alone led to a decrease in the number of tyrosine hydroxylase-positive neurons, whereas nondopaminergic neurons were not affected (*see* Fig. 2B). Very recently, phosphorylation of tyrosine hydroxylase by CDK5 has been shown to increase enzyme stability and activity *(11)*, which may underlie the neurotoxic effect of the CDK5 inhibitor on cultured dopaminergic neurons. On the other hand, the pan-CDK inhibitor flavopiridol attenuated the loss of dopaminergic neurons in rats injected with 1-methyl-4-phenyl-1,2,3,6-tetrahydropyridine *(12)*. Thus, we cannot exlude that a compound-related effect may contribute to the selective neurotoxicity of Indolinone A in cultured dopaminergic neurons seen in our studies. Indolinone A was also active in vivo. Intravitreal injections promoted survival of retinal ganglion cells after transection of the optic nerve, and intracerebroventricular infusion significantly reduced infarct volume following transient focal cerebral ischemia in rats (Table 1) *(8)*. Most importantly, compounds that inhibit mitotic CDKs but not CDK5 did not show neuroprotective effectiveness in any paradigm (Table 1), strongly suggesting that CDK5 represents an attractive target for neuroprotection.

3. Point of Intervention

Studies during the last decade have shown that different cell death pathways converge on mitochondria causing the release of various cell death-promoting proteins *(13)*. A therapeutic intervention at the early steps in the cell death cascade seems a prerequisite for long-term functional neuroprotection.

In our studies using cultured neurons, release of cytochrome *c* from mitochondria during staurosporine-induced apoptosis was prevented by Indolinone A, as assessed by immunoblotting *(8)*. More importantly, the CDK5 inhibitor preserved mitochondrial transmembrane potential after glutathione depletion by BSO, while mitochondria were completely depolarized in cultures treated with the pan-caspase inhibitor zVAD-fmk (*see* Fig. 3A).

In order to determine whether Indolinone A directly acts on mitochondria, two experiments were performed. First, cytochrome *c* release and subsequent neuronal cell death was triggered directly at the mitochondrial level using HA14-1, a small molecule inhibitor of the antiapoptotic mitochondrial Bcl-2 protein *(14)*. HA14-1 binds to the surface pocket of Bcl-2, thereby preventing heterodimerization with cell death-promoting

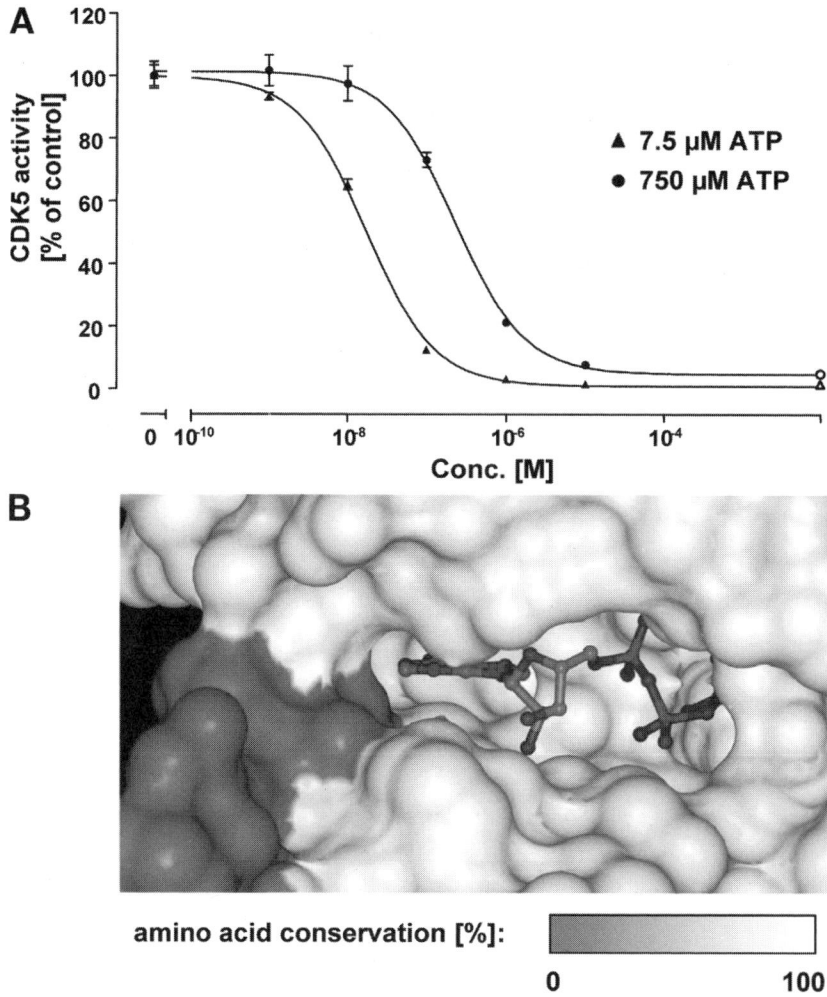

Fig. 1. (A) Inhibition of recombinant CDK5/p25 by Indolinone A in a cell-free assay. At higher ATP concentrations (circles) dose-response curve is shifted to the right indicating an ATP-competitive mode of action. Empty circle/triangle indicates absence of CDK5 substrate histone H1. **(B)** Molecular model showing amino acid conservation between mitotic CDK2 and neuronal CDK5 in a region encompassing the ATP binding site. ATP bound within its pocket is depicted.

Bcl-2 family members, like Bax, Bad, or Bak. Bax oligomers then insert into the outer mitochondrial membrane mediating the release of cytochrome *c*. In cultured cerebellar granule neurons, HA14-1 dose-dependently increased mitochondrial dysfunction and cell death which was not reduced by cotreatment with the CDK5 inhibitor (*see* Fig. 4A). Secondly, the compounds were administered directly to mitochondria that had been isolated from mouse forebrains. Treatment of density gradient-purified mitochondria with high concentrations of calcium led to a decrease in transmembrane potential. Mitochondrial depolarization was not prevented by coadministration of Indolinone A, although a small amount of CDK5 copurified with mitochondria (*see* Fig. 4B). These findings indicate that Indolinone A exerts its neuroprotective effects upstream of the Bcl-2/Bax family, mitochondrial depolarization, and cytochrome *c* release.

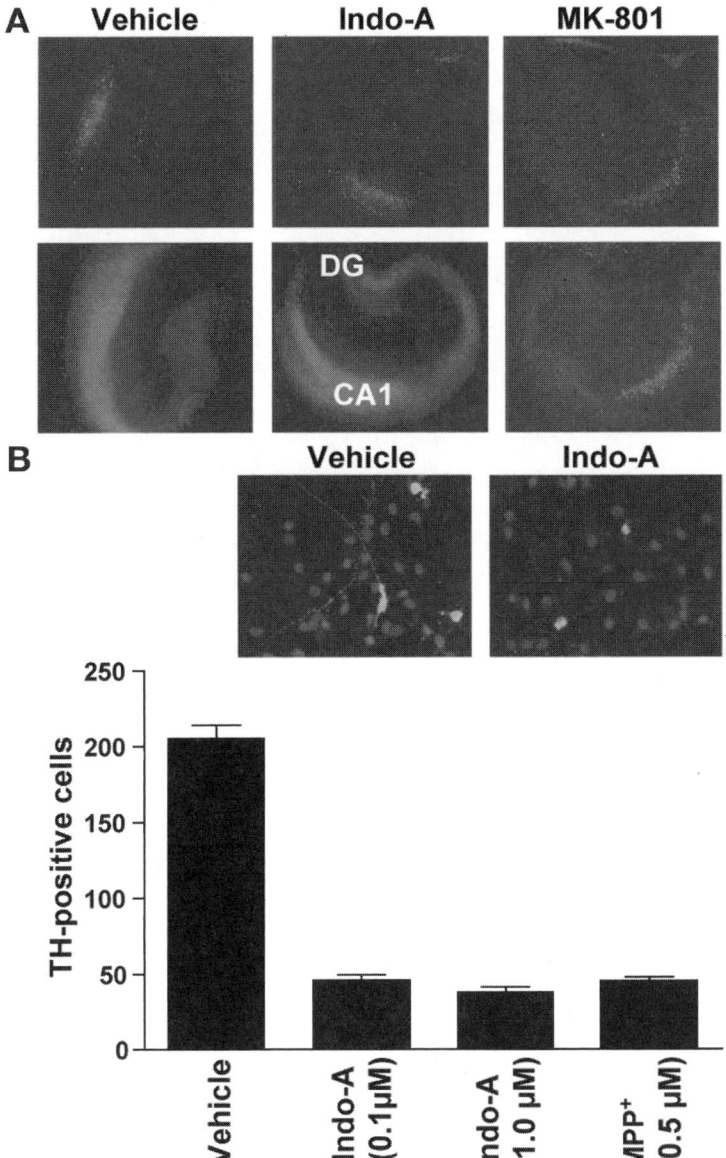

Fig. 2. (A) The CDK5 inhibitor Indolinone A does not prevent excitotoxic death of hippocampal neurons induced by glutamate. Rat organotypic hippocampal slice cultures were stained with propidium iodide and monitored 6 h before (upper row) and 16 h after (lower row) glutamate exposure (50 μ*M* for 1h), respectively. Uptake of prodium iodide (fluorescence) indicates neuronal cell death in the granule cell layer of the dentate gyrus (DG) and the pyramidal cell layer of the cornu ammonis (CA). Neurodegeneration is prevented by pretreatment with the glutamate receptor antagonist MK-801, but not with the CDK5 inhibitor Indolinone A (Indo-A). **(B)** Treatment with Indolinone A causes a decline in the number of tyrosine hydroxylase (TH) immunopositive neurons in vitro. Neuronal cell cultures from the embryonic rat midbrain were incubated with either the neurotoxin 1-methyl-4-phenylpyridinium (MPP⁺) or Indolinone A (Indo-A). Dopaminergic neurons were visualized by tyrosine hydroxylase immunocytochemistry and healthy neurons with long neurites (inset) were counted. Values represent mean ± SD, *n* = 4.

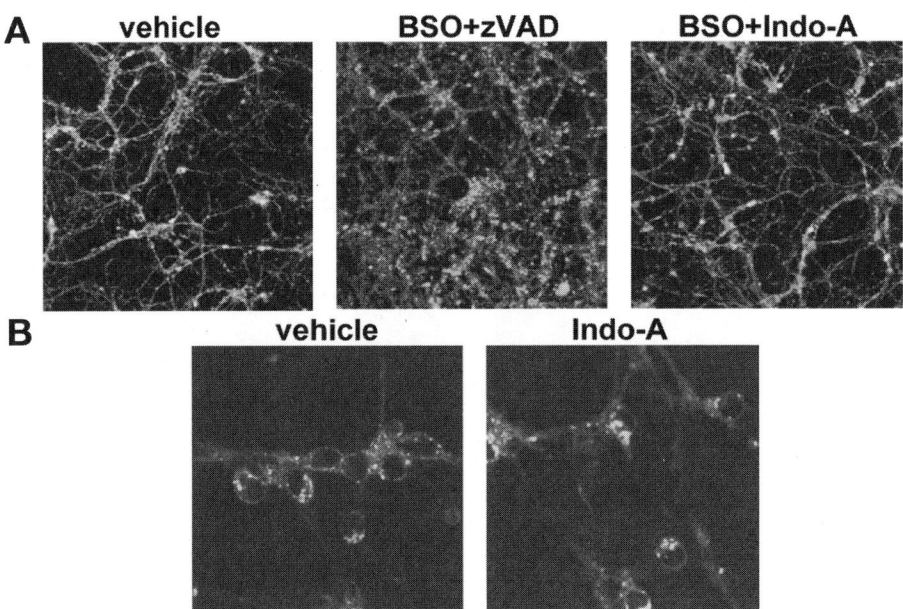

Fig. 3. (A) Indolinone A preserves mitochondrial transmembrane potential in cultured rat cerebellar granule neurons during free radical stress. Neuronal cell cultures were treated with BSO which inhibits glutathione biosynthesis, thereby increasing free radical generation and cell death. Hyperpolarized, functional mitochondria exhibit red fluorescence following administration of the potential-sensitive dye JC-1, whereas depolarized mitochondria show green fluorescence. An overlay of confocal lasercan microscope images is presented. Complete depolarization of mitochondria is visible in cultures coincubated with BSO and the pan-caspase inhibitor zVAD-fmk (zVAD, 100 μ*M*), whereas Indolinone A (Indo-A, 0.3 μ*M*) prevents loss of mitochondrial transmembrane potential. **(B)** Indoline A does not affect synaptic vesicle recycling in rat cerebellar granule neurons. Cultured neurons were co-incubated with compound and FM 1–43, a fluorescent tracer of synaptic vesicle endocytosis. Confocal laser scan microscope images show similar signal intensity compared to vehicle-treated control cultures.

Activity of both antiapoptotic and proapoptotic Bcl-2 family members can be modulated by phosphorylation and various kinase signaling pathways influence cell survival *(15)*. In neuronal cell culture, activity of antiapoptotic extracellular signal-regulated kinase (ERK) versus proapoptotic c-Jun N-terminal kinase/p38 mitogen-activated protein kinase (JNK/p38 MAPK) has been shown to determine survival *(16)*. Studies in *CDK5*-deficient mice point to a crosstalk between CDK5 and ERK/JNK signaling pathways *(17,18)*. As shown in Fig. 5, incubation of cerebellar granule neurons with Indolinone A caused a rapid increase in phosphorylation of both ERK and c-Jun indicating activation of ERK/JNK signaling. By contrast, Indolinone A treatment did not modulate phosphorylation/activation of Akt kinase or p38 MAPK.

To analyze whether activation of anti-apoptotic ERK by Indolinone A underlies its neuroprotective effectiveness, we co-incubated cerebellar granule neurons with Indolinone A and the ERK inhibitor U0126 *(19)*. Whereas U0126 (10 μ*M*) completely blocked the increase in ERK phosphorylation by Indolinone A, the compound did not significantly affect neuroprotection by Indolinone A against BSO-induced cell death. Similarly, treatment of cerebellar granule neurons with the JNK inhibitor SP600125 (30 μ*M*) *(20)* plus

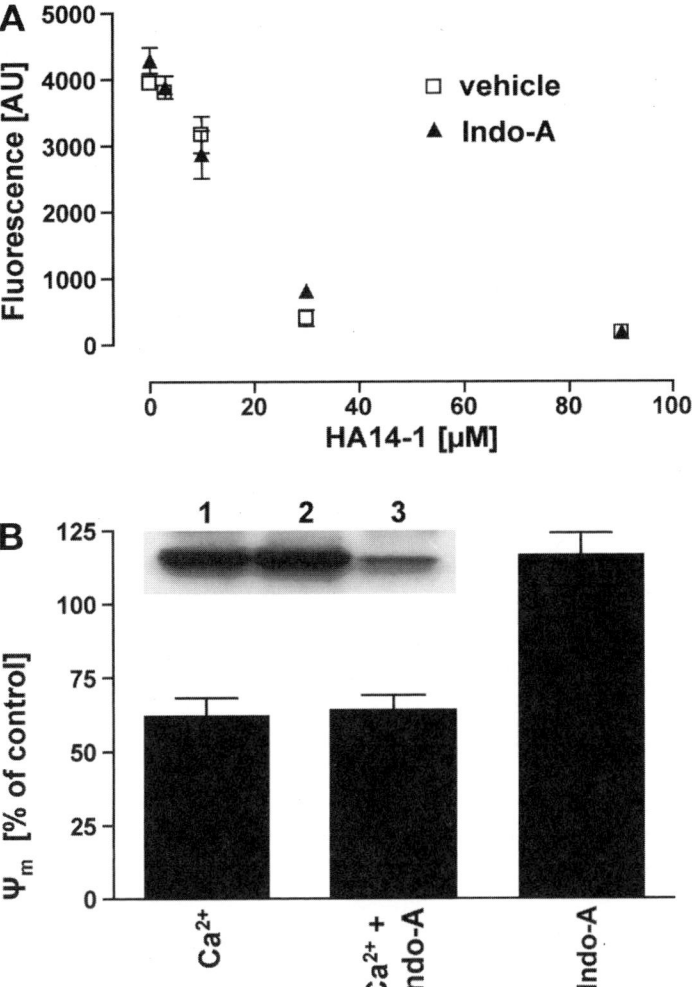

Fig. 4. (**A**) The CDK5 inhibitor Indolinone A does not block the cell death program, if it is triggered directly at the mitochondrial level. Administration of HA14-1, a small molecule inhibitor of antiapoptotic Bcl-2 protein, dose-dependently induces cell death in cultured rat cerebellar granule neurons as assessed by Alamar Blue assay. Coadministration of Indolinone A (0.3 μM) does not promote survival. (**B**) Indolinone A has no effect on isolated mitochondria. Metabolically active mitochondria were purified from mouse forebrains. Mitochondrial transmembrane potential (Ψ_m) was measured in a microtiter plate reader following administration of the potential-sensitive dye JC-1. Preincubation with Indo-A does not prevent mitochondrial depolarization induced by calcium overload (100 nmol/mg). Values represent mean ± SD, $n = 4$-5. (Inset) Immunoblot analysis demonstrates that a small fraction of CDK5 co-purifies with brain mitochondria. *Lane 1*: homogenate; *lane 2*: cytosol; *lane 3*: purified mitochondria.

Indolinone A prevented the accumulation of phospho-c-Jun, however, blockade of BSO-induced cell death by Indolinone A remained unchanged. Thus, activation of ERK/JNK signaling is dispensable for Indolinone A-mediated neuroprotection against free radical stress. In contrast, cultured cortical neurons from *CDK5* knockout mice showed increased JNK activity and c-Jun phosphorylation, as well as a faster rate of apoptotic cell death following ultraviolet (UV) irradiation (*17*).

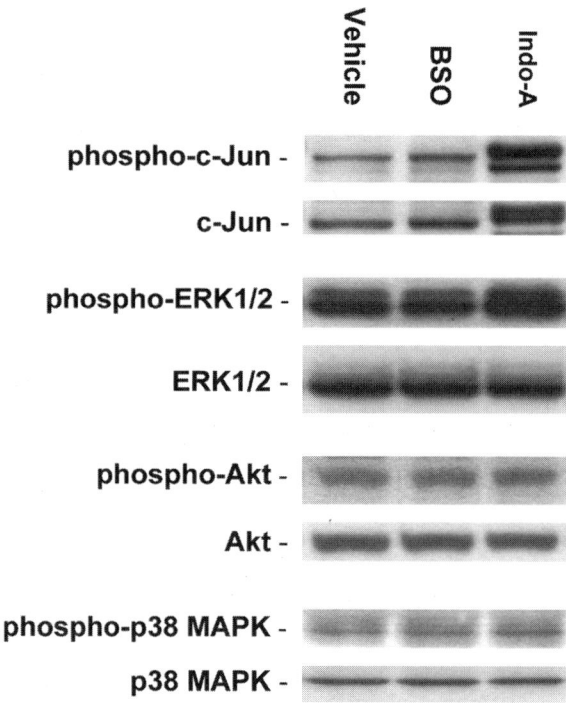

Fig. 5. Crosstalk between CDK5 and ERK/JNK signaling cascades. Rat cerebellar granule neurons were cultured in 60-mm Petri dishes and treated with the CDK5 inhibitor Indolinone A (Indo-A, 0.3 µM) or BSO (500 µM) for 3 h. Proteins were detected in cell lysates by immunoblot analysis using the following primary antibodies: anti phospho-c-Jun(Ser73), anti total c-Jun, anti phospho-ERK1/2(Thr202/Tyr204), anti total ERK1/2, anti phospho-Akt(Thr308), anti total Akt, anti phospho-p38 MAPK(Thr180/Tyr182), and anti total p38 MAPK. Inhibition of CDK5 causes a selective increase in phosphorylation of ERK1/2 and the JNK substrate c-Jun indicating activation of ERK and JNK signaling, respectively. The total c-Jun immunoreactive band is shifted suggesting that c-Jun becomes phosphorylated at multiple sites.

4. Phosphoproteome Analysis

Following proteolytic cleavage of its activatory subunit p35 to p25, CDK5/p25 translocates from the plasma membrane to the cytosol and nucleus where novel substrates become hyperphosphorylated leading to neuronal cell death. Toxic substrates such as tau and neurofilament proteins have already been identified in Alzheimer's disease and amyotrophic lateral sclerosis *(1,3,4)*. Incubation of cerebellar granule neurons with Indolinone A lead to a complete dissappearance of basal phospho-tau(Ser202/Thr205) immunoreactivity, as assessed by immunoblotting *(21)*. However, we could not detect an increase in tau phosphorylation in our models of neurodegeneration and thus, other pathogenic substrates remain to be identified.

In order to identify novel CDK5 substrates in brain lysates, we performed a pilot study using KESTREL (kinase substrate tracking and elucidation) *(22)*. Pig brains were homogenized, purified by gel filtration chromatography, and fractionated by heparin chromatography. An aliquot from each fraction was incubated with recombinant CDK5/p25 and radiolabeled ATP. In addition to the CDK5/p25 autophosphorylation

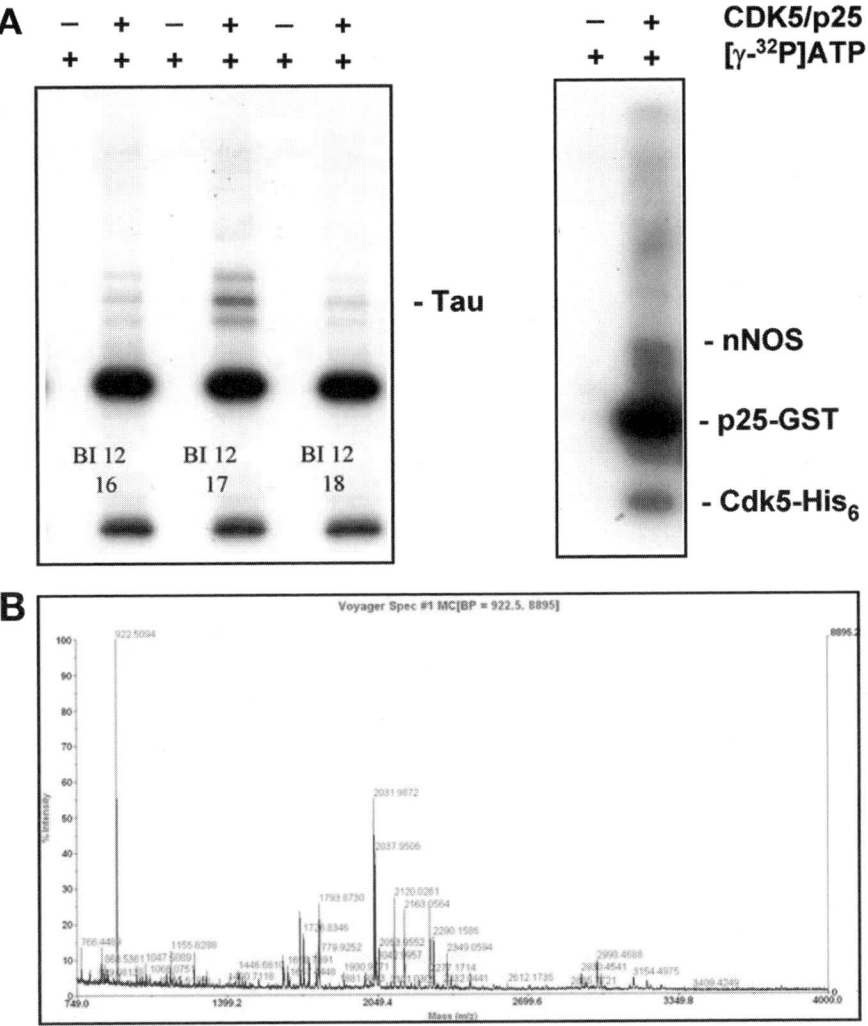

Fig. 6. (A) Identification of novel CDK5 substrates by KESTREL. Protein extracts from pig brains were fractioned by chromatography. An aliquot of each fraction was supplemented with recombinant CDK5/p25 and [γ-^{32}P]ATP. Reactions were analyzed by sodium dodecyl sulfate-polyacrylamide gel electrophoresis (SDS-PAGE) and autoradiography. Radiolabeled bands were excised, proteins were digested with trypsin and identified by peptide mass fingerprinting. *Left*: phosphorylation of the known CDK5 substrate tau (three isoforms). *Right:* autophosphorylation of recombinant CDK5/p25 and phosphorylation of neuronal nitric oxide synthase (nNOS). **(B)** Peptide mass fingerprint of collapsin response mediator protein-4. Rat cerebellar granule neurons were coincubated with ^{32}P orthophosphoric acid and the CDK5 inhibitor Indolinone A (or vehicle). Proteins were separated by two-dimensional gel electrophoresis. Differentially radiolabeled gel spots were digested with trypsin and analyzed by MALDI-TOF mass spectrometry.

bands, several ^{32}P-labeled proteins of varying molecular weight were detected (*see* Fig. 6A). The fractions containing these potential CDK5 substrates were further purified by ion exchange and gel filtration chromatography. The bands were excised from the gel, digested with trypsin, and identified by peptide mass fingerprinting. The peptide masses were matched with tau protein (accession number P29172, three isoforms) and neuronal nitric oxide synthase (accession number Q9Z0J4, N-terminal fragment), respectively. Neuronal nitric oxide synthase exhibits several CDK5 consensus phosphorylation sites (S/TPXK/H/R) and potentially represents a novel CDK5 substrate. About 40 substrates for CDK5 have been published *(4)*, however, phosphorylation under physiological conditions has not been demonstrated for all of them. Using KESTREL we could identify only tau and p25. Some CDK5 substrates are transmembrane proteins which are difficult to solublize, others are low-abundant proteins which are difficult to detect. Preliminary results indicate that additional radiolabeled bands become detectable following modification of KESTREL (A. Knebel, personal communication).

We also used metabolic labeling followed by two-dimensional (2D) gel electrophoresis and matrix assisted laser desorption/ionization-time of flight (MALDI-TOF) mass spectrometry to analyze global changes in protein phosphorylation in cerebellar granule neurons in vitro. In Indolinone A-treated cultures, a significant increase in ^{32}P incorporation was observed in cofilin (accession number P45592), tubulin β (P04691), α-internexin (Q07803), syndapin I (Q9Z0W5), heterogeneous nuclear ribonucleoprotein K (Q07244), and elongation factor G (Q07803) *(21)*. Neuronal proteins showing a Indolinone A-induced loss of ^{32}P incorporation included dynein light intermediate chain 2 (O43237), collapsin response mediator protein-2 (P47942), and collapsin response mediator protein-4 (Q62952) *(see* Fig. 6B). Corresponding protein spots on silver-stained gels did not show significant changes in staining intensity. Prediction of potential phosphorylation sites was performed using the computer program NetPhos *(www.cbs.dtu.dk/services/NetPhos).* Phosphorylation at serine, threonine or tyrosine residues was predicted with a score greater than 0.9 for all radiolabeled proteins.

Both dynein light intermediate chain 2 and collapsin response mediator proteins-2/-4 exhibit conserved CDK5 consensus phosphorylation site [(S/T)PX(K/H/R)] and phosphorylation of collapsin response mediator protein-2 by CDK5 has recently been confirmed by others *(23)*. Collapsin response mediator proteins regulate axonal elongation and phosphorylation by Rho-associated kinase causes growth cone collapse. In our study, incubation of cultured chicken dorsal root ganglion neurons with the CDK5 inhibitor Indolinone A significantly increased neurite length *(21)*. Consistently, growth cone collapse was blocked in rat dorsal root ganglion neurons by the CDK5 inhibitor roscovitine or by transfection of mutant collapsin response mediator protein-2(S522A) that cannot be phosphorylated by CDK5 *(23)*. Similar results were obtained in dorsal root ganglion neurons from *CDK5*-deficient mice *(24)*. It should be noted however, that conflicting data have been reported using cultured chicken sympathetic ganglia, where both roscovitine and a dominant-negative CDK5 mutant lead to a marked reduction in neurite outgrowth *(25)*.

CDK5 has been shown to modulate the activity of various protein kinases and protein phosphatases (reviewed in ref. *4)* which may help explain why numerous proteins become phosphorylated in cerebellar granule neurons following CDK5 inhibitor treatment. Cofilin is best known for its destabilizing effects on the actin cytoskeleton. Recently, however, it has been shown that following different apoptotic stimuli dephosphorylated cofilin rapidly translocates to mitochondria where it induces cytochrome *c* release and cell death. Phosphorylation of cofilin inhibited mitochondrial transloction and cytochrome *c* release *(26)*. Because Indolinone A prevents mitochondrial release of cytochrome *c* and enhances ^{32}P incorporation in cofilin, we analyzed, whether this mechanism might contribute to Indolinone A-mediated neuroprotection. Following serum/potassium deprivation in cultured cerebellar granule neurons, we detected a five-fold accumulation of cofilin in the mitochondrial fraction that was significantly reduced by administration of the CDK5 inhibitor *(21)*. Consistently, Indolinone A also lead to a twofold increase in the ratio of phospho-cofilin(Ser-3) to total cofilin in cell lysates as assessed by immunoblotting. It may be hypothesized that early in the cell death cascade, CDK5 inactivates/activates the kinase/phosphatase acting on cofilin(Ser-3).

Finally, treatment of rat cerebellar granule neurons with Indolinone A significantly increased ^{32}P incorporation into elongation factor G and heterogeneous nuclear ribonucleoprotein K. Elongation factor G catalyzes the translocation of ribosomes during protein synthesis and heterogeneous nuclear ribonucleoprotein K couples extracellular signals to mRNA transcription and translation *(27)* suggesting that CDK5 inhibitors might also influence gene expression.

5. Gene Expression Profiling

In several studies nuclear localization of CDK5 and association with nuclear proteins (e.g., SET protein) has been described *(28)*. More importantly, CDK5 has been shown to phosphorylate several transcription factors [retinoblastoma protein, p53, myocyte enhancer factor 2 (MEF2), signal transducer and activator of transcription 3 (STAT3), suppressor of defective silencing 3 Sds3)] and to modulate transcriptional activity *(4,29–31)*. CDK5-mediated phosphorylation/inactivation of the survival-promoting transcription factor MEF2 contributes to neuronal cell death following free radical stress. We have therefore included gene expression profiling to investigate the influence of the neuroprotective CDK5 inhibitor Indolinone A on neuronal gene expression *(21)*.

cDNA microarray analysis demonstrated that numerous mRNAs were either upregulated or downregulated more than or equal to twofold already 3 h after compound administration (Table 2A,B). Phosphorylation by CDK5 inhibits MEF2 transcriptional activity and enhances Sds3-mediated transcriptional repression *(29,31)*, whereas transcription of STAT3 target genes is increased *(30)*. Consistently, Indolinone A treatment lead to a decrease in mRNA of FBJ osteosarcoma oncogene (c-fos), a well known target gene for STAT3, and sequence analysis revealed that several up-regulated genes contain binding consensus sequences for MEF2 [CT(A/t)(a/t)AAATAG] in their regulatory regions *(21)*. Interestingly, expression of MEF2D mRNA is rapidly downregulated following MEF2 disinhibition by the CDK5 inhibitor. Additionally, there seems to be a compensatory downregulation of several protein phospatases (Table 2B). Similar to phosphoproteome analysis, gene expression profiling revealed that several components of the basic gene transcription machinery (e.g., RNA polymerase, nuclear ribonucleoproteins) are modulated following Indolinone A treatment. This raises the possibility that Indolinone A might also inhibit CDK7, -8, or -9, which are involved in regulation of RNA transcription *(32)*. Similar to our data however, chemical inhibition of Pho85 kinase in yeast (the functional homolog to mammalian CDK5) caused induction/repression of more than 300/500 genes *(33)*. Notably, changes in gene expression were transient and only detectable after the rapid loss of Pho85 activity caused by chemical inhibition, but not in *Pho85* gene-deficient yeast mutants.

Table 2A
Gene Induction Following Indolinone A Treatment

ID	Acc.No.	Gene	Title	fold induction [log(2)]
906E12	AU020524	Hmgcs1	3-hydroxy-3-methylglutaryl-Coenzyme A synthase 1	2.263358890
854D11	BG078400	Ap1m2	adaptor protein complex AP-1, mu 2 subunit	1.509966312
322B4	BE860148	Alex3-pending	ALEX3 protein	1.136517749
837F10	AW538795	Asb6	ankyrin repeat and SOCS box-containing protein 6	1.003616405
968H11	BG087691	Amhr2	anti-Mullerian hormone type 2 receptor	1.103956803
983F1	AW558347	Caskin2	cask-interacting protein 2	1.520218169
834A4	BG076855	Cd2bp2	CD2 antigen (cytoplasmic tail) binding protein 2	1.230781176
975H4	AW556496	BC003940	cDNA sequence BC003940	1.152310219
837H1	BG077109	BC021530	cDNA sequence BC021530	1.487676484
979A9	BG088348	BC032200	cDNA sequence BC032200	1.212097459
987E2	NM_013654	Ccl7	chemokine (C-C motif) ligand 7	1.896125534
969E2	AW555109	Chd1	chromodomain helicase DNA binding protein 1	1.474401986
966B2	AW554252	Csk	c-src tyrosine kinase	1.034350284
332D12	BE859476	Cnp1	cyclic nucleotide phosphodiesterase 1	1.381581172
233B7	AI587800	Cdkn2c	cyclin-dependent kinase inhibitor 2C (p18, inhibits CDK4)	1.187476618
835A7	BI076508	Dhx36	DEAH (Asp-Glu-Ala-His) box polypeptide 36	1.544586027
977B7	AW556827	D12Ertd748e	DNA segment, Chr 12, ERATO Doi 748, expressed	1.228579753
984B10	BG088755	Dnaja1	DnaJ (Hsp40) homolog, subfamily A, member 1	1.109739733
840H1	AW540988	Dnaja3	DnaJ (Hsp40) homolog, subfamily A, member 3	1.044360218
980A10	BG088432	Drap1	Dr1 associated protein 1 (negative cofactor 2 alpha)	1.155361719
332E3	BE651854	Dnm	dynamin	1.274349029
857G7	BG078688	Dtnb	dystrobrevin, β	1.267983758
324A4	BE651921	Enpp2	ectonucleotide pyrophosphatase/phosphodiesterase 2	1.010581079
941A8	BG085501	Eif1a	eukaryotic translation initiation factor 1A	1.811237612
915D3	AU016279	AF013969	expressed sequence AF013969	1.386536154
319G9	BE651714	AI426686	expressed sequence AI426686	1.156323951
322A3	BE652683	AI426938	expressed sequence AI426938	1.312695745

344G8	BE860139	AI848100	expressed sequence AI848100	0.959912505
973B6	BG087891	AU020206	expressed sequence AU020206	1.264458751
335D9	BE652725	Foxo1	forkhead box O1	1.123737896
988D10	NM_010235	Fosl1	fos-like antigen 1	1.503112887
6A3	AI385474	Gpr14	G protein-coupled receptor 14	1.189622888
980F3	BG088479	Gtlf3b	gene trap locus F3b	1.289962938
842E1	BG077445	Golph2	golgi phosphoprotein 2	1.511070409
963G12	BG087208	H1f0	H1 histone family, member 0	1.243853116
956B8	BG086671	H19	H19 fetal liver mRNA	1.489823764
335D6	BE860210	Hbb-b1	hemoglobin, beta adult major chain	1.033707681
955A11	AW551778	Hnrpc	heterogeneous nuclear ribonucleoprotein C	1.831768534
929E10	BG084577	Hs1bp3-pending	HS1 binding protein 3	1.498311536
953B11	BG086436	C130006E23	hypothetical protein C130006E23	1.458076364
349A12	BE861346	MGC38046	hypothetical protein MGC38046	1.414133201
981E2	BG088546	MGC47001	hypothetical protein MGC47001	1.253493232
762E2	NM_153175	Ian6	immune associated nucleotide 6	1.528882179
252E11	AA184823	Ifi47	interferon gamma inducible protein	1.093851904
867F11	BG066248	Ifi30	interferon gamma inducible protein 30	1.102862378
988D11	NM_010591	Jun	Jun oncogene	1.365047586
936G6	BI076450	Ltf	lactotransferrin	1.168856547
343B4	BE859813	Lamb1-1	laminin B1 subunit 1	1.421635857
345F12	BE652510	Ly6c	lymphocyte antigen 6 complex, locus C	1.681490503
987H8	NM_008529	Ly6e	lymphocyte antigen 6 complex, locus E	1.086073304
842H8	BG064137	Lyric-pending	lyric	1.192396335
977B4	AW556823	Lamp1	lysosomal membrane glycoprotein 1	1.502351183
949H7	BG086197	Mgat1l-pending	mannoside acetylglucosaminyltransferase 1-like	0.968370336
335F9	BE652732	Mbd3	methyl-CpG binding domain protein 3	1.280463224
891C9	BG068329	Mybl1	myeloblastosis oncogene-like 1	1.103594652
952F4	AW550881	Nif3l1	Ngg1 interacting factor 3-like 1 (S. pombe)	1.257796240
961H12	AW553287	Osf2-pending	osteoblast specific factor 2 (fasciclin I-like)	1.123708527
955E2	BG086615	Ovcov1	ovarian cancer overexpressed 1	1.074767077
856F1	BG078591	Perp-pending	p53 apoptosis effector related to Pmp22	1.059799816

(Continued)

Table 2A (*Continued*)

ID	Acc.No.	Gene	Title	fold induction [log(2)]
342F2	BE860247	Pknox1	Pbx/knotted 1 homeobox	1.030895696
983E4	BG088703	Hspg2	perlecan (heparan sulfate proteoglycan 2)	1.444840704
977B6	AW556825	Pla2g7	phospholipase A2, group VII (platelet-activating factor acetylhydrolase, plasma)	1.295882781
978C6	BG088297	Parg1-pending	PTPL1-associated RhoGAP 1	1.374483858
836H4	BG077040	Rad50	RAD50 homolog (S. cerevisiae)	1.337533724
833B7	BG063350	Ranbp1	RAN binding protein 1	1.152857066
841E4	BG077353	Rfxank	regulatory factor X-associated ankyrin-containing protein	1.074238637
762E4	NM_011269	Rhag	Rhesus blood group-associated A glycoprotein	1.114459133
958E1	BG086862	Rbm8	RNA binding motif protein	2.009002422
330A10	BE859664	S100a1	S100 calcium binding protein A1	0.945521522
986E1	BG088931	Sall2	sal-like 2 (Drosophila)	1.890928054
975B7	BG088059	Stk31	serine threonine kinase 31	1.825952339
956G1	AW552084	Sh3glb1	SH3-domain GRB2-like B1 (endophilin)	0.983726642
977A8	AW556781	Slit2	slit homolog 2 (Drosophila)	1.234985595
331A6	BE652402	Slc1a3	solute carrier family 1, member 3	1.186666453
978E11	BG088317	Shc1	src homology 2 domain-containing transforming protein C1	1.136748257
946A3	AW548440	Syn1	synapsin I	1.807671696
833H4	BG076846	Sdc1	syndecan 1	1.038927867
965F12	BG087362	Sdc4	syndecan 4	1.003555938
946F2	BG085915	Tcfl1	transcription factor-like 1	1.123970670
911H5	BG083135	Tmod3	tropomodulin 3	0.915977918
844G1	BG077629	Twistnb	TWIST neighbor	1.454904838
953A11	AW551034	Usmg5	upregulated during skeletal muscle growth 5	1.220283386
978G6	BG075788	Xpc	xeroderma pigmentosum, complementation group C	1.778736657
762H8	NM_011756	Zfp36	zinc finger protein 36	1.363965617
340C2	BE859763	Zfp91	zinc finger protein 91	1.069518743
964A6	BG087226	Zp2	zona pellucida glycoprotein 2	1.936225234

Table 2B
Gene Repression Following Indolinone A Treatment

ID	Acc.No.	Gene	Title	fold induction [log(2)]
951E6	BG086322	Atf4	activating transcription factor 4	-1.254958598
357E5	AI845557	Arl4	ADP-ribosylation factor-like 4	-1.193286778
339F11	BE859750	Aars	alanyl-tRNA synthetase	-1.082206263
317G4	BE859535	Alas1	aminolevulinic acid synthase 1	-1.405198655
949E1	BG086158	Amotl2	angiomotin like 2	-1.908459702
349D8	BE653450	Atip1-pending	angiotensin II AT2 receptor-interacting protein 1	-1.119802615
844A10	BG077577	Arcn1	archain 1	-1.041678675
932G9	BG084937	Arntl	aryl hydrocarbon receptor nuclear translocator-like	-1.144971372
941D11	BG085533	Btg1	B-cell translocation gene 1, antiproliferative	-2.373969904
344A12	BE860100	Bub3	budding uninhibited by benzimidazoles 3 homolog (S. cerevisiae)	-1.192025656
2E9	AI326507	Cdh1	cadherin 1	-1.197291971
330C8	BE859630	Clp1-pending	cardiac lineage protein 1	-1.513510187
855G12	BG078527	Cir-pending	CBF1 interacting corepressor	-0.917509090
903H8	BG082503	Cited4	Cbp/p300-interacting transactivator, with Glu/Asp-rich C-terminal domain, 4	-1.509696485
233E9	AI587885	Clk4	CDC like kinase 4	-1.030674319
351H5	BE860008	BC003251	cDNA sequence BC003251	-1.071808008
975A5	BG088047	Cmkor1	chemokine orphan receptor 1	-1.268246239
871B6	BG079703	Cstf3	cleavage stimulation factor, 3′ pre-RNA, subunit 3	-0.918266049
253D9	AI451894	Ccng2	cyclin G2	-1.821473007
3E7	AI528658	Ddx24	DEAD (Asp-Glu-Ala-Asp) box polypeptide 24	-1.750049304
850G8	BG078113	Ddx5	DEAD (Asp-Glu-Ala-Asp) box polypeptide 5	-0.908876539
347E8	BE653185	Ddit3	DNA-damage inducible transcript 3	-1.085510777
870F5	BG066495	Dnajb1	DnaJ (Hsp40) homolog, subfamily B, member 1	-1.358385916
9H10	AI386158	Dusp1	dual specificity phosphatase 1	-1.202402209
939D7	BG085362	Dusp14	dual specificity phosphatase 14	-1.924781361
958A7	BG086826	Ech1	enoyl coenzyme A hydratase 1, peroxisomal	-1.622216124
902C5	BG082350	Eif5	eukaryotic translation initiation factor 5	-1.255651683

(Continued)

Table 2B (*Continued*)

ID	Acc.No.	Gene	Title	fold induction [log(2)]
335F5	AI841931	AI256456	expressed sequence AI256456	-0.946885939
344E6	BE860123	AI837181	expressed sequence AI837181	-1.702047103
965H5	BG087378	AI837181	expressed sequence AI837181	-1.051245654
987C7	NM_010234	Fos	FBJ osteosarcoma oncogene	-1.149800521
949C12	BG086146	Fbxo18	F-box only protein 18	-1.122402415
218A7	W48154	Fbxo2	F-box only protein 2	-0.962021583
342B4	BE860233	Gabra6	gamma-aminobutyric acid (GABA-A) receptor, subunit alpha 6	-1.211661495
337E12	AI840450	Gtf2b	general transcription factor IIB	-1.239747183
340E8	BE652219	Gars	glycyl-tRNA synthetase	-1.107935366
881C2	BG080485	Gadd45g	growth arrest and DNA-damage-inducible 45 gamma	-1.36646492
350F2	BE653223	H3f3b	H3 histone, family 3B	-1.649702245
960H1	BG087043	Hspa8	heat shock protein 8	-1.075527705
318A8	BE859389	Herpud1	homocysteine-inducible, ER stress-inducible, ubiquitin-like domain member 1	-2.341701509
881D5	BG080499	Hspbap1	Hspb associated protein 1	-0.922325776
367E3	AI847367	LOC224093	hypothetical protein LOC224093	-1.309299707
987D11	NM_008329	Ifi204	interferon activated gene 204	-1.543705128
963G1	BG087199	Jtv1-pending	JTV1 gene	-1.183824067
219E2	W64940	Mab2l2	mab-21-like 2 (*C. elegans*)	-0.913480060
348F1	BE859961	Magee1	melanoma antigen, family E, 1	-0.920709865
953H5	BG086491	Mat2a	methionine adenosyltransferase II, α	-1.141585399
360B8	BE860478	Mrpl15	mitochondrial ribosomal protein L15	-1.015346490
221D12	AI430437	Mln51-pending	MLN51 protein	-1.122206799
320G6	AI836493	Myc	myelocytomatosis oncogene	-1.208928111
876E12	BG080094	Mcl1	myeloid cell leukemia sequence 1	-1.121497938
974A11	BG087970	Mef2d	myocyte enhancer factor 2D	-1.140153304
318E9	AI835157	Neurod1	neurogenic differentiation 1	-1.828387573
949B11	BG086134	Nipa-pending	nuclear interacting partner of anaplastic lymphoma kinase (Alk)	-1.196916743
852E3	BG064932	Ncl	nucleolin	-0.930569055
987A9	NM_008781	Pax3	paired box gene 3	-1.280875248

360A11	BE654231	Ptch	patched homolog	-1.264757560
986C3	BG088910	Pde4b	phosphodiesterase 4B, cAMP specific	-1.895407179
339B2	BE652144	Psat1	phosphoserine aminotransferase 1	-1.547179873
217E1	W34773	Pl6-pending	Pl6 protein	-1.145820395
3F7	AI528709	Vpreb3	pre-B lymphocyte gene 3	-4.258411262
337D2	AI839158	Psmd8	proteasome (prosome, macropain) 26S subunit, non-ATPase, 8	-1.438979538
215A8	W15071	Ppp1r1c	protein phosphatase 1, regulatory (inhibitor) subunit 1C	-1.803774360
983C3	BG076151	Ppm1d	protein phosphatase 1D magnesium-dependent, delta isoform	-1.658182946
935E5	BG085090	Rasd1	RAS, dexamethasone-induced 1	-1.538476917
318B5	BE859393	Rraga	Ras-related GTP binding A	-1.084709354
893F1	BG081562	Rgs2	regulator of G-protein signaling 2	-1.867596218
341B12	BE652245	Rlucl-pending	ribosomal large subunit pseudouridine synthase C like	-1.161114858
345B12	BE652489	Rpl15	ribosomal protein L15	-1.333707849
949C11	BG086145	Rbm16	RNA binding motif protein 16	-0.926581565
876D9	BG066998	Rpo1-1	RNA polymerase 1-1	-0.912932972
337A12	BE860038	Sqstm1	sequestosome 1	-1.153547323
838D7	BG077146	Sars1	seryl-aminoacyl-tRNA synthetase 1	-0.992646725
344A7	BE860098	Spc18-pending	signal peptidase complex	-0.948858542
943G5	BG085725	Snrpd1	small nuclear ribonucleoprotein D1	-1.763272576
350F11	AI841348	Snrpn	small nuclear ribonucleoprotein N	-1.133172974
337F1	BE860015	Slc3a2	solute carrier family 3, member 2	-1.357090401
945C3	BG085816	Spna2	spectrin alpha 2	-1.094000859
829B5	AA409665	Sfrs7	splicing factor, arginine/serine-rich 7	-1.241856383
965D3	BG087332	Tbc1d8	TBC1 domain family, member 8	-0.930975758
337B3	BE860040	Tm4sf7	transmembrane 4 superfamily member 7	-1.542375979
947C7	BG085973	Trb1-pending	tribbles homolog 1 (Drosophila)	-2.016653846
962D1	BG087083	Tde1	tumor differentially expressed 1	-1.088542661
344D6	BE860116	Yars	tyrosyl-tRNA synthetase	-1.204260312
348C7	BE859951	Ubc	ubiquitin C	-1.401802889
970G1	BG075137	B4galt3	UDP-Gal:betaGlcNAc beta 1,4-galactosyltransferase, polypeptide 3	-1.167223585
927A1	BG084364	Zfp216	zinc finger protein 216	-2.022021400
5F12	W41117	Zfp334	zinc finger protein 334	-1.009478171

The physiological relevance of the genes listed in Table 2 remains to be determined. We could not detect changes in the amount of mRNAs encoding anti-/proapoptotic proteins following administration of neuroprotective Indolinone A. However, an effect of the CDK5 inhibitor on synaptic function is indicated by several transcripts encoding synaptic proteins. CDK5 plays a central role in synaptic vesicle endocytosis by phosphorylating amphiphysin, dynamin and synaptojanin, which regulates interaction with other components (e.g., endophilin) *(34–36)*. In our study, mRNA expression of dynamin and endophilin increases rapidly following CDK5 inhibitor administration. Compound treatment also enhanced expression of synapsin mRNA and phosphorylation of syndapin protein which are involved in synaptic vesicle recycling. To determine whether expression of these genes is modulated as a consequence of functional alterations in synaptic signaling induced by CDK5 blockade, we coincubated cerebellar granule neurons under depolarizing conditions with Indolinone A and FM 1-43, a fluorescent tracer of synaptic vesicle endocytosis *(37,38)*. Internalization of FM 1–43 into recycling vesicles was monitored using a Leica confocal laserscan microscope or a Millipore CytoFluor plate reader. In our study, signal intensity did not differ between Indolinone A-treated and vehicle-treated neuronal cell cultures *(see* Fig. 3B). In a more detailed analysis by Tan et al. *(35)* however, both the CDK5 inhibitor roscovitine and a dominant-negative CDK5 mutant reduced synaptic vesicle endocytosis in cultured rat cerebellar granule neurons.

6. Conclusions

Taken together, our data indicate that inhibition of mitotic CDKs is not sufficient for neuroprotection in numerous cell death paradigms, whereas coinhibition of CDK5 by some Indolinones preserves neuronal structure and function. High-throughput screening of compound libraries for CDK5 inhibitors may be biased towards identification of potent, but nonselective compounds acting via ATP competition. However, crystal structure analysis of CDK5/p25 revealed clear differences in kinase activation and substrate recognition compared with mitotic CDKs *(39)* that may be exploited to design more selective inhibitors. On the other hand, CDKs and other cell death promoting kinases become coactivated in some paradigms of neurodegeneration *(1)* suggesting that nonselective inhibitors may be superior for neuroprotection. Additionally, we cannot exclude that neuroprotection following CDK5 inhibitor administration is achieved by inhibition of an unknown kinase, since more than 500 putative protein kinase genes have been identified in the human genome *(40)*.

Phosphorylation of several proteins and expression of nearly 200 genes is modulated 3 h after compound administration. Modulation of synaptic proteins is indicative of an interference with the physiological function of CDK5 in neurotransmission which may preclude long-term administration in chronic neurodegenerative diseases. Nevertheless, the CDK5-inhibiting Indolinones promote both neuronal survival and neurite outgrowth in various models suggesting that small molecule CDK5 inhibitors may favour functional survival and regeneration following acute neurological insults.

Note added in proof: Novel CDK5 inhibitors have been published by Ahn and colleagues (Chem Biol 2005;12:811–823). Effect of p35 gene deletion in rodent models of neurodegeneration has been described by Hallows et al. (J Neurosci 2006; 26:2738–2744).

Acknowledgments

Data presented were generated in collaboration with Dr. Axel Knebel (Kinasource, Dundee), Dr. Peter Steinlein (Research Institute of Molecular Pathology, Vienna), Dr. Christopher Gerner (University Vienna), Prof. Klaus Unsicker (University Heidelberg), Dr. Jochen Weishaupt (University Hospital Göttingen), Dr. Armin Heckel (Boehringer Ingelheim), Lothar Kussmaul (Boehringer Ingelheim), and Dr. Alexander Pautsch (Boehringer Ingelheim).

References

1. Nguyen MD, Mushynski WE, Julien J-P. Cycling at the interface between neurodevelopment and neurodegeneration. Cell Death Differ 2002;9:1294–1306.
2. O'Hare M, Wang F, Park DS. Cyclin-dependent kinases as potential targets to improve stroke outcome. Pharmacol Therap 2002;93:135–143.
3. Dhavan R, Tsai L-H. A decade of CDK5. Nature Rev Mol Cell Biol 2001;2:749–759.
4. Shelton SB, Johnson GVW. Cyclin-dependent kinase-5 in neurodegeneration. J Neurochem 2004;88:1313–1326.
5. Cohen P. Protein kinases – the major drug targets of the twenty-first century? Nature Rev Drug Discovery 2002;1:309–315.
6. Noble MEM, Endicott JA, Johnson LN. Protein kinase inhibitors: insights into drug design from structure. Science 2004;303:1800–1805.
7. Wüllner U, Seyfried J, Groscurth P, et al. Glutathione depletion and neuronal cell death: the role of reactive oxygen intermediates and mitochondrial function. Brain Res 1999;826:53–62.
8. Weishaupt JH, Kussmaul L, Grötsch P, et al. Inhibition of CDK5 is protective in necrotic and apoptotic paradigms of neuronal cell death and prevents mitochondrial dysfunction. Mol Cell Neurosci 2003;24:489–502.
9. Wang J, Liu S, Fu Y, Wang JH, Lu Y. Cdk5 activation induces hippocampal CA1 cell death by directly phosphorylating NMDA receptors. Nat Neuroscience 2003;6:1039–1047.
10. Lotharius J, O'Malley KL. The parkinsonism-inducing drug 1-methyl-4-phenylpyridinium triggers intracellular dopamine oxidation. J Biol Chem 2000;275:38,581–38,588.
11. Moy LY, Tsai LH. Cyclin-dependent kinase 5 phosphorylates serine 31 of tyrosine hydroxylase and regulates its stability. J Biol Chem 2004;279:54,487–54,493.
12. Smith PD, Crocker SJ, Jackson-Lewis V, et al. Cyclin-dependent kinase 5 is a mediator of dopaminergic neuron loss in a mouse model of Parkinson's disease. Proc Natl Acad Sci USA 2003;100:13,650–13,655.
13. Polster BM, Fiskum G. Mitochondrial mechanisms of neuronal cell apoptosis. J Neurochem 2004;90:1281–1289.
14. An J, Chen Y, Huang Z. Critical upstream signals of cytochrome c release induced by a novel Bcl-2 inhibitor. J Biol Chem 2004;279:19,133–19,140.
15. Yuan J, Yankner BA. Apoptosis in the nervous system. Nature 2000;407:802–809.
16. Xia Z, Dickens M, Raingeaud J, Davis RJ, Greenberg ME. Opposing effects of ERK and JNK-p38 MAP kinases on apoptosis. Science 1995;270:1326–1331.
17. Li BS, Zhang L, Takahashi S, Ma W, Jaffe H, Kulkarni AB, Pant HC. Cyclin-dependent kinase 5 prevents neuronal apoptosis by negative regulation of c-Jun N-terminal kinase 3. EMBO J 2002;21:324–333.
18. Sharma P, Veeranna D, Sharma M, et al. Phosphorylation of MEK1 by cdk5/p35 down-regulates the mitogen-activated protein kinase pathway. J Biol Chem 2002; 277:528–534.
19. Favata MF, Horiuchi KY, Manos EJ, et al. Identification of a novel inhibitor of mitogen-activated protein kinase kinase. J Biol Chem 1998;273:18,623–18,632.

20. Bennett BL, Sasaki DT, Murray BW, et al. SP600125, an anthrapyrazolone inhibitor of Jun N-terminal kinase. Proc Natl Acad Sci USA 2001;98:13,681–13,686.

21. Gillardon F, Steinlein P, Bürger E, Hildebrandt T, Gerner C. Phosphoproteome and transcriptome analysis of the neuronal response to a CDK5 inhibitor. Proteomics 2005;5: 1299–1307

22. Knebel A, Morrice N, Cohen P. A novel method to identify protein kinase substrates: eEF2 kinase is phosphorylated and inhibited by SAPK4/p38δ. EMBO J 2001;20:4360–4369.

23. Brown M, Jacobs T, Eickholdt B, et al. α2-chimaerin, cyclin-dependent kinase 5/p35, and its target collapsin response mediator protein-2 are essential components in semaphorin 3A-induced growth-cone collapse. J Neurosci 2004;24:8994–9004.

24. Sasaki Y, Cheng C, Uchida Y, et al. Fyn and Cdk5 mediate semaphorin-3A signaling, which is involved in regulation of denrite orientation in cerebral cortex. Neuron 2002;35:907–920.

25. Ledda F, Paratcha G, Ibáñez CF. Target-derived GFRα1 as an attractive guidance signal for developing sensory and sympathetic axons via activation of Cdk5. Neuron 2002;36:387–401.

26. Chua BT, Volbracht C, Tan KO, Li R, Yu VC, Li P. Mitochondrial translocation of cofilin is an early step in apoptosis induction. Nat Cell Biol 2003;5:1083–1089.

27. Bomsztyk K, Denisenko O, Ostrowski J. hnRNP K: one protein multiple processes. Bioessays 2004;26:629–638.

28. Qu D, Li Q, Lim HY, Cheung NS, Li R, Wang JH, Qi RZ. The protein SET binds the neuronal Cdk5 activator p35^{nck5a} and modulates Cdk5/p35^{nck5a} activity. J Biol Chem 2002;277: 7324–7332.

29. Gong X, Tang X, Wiedmann M, et al. Cdk5-mediated inhibition of the protective effects of transcription factor MEF2 in neurotoxicity-induced apoptosis. Neuron 2003;38:33–46.

30. Fu AKY, Fu WY, Ng AKY, et al. Cyclin-dependent kinase 5 phosphorylates signal transducer and activator of transcription 3 and regulates its transcriptional activity. Proc Natl Acad Sci USA 2004;101:6728–6733.

31. Li Z, David G, Hung KW, DePinho RA, Fu AKY, Ip NY. Cdk5/p35 phosphorylates mSds3 and regulates mSds3-mediated repression of transcription. J Biol Chem 2004;279: 54,438–54,444.

32. Sausville EA. Complexities in the development of cyclin-dependent kinase inhibitor drugs. Trends Mol Med 2002;8:S32–S37.

33. Carroll AS, O'Shea EK. Pho85 and signaling environmental conditions. Trends Biochem Sci 2002;27:87–93.

34. Samuels BA, Tsai LH. Cdk5 is a dynamo at the synapse. Nature Cell Biol 2003;5:689–690.

35. Tan TC, Valova VA, Malladi CS, et al. Cdk5 is essential for synaptic vesicle endocytosis. Nature Cell Biol 2003;5:701–710.

36. Lee SY, Wenk MR, Kim Y, Nairn AC, De Camill P. Regulation of synaptojanin 1 by cyclin-dependent kinase 5 at synapses. Proc Natl Acad Sci USA 2004;101:546–551.

37. Yamagishi S, Fujikawa N, Kohara K, Tominaga-Yoshino K, Ogura A. Increased exocytotic capability of rat cerebellar granule neurons cultured under depolarizing conditions. Neuroscience 1999;95:473–479.

38. Ryan TA. Presynaptic imaging techniques. Curr Opin Neurobiol 2001;11:544–549.

39. Tarricone C, Dhavan R, Peng J, Areces LB, Tsai LH, Musacchio A. Structure and regulation of the CDK5-p25^{nck5a} complex. Mol Cell 2001;8:657–669.

40. Manning G, Whyte DB, Maritnez R, Hunter T, Sudarsanam S. The protein kinase complement of the human genome Science 2002;298:1912–1934.

5

Critical Roles of the Raf/MEK/ERK Pathway in Apoptosis and Drug Resistance

James A. McCubrey, Fred E. Bertrand, Linda S. Steelman, Fumin Chang, David M. Terrian, and Richard A. Franklin

Summary

The Ras/Raf/MEK/ERK pathway plays a critical role in the transmission of signals from growth factor receptors to the nucleus to regulate gene expression. Components of this pathway (e.g., Ras and B-Raf) are frequently mutated in human cancer. Mutations at upstream receptors (e.g., epidermal growth factor receptor [EGFR] and Flt-3) and chimeric chromosomal translocations (e.g., BCR-ABL), which transmit their signals through the Ras/Raf/MEK/ERK cascade were also frequently observed in human cancer. This pathway also interacts with other signaling pathways (e.g., PI3K/PTEN/Akt) to regulate cell growth. In some cells, mutation of PTEN may contribute to suppression of Raf/MEK/ERK because of the ability of Akt to phosphorylate and inactivate Raf-1. Other regulatory components of the Raf/MEK/ERK pathway (e.g., the Raf Kinase Inhibitor Protein, RKIP) may display altered expression during metastasis, which lead to activation of the pathway. This chapter describes the roles of the Raf/MEK/ERK pathway in signal transduction, prevention of apoptosis and induction of drug resistance.

Key Words: Raf; MAPK; apoptosis; cell cycle; signal transduction; drug resistance.

1. Overview of Ras/Raf/MEK/ERK Signaling and its Role in Apoptosis and Drug Resistance

The Ras/Raf/MEK/ERK cascade couples signals from cell surface receptors to transcription factors, which regulate gene expression. A diagrammatic overview of the Ras/Raf/MEK/ERK pathway is presented in Fig. 1. This pathway is often activated by mutations or overexpression in upstream molecules such as BCR-ABL and epidermal growth factor receptor (EGFR) in certain tumors. The Raf/MEK/ERK pathway also has profound effects on the regulation of apoptosis by the post-translational phosphorylation of apoptotic regulatory molecules including Bad, caspase 9, and Bcl-2. Depending upon the stimulus and cell type, this pathway can transmit signals, which regulate apoptosis and cell-cycle progression (1–5). A survey of the literature reveals the daily increase in complexity in this pathway, as there are multiple members of the kinase, transcription factor, apoptotic regulator, and caspase executioner families, which can be activated or inactivated by protein phosphorylation. Raf, either through downstream MEK and ERK, or independently of MEK and ERK, can induce the phosphorylation of proteins, which control apoptosis. The diversity of signals transduced by this pathway is further increased as different Raf family members heterodimerize to transmit different signals. Furthermore, additional signal transduction pathways interact with the Raf/MEK/ERK pathway to positively or negatively regulate its activity. Abnormal

From: *Apoptosis, Cell Signaling, and Human Diseases: Molecular Mechanisms, Volume 2*
Edited by R. Srivastava © Humana Press Inc., Totowa, NJ

activation of this pathway occurs in human cancer as a result of mutations at Ras and B-Raf as well as genes in other pathways (e.g., PI3K, PTEN, Akt), which serve to regulate Raf activity. The Raf/MEK/ERK pathway also influences chemotherapeutic drug resistance as ectopic activation of Raf induces resistance to doxorubicin and paclitaxel in breast cancer cells and some mutations at B-Raf have been detected in breast cancers. For all the above reasons, the Raf/MEK/ERK pathway is an important pathway to target for therapeutic intervention. Inhibitors of Ras, Raf, MEK and some downstream targets have been developed and many are currently in clinical trials. This chapter will summarize our current understanding of the Ras/Raf/MEK/ERK signal transduction pathway and other interacting signaling and apoptotic pathways.

2. Ras and its Role in the Raf/MEK/ERK Kinase Cascade

Ras is a small GTP-binding protein, which is the common upstream molecule of several signaling pathways including Raf/MEK/ERK, PI3K/Akt and RalEGF/Ral *(6)*. So far three Ras proteins have been identified, namely Ha-Ras, Ki-Ras, and N-Ras. Ras proteins show varying abilities to activate the Raf/MEK/ERK and PI3K/Akt cascades. For example, Ki-Ras has been associated with the Raf/MEK/ERK pathway whereas Ha-Ras is associated with PI3K/Akt activation *(7)*. Different mutation frequencies have been observed between Ras genes in human cancer (Ki-Ras > Ha-Ras).

For Ras to be targeted to the cell membrane, it must be farnesylated by farnesyl transferase (Ha-, Ki-, and N-Ras) or geranylgeranylated by geranylgeranyl transferase (N-and Ki-Ras). Farnesylation and geranylgeranylation both occur on the same cysteine residue. Ras preferentially undergoes farnesylation, however, in the presence of farnesylation inhibitors N-Ras and Ki-Ras can undergo gernylgernylation. Farnesylation and geranylgeranylation are important for targeting Ras to the cell membrane. Ha-Ras and N-Ras can also undergo palmitoylation with Ha-Ras having two palmitoylation sites and N-Ras having one palmitoylation site. Ki-Ras appears to lack a palmitylation site. It is believed that palmitoylation plays a role in plasma membrane microlocalization. Following binding of cytokines, growth factors or mitogens to their appropriate receptors activation of the coupling complex Shc/Grb2/Sos occurs. Upon stimulation by Shc/Grb2/SOS, the inactive Ras exchanges GDP for GTP and undergoes a conformational change and becomes active. The GTP bound active Ras can then recruit Raf to cell membrane (*see* Fig. 1).

3. Raf Activation and Inhibition

Raf is a serine/threonine (S/T) kinase and is normally activated by a complex series of events including: (1) recruitment to the plasma membrane mediated by an interaction with Ras *(7)*; (2) dimerization of Raf proteins *(8)*; (3) phosphorylation/dephosphorylation on different domains *(9)*; (4) disassociation with the Raf kinase inhibitory protein (RKIP) and, (5) association with scaffolding complexes (e.g., kinase suppressor of Ras, [KSR]) (*see* Fig. 2). Raf activity is further modulated by adaptor proteins including Bag1 and 14-3-3 *(10)*. Inhibition of Ras activity hinders Raf activation.

4. Regulation of Raf Activity by Phosphorylation

The mammalian Raf family consists of A-Raf, B-Raf, and Raf-1 (C-Raf). These proteins have three conserved regions, which have been termed CR1, CR2, and CR3 *(4)*. Among these three domains, CR1 contains the Ras binding site; CR2 contains the regulatory domain and CR3 contains the kinase domain.

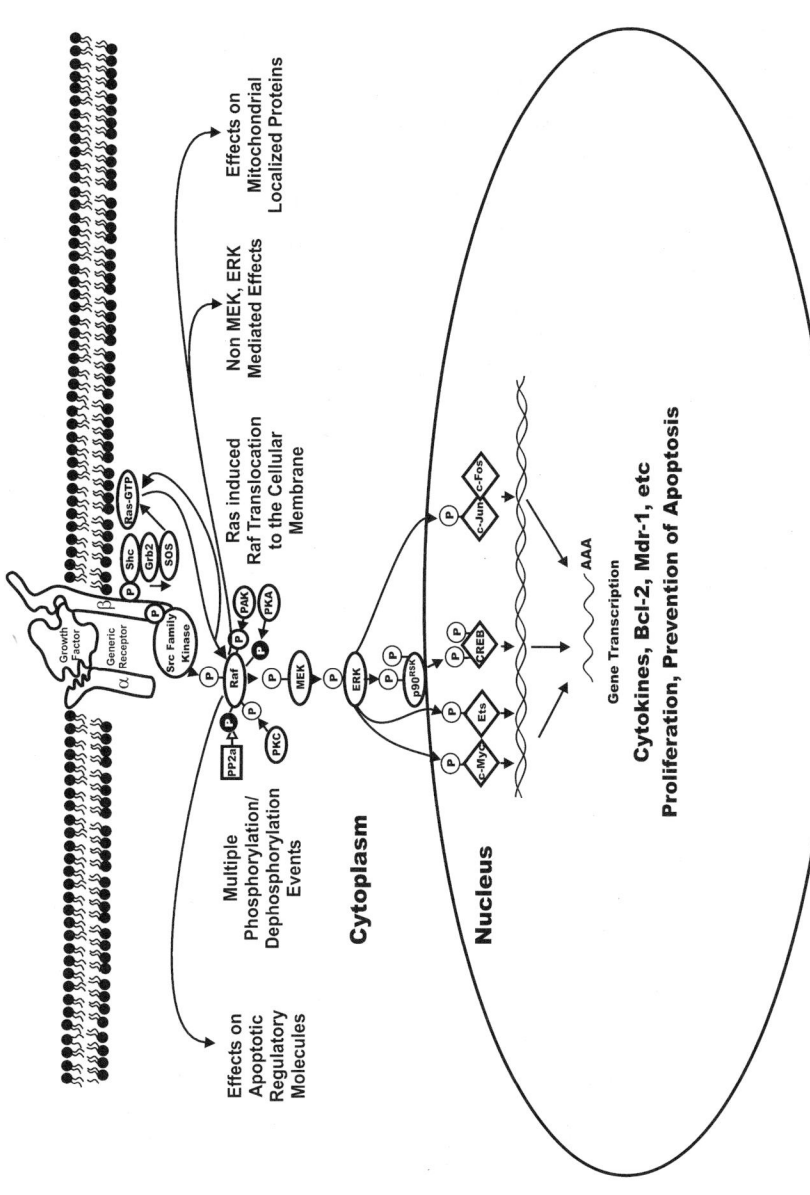

Fig. 1. Overview of Raf/MEK/ERK Pathway. This figure illustrates how the Raf/MEK/ERK pathway is regulated by Ras as well as various kinases, which serve to phosphorylate S/T and Y residues on Raf. Some of these phosphorylation events serve to enhance Raf activity (shown by a black P in a white circle) whereas others serve to inhibit Raf activity (shown by a white P in a black circle. Moreover there are phosphatases such as PP2A, which remove phosphates on certain regulatory residues. The downstream transcription factors regulated by this pathway are indicated in diamond shaped outlines.

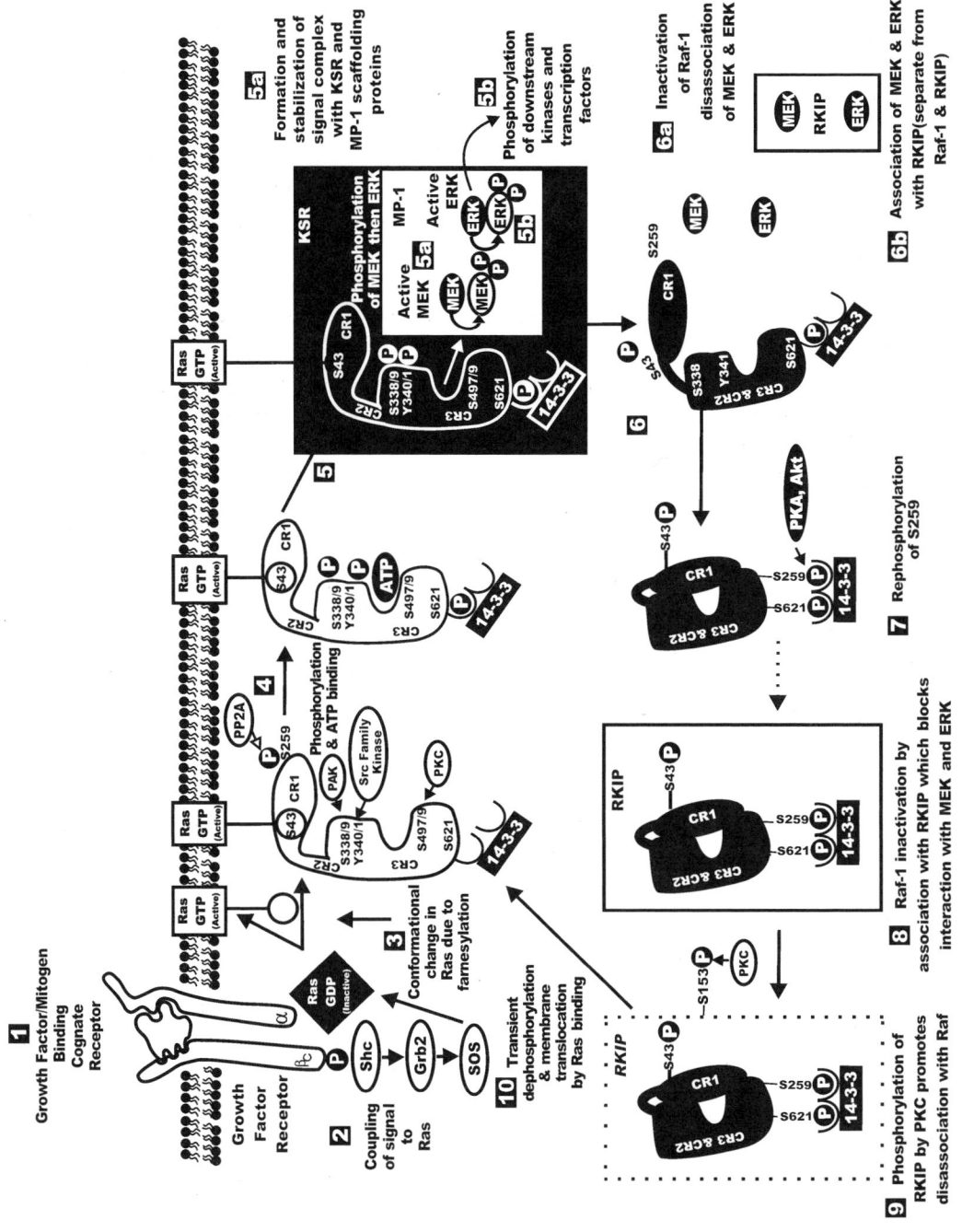

There are at least thirteen regulatory phosphorylation sites on Raf-1 *(1,10,11)*. Some of these sites (e.g., S43, S259 and S621) are phosphorylated when Raf-1 is inactive. This allows the 14-3-3 chaperonin proteins to bind Raf-1 and confer a formation which is inactive *(see* Fig. 2). Upon cell stimulation, S621 becomes transiently dephosphorylated by a phosphatase not yet identified. Phosphatases such as protein phosphatase 2A (PP2A) dephosphorylate S259 *(11)*. 14-3-3 then disassociates from Raf-1. This allows Raf-1 to be phosphorylated at S338, Y340, and Y341, which renders Raf-1 active. A Src family kinase is likely responsible for phosphorylation at Y340 and Y341 *(10–12)*. The phosphatases, which dephosphorylate S621 and other Raf phosphorylation sites, excluding S259, are unknown.

Y340 and Y341 on Raf-1, the phosphorylation targets of Src family kinases, are conserved in A-Raf (Y299 and Y300), but are replaced with aspartic acid (D) at the corresponding positions in B-Raf (D492 and D493) *(10,13)*. The negatively charged aspartic acid residues mimic activated residues, which makes B-Raf highly active. Maximal activation of Raf-1 and A-Raf requires both Ras and Src activity whereas B-Raf activation is Src-independent *(14)*. Interestingly, as will be discussed later, more mutations at B-Raf, than either Raf-1 or A-Raf, have been detected in human cancer. This may have resulted from a simpler mode of activation and selection of cells containing B-Raf mutations than either Raf-1 or A-Raf mutations.

The S338 residue present in Raf-1 is conserved among the three Raf isoforms, however, in B-Raf (S445), this corresponding site is constitutively phosphorylated *(15)*. S338 phosphorylation on Raf-1 is stimulated by Ras and is dependent on p21-activated protein kinase (PAK) *(16)*. Other phosphorylation sites in Raf-1 that may modulate its activity include: S43, S339, T491, S494, S497, S499, S619, and S621. Protein kinase C (PKC) has been shown to activate Raf and induce crosstalk between PKC and Raf/MEK/ERK signaling pathways *(17)*. S497 and S499 were identified as the target residues on Raf-1 for PKC phosphorylation *(17)*. However, other studies suggest that these sites are not necessary for Raf-1 activation *(18)*.

Fig. 2. (*Opposite page*) Regulation of Raf Activity by Phosphorylation, Ras Binding and Complex Formation. Activation of Ras occurs after receptor ligation and results in the posttranslational modification of Ras and membrane translocation. Ras recruits Raf-1 to the membrane by binding the Ras binding domain present on Raf-1. Also occurring at this time is the transient dephosphorylation of S621 present on Raf-1. The S259 present on Raf-1 is then dephosphorylated by PP2A. This allows Raf-1 to be phosphorylated and activated by other kinases (PAK, Src family kinases, and potentially PKC). Raf-1 then binds ATP and phosphorylates MEK, which in turn phosphorylates ERK. Raf-1 is then inactivated by protein dephosphorylation and binds 14-3-3. This results in a conformational change and Raf-1 is translocated to the cytoplasm and is inactive. Raf-1 also associates with RKIP which results in its inactivation. MEK and ERK also interact with RKIP which prevents this interaction with Raf-1. RKIP is phosphorylated by PKC which results in the disassociation of Raf-1 from RKIP and activation of Raf-1. The phosphorylation/dephosphorylation events alter the configuration of Raf-1 and can result in the disassociation of the 14-3-3 protein, which unlocks the Raf-1 protein allowing it to be phosphorylated by other kinases. The binding of the chaperonin protein 14-3-3 and the subsequent conformational changes and translocation to the cytoplasm are indicated. This figure is based in part by the models proposed by Dr. Walter Kolch *(11,17,20,23)*.

Raf activity is also negatively regulated by phosphorylation on the CR2 regulatory domain. Akt and protein kinase A (PKA) phosphorylate S259 on Raf-1 and inhibit its activity *(19,20)*. Furthermore, Akt or the related serum/glucocorticoid regulated kinase (SGK) phosphorylate B-Raf on S364 and S428 and inactivate its kinase activity *(21,22)*. These S-phosphorylated Rafs associate with 14-3-3 and become inactive. This inhibitory effect of Akt on Raf activity is cell type-specific and may depend on the differentiation state of the cells *(19)*. It was suggested that some differentially expressed mediators are essential for the association between Akt and Raf.

Recently a scaffolding protein, Raf Kinase Inhibitory Protein (RKIP) has been shown to inhibit Raf-1 activity *(23)*. RKIP is a member of the phosphatidylethanolamine-binding protein (PEBP) family. This multi-gene family is evolutionarily conserved and has related members in bacteria, plants and animals *(24)*. Interesting RKIP can bind Raf or MEK/ERK complex but not to Raf, MEK and ERK together. Various isoforms of PKC have been shown to phosphorylate RKIP on S153 that results in the disassociation of Raf and RKIP. Subsequently RKIP binds the G-protein-coupled receptor kinase 2 (GRK2) and inhibits its activity *(24)*. The role of RKIP in metastasis will be discussed in Section 17.

The importance of Raf-1 in the Raf/MEK/ERK signal transduction pathway has come into question as a result of the discovery that B-Raf was a much more potent activator of MEK compared with Raf-1 and A-Raf. Many of the "functions" of Raf-1 still persist in Raf-1 knock-out mice likely as a result of function of endogenous B-Raf *(25)*. Interestingly and controversially it has recently been proposed that B-Raf is not only the major activator of MEK1, but B-Raf is also required for Raf-1 activation. Furthermore, B-Raf may be temporally activated before Raf-1. However, there may be different subcellular localizations of B-Raf and Raf-1 within the cell that exert different roles in signaling and apoptotic pathways *(26)*. In some cases, B-Raf may transduce its signal through Raf-1. The reasons for this added step in the kinase cascade are not obvious but may represent another layer of fine tuning.

Raf-1 plays important roles in promoting cell-cycle progression and preventing apoptosis. Recently, Raf-1 has been postulated to have nonenzymatic functions and serve as a docking protein. An excellent summary of these non-MEK/ERK related activities of Raf-1 is presented in the review by Hindley and Kolch *(27)*. Raf-1 has been proposed to have important functions at the mitochondrial membrane. Expression of membrane targeted Raf was shown to complement a BCR-ABL mutant in abrogating the cytokine-dependence of hematopoietic cells *(28)*. In these mutant BCR-ABL transfected cells, Bad was expressed in the hyperphosphorylated inactive form and released from the mitochondria into the cytosol. In contrast, in the cells containing the BCR-ABL mutant but lacking the membrane-targeted Raf-1, which were not cytokine-dependent, Bad was hypophosphorylated and present in the mitochondrial fraction. BCR-ABL may interact with mitochondrial targeted Raf-1 to alter the phosphorylation of Bad at the mitochondrial membrane and hence regulate (prevent) apoptosis in hematopoietic cells containing the BCR-ABL chromosomal translocation *(29)*. This survival mechanism is independent of MEK and ERK.

5. Downstream of Raf Lies MEK1

Mitogen-activated protein kinase/ERK kinase (MEK1) is a tyrosine (Y-) and S/T-dual specificity protein kinase *(30,31)*. Its activity is positively regulated by Raf phosphorylation on S residues in the catalytic domain. All three Raf family members are able to phosphorylate and activate MEK but different biochemical potencies have been observed (B-Raf > Raf-1>> A-Raf) *(30,31)*. Although MEK1 is a key kinase in the Raf/MEK/ERK cascade, it does not appear to be frequently mutated in human cancer. Activated mutants can be constructed which will abrogate the cytokine-dependence of hematopoietic cells and morphologically transform NIH-3T3 cells *(1)*. Another interesting aspect regarding MEK1 is that it's predominate downstream target is ERK. In contrast, both upstream Raf and downstream ERK appear to have multiple targets. Thus, therapeutic targeting of MEK1 is relatively specific.

6. Downstream of MEK1 Lies ERK

Extracellular-signal-regulated kinases 1,2 (ERK), are S/T kinases and their activities are positively regulated by phosphorylation mediated by MEK1 and MEK2. ERKs can directly phosphorylate many transcription factors including Ets-1, c-Jun and c-Myc. ERK can also phosphorylate and activate the 90 kDa ribosomal six kinase (p90[Rsk]), which then leads to the activation of the transcription factor CREB *(32)* (*see* Fig. 1). Moreover, through an indirect mechanism, ERK can lead to activation of the NF-κB transcription factor (nuclear factor immunoglobulin κ-chain enhancer-B cell) by phosphorylating and activating inhibitor κB kinase (IKK) (see below). ERK1 and ERK2 are differentially regulated. ERK2 has been positively associated with proliferation while ERK1 may inhibit the effects of ERK2 in certain cells *(33)*.

7. The Proliferative and Anti-Proliferative Effects of Ras/Raf/MEK/ERK Signaling

Amplification of *ras* proto-oncogenes and activating mutations that lead to the expression of constitutively-active Ras proteins are observed in approximately 30% of all human cancers *(34,35)*. B-Raf is mutated in approx 7% of all cancers *(36)*. The effects of Ras on proliferation and tumorigenesis have been documented in immortal cell lines *(37)*. However, antiproliferative responses of oncogenic Ras have also been observed in nontransformed fibroblasts, primary rat Schwann cells and primary fibroblast cells of human and murine origins *(38)*. Ras and its downstream effector molecules affect the expression of many molecules which regulate cell cycle including p16[Ink4a], p15[Ink4b], and p21[Cip1], and can lead to premature cell-cycle arrest at the G1 phase. This p15[Ink4b]/p16[Ink4a] or p21[Cip1]-mediated premature G_1 arrest and subsequent senescence is dependent on the Raf/MEK/ERK pathway *(39–41)*.

Overexpression of activated Raf proteins is associated with such divergent responses as cell growth, cell-cycle arrest or even apoptosis *(42–45)*. The fate of the cells depends on the level and isoform of Raf kinase expressed. Ectopic overexpression of Raf proteins is associated with cell proliferation in cells including hematopoietic cells *(42)* erythroid progenitor cells *(46)* and A10 smooth muscle cells *(47)*. However, overexpression of activated Raf proteins is associated with cell-cycle arrest in rat Schwann cells, mouse PC12 cells, human promyelocytic leukemia HL-60 cells, small cell lung cancer cell lines, and some hematopoietic cells *(48–50)*. Depending on the Raf isoform, overexpression of Raf can lead to cell proliferation or cell growth arrest in NIH-3T3 fibroblast and FDC-P1 hematopoietic cells. It is not clear why overexpression of the Raf gene can lead to such conflicting results, but it has been suggested that the opposite outcomes may be determined by the amount or activity of the particular Raf oncoprotein *(43,45)*.

NIH-3T3 cells have been transfected with the three different Raf genes. The introduced A-Raf molecule was able to upregulate the expression of cyclin D_1, cyclin E, Cdk2, and Cdk4 and down-regulate the expression of Cdk inhibitor p27^{Kip1} *(51)*. These changes induced the cells to pass through G_1 phase and enter S phase. It should be remembered that A-Raf is the weakest Raf kinase and its role in cell proliferation is not clear. However, in B-Raf and Raf-1 transfected NIH-3T3 cells, there was also a significant induction of p21^{Cip1}, which led to G_1 arrest. Using cytokine-dependent FDC-P1 hematopoietic cells transfected with conditionally-active mutant Raf-1, A-Raf, and B-Raf genes as a model, we have demonstrated that moderate Raf activation, such as A-Raf and Raf-1, led to cell proliferation, which was associated with the induction of cyclin expression and Cdks activity. However, ectopic expression of the much more potent B-Raf could lead to apoptosis *(42–45)*.

An alternative explanation for the diverse proliferative results obtained with the three Raf genes is the different biological effects of A-Raf, B-Raf, and Raf-1. The individual functions of these three different Raf proteins are not fully understood. Even though it has been shown that all three Raf proteins are activated by oncogenic Ras *(14,52–56)* target the same downstream molecules (i.e., MEK1 and MEK2) *(21,57–59)*, and use the same adapter protein (14-3-3) for conformational stabilization *(60)*, different biological and biochemical properties among them have been reported and their functions are not always compensatable *(59,61–66)*. It is safe to say that even as we learn more about the intricacies of these Raf molecules, we discover that there are more questions regarding their specificities and mechanism of activation.

The cellular distribution of the different Raf proteins in mice is very diverse. The Raf-1 protein is expressed ubiquitously whereas A-Raf is predominately expressed in urogenital tissues. B-Raf was originally shown to be expressed in neuronal tissues and hematopoietic cells *(63,65–67)*. However, as stated previously, B-Raf is now thought to be the most important activator of the Raf/MEK/ERK cascade and recognition of its cellular distribution in normal and neoplastic samples has increased *(36)*. This is most likely the result of improved ability to detect expression and mutation of the different Raf isoforms.

Knockout mice with a homologous deletion of these Raf genes have revealed very different phenotypes *(64,66,67)*. B-Raf$^{-/-}$ mice died embryonically with serious defects in vascular endothelial cell survival and differentiation indicated by an increased number of endothelial cell precursors and apoptotic cells in vascular endothelium *(66)*. A-Raf$^{-/-}$ mice showed gastro-intestinal and neurological defects and died shortly after birth *(63)*. Raf-1$^{-/-}$ mice also died embryonically and showed defects in the development of skin, lung, and placenta *(66)*. However, in both A-Raf$^{-/-}$ and Raf-1$^{-/-}$ mice no significant apoptotic cells were observed, likely resulting from the presence of functional B-Raf in these cells *(57)*.

The abilities of the Ras isoforms to activate the three Raf molecules are different *(14,54,56,68)*. Ras and Src stimulate the kinase activity of both Raf-1 and A-Raf. However, B-Raf kinase activity is controlled by both Ras and other members of the small G-protein family, such as Rap1 *(14,69,70)* or TC21 *(14,70)*. Raf-isoform-specific interaction partners have been identified using three different Raf molecules as bait in yeast two-hybrid screens. PA28α, a subunit of the 11 S regulator of proteasomes, binds with B-Raf but not Raf-1 or A-Raf *(71)*. Both CK2β (the regulatory subunit of protein kinase CK2) and pyruvate kinase M2 are A-Raf-specific interaction partners *(67,72)*. RKIP may display differential abilities to bind Raf-1, B-Raf and A-Raf. Although RKIP binds and inhibits Raf-1, it is not clear whether RKIP binds and inhibits B-Raf and A-Raf. Recently it was shown by proteomic analysis of Raf-1 signaling complexes that the <u>M</u>ammalian <u>S</u>terile 20-like kinase (MST-2) binds Raf-1 but not B-Raf *(73)*. Thus although the Raf proteins are related and often called "isoforms," they exhibit different modes of activation and different interaction patterns. Hence, caution must be applied when using the word "isoform."

8. Downstream Transcription Factor Targets of the Ras/Raf/MEK/ERK Pathway

The Ras/Raf/MEK/ERK signaling pathway can exert proliferative or antiproliferative effects through downstream transcription factor targets including NF-κB, CREB, Ets-1, AP-1, and c-Myc. ERKs can directly phosphorylate Ets-1, AP-1, and c-Myc, which lead to their activation *(74–142)* (*see* Fig. 3). These transcription factors induce the expression of genes important for cell-cycle progression, prevention of apoptosis and regulation of drug resistance. However, under certain circumstances, strong Raf signaling has been shown to result in the inactivation of downstream transcription factors *(99)*, including NF-κB and c-Myc, which may account for the Raf-induced antiproliferative responses observed in some studies.

9. Raf Signaling and NF-κB

NF-κB is a dimeric transcription factor comprised of members of the Rel gene family of DNA-binding proteins including RelA, RelB, p105/NF-κB1, and p100/NF-κB2. Many target genes of NF-κB play important roles in proliferation, prevention of apoptosis, angiogenesis, metastasis, and immune responses *(74–80)*. NF-κB can modulate the expression of its target genes encoding: cytokines, growth factors, other transcription factors, and regulators of cell proliferation and apoptosis by binding to the κB cis-acting element contained in their promoter regions *(74–76)*.

Raf activation induces the expression of reporter genes driven by the NF-κB promoter *(79,80)*. Raf activates NF-κB through two pathways. First, Raf can activate NF-κB in a rapid and direct fashion, which involves the activation of <u>M</u>itogen-activated protein kinase/<u>E</u>RK <u>K</u>inase <u>K</u>inase-1 (MEKK1) and it's target the <u>I</u>nhibitor of <u>κ</u>B <u>K</u>inase β (IKKβ) *(80–83)*. Second, Raf may also activate NF-κB through an autocrine loop in certain cell types *(42,80,83)*. Earlier work had suggested a possible role of p90[Rsk] in the activation of NF-κB because in vitro studies showed that p90[Rsk] could phosphorylate IκB on S32, which lead to IκB degradation *(84,85)*. Further studies indicated that Raf-mediated NF-κB activation is independent of the MEK/ERK/RSK pathway, but dependent on MEKK1 *(81)*. This is another example of a Raf-dependent event, which is independent of MEK/ERK activity. Raf-induced MEKK1 preferentially activates IKKβ and has little effect on IKKα *(81,86)*. This activation is possibly through direct phosphorylation *(87)*. However, Raf inhibits NF-κB activity in the human glioblastoma cell line, T98G, through the MEK/ERK pathway. So, the Raf/MEK/ERK pathway may in some circumstances exert a negative regulatory role in NF-κB activity *(88)*.

10. Raf Signaling and CREB

The <u>C</u>yclic AMP-<u>R</u>esponsive <u>E</u>lement-<u>B</u>inding protein (CREB) was initially identified as a leucine zipper transcription factor that functions as a regulatory effector of the cAMP signaling pathway. CREB also plays important roles in signal transduction initiated

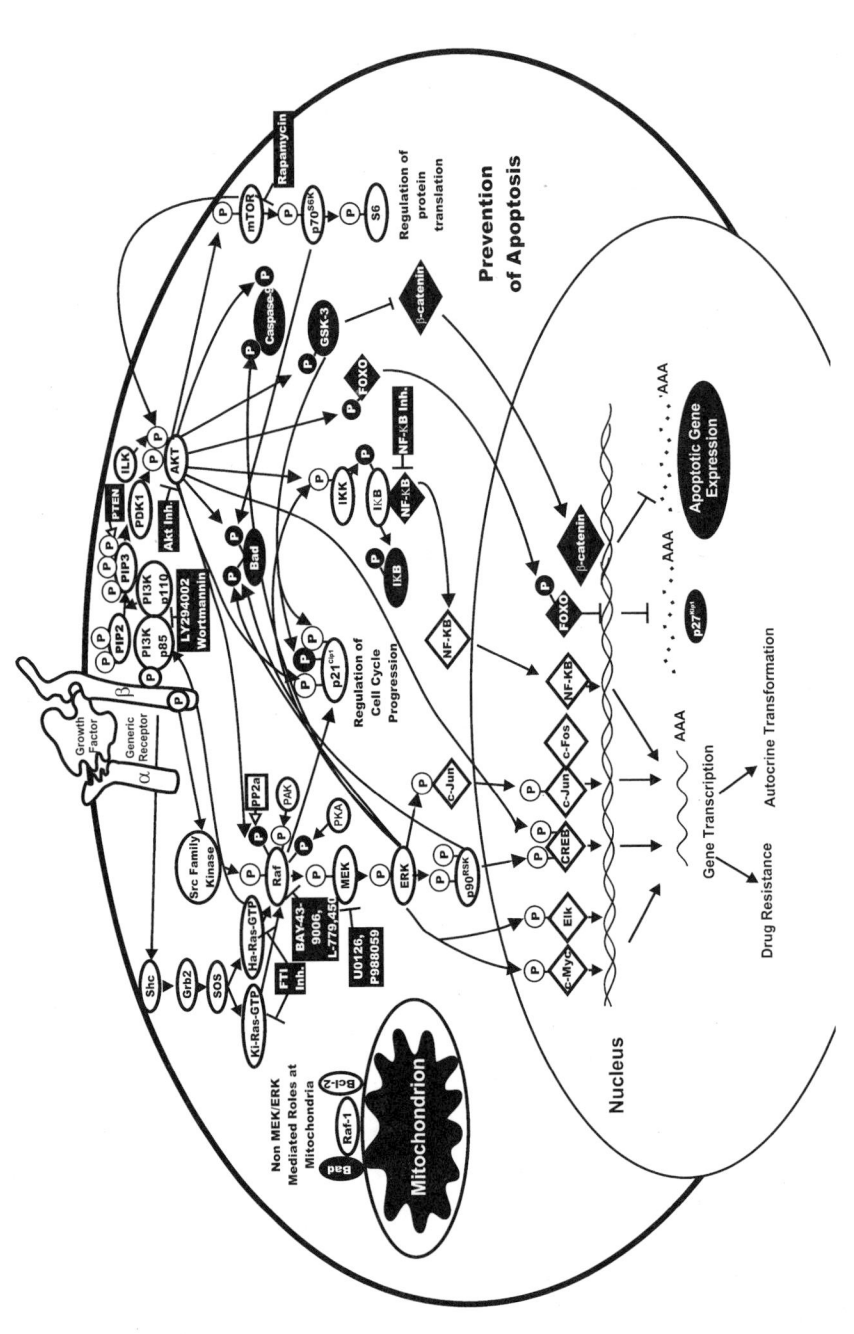

Fig. 3. Interactions between the Raf/MEK/ERK and PI3K/Akt pathways and sites of action of small molecular weight signal transduction pathway inhibitors. Regulatory signals are transduced to either or both Ki-Ras and Ha-Ras, which results in the activation of Raf or PI3K or both. Akt can regulate the activity of both Raf-1 and B-Raf. This can result in the inactivation of both of these kinases, which would suppress their affects on p53 and p21[Cip1], and the induction of cell-cycle arrest. In normal cells, cell proliferation is tightly balanced with apoptosis. Uncontrolled cell proliferation and prevention of apoptosis results in breakdown of this balance. Some of the sites targeted by small molecular weight inhibitors are illustrated.

by Ca^{++}, growth factors, and stress signals *(89,90)*. CREB regulates angiotensins, cell growth, long-term memory, and immune functions. CREB/ATF is a gene family of basic leucine zipper DNA binding proteins, including CREB, CREM, ATF-1, ATF-2, ATF-3, and ATF-4, that can form homodimers, heterodimers, or even in some cases cross-family heterodimers with other leucine zipper DNA binding transcription factors in the Jun family *(91,92)*.

Protein kinases including PKA, Ca^{2+}-calmodulin-dependent kinase IV (CaMKIV), $p70^{S6K}$ and $p90^{Rsk}$ phosphorylate *CREB (93–95)*. All these kinases target CREB on S133 to activate CREB. Raf signaling activates CREB in several cell types with different treatments that lead to a mitogenic effect. This is believed to result from the direct phosphorylation of CREB by RSKs. RSK2 was shown to mediate growth factor induction of phosphorylation of CREB on S133 both in vivo and in vitro.

Genes regulated by the CREB/ATF transcription factors include cyclin A *(96)*, cyclin D_1 *(97)*, c-Fos *(98)*, and Bcl-2 *(99)*. Cell-cycle arrest mediated by transforming growth factor (TGF)-β occurs through the down-regulation of cyclin A via CREB inactivation *(96)*. TGF-β reduced CREB protein levels and stimulated its dephosphorylation. Transcription factors such as serum response factor (SRF) and Elk-1 contribute to the c-Fos induction mediated by growth factors *(100)*.

11. Raf Signaling and Ets-1

Ets is a family of transcription factors, which include Ets-1, Ets-2, Elk-1, SAP1, SAP2, E1AF, PEA3, PU1, ERF, NET, YAN, and TEL *(101)*. They share an 85 amino acid sequence called the Ets DNA binding domain. The Ets transcription factors regulate many genes including: transcription factor [p53 *(102)*, c-Fos *(103)*, and NF-κB *(104)*]; cell-cycle regulatory [cyclin D_1 *(105)*, Rb *(106)*, and p21^{Cip1} *(107)*]; apoptosis-related [Bcl-2 *(108)*, Bcl-X_L *(109)*, and Fas *(110)*]; cytokine [GM-CSF and IL-3 *(4,111)*]; and growth factor [platelet-derived growth factor (PDGF) *(112)* and heparin-binding epidermal growth factor (HB-EGF) *(113)*].

Raf-signaling regulates Ets through direct phosphorylation by ERK *(69,96,114,115)*. Elk-1 is a target for all three MAPK pathways (i.e., p38^{MAPK}, JNK and ERK pathways). However, different residues of Elk-1 are phosphorylated by the three different kinases. ERK-induced Elk-1 phosphorylation leads to enhanced DNA binding and transcriptional activation *(100,116–118)*. One of the best-characterized target genes for Elk-1 is c-Fos. Elk-1, after phosphorylation by ERK, binds to the SRE cis-acting element in the promoter region of *c-fos* and induces its transcription *(97,119,120)*.

ERK can physically associate with ETS2 repressor factor (ERF) and phosphorylate ERF on multiple sites *(114,121)*. Phosphorylated ERF is exported from the nucleus into the cytoplasm and becomes inactive. The transportation-inactivation of ERF is eliminated when ERK activity is inhibited *(114)*. ERF mutants with the ERK phosphorylation sites mutated to alanine (A) are insensitive to ERK activation and block Ras-induced transformation of NIH-3T3 cells *(114)*. Interestingly, the TEL gene, which is engaged in many chromosomal translocations (e.g., TEL-JAK, TEL-PDGFR and TEL-AML), is a member of the ERF family.

Ets-binding sequences have been identified in the promoter regions of many growth factor genes including granulocyte macrophage-colony stimulating factor (GM-CSF), interleukin (IL)-3, PDGF, and HB-EGF. This suggests the autocrine loop induced by Raf signaling may be mediated by Ets transcription factors. However, Ets binding sequences are also present in the promoter regions of genes that may inhibit cell-cycle progression such as the Cdk Kinase Inhibitors (CKI) p21[Cip1] and p16[Ink4a] *(107,121,122)* and the tumor suppressor gene p53 *(102)*. Thus the ETS transcription factors may stimulate or prevent cell-cycle progression depending on the presence of specific cellular signals.

12. Raf Signaling and AP-1

Regulation of c-Fos activity by Raf signaling is mainly at the transcriptional level through the activation of transcription factors Elk-1 and CREB. Phosphorylated Elk-1 and CREB bind to the SRE and CRE cis-acting elements in the c-*fos* promoter to induce its transcription *(119,120,123)*.

Regulation of c-Jun activity by Raf signaling occurs at both the transcriptional and post-translational levels. Cytokine induction of c-*jun* transcription is ERK-dependent *(124–126)*. c-Fos forms heterodimers with c-Jun and bind the AP-1 site on the c-*jun* promoter region inducing its transcription *(124)*. c-Jun activity is regulated by phosphorylation on the amino-terminal transactivating domain and the carboxyl-terminal DNA binding domain. Phosphorylation on the carboxyl-terminal domain significantly reduces its DNA binding ability and inactivates c-Jun transactivation ability. In contrast, phosphorylation at S63 and S73 in the transactivation domain at the amino-terminal activates c-Jun. Activation of c-Jun can occur after phosphorylation mediated by Jun N-terminal Kinase (JNK) *(127)*. ERK also phosphorylates Jun in the transactivating domain, and also S243 in the carboxyl DNA binding domain of c-Jun to positively or negatively regulate AP-1 activity *(127–130)*.

13. Raf Signaling and c-Myc

c-Myc is a transcription factor that regulates genes whose functions are associated with growth *(131–136)*. Deregulated c-Myc expression can either stimulate proliferation or apoptosis in response to growth promoting or inhibitory signals respectively *(131,136)*. Moderate c-Myc expression leads to cell-cycle progression while strong c-Myc expression can lead to apoptosis *(131–136)*.

c-Myc can be phosphorylated by ERK and other kinases *(129,137–140)*. ERK can phosphorylate c-Myc on two residues S62 or T58. Phosphorylation at S62 leads to increased c-Myc transactivation activity *(134,137–139)*. T58 phosphorylation facilitates rapid c-Myc proteolysis through the ubiquitin-proteosome pathway *(140–142)*. Phosphorylation on S62 significantly increases the c-Myc half-life *(137)*. In summary, ERK phosphorylates c-Myc at S62 or T58 to positively or negatively regulate its transactivation activity.

14. Raf Signaling and the Suppression of Apoptosis

Clearly Raf has many roles in kinase cascades and downstream transcription factors which regulate apoptosis. The Raf/MEK/ERK cascade also has direct roles on key molecules involved in the prevention of apoptosis. For many years now, it has been known that the Raf/MEK/ERK pathway can phosphorylate Bad on S112 which contributes to its inactivation and subsequent sequestration by 14-3-3 proteins *(143)*. This allows Bcl-2 to form homodimers and an antiapoptotic response is generated. Recently it has been shown that the Raf/MEK/ERK cascade can phosphorylate caspase 9 on residue T125 which contributes to the inactivation of this protein *(144)*. Interesting, both Bad and caspase 9 are also phosphorylated, on different residues, by the Akt pathway indicating that the Raf/MEK/ERK and PI3K/Akt pathways can crosstalk and result in the prevention of apoptosis *(145)*. More controversially, Bcl-2 is also phosphorylated by the Raf/MEK/ERK cascade on certain residues, in the loop region, which have also been associated with enhanced antiapoptotic activity *(146,147)*. As noted earlier, Raf has MEK- and ERK- independent functions at the mitochondrial membrane. For example, mitochondrial localized Raf can phosphorylate Bad, which results in its disassociation from the mitochondrial membrane *(28,29)*.

Recently Raf-1 was shown to interact with Mammalian Sterile 20-like kinase (MST-2) and prevent its dimerization and activation *(73)*. MST-2 is a kinase, which is activated by proapoptotoic agents such as staurosporine and Fas ligand. Raf-1 but not B-Raf binds MST-2 and depletion of MST-2 from Raf-1$^{-/-}$ cells abrogated sensitivity to apoptosis and overexpression of MST-2 increased sensitivity to apoptosis. It was proposed that Raf-1 might control MST-2 by sequestering it into an inactive complex. This complex of Raf-1:MST-2 is independent of MEK and downstream ERK. Raf-1 can also interact with the Apoptosis Signal Related Kinase (ASK1) to inhibit apoptosis *(148)*. ASK1 is a general mediator of apoptosis and it is induced in response to a variety of death signals. It is induced in response to various cytotoxic stresses including TNF, Fas and reactive oxygen species (ROS). ASK1 appears to be involved in the activation of the JNK and p38 MAP kinases. This appears to be an interaction which is independent of MEK and ERK.

15. Raf-Induced Growth Factor Expression

A common feature of cells transformed by Raf is the expression of growth factors, which often have an autocrine effect. NIH-3T3 cells transformed by activated Raf secrete Heparin Binding Epidermal Growth Factor (hbEGF) *(83)*. FDC-P1 and TF-1 hematopoietic cells transformed by activated Raf genes often express GM-CSF, which has autocrine growth factor effects *(42,53,83,113,149–154)*. Kaposi's sarcoma transformed B-cells, which express elevated levels of B-Raf, express high levels of Vascular Endothelial Growth Factor (VEGF) *(155)*. Recently it has been shown that B-Raf increases the infectivity of Kaposi's Sarcoma Virus *(156,157)*. One mechanism responsible for this enhancement of viral infection by B-Raf is its ability to induce VEGF expression *(158,159)*. Many growth factor genes contain in their promoter regions binding sites for transcription factors phosphorylated by the Raf/MEK/ERK pathway *(4)*. Thus aberrant Raf expression may establish an autocrine loop, which results in the continuous stimulation of cell growth. Alternatively, the VEGF expression induced by Raf can promote angiogenesis. Raf-induced growth factor expression will contribute to both the prevention of apoptosis as well as chemotherapeutic drug resistance as growth factor expression has been associated with both the prevention of apoptosis and drug resistance *(161)*.

16. B-Raf and Human Cancer

For many years, the Raf oncogenes were not thought to be frequently mutated in human cancer and all attention to abnormal activation of this pathway was dedicated to Ras mutations. However, recently it was shown that B-Raf is frequently mutated in certain types of cancer, especially melanoma (27–70%), papillary thyroid cancer (36–53%), colorectal cancer (5–22%) and ovarian cancer (30%) *(36,162–166)*. The reasons for mutation at B-Raf and not Raf-1 or A-Raf are not entirely clear. Based on the mechanism of activation of B-Raf, it may be easier to select for B-Raf than either Raf-1 or A-Raf mutations. Activation of B-Raf would require one genetic mutation while activation Raf-1 and A-Raf would require two genetic events. It has been proposed recently that the structure of B-Raf, Raf-1 and A-Raf may dictate the ability of mutations to occur at these genes, which can permit the selection of activated oncogenic forms *(36,167)*. These predictions have arisen from determining the crystal structure of B-Raf *(168)*. Like many enzymes, B-Raf is proposed to have small and large lobes, which are separated, by a catalytic cleft *(168)*. The structural and catalytic domains of B-Raf and the importance of the size and positioning of the small lobe, may be critical in its ability to be stabilized by certain activating mutations. In contrast, the precise substitutions in A-Raf and Raf-1 are not predicted to result in small lobe stabilization thus preventing the selection of mutations at A-Raf and Raf-1, which would result in activated oncogenes *(36,166)*.

The most common B-Raf mutation is a change at nucleotide 1796 that converts a valine to a glutamic acid (V599E) *(36)*. This B-Raf mutation accounts for over 90% of the B-Raf mutations found in melanoma and thyroid cancer. It has been proposed that B-Raf mutations may occur in certain cells, which express high levels B-Raf as a result of hormonal stimulation. Certain hormonal signaling will elevate intracellular cAMP levels, which result in B-Raf activation, which leads to proliferation. Melanocytes and thyrocytes are two such cell types that have elevated B-Raf expression as they are often stimulated by the appropriate hormones *(36,169)*. Moreover, it has been proposed recently that B-Raf is the more important kinase in the Raf/MEK/ERK cascade *(1,36)*, thus mutation at B-Raf activates downstream MEK and ERK. In some models wild-type and mutant B-Raf activates Raf-1, which in turn activates MEK and ERK *(36)*.

In some cells, B-Raf mutations are believed to be initiating events and not sufficient for full-blown neoplastic transformation *(170,171)*. Moreover, there appears to be cases where certain B-Raf mutations (V599E) and Ras mutations are not permitted in the transformation process as they might result in hyperactivation of Raf/MEK/ERK signaling and expression, which leads to cell-cycle arrest *(162)*. In contrast, there are other situations, which depend on the particular B-Raf mutation and require both B-Raf and Ras mutations for transformation. The B-Raf mutations in these cases result in weaker levels of B-Raf activity *(162,171)*.

Different B-Raf mutations have been mapped to various regions of the B-Raf protein. However, the mutations at 599 residue in B-Raf appear to be the most common. This mutation (V599E) results in activation of B-Raf and downstream MEK and ERK. Some of the other B-Raf mutations are believed to result in B-Raf molecules with impaired B-Raf activity, which must signal through Raf-1 *(36)*. Others mutations, such as D593V, may activate alternative signal transduction pathways *(36)*.

17. RKIP, Raf, and Prostate Cancer

Recently, a role for RKIP in cancer was hypothesized. Certain advanced prostate cancers express lower amounts of RKIP than less malignant prostate cancer specimens *(172–175)*. Then it was determined that inhibition of RKIP expression made certain prostate cells more metastatic *(175)*. The mechanism responsible for this increase in metastasis is believed to result from the enhanced activity of the Raf/MEK/ERK signaling pathways. After RKIP disassociates from Raf, it is believed to bind and inhibit

GRK2 activity. RKIP is not thought to alter the tumorigenic properties of prostate cancer cells; rather it is thought to be a suppressor of metastasis and may function by decreasing vascular invasion *(175)*.

The role of the Raf/MEK/ERK pathway in prostate cancer remains controversial. The studies with RKIP would suggest that increasing Raf activity, by inhibition of RKIP after phosphorylation by PKC, is somehow linked with metastisis in prostate cancer *(24,172–175)*. However, the Raf/MEK/ERK pathway may be shut off in advanced prostate cancer as a result of deletion of the PTEN gene, which normally regulates the activity of Akt by counterbalancing PI3K activity *(176–178)*. In some cells, Akt may inhibit Raf-1 activity by phosphorylation of Raf-1 on S259. Interesting, we have observed a very low expression of the Raf/MEK/ERK pathway in prostate cancer cells, which often have overexpression of activated Akt caused by deletion or mutation of the PTEN phosphatase, which regulates the activity of PI3K *(177)*. Although the Raf/MEK/ERK genes are present in these cells and Raf/MEK/ERK can be induced upon treatment with chemotherapeutic drugs such as paclitaxel *(179–182)*, their expression under normal growth conditions are very low. This may indicate that elevated Akt expression shuts off the Raf/MEK/ERK pathway, which may normally induce the terminal differentiation of prostate cells (*see* Fig. 4). In addition, we have observed that ectopic expression of Akt will suppress the levels of Raf/MEK/ERK activation in breast cancer cells. Others have also speculated that rather than Raf being overexpressed in prostate cancer, it may be expressed at low levels or not at all. Therapies aimed at increasing Raf expression might be more effective in the treatment of prostate cancer as they may induce the terminal differentiation of the cells *(179,181)*. To reconcile these differences between low Raf expression in prostate cancer cells in culture and the studies with RKIP and Raf expression during prostate cancer metastasis in vivo, it could result from a difference between cellular growth and metastasis as Raf may be expressed at low levels during growth, but at higher levels in cells that have metastized.

18. Raf and Chemotherapeutic Drug Resistance

In certain cancer types, expression of the Raf/MEK/ERK pathway will modulate the expression of drug pumps and antiapoptotic molecules such as Bcl-2 *(183–186)*. We have observed that ectopic expression of Raf will increase the levels of both the Mdr-1 drug pump and the antiapoptotic Bcl-2 protein in breast cancer cells *(184,186)*. This increased expression of Mdr-1 and Bcl-2 most likely occurs by a transcriptional mechanism by downstream target kinases of the Raf/MEK/ERK pathway inducing the phosphorylation of transcription factors which bind the promoter regions of Mdr-1 and Bcl-2 and stimulate transcription *(187,188)*. The increased expression of Mdr-1 and Bcl-2 has been associated with the drug resistance of these breast cancer cells *(184,186)*.

The Raf/MEK/ERK pathway may also induce the phosphorylation of Bcl-2 itself *(146,147)*. Certain phosphorylation events on Bcl-2 have been associated with prolonged activation *(146,147)*, whereas other phosphorylation events on Bcl-2 induced by chemotherapeutic drugs such as paclitaxel have been associated with inactivation of Bcl-2 *(182,189,190)*. Likewise, phosphorylation of ERK is normally associated with a proliferative response. However, certain chemotherapeutic drugs such as paclitaxel can induce the phosphorylation of ERK at the same residues, which is associated with the prevention of apoptosis *(146,147,189,190)*.

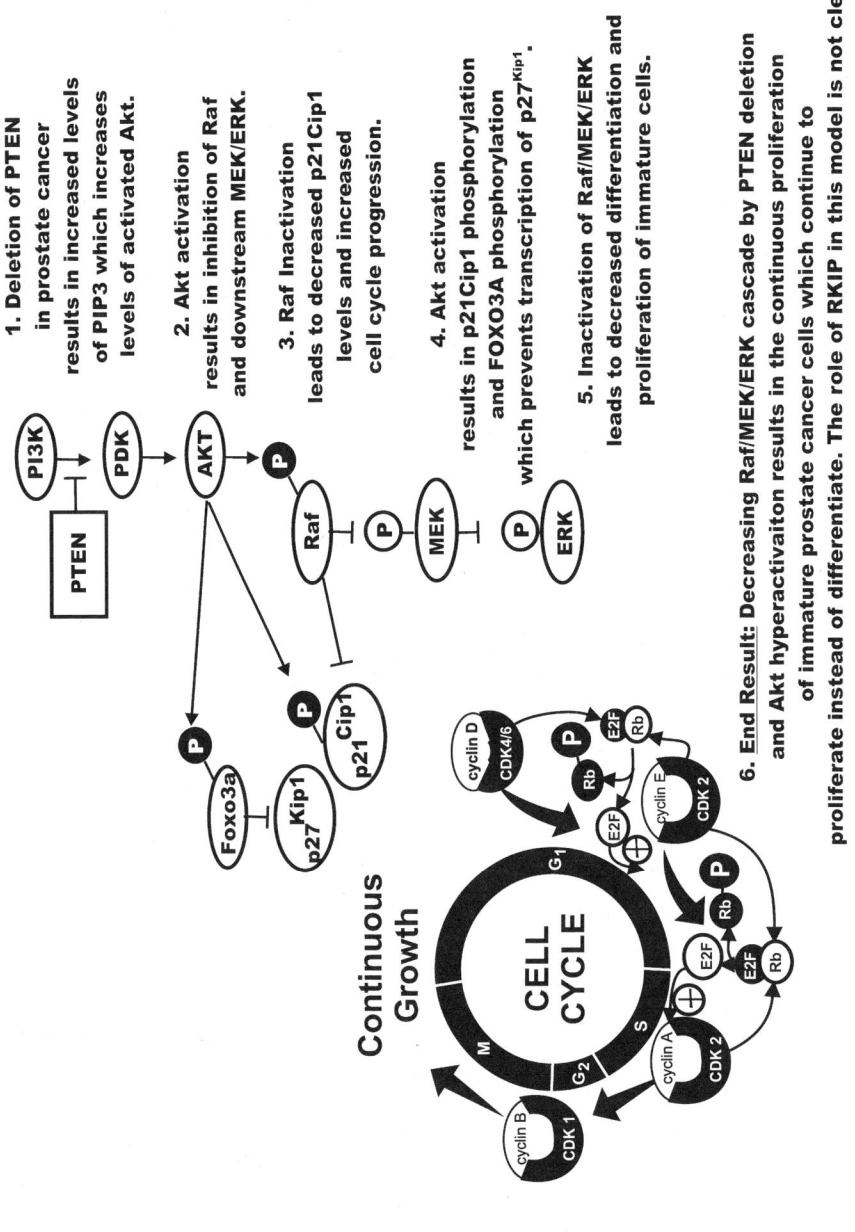

Fig. 4. Effects of PTEN deletion on PI3K/Akt and Raf/MEK/ERK activation in prostate cancer. PTEN deletion results in Akt activation. Akt activation can result in the phosphorylation and inactivation of Raf. This decrease in downstream MEK and ERK activation may lead to the loss of differentiation. Lack of Raf activation blocks p21[Cip1] activation that may result in cell-cycle progression. Akt activation can also result in the phosphorylation and inactivation of p21[Cip1] and Foxo3a. Foxo3a normally leads in the transcription of p27[Kip1] which normally acts to inhibit cell-cycle progression.

19. The Role of the Ras/RAF/MEK/ERK Pathway in Drug Resistance to Therapies Based on Oxygen Radicals

Many of the current therapies for cancer result in oxygen radical production. For example, doxorubicin (adriamycin) a drug that is commonly used to treat breast cancer and leukemia patients generates oxygen radicals within the cells *(191)*. These oxygen radicals can result in DNA adduct formation and the eventual death of that cell. γ-irradiation, another commonly used treatment for cancer, also results in the production of oxygen radicals *(192)*. Recently, ischemia reperfusion, which is well known to result in oxygen radical production, was used to treat individuals suffering with hepatic carcinomas *(193)*. Photodynamic therapy (PDT) is a three-component treatment that is used for cancer treatment *(194)*. In PDT, a photosensitizer is used that accumulates preferentially in abnormal tissue, including tumors. The photosenthesizer can then be activated by a laser and in the presence of molecular oxygen gives rise to reactive oxygen intermediates *(195)*. The production of oxygen radicals by these different cancer treatments appears to be important for their therapeutic effects because antioxidants, such as *n*-acetyl cysteine (NAC), inhibit cell death in response to chemotherapy, photodynamic therapy, and irradiation *(191,192,196–198)*.

Cells, however, can be exposed to oxygen radicals from many other sources as well. Electrons will leak from the electron transport chain leading to superoxide production. Superoxide can undergo dismutation and form hydrogen peroxide. Hydrogen peroxide can be further reduced to form the highly reactive hydroxyl radical. Furthermore, cells can be exposed to intracellularly produced oxygen radicals produced by the triggering of certain cell surface receptors such as the TNF and EGF receptors. In addition, cells may be found in an environment in which oxidative stress is high such as in tumors and inflammatory sites. It appears that over the course of time, cells have evolved to protect themselves from oxygen radicals. Cells express enzymes such as catalase and superoxide dismutase to help detoxify the oxygen radicals. In addition, cells have several antioxidant buffering systems such as glutathione and thioredoxin that help scavenge oxygen free radicals.

It appears that cells have also adapted to oxygen radical-induced stress by activating cellular signaling pathways that are thought to promote cell survival such as the Akt, NF-κB and Ras/Raf/MEK/ERK signaling pathways. Thus oxygen radicals not only activate stress kinase signaling pathways which are presumed to lead to apoptosis *(199)*, but also additional signaling pathways which are presumed to promote cell survival *(200,201)*.

What can be seen clearly from a number of studies is that ROS (which can be toxic to cells) can also lead to the activation of signaling pathways that are involved with cell growth or the prevention of apoptosis. In other words, the cancer therapies themselves can activate pathways that give rise to drug resistance. Thus, it should be possible to cause cancer cells to be more susceptible to the treatment by inhibiting some of the life saving pathways which are turned on in response to oxygen intermediates. The first step to doing this is to identify potential targets, which are activated by ROI.

Reactive oxygen intermediates are known to result in the activation of the MAP kinases in a number of different cell types *(202–205)*. It is thought that this activation can be mediated in part by acting directly on Ras. Hoyos et al. reported that ultraviolet (UV) light could induce the activation of Raf-1 *(206)*. They reported that the primary event in this reaction was the oxidation of cysteine residues in the cysteine rich domain

(CRD) of Raf-1. Similar to UV light, hydrogen peroxide treatment also induced oxidation of the cysteine residues in the CRD. The activation of Ras by UV light was Ras-dependent and also resulted in the phosphorylation of MEK.

Other mechanisms for oxygen radical-induced activation of the Ras/Raf/MEK/ERK pathway also exist. Both Fyn and Jak2 can be activated by Reactive Oxygen Species (ROS) and the activation of these two kinases is thought to promote Ras activity under these conditions *(207)*. This activation of Ras would appear to be productive as these same authors found in a subsequent study that oxygen radical-induced p90[Rsk] activation occurs in a Ras- and Fyn-dependent manner *(208)*. Accorsi and coworkers found that oxidizing agents such as hydrogen peroxide also regulate Ras activation and down-regulated GAP activity and a GDP/GTP exchange factor (GEF) *(209)*. The authors suggest that, depending on the activating agent the balance between GEF and GAP inactivation could lead to Ras activation or inactivation. Activation of PKC by the pharmacological agent PMA is well known to transiently induce ERK activation in virtually all cell types. PMA is also known to stimulate intracellular production of ROS *(210)* and oxygen radicals are capable of activating PKC *(211)*.

Oxidants may directly intervene on the Ras/Raf/MEK/ERK pathway and may also activate receptors that induce the activation of this pathway. For example, ROS have been shown to induce EGF receptor dimerization and phosphorylation in the absence of its ligand. The fact that the Ras/Raf/MEK/ERK pathway can be activated by oxidants and that this pathway has been shown to have antiapoptotic effects, suggests that the activation of this pathway by chemotherapeutic drugs and other cancer treatments can give rise to drug resistance in the absence of any abnormal functioning of the Ras/Raf/MEK/ERK pathway. In cases where Ras, Raf, MEK, or ERK may be over-expressed, the balance of the antiapoptotic and apoptotic activity may be shifted to favor life over death.

20. Targeting the Ras/Raf/MEK/ERK Pathway for Therapeutic Intervention

Mutations at three different Ras codons (12, 13, and 61) convert the Ras protein into a constitutively-active protein *(1,34,35)*. These point mutations can be induced by environmental mutagens as well as errors in DNA synthesis during cellular division. Given the high level of mutations that have been detected in Ras, this pathway has long been considered a key target for therapeutic intervention. Ras mutations are frequently observed in certain hematopoietic malignancies including myelodysplastic syndromes *(34,35)*. Ras mutations are also detected in many other types of tumors and often occur in the same types of tumors that B-Raf mutations occur *(36)*. These observations suggest that activation of the Ras/Raf/MEK/ERK pathway may contribute to abnormal proliferation in certain types of tumors.

Ras mutations are often one step in tumor progression and mutations at other genes (chromosomal translocations, gene amplification, tumor suppressor gene inactivation/deletion) have to occur for the full malignant phenotype to manifest. Pharmaceutical companies have developed many FT inhibitors, which suppress the farnesylation of Ras, precluding it from localizing to the cell membrane.

The biochemical differences between these Ras proteins have remained elusive. Ki-Ras mutations have been more frequently detected in human neoplasia than Ha-Ras mutations *(212)*. Ras has been shown to activate both the Raf/MEK/ERK and the PI3K/Akt pathways. Thus mutations at Ras should theoretically activate both pathways simultaneously. Ras mutations have a key role in malignant transformation as both of these pathways can prevent apoptosis as well as regulate cell-cycle progression. There is specificity in terms of the ability of Ki-Ras and Ha-Ras to induce the Raf/MEK/ERK and PI3K/Akt pathways *(7)*. Ki-Ras preferentially activates the Raf/MEK/ERK pathway whereas Ha-Ras preferentially activates the PI3K/Akt pathway *(7)*. Therefore, if Ras inhibitors could be developed which would specifically inhibit one particular Ras protein, it might be possible to inhibit one of the downstream pathways as opposed to inhibiting both. This might under certain circumstances be advantageous, as Raf has cell-cycle inhibitory effects under certain conditions. Furthermore, decreases in ERK expression may affect differentiation. In advanced prostate cancer, PTEN is often deleted resulting in higher levels of Akt activity, which suppresses Raf/MEK/ERK activation. This is a case where therapeutic inhibition of Ha-Ras may be more important than inhibition of Ki-Ras. Thus in certain tumors, it might be desirable to inhibit the effects that Ras has on the PI3K/Akt pathway as opposed to the effects Ras has on Raf. Targeting of Ha-Ras as opposed to Ki-Ras might inhibit apoptosis suppression by Ha-Ras but not inhibit the effects Ki-Ras has on inhibition of cell-cycle progression or differentiation.

Overexpression of the Raf/MEK/ERK cascade is frequently observed in human neoplasia. A prime consequence of this activation may be the increased expression of growth factors which can potentially further activate this cascade by autocrine loops. Many cytokine and growth factor genes contain transcription factor binding sites, which are bound by transcription factors whose activity are often regulated by the Raf/MEK/ERK cascade. Identification of the mechanisms responsible for activation of this cascade in human cancer has remained elusive, as there are many points for activation.

Recently B-Raf mutations have been detected in hematological malignancies *(163–166)*. A low frequency of B-Raf mutations was detected in 4/164 non-Hodgkin's lymphoma *(163)*. The mutations detected in these four non-Hodgkin's lymphomas G468A, two G468R and one D593G are not the common mutation (V599E) detected in melanomas. B-Raf mutations have also been detected in Acute Myeloid Leukemia (AML) *(164)*. Approximately 4% of AML surveyed in one study have mutations at B-Raf. Interesting the mutation detected in AML and non-Hodgkin's lymphoma are not the same as the most common mutation detected in over 70% of melanoma (V599E), but some of these mutations detected in hematological cancer result in the constitutive activation of B-Raf activity. At least two additional groups have detected mutations at B-Raf in hematological malignancies, which were presented at the ASH Annual Meeting in 2004 *(165,166)*. Some B-Raf mutations have been observed in breast cancer *(36)*. Although these mutations in B-Raf may be relatively rare, they do point to the fact that they occur in other types of cancer besides melanomas, thyroid and colonic cancers.

Raf inhibitors have been developed and some are being evaluated in clinical trials *(212–218)*. As stated previously, Raf-1 has at least thirteen regulatory phosphorylation residues. Inhibition of Raf is a complicated affair as certain phosphorylation events stimulate Raf activity whereas others inhibit Raf activity and promote Raf association with 14-3-3 proteins, which render it inactive and present in the cytoplasm. Certain Raf inhibitors were developed which inhibit the Raf kinase activity as determined by assays with purified Raf proteins and substrates (MEK). These inhibitors (e.g., L-779,450, ZM 336372, Bay 43-9006) bind the Raf kinase domain and therefore prevent its activity. Some Raf inhibitors may affect a single Raf isoform (e.g., Raf-1), others may affect Raf proteins, which are more similar (Raf-1 and A-Raf), whereas other Raf inhibitors may affect all three Raf proteins (Raf-1, A-Raf and B-Raf). We have observed that the L-779,450 inhibitor suppresses the effects of A-Raf and Raf-1 more than the effects of B-Raf. *(217)*. Like many Raf inhibitors, L-779,450 is not specific for Raf; it also inhibits the closely related p38^{MAPK}. Likewise, Bay49-9006 inhibits other kinases besides Raf (see below). Knowledge of the particular Raf gene mutated or overexpressed in certain tumors may provide critical information regarding how to treat the patient as some cancers which overexpress a particular Raf gene may be more sensitive to inhibition by agents which target that particular Raf protein. Inhibition of certain Raf genes might prove beneficial while suppression of other Raf genes under certain circumstances might prove detrimental. Thus the development of unique and broad-spectrum Raf inhibitors may prove useful in human cancer therapy.

Chaperonin proteins such as 14-3-3 and hsp-90 regulate Raf activity *(218)*. Raf activity is regulated by dimerization. These biochemical properties result in Raf activity being sensitive to drugs, which block protein–protein interactions such as geldanamycin *(220)*. Geldanamycin and its 17-allylamino-17-demethoxy analog (17-AAG), are nonspecific Raf inhibitors as they also affect the activity of many proteins including the BCR-ABL oncoprotein, which has a critical role in the etiology of chronic myelogenous leukemia. Geldanamycin and 17-AAG are currently in clinical trials *(220,221)*. We often think of a single Raf protein carrying out its biochemical activity. However, Raf isoforms dimerize with themselves and other Raf isoforms to become active. Drugs such as coumermycin, which inhibit Raf dimerization and others such as geldanmycin, which prevent interaction of Raf with Hsp90 and 14-3-3 proteins suppress Raf activity *(218,222)*.

Some of the studies performed with the Raf inhibitor Bay 43-9006 have shown significant promise in the treatment of diverse cancers which have been difficult to treat (e.g., colorectal, ovarian) *(223–227)*. As with many inhibitors, the true target remains controversial. Recently Bay 43-9006 has been shown to have additional targets such as PDGF-R and the VEGF-II receptor KDR that is important in the control of angiogenesis *(227)*.

Prevention of Raf activation by targeting kinases (e.g., Src, PKC, PKA, PAK, or Akt) and phosphatases (e.g., PP2A) involved in Raf activation may be a mechanism to inhibit/regulate Raf activity. It is worth noting that some of these kinases normally inhibit Raf activation (Akt, PKA). A major limitation of this approach would be that these kinases and phosphatases could result in activation or inactivation of other proteins and would have other effects on cell physiology.

Currently it is believed that MEK1 is not frequently mutated in human cancer. However, aberrant expression of MEK1 is observed in many different cancers because of the activation of the Raf/MEK/ERK pathway by upstream kinases (e.g., BCR-ABL) and growth factor receptors (e.g., EGFR). Specific inhibitors to MEK have been developed (PD98059, U0126 and PD184352 a.k.a., CI1040). Second generation MEK inhibitors such as PD184352 have been developed and are currently in clinical trials *(228–231)*. Other MEK inhibitors inactivate MEK1, MEK2 and the closely related MEK5 (PD98059) while others are more specific and only inhibit MEK1 and MEK2 (U0126). The successful development of MEK inhibitors may result from the relatively few phosphorylation sites on MEK involved in activation/inactivation. An advantage of targeting the Raf/MEK/ERK cascade is that it can be targeted without knowledge of the precise genetic mutation, which results in its aberrant activation. This is important as the nature of the critical mutation(s), which leads to the malignant growth of at least 50% of AMLs and other cancers, is not currently known.

To our knowledge, no small molecular weight ERK inhibitors have been developed yet, however, inhibitors to ERK could prove very useful as ERK can phosphorylate many targets (e.g., Rsk, c-Myc, Elk, and so on). There are at least two ERK molecules regulated by the Raf/MEK/ERK cascade, ERK1 and ERK2. Little is known about the different in vivo targets of ERK1 and ERK2. However ERK2 has been postulated to have proproliferative effects whereas ERK1 has antiproliferative effects *(232–234)*. Development of specific inhibitors to ERK1 and ERK2 might eventually prove useful in the treatment of certain diseases.

The MAP kinase phosphatase-1 (MKP-1) removes the phosphates from ERK. MKP-1 is mutated in certain tumors and could be considered a tumor suppressor gene *(235,236)*. An inhibitor of this phosphatase has been developed (Ro-31-8220). Therapeutic targeting of phosphatases has lagged behind targeting of kinases as it is often more difficult to precisely target them. However, the development of phosphatase inhibitors is a novel area, which may yield many important inhibitors.

21. Conclusions

Over the past 25 yr, there has been much progress in elucidating the involvement of the Ras/Raf/MEK/ERK cascade in etiology of human neoplasia. From initial studies with defining the oncogenes present in avian and murine oncogenes, we learned that ErbB, Ras, Src, Abl, Raf, Jun, Fos, Ets, and NF-κB were originally cellular genes that were captured by retroviruses. Biochemical studies defined and continue to elucidate the roles that the cellular and viral oncogenes had in cellular transformation. We have learned that many of these oncogenes are connected to the Ras/Raf/MEK/ERK pathway and either feed into this pathway (e.g., BCR-ABL, ErbB) or are downstream targets, which regulate gene expression (e.g., Jun, Fos, Ets, and NF-κB).

The Ras/Raf/MEK/ERK pathway has what often appear to be conflicting roles in cellular proliferation, differentiation, and the prevention of apoptosis. Classical studies have indicated that Ras/Raf/MEK/ERK can promote proliferation and malignant transformation in part because of the prevention of apoptosis, which may be mediated by phosphorylation of key components of the apoptotic pathway. Indeed, the latest "hot" area of the Ras/Raf/MEK/ERK pathway is the discovery of B-Raf mutants, which promote proliferation and transformation *(36)*. However, it should be remembered that only a few years ago, hyperactivtion of B-Raf and Raf-1 was proposed to promote cell-cycle arrest *(48–52)*. Thus it is probably fine-tuning of these mutations, which dictates whether there is cell-cycle arrest or malignant transformation.

Initially it was thought that Raf-1 was the most important Raf isoform. Raf-1 was the earliest studied Raf isoform and homologous genes are present in both murine and avian transforming retroviruses. Originally it was shown that Raf-1 was ubiquitously expressed, indicating a more general and important role while B-Raf and A-Raf had more limited patterns of expression. However, it is now believed that not only B-Raf is the more important activator of the Raf/MEK/ERK cascade but also in some cases the activation of Raf-1 may require B-Raf. The role of A-Raf remains poorly defined yet it is an interesting isoform. It is the weakest Raf kinase, yet it can stimulate proliferation without having the negative effects on cell proliferation that B-Raf and Raf-1 have *(42–45,51)*.

Activation of the Raf proteins is very complex as there are many phosphorylation sites on Raf. Phosphorylation at different sites can lead to either activation or inactivation. Clearly there are many kinases and phosphatases that regulate Raf activity and the state of phosphorylation will determine where Raf is active or inactive. While the kinases involved in regulation in Raf/MEK/ERK have been extensively studied, we have only a very limited knowledge of the specific phosphatases involved in these regulatory events.

Raf-1 has many roles, which are apparently independent of downstream MEK/ERK. Some of these roles occur at the mitochondria and are intimately associated with the prevention of apoptosis. Raf-1 may function as a scaffolding molecule to inhibit the activity of kinases which promote apoptosis.

The Raf/MEK/ERK pathway is both positively (KSR, MP-1) and negatively (RKIP, 14-3-3, hsp-90) regulated by association with scaffolding proteins. The expression of some of the scaffolding proteins is altered in human cancer (e.g., RKIP) in some cases. Some of these scaffolding proteins (e.g., hsp-90) are being evaluated as potential therapeutic targets (geldanamycin).

The Raf/MEK/ERK pathway is intimately linked with the PI3K/PTEN/Akt pathway. Ras can regulate activation of both pathways. Furthermore, in some cell types, Raf activity is negatively regulated by Akt, indicating crosstalk between the two pathways. Both pathways may result in the phosphorylation of many downstream targets and impose a role in the regulation of cell survival and proliferation.

Although we often think of phosphorylation of these molecules as being associated with the prevention of apoptosis and the induction of gene transcription, this view is oversimplified. For example, in certain situations such as in advanced prostate cancer, the Raf/MEK/ERK pathway may be inhibited; hence the phosphorylation of Bad and CREB normally mediated by the Raf/MEK/ERK cascade, which is associated with the prevention of apoptosis, will be inhibited.

Although it has been known for many years that the Raf/MEK/ERK pathway can effect differentiation, this is probably one of the weakest research areas in the field resulting from the often cell-lineage specific effects which must be evaluated in each cell type. An intriguing aspect of human cancer therapy is that in some cases stimulation of the Raf/MEK/ERK pathway may be desired to promote terminal differentiation, while in other types of malignant cancer cells which proliferate in response to Raf/MEK/ERK activity, inhibition of the Raf/MEK/ERK pathway may be desired to suppress proliferation. Thus we must be flexible in dealing with the Raf/MEK/ERK pathway. As we learn more, our conceptions continue to change.

Acknowledgments

JAM has been supported in part by a grant from the NIH (R01098195).

References

1. Steelman LS, Pohnert SC, Shelton JG, Franklin RA, Bertrand FE, McCubrey JA. JAK/STAT, Raf/MEK/ERK, PI3K/Akt and BCR-ABL in cell cycle progression and leukemogenesis. Leukemia 2004;18:189–218.
2. Lee JT, Jr, McCubrey JA. The Raf/MEK/ERK Signal Transduction Cascade as a Target for Chemotherapeutic Intervention. Leukemia 2002;16:486–507.
3. Chang F, Steelman LS, Lee JT, et al. Signal transduction mediated by the Ras/Raf/MEK/ERK pathway from cytokine receptors to transcription factors: potential targeting for therapeutic intervention. Leukemia 2003;17:1263–1293.

4. Blalock WL, Weinstein-Oppenheimer C, Chang F, et al. Signal transduction, cell cycle regulatory, and anti-apoptotic pathways regulated by IL-3 in hematopoietic cells: possible sites for intervention with anti-neoplastic drugs. Leukemia 1999;13:1109–1166.

5. Chang F, Lee JT, Navolanic PM, et al. Involvement of PI3K/Akt pathway in cell cycle progression, apoptosis, and neoplastic transformation: a target for cancer chemotherapy. Leukemia 2003;17:590–603.

6. Peyssonnaux C, Provot S, Felder-Schmittbuhl MP, Calothy G, Eychéne A. Induction of postmitotic neuroretina cell proliferation by distinct Ras downstream signaling pathways. Mol Cell Biol 2000;20:7068–7079.

7. Yan J, Roy S, Apolloni A, Lane A, Hancock JF. Ras isoforms vary in their ability to activate Raf-1 and phosphoinositide 3-kinase. J Biol Chem 1998;273:24,052–24,056.

8. Luo Z, Tzivion G, Belshaw PJ, Vavvas D, Marshall M, Avruch J. Oligomerization activates c-Raf-1 through a Ras-dependent mechanism. Nature 1996;383:181–185.

9. Hagan S, Garcia R, Dhillon A, Kolch W. Raf kinase inhibitor protein regulation of raf and MAPK signaling. Methods Enzymol 2005;407:248–259.

10. Fabian JR, Daar IO, Morrison DK. Critical tyrosine residues regulate the enzymatic and biological activity of Raf-1 kinase. Mol Cell Biol 1993;13:7170–7179.

11. Dhillon AS, Meikle S, Yazici Z, Eulitz M, Kolch W. Regulation of Raf-1 activation and signaling by dephosphorylation. EMBO J 2002;21:64–71.

12. Marais R, Light Y, Paterson HF, Marshall CJ. Ras recruits Raf-1 to the plasma membrane for activation by tyrosine phosphorylation. EMBO J 1995;14:3136–3145.

13. Lee JE, Beck TW, Wojnowski L, Rapp UR. Regulation of A-Raf expression. Oncogene 1996;12:1669–1677.

14. Marais R, Light Y, Paterson HF, Mason CS, Marshall CJ. Differential regulation of Raf-1, A-Raf, and B-Raf by oncogenic ras and tyrosine kinases. J Biol Chem 1997;272: 4378–4383.

15. Mason CS, Springer CJ, Cooper RG, Superti-Furga G, Marshall CJ, Marais R. Serine and tyrosine phosphorylations cooperate in Raf-1, but not B-Raf activation. EMBO J 1999;18: 2137–2148.

16. Diaz B, Barnard D, Filson A, MacDonald S, King A, Marshall M. Phosphorylation of Raf-1 serine 338-serine 339 is an essential regulatory event for Ras-dependent activation and biological signaling. Mol Cell Biol 1997;17:4509–4516.

17. Kolch W, Heidecker G, Kochs G, et al. Protein kinase Cα activates RAF-1 by direct phosphorylation. Nature 1993;364:249–252.

18. Barnard D, Diaz B, Clawson D, Marshall M. Oncogenes, growth factors and phorbol esters regulate Raf-1 through common mechanisms. Oncogene 1998;17:1539–1547.

19. Rommel C, Clarke BA, Zimmermann S, et al. Differentiation stage-specific inhibition of the Raf-MEK-ERK pathway by Akt. Science 1999;286:1738–1741.

20. Dhillon AS, Pollock C, Steen H, Shaw PE, Mischak H, Kolch W. Cyclic AMP-dependent kinase regulates Raf-1 kinase mainly by phosphorylation of serine 259. Mol Cell Biol 2002;22:3237–3246.

21. Guan KL, Figueroa C, Brtva TR, Zhu T, Taylor J, Barber TD, Vojtek AB. Negative regulation of the serine/threonine kinase B-Raf by Akt. J Biol Chem 2000;275:27,354–27,359.

22. Zhang BH, Tang ED, Zhu T, Greenberg ME, Vojtek AB, Guan KL. Serum- and glucocorticoid-inducible kinase SGK phosphorylates and negatively regulates B-Raf. J Biol Chem 276:2001;31,620–31,626.

23. Yeung K, Seitz T, Li S, et al. Suppression of Raf-1 kinase activity and MAP kinase signaling by RKIP. Nature 1999;401:173–177.

24. Corbit KC, Trakul N, Eves EM, Diaz B, Marshall M, Rosner MR. Activation of Raf-1 signaling by protein kinase C through a mechanism involving Raf kinase inhibitory protein. J Biol Chem 2003;278:13,061–13,068.

25. Mercer KE, Pritchard CA. Raf proteins and cancer: B-Raf is identified as a mutational target. Biochimica et Biophysica Acta 2003;1653:25–40.

26. Brummer T, Shaw PE, Reth M, Misawa Y. Inducible gene deletion reveals different roles for B-Raf and Raf-1 in B-cell antigen receptor signalling. EMBO J 2002;21:5611–5622.

27. Hindley A, Kolch W. Extracellular signal regulated kinase (ERK)/mitogen activated protein kinase (MAPK)-independent functions of Raf kinases. J Cell Sci 2002;115:1575–1581.

28. Salomoni P, Wasik MA, Riedel RF, et al. Expression of constitutively active Raf-1 in the mitochondria restores antiapoptotic and leukemogenic potential of a transformation-deficient BCR/ABL mutant. J Exp Med 1998;187:1995–2007.

29. Neshat MS, Raitano AB, Wang HG, Reed JC, Sawyers CL. The survival function of the Bcr-Abl oncogene is mediated by Bad-dependent and -independent pathways: roles for phosphatidylinositol 3-kinase and Raf. Mol Cell Biol 2000;20:1179–1186.

30. Alessi DR, Saito Y, Campbell DG, et al. Identification of the sites in MAP kinase kinase-1 phosphorylated by p74raf-1. EMBO J 1994;13:1610–1619.

31. Wu X, Noh SJ, Zhou G, Dixon JE, Guan KL. Selective activation of MEK1 but not MEK2 by A-Raf from epidermal growth factor-stimulated Hela cells. J Biol Chem 1996;271:3265–3271.

32. Blenis J. Signal transduction via the MAP kinases: proceed at your own RSK. Proc Natl Acad Sci USA 1993;90:5889–5892.

33. Pouyssegur J, Volmat V, Lenormand P. Fidelity and spatio-temporal control in MAP kinase (ERKs) signaling. Biochem Pharm 2002;64:755–763.

34. Flotho C, Valcamonica S, Mach-Pascual S, et al. RAS mutations and clonality analysis in children with juvenile myelomonocytic leukemia (JMML). Leukemia 1999;13:32–37.

35. Stirewalt DL, Kopecky KJ, Meshinchi S, et al. FLT3, RAS, and TP53 mutations in elderly patients with acute myeloid leukemia. Blood 2001;97:3589–3595.

36. Garnett MJ, Marais R. Guitly as charged: B-Raf is a human oncogene. Cancer Cell 2004;6: 313–319.

37. McCubrey JA, Steelman LS, Wang X-Y, et al. Differential effects of viral and cellular oncogenes on the growth factor-dependency of hematopoietic cells. Int J Oncol 1995;7: 295–310.

38. Serrano M, Lin AW, McCurrach ME, Beach D, Lowe SW. Oncogenic ras provokes premature cell senescence associated with accumulation of p53 and p16[INK4a]. Cell 1997;88:593–602.

39. Malumbres M, Castro IPD, Hernández MI, Jiménez M, Corral T, Pellicer A. Cellular response to oncogenic ras involves induction of the Cdk4 and Cdk6 inhibitor p15[INK4b]. Mol Cell Biol 2000;20:2915–2925.

40. Hirakawa T, Ruley HE. Rescue of cells from ras oncogene-induced growth arrest by a second complementing oncogene. Proc Natl Acad Sci USA 1988;85:1519–1523.

41. Adnane J, Jackson RJ, Nicosia SV, Cantor AB, Pledger WJ, Sebti SM. Loss of p21[WAF1/CIP1] accelerates Ras oncogenesis in a transgenic/knockout mammary cancer model. Oncogene 2000;19:5338–5347.

42. Hoyle PE, Moye PW, Steelman LS, et al. Differential abilities of the Raf family of protein kinases to abrogate cytokine dependency and prevent apoptosis in murine hematopoietic cells by a MEK1-dependent mechanism. Leukemia 2000;14:642–656.

43. Chang F, McCubrey JA. P21[Cip1] induced by Raf is associated with increased Cdk4 activity in hematopoietic cells. Oncogene 2001;20:4354–4364.

44. Chang F, Steelman LS, McCubrey JA. Raf-induced cell cycle progression in human TF-1 hematopoietic cells. Cell Cycle 2002;1:220–226.

45. Shelton JG, Chang F, Lee JT, Franklin RA, Steelman LS, McCubrey JA. B-Raf and insulin synergistically prevent apoptosis and induce cell cycle progression in hematopoietic cells. Cell Cycle 2004;3:189–196.

46. Sanders MR, Lu H, Walker F, Sorbad S, Dainiak N. The Raf-1 protein mediates insulin-like growth factor-induced proliferation of erythroid progenitor cells. Stem Cells 1998;16:200–207.

47. Cioffi CL, Garay M, Johnston JF, et al. Monia BP. Selective inhibition of A-Raf and C-Raf mRNA expression by antisense oligodeoxynucleotides in Rat vascular smooth muscle cells: role of A-Raf and C-Raf in serum-induced proliferation. Mol Pharmacol 1997;51:383–389.

48. Ravi RK, Weber E, McMahon M, et al. Activated Raf-1 causes cell cycle arrest in small cell lung cancer cells. J Clin Invest 1998;101:153–159.

49. Lloyd AC, Obermuller F, Staddon S, Barth CF, McMahon M, Land H. Cooperating oncogenes converge to regulate cyclin/Cdk complexes Genes Devel 1997;11:663–677.

50. Yen A, Williams M, Platko JD, Der C, Hisaka M. Expression of activated Raf accelerates cell differentiation and RB protein down-regulation but not hypophosphorylation. Eur J Cell Biol 1994;65:103–113.

51. Woods D, Parry D, Cherwinski H, Bosch E, Lees E, McMahon M. Raf-induced proliferation or cell cycle arrest is determined by the level of Raf activity with arrest mediated by p21^{Cip}. Mol Cell Biol 1997;19:5598–5611.

52. Chang F, Steelman LS, Shelton JG, et al. McCubrey JA. Regulation of cell cycle progression and apoptosis by the Ras/Raf/MEK/ERK pathway. Int J Oncol 2003;22:469–480.

53. McCubrey JA, Lee JT, Steelman LS, et al. Interactions between the PI3K and Raf signaling pathways can result in the transformation of hematopoietic cells. Cancer Detection and Prevention 2001;25:375–393.

54. Kuroda S, Ohtsuka T, Yamamori B, Fukui K, Shimizu K, Takai Y. Different effects of various phospholipids on Ki-Ras-, Ha-Ras-, and Rap1B-induced B-Raf activation. J Biol Chem 1996;271:14,680–14,683.

55. Voice JK, Klemke RL, Le A, Jackson JH. Four human ras homologs differ in their abilities to activate Raf-1, induce transformation, and stimulate cell motility. J Biol Chem 1999;274:17,164–17,170.

56. Weber CK, Slupsky JR, Hermann C, Schuler M, Rapp UR, Block C. Mitogenic signaling of Ras in regulated by differential interaction with Raf isozymes. Oncogene 2000;19:169–176.

57. Pritchard CA, Samuels ML, Bosch E, McMahon M. Conditionally oncogenic forms of the A-Raf and B-Raf protein kinases display different biological and biochemical properties in NIH-3T3 cells. Mol Cell Biol 1995;15:6430–6442.

58. Kolch W. To be or not to be: a question of B-Raf. Trends Neurosci 2001;21:498–500.

59. Fantl WJ, Muslin AJ, Kikuchi A, et al. Activation of Raf-1 by 14-3-3 proteins. Nature 1994;371:612–614.

60. Freed E, Symons M, MacDonald SG, McCormick F, Ruggieri R. Binding of 14-3-3 proteins to the protein kinase Raf and effects on its activation. Science 1994;265:1713–1716.

61. Irie K, Gotoh Y, Yashar BM, Errede B, Nishida E, Matsumoto K. Stimulatory effects of yeast and mammalian 14-3-3 proteins on the Raf protein kinase. Science 1994;265:1716–1719.

62. Papin C, Denouel A, Calothy G, Eychene A. Identification of signaling proteins interacting with B-Raf in the yeast two-hybrid system. Oncogene 1996;12:2213–2221.

63. Pritchard CA, Bolin L, Slattery R, Murray R, McMahon M. Post-natal lethality and neurological and gastrointestinal defects in mice with targeted disruption of the A-Raf protein kinase gene. Curr Biol 1996;6:614–617.

64. Shinkai M, Masuda T, Kariya K, et al. Difference in the mechanism of interaction of Raf-1 and B-Raf with H-Ras. Biochem Biophys Res Commun 1996;22:729–734.

65. Wadewitz AG, Winer MA, Wolgemuth DJ. Developmental and cell lineage specificity of raf family gene expression in mouse. Oncogene 1993;8:1055–1062.

66. Wojnowski L, Zimmer AM, Bec TW, et al. Endothelial apoptosis in Braf-deficient mice. Nat Genet 1997;16:293–297.

67. Hagemann C, Rapp UR. Isotype-specific functions of Raf kinases. Exp Cell Res 1999;253: 34–46.

68. Vossler MR, Yao H, York RD, Pan M-G, Rim CS, Stork PJS. CAMP activates MAP kinase and Elk-1 through a B-Raf and Rap1-dependent pathway. Cell 1997;89:73–82.

69. Shinkai M, Masuda T, Kariya K, et al. Difference in the mechanism of interaction of Raf-1 and B-Raf with H-Ras. Biochem Biophys Res Commun 1996;223:729–734.

70. Rosario M, Paterson HF, Marshall CJ. Activation of the Raf/MAP kinase cascade by the Ras-related protein TC21 is required for the TC21-mediated transformation of NIH-3T3 cells. EMBO J 1999;18:1270–1279.

71. Kalmes A, Hagemann C, Weber CK, Wixler L, Schuster T, Rapp UR. Interaction between the protein kinase B-Raf and the alpha-subunit of the 11S proteasome regulator. Cancer Res 1998;58:2986–2990.

72. Hagemann C, Kalmes A, Wixler V, Wixler L, Schuster T, Rapp UR. The regulatory subunit of protein kinase CK2 is a specific A-Raf activator. FEBS Lett 1997;403:200–202.

73. O'Neill E, Rushworth L, Baccarini M, Kolch W. Role of the kinase MST2 in suppression of apoptosis by the proto-oncogene product Raf-1. 2004;306:2267–2270.

74. Yamamoto Y, Gaynor RB. Therapeutic potential of inhibition of the NF-κB pathway in the treatment of inflammation and cancer. J Clin Invest 2001;107:135–142.

75. Chen FE, Ghosh G. Regulation of DNA binding by Rel/NF-kappaB transcription factors: structural views. Oncogene 1999;18:6845–6852.

76. Wang CY, Mayo MW, Komeluk RG, Goeddel DV, Baldwin AS, Jr. NF-κB antiapoptosis: induction of TRAF1 and TRAF2 and c-IAP1 and c-IAP2 to suppress caspase-8 activation. Science 1998;281:1680–1683.

77. Govind S. Control of development and immunity by Rel transcription factors in Drosophila. Oncogene 1999;18:6875–6887.

78. Baldwin AS. Control of oncogenesis and cancer therapy resistance by the transcription factor NF-κB. J Clin Invest 2001;107:241–246.

79. Finco TS, Westwick JK, Norris JL, Beg AA, Der CJ, Baldwin AS, Jr. Oncogenic Ha-Ras-induced signaling activates NF-κB transcriptional activity, which is required for cellular transformation. J Biol Chem 1997;272:24,113–24,116.

80. Norris JL, Baldwin AS, Jr. Oncogenic Ras enhances NF-κB transcriptional activity through Raf-dependent and Raf-independent mitogen-activated protein kinase signaling pathways. J Biol Chem 1999;274:13,841–13,846.

81. Baumann B, Weber CK, Troppmair J, Whiteside S, Israel A, Rapp UR. Raf induces NF-κB by membrane shuttle kinase MEKK1, a signaling pathway critical for transformation. Proc Natl Acad Sci USA 2000;97:4615–4620.

82. Ludwig L, Kessler H, Wagne, M, et al. Nuclear factor-κB is constitutively active in C-cell carcinoma and required for RET-induced transformation. Cancer Res 2001;61: 4526–4535.

83. McCarthy SA, Samuels ML, Pritchard CA, Abraham JA, McMahon M. Rapid induction of heparin-binding epidermal growth factor/diphtheria toxin receptor expression by Raf and Ras oncogenes. Genes Dev 1995;9:1953–1964.

84. Schouten GJ, Vertegaal AC, Whiteside ST, et al. IκBα is a target for the mitogen-activated 90 kDa ribosomal S6 kinase. EMBO J 1997;16:3133–3144.

85. Ghoda L, Lin X, Greene WC. The 90-kDa ribosomal S6 kinase (pp90rsk) phosphorylates the N-terminal regulatory domain of IκBα and stimulates its degradation *in vitro*. J Biol Chem 1997;272:21,281–21,288.

86. DiDonato J, Mercurio F, Rosette C, et al. Mapping of the inducible IκB phosphorylation sites that signal its ubiquitination and degradation. Mol Cell Biol 1996;16:1295–1304.

87. Lee FS, Peters RT, Dang LC, Maniatis T. MEKK1 activates both IκB kinase α and IκB kinase β. Proc Natl Acad Sci USA 1998;95:9319–9324.

88. Funakoshi M, Tago K, Sonoda Y, Tominaga S, Kasahara T. A MEK inhibitor, PD98059 enhances IL-1-induced NF-κB activation by the enhanced and sustained degradation of IκBα. Biochem Biophys Res Comm 2001;283:248–254.

89. Muthusamy N, Leiden JM. A protein kinase C-, Ras-, and RSK2-dependent signal transduction pathway activates the cAMP-responsive element-binding protein transcription factor following T cell receptor engagement. J Biol Chem 1998;273:22,841–22,847.

90. Haus-Seuffert P, Meisterernst M. Mechanisms of transcriptional activation of cAMP-responsive element-binding protein CREB. Mol Cell Biochem 2000;212:5–9.

91. Boehlk S, Fessele S, Mojaat A, et al. ATF and Jun transcription factors, acting through an Ets/CRE promoter module, mediate lipopolysaccharide inducibility of the chemokine RANTES in monocytic Mono Mac 6 cells. Eur J Immunol 2000;30:1102–1112.

92. Feuerstein N, Firestein R, Aiyar N, He X, Murasko D, Cristofalo V. Late induction of CREB/ATF binding and a concomitant increase in cAMP levels in T and B lymphocytes stimulated via the antigen receptor. J Immunol 1996;156:4582–4593.

93. Xing J, Ginty DD, Greenberg ME. Coupling of the RAS-MAPK-pathway to gene activation by RSK2, a growth factor-regulated CREB kinase. Science 1996;273:959–963.

94. Matthews RP, Guthrie CR, Wailes LM, Zhao X, Means AR, McKnight GS. Calcium/calmodulin-dependent protein kinase types II and IV differentially regulate CREB-dependent gene expression. Mol Cell Biol 1994;14:6107–6116.

95. De Groot R, Ballou LM, Sassone-Corsi P. Positive regulation of the cAMP-responsive activator CREM by the p70 S6 kinase: an alternative route to mitogen-induced gene expression. Cell 1994;79:81–91.

96. Shimizu M, Nomura Y, Suzuki H, et al. Activation of the Rat cyclin A promoter by ATF2 and Jun family members and its suppression by ATF4. Exp Cell Res 1998;239:93–103.

97. Nagata D, Suzuki E, Nishimatsu H, et al. Transcriptional actiavtion of the cyclin D_1 gene is mediated by multiple cis-elements, including SP1 sites and a cAMP-responsive element in vascular endothelial cells. J Biol Chem 2001;276:662–669.

98. Vanhoutte P, Barrier J, Gilbert B, et al. Glumate induces phosphorylation of Elk-1 and CREB, along with c-fos activation, via an extracellular signal-regulated kinase-dependent pathway in brain slices. Mol Cell Biol 1999;19: 136–146.

99. Arcinas M, Heckman CA, Mehew JW, Boxer LM. Molecular mechanisms of transcriptional control of bcl-2 and c-myc in follicular and transformed lymphoma. Cancer Res 2001;61:5202–5206.

100. Treisman R. Regulation of transcription by MAP kinase cascades. Curr Opin Cell Biol 1996;8:205–215.

101. Wasylyk B, Hahn SL, Giovane A. The Ets family of transcription factors. Eur J Biochem 1993;211:7–18.

102. Vananzoni MC, Robinson LR, Hodge DR, Kola I, Seth A. ETS1 and ETS2 in p53 regulatin: spatial separation of ETS binding sites (EBS) modulate protein:DNA interaction. Oncogene 1996;12:1199–1204.

103. Liu SH, Ng SY. Serum response factor associated ETS proteins: ternary complex factors and PEA3-binding factor. Biochem Biophys Res Commun 1994;201:1406–1413.

104. Lambert PE, Ludford-Menting MJ, Deacon NJ, Kola I, Doherty RR. The nfκb1 promoter is controlled by proteins of the Ets family. Mol Biol Cell 1997;8:313–323.

105. Albanese C, Johnson J, Watanabe G, et al. Transforming p21ras mutants and c-Ets-2 activate the cyclin D_1 promoter through distinguishable regions. J Biol Chem 1995;270: 23,589–23,597.

106. Tamir A, Howard J, Higgins RR, et al. Fil-1, an Ets-related transcription factor, regulates erythropoietin-induced erythroid proliferation and differentiatin evidence for direct transcriptional repression of the Rb gene during differentiation. Mol Cell Biol 1999;19: 4452–4464.

107. Beier F, Taylor AC, LuValle P. The Raf-1/MEK/ERK pathway regulates the expression of the p21$^{Cip1/Waf1}$ gene in chondrocytes. J Biol Chem 1999;274:30,273–30,279.

108. Frampton J, Ramqvist T, Graf T. v-Myb of E26 leukemia virus up-regulates bcl-2 and suppress apoptosis in myeloid cells. Genes Dev 1996;10:2720–2731.

109. Sevilla L, Aperlo C, Dulic V, et al. The Ets2 transcription factor inhibits apoptosis induced by colony stimulating factor 1 deprivation of macrophages through a Bcl-xL-dependent mechanism. Mol Cell Biol 1999;19:2624–2634.

110. Li XR, Chong AS, Wu J, et al. Transcriptional regulation of Fas gene expression by GA-binding protein and AP-1 in T cell antigen receptor. CD3 complex-stimulated T cells. J Biol Chem 1999;274:35,203–35,210.

111. Nimer S, Zhang J, Avraham H, Miyazaki Y. Transcriptional regulation of interleukin-3 expression in megakaryocytes. Blood 1996;88:66–74.

112. Maul RS, Zhang H, Reid JDT, Pedigo NG, Kaetzel DM. Identification of a cell type-specific enhancer in the distal 5′-region of the platelet-derived growth factor A-chain gene. J Biol Chem 1998;273:33,239–33,246.

113. McCarthy SA, Chen D, Yang BS, et al. Rapid phosphorylation of Ets-2 accompanies mitogen-activated protein kinase activation and the induction of heparin-binding epidermal growth factor gene expression by oncogenic Raf-1. Mol Cell Biol 1997;17: 2401–2412.

114. Le Gallic L, Sgouras D, Beal G, Jr, Mavrothalassitis G. Transcriptional repressor ERF is a Ras/Mitogen-Activated Protein Kinase target that regulates cellular proliferation. Mol Cell Biol 1999;19:4121–4133.

115. Sgambato V, Vanhoutte P, Pages C, Rogard M, Hipskind R, Besson MJ, Caboche J. *In vivo* expression and regulation of Elk-1, a target of the extracellular-regulated kinase signaling pathway, in the adult rat brain. J Neurosci 1998;18:214–226.

116. Yang B, Hauser CA, Henkel G, et al. Ras-mediated phosphorylation of a conserved threonine residue enhances the transactivation activities of c-Ets1 and c-Ets2. Mol Cell Biol 1996;16:538–547.

117. Yang SH, Yates PR, Whitmarsh AJ, Davis RJ, Sharrocks AD. The Elk-1 ETS-domain transcription factor contains a mitogen-activated protein kinase targeting motif. Mol Cell Biol 1998;18:710–720.

118. Binetruy B, Smeal T, Karin M. Ha-Ras augments c-Jun activity and stimulates phosphorylation of its activation domain. Nature 1991;351:122–127.

119. Meyer-Vehn T, Covacci A, Kist M, Pahl HL. Helicobacter pylori activates mitogen-activated protein kinase cascades and induces expression of the proto-oncogenes c-fos and c-jun. J Biol Chem 2000;275:16,064–16,072.

120. Premkumar DR, Adhikary G, Overholt JL, Simonson MS, Cherniack NS, Prabhakar NR. Intracellular pathways linking hypoxia to activation of c-fos and AP-1. Adv Exp Med Biol 2000;475:101–109.

121. Sgouras DN, Athanasiou MA, Beal GJ, Jr, Fisher RJ, Blair DG, Mavrothalassitis GJ. ERF: an ETS domain protein with strong transcriptional repressor activity, can suppress ets-associated tumorigenesis and is regulated by phosphorylation during cell cycle and mitogenic stimulation. EMBO J 1995;14:4781–4793.

122. Lallemand D, Spyrou G, Yaniv M, Pfarr CM. Variations in Jun and Fos protein expressin and AP-1 activity in cycling, resting and stimulated fibroblasts. Oncogene 1997;14:819–830.

123. Wang D, Westerheide SD, Hanson JL, Baldwin AS, Jr. TNFα-induced phosphorylation of RelA/p65 on ser529 is controlled by casein kinase II. J Biol Chem 2000;275: 32,592–32,597.

124. Angel P, Hattori K, Smeal T, Karin M. The jun proto-oncogene is positively autoregulated by its product, Jun/AP-1. Cell 1988;55:875–885.

125. Han TH, Prywes R. Regulatory role of MEF2D in serum induction of the c-jun promoter. Mol Cell Biol 1995;15:2907–2915.

126. Chen J, Fujii K, Zhang L, Roberts T, Fu H. Raf-1 promotes cell survival by antagonizing apoptosis signal-regulating kinase 1 through a MEK-ERK independent mechanism. Proc Natl Acad Sci USA 2001;98:7783–7788.

127. Weitzman JB. Quick guide, Jnk. Curr Biol 2000;10:R290.

128. Smeal T, Binetruy B, Mercola DA, Grover-Bardwick A, Heidecker G, Rapp UR, Karin M. Oncoprotein-mediated signaling cascade stimulates c-Jun activity by phosphorylation of serine 63 and 73. Mol Cell Biol 1992;12:3507–3513.

129. Alvarez E, Northwood IC, Gonzalez FA, et al. Pro-Leu-Ser/Thr-Pro is a consensus primary sequence for substrate protein phosphorylation. Characterization of the phosphorylation of c-myc and c-jun proteins by an epidermal growth factor receptor threonine 669 protein kinase. J Biol Chem 1991;266:15,277–15,285.

130. Chou SY, Baichwal V, Ferrell JE, Jr. Inhibition of c-Jun DNA binding by mitogen-activated protein kinase. Mol Biol Cell 1992;3:1117–1130.

131. Nasi S, Ciarapica R, Jucker R, Rosati J, Soucek L. Making decisions through Myc. FEBS Lett 2001;490:153–162.

132. Amati B, Frank SR, Donjerkovic D, Taubert S. Function of the c-Myc oncoprotein in chromatin remodeling and transcription. Biochimica et Biophysica Acta 2001;1471: M135–M145.

133. Prendergast GC. Mechanisms of apoptosis by c-Myc. Oncogene 1999;18:2967–2987.

134. Dang CV. c-Myc target genes involved in cell growth, apoptosis, and metabolism. Mol Cell Biol 1999;19:1–11.

135. Iritani BM, Eisenman RN. c-Myc enhances protein synthesis and cell size during B lymphocyte development. Proc Natl Acad Sci USA 1999;96:13,180–13,185.

136. Schmidt EV. The role of c-myc in cellular growth control. Oncogene 1999;18:2988–2996.

137. Gupta S, Davis R. MAP kinase binds to the NH2-terminal activation domain of c-Myc. FEBS Lett 1994;353:281–285.

138. Henriksson M, Barkardjiev A, Klein G, Luscher B. Phosphorylation sites mapping in the N-terminal domain of c-myc modulate its transforming potential. Oncogene 1993;8:3199–3209.

139. Seth A, Alvarez E, Gupta S, Davis RJ. A phosphorylation site located in the NH_2-terminal domain of c-Myc increases transactivation of gene expression. J Biol Chem 1991;266: 23,521–23,524.

140. Gregory MA, Hann SR. c-Myc proteolysis by the ubiquitin-proteosome pathway: stabliaztion of c-Myc in Burkitt's lymphoma cells. Mol Cell Biol 2000;20:2423–2435.

141. Salghetti ES, Kim SY, Tansey WP. Destruction of Myc by ubiquitin-mediated proteolysis:cancer-associated and transforming mutations stablize Myc. EMBO J 1999;18: 717–726.

142. Niklinski J, Claassen G, Meyers C, et al. Disruption of Myc-tubulin interaction by hyperphosphorylation of c-Myc during mitosis or by constitutive hyperphosphorylation of mutant c-Myc in Burkitt's lymphoma. Mol Cell Biol 2000;20:5276–5284.

143. Zha J, Harada H, Yang E, Jockel J, Korsmeyer SJ. Serine phosphorylation of death agonist BAD in response to survival factor results in binding to 14-3-3 not Bcl-XL. Cell 1996;87: 589–592.

144. Allan LA, Morrice N, Brady S, Magee G, Pathak S, Clarke PR. Inhibition of caspase-9 through phosphorylation at Thr 125 by ERK MAPK. Nature Cell Biol 2003;7:647–654.

145. Crdone MH, Roy N, Stennicke HR, et al. Regulation of cell death protease caspase 9 by phosphorylation. Science 1998;282:1318–1321.

146. Deng X, Ruvolo P, Carr B, May WS, Jr. Survival function of ERK1/2 as an IL-3 activated, staurosporine-resistant Bcl2 kinases. Proc Nat Acad USA 2000;97:1578–1583.

147. Deng X, Kornblau SM, Ruvulo PP, May WS, Jr. Regualtion of Bcl2 phosphorylation and potential significance for leukemic cell chemoresistance. JNCI Monographs 2001;28:30–37.

148. Du J, Cai SH, Shi Z, Nagase F. Binding activity of H-Ras is nessary for in vivo inhibition of ASK1 activity. Cell Research. 14:2004;148–154.

149. McCubrey JA, Steelman LS, Hoyle PE, et al. Differential abilities of activated Raf onco-proteins to abrogate cytokine-dependency, prevent apoptosis and induce autocrine growth factor synthesis in human hematopoietic cells. Leukemia 1998;12:1903–1929.

150. Blalock WL, Pearce M, Steelman LS, et al. A conditionally-active form of MEK1 results in autocrine transformation of human and mouse hematopoeitic cells. Oncogene 2000;19: 526–536.

151. Blalock WL, Moye PW, Chang F, et al. Combined effects of aberrant MEK1 activity and BCL2 overexpression on relieving the cytokine-dependency of human and murine hemato-poietic cells. Leukemia 2000;14:1080–1096.

152. Moye PW, Blalock WL, Hoyle PE, et al. Synergy between Raf and BCL2 in abrogating the cytokine-dependency of hematopoietic cells. Leukemia 2000;14:1060–1079.

153. Blalock WL, Pearce M, Chang F, et al. Effects of inducible MEK1 activation on the cytokine-dependency of lymphoid cells. Leukemia 2001;15:794–807.

154. Shelton JG, Steelman LS, Lee JT, et al. Effects of the Raf/MEK/ERK and PI3K signal transduction pathways on the abrogation of cytokine dependence and prevention of apop-tosis in hematopoietic cells. Oncogene 2003;24:2478–2492.

155. Akula SM, Ford PW, Whitman AG, et al. B-Raf dependent expression of vascular endothe-lial growth factor-A in Kaposi's sarcoma-associated herpesvirus infected human B cells. Blood 2005;105:4516–4522.

156. Akula SM, Ford PW, Whitman AG, Hamden KH, Shelton JG, McCubrey JA. Raf promotes human herpesvirus-8 (HHV-8/KSHV) infection. Oncogene 2004;23:5227–5241.

157. Whitman AG, Hamden KH, Ford PW, McCubrey JA. Role for Raf in the entry of viruses associated with AIDS. Int J Oncol 2004;469–480.

158. Ford PW, Hamden KE, Whitman AG, McCubrey JA, Akula SM. Vascular endothelial growth factor augments human Herpesvirus-8 (HHV-8/KSHV) infection. Cancer Biol Ther 2004;3:876–881.

159. Hamden KE, Ford PW, Whitman AG, McCubrey JA, Akula SM. Raf induced Vascular Endothelial Growth Factor Augments Kaposi's Sarcoma-Associated Herpesvirus (KSHV/HHV-8) Infection. J. Virol. 2004;78:13,381–13,390.

160. Hamden KE, Ford PW, Whitman AG, Shelton JG, McCubrey JA, Akula SM. Raf and VEGF: Emerging Therapeutic targets in Kaposi's Sarcoma-Associated Herpesviurs Infection and Angiogenesis in Hematopoietic and Non-Hematopoietic Tumors. Leukemia 2005;19: 18–26.

161. Alexia C, Fallot G, Lasfer M, Schweizer-Groyer G, Groyer A. An evaluation of the role of insulin-like growth factors (IGF) and of type-I IGF receptor in signaling in hepatocarcino-genesis and in the resistance of hepatocarcinoma cells against drug-induced apoptosis. Biochemical Pharmacology 2004;68:1003–1015.

162. Davies H, Bignell GR, Cox C, et al. Mutations of the BRAF gene in human cancer. Nature 2002;417:949–954.

163. Lee JW, Yoo NJ, Soung YH, et al. BRAF mutations in non-Hodgkin's lymphoma. Br J Cancer 2003;89:1958–1960.

164. Lee JW, Soung YH, Park WS, et al. BRaf mutations in acute leukemias. Leukemia 2004; 18:170–172.

165. Zebisch A, Staber PB, Fischereder K, et al. Two novel activating germiline mutations of the c-Raf proto-oncogene predisposing to solid tumors and therapy related acute myeloid leukemia. Blood Supplement. 2004;11:920.

166. Pedersen-Bjergaad J, Christiansen DH, Andersen MK. Genetic pathways in t-MDS and t-AML, a revised model. Blood Supplement. 2004;11:818.

167. Fransen K, Klinntenas M, Osterstrom A, Dimberg J, Monsteis HJ, Soderkvist P. Mutation analysis of the B-Raf, A-Raf and Raf-1 genes in human colorectal adenocarcinomas. Carcinogensis 2004;25:527–533.

168. Wan PT, Garnett MJ, Ros SM, et al. Mechanism of activation of the Raf-MEK signaling pathway by oncogenic mutations of B-Raf. Cell 2004;116:856–867.

169. Busca R, Abbe P, Mantous F, et al. Ras mediates the cAMP-dependent activation of extracellular signal-regulated in melanocytes. EMBO J. 2000;19:2900–2910.

170. Rajagopalan H, Bordelli A, Lengauer C, Kinzler KN, Vogelstein B, Velculesccu VE. Tumorigenesis: Raf/Ras oncogenes and mismathch-repair status. Nature 2002;418:934.

171. Yuen ST, Davies H, Chan TL, et al. Similarity of the phenotypic patterns associated with B-Raf and KRAS mutations in colorectal neoplasia. Cancer Research 2002; 62:6451–6455.

172. Keller ET, Fu Z, Brennan M. The role of Raf kinase inhibitor protein (RKIP) in helath and disease. Biochemical Pharmacology 2004;68:1049–1053.

173. Keller ET. Metastasis suppressor genes: a role for raf kinase inhibitor protein (RKIP). Anti-Cancer Drugs 2004;15:663–669.

174. Keller ET, Fu Z, Yeung K, Brennan M. Raf kinase inhibitor protein: a prostate cancer metastasis suppressor gene. Cancer Letters 2004;207:131–137.

175. Fu Z, Smith PC, Zhang L, et al. Effects of raf kinase inhibitor protein on suppression of prostate cancer metastasis. JNCI 2003;95:878–889.

176. Steelman LS, Bertrand FE, McCubrey JA. The complexity of PTEN: mutation, marker and potential target for therapeutic intervention. Expert Opinion Therapeutic Targets. 2004;8:537–550.

177. Lee JT, Steelman, LS, McCubrey JA. Modulation of Raf/MEK/ERK pathway in prostate cancer drug resistance. Int J Oncol 2005;26(6):1637–1645.

178. Lee JT, Steelman LS, McCubrey JA. PI3K Activation leads to MRP1 expression and subsequent chemoresistance in advanced prostate cancer cells. Cancer Res 2004;64: 8397–8404.

179. Blagosklonny MV. The mitogen-activated protein kinase pathway mediates growth arrest or E1A-dependent apoptosis in SKBR3 human breast cancer cells. Int J Cancer 1998;78: 511–517.

180. Blagosklonny MV, Prabhu NS, El-Deiry WS. Defects in p21WAF1/CIP1, Rb and c-myc signaling in phorbol ester-resistant cancer cells. Cancer Res 1997;57:320–325.

181. Ravi RK, McMahon M, Yangang Z, et al. Raf-1-induced cell cycle arrest in LNCap human prostate cancer cells. J Cell Biochem. 1999;72:458–469.

182. Blagosklonny MV, Schulte T, Nguyen P, Trepel J, Neckers LM. Taxol-induced apoptosis and phosphorylation of Bcl-2 protein involves c-Raf-1 and represents a novel c-Raf-1 signal transduction pathway. Cancer Res 1996;56:1851–1854.

183. Blagosklonny MV. Drug-resistance enables selective killing of resistant leukemia cells: exploiting of drug resistance instead of reversal. Leukemia 1999;13:2031–2035.

184. Weinstein-Oppenheimer CR, Henríquez-Roldán CF, Davis J, et al. Role of the Raf signal transduction cascade in the in vitro resistance to the anticancer drug doxorubicin. Clinical Cancer Res 2001;7:2892–2907.

185. Weinstein-Oppenheimer CR, Burrows C, Steelman LS, McCubrey JA. The effects of β-estradiol on Raf activity, cell cycle progression and growth factor synthesis in the MCF-7 breast cancer cell line. Cancer Biol Ther 2002;1:254–260.

186. Davis JM, Weinstein-Oppenheimer CR, Steelman LS, et al. Raf-1 and Bcl-2 induce distinct and common pathways which contribute to breast cancer drug resistance. Clinical Cancer Res 2003;9:1161–1170.

187. Kim SH, Lee SH, Kwak NH, Kang CD, Chung BS. Effect of the activated Raf protein kinase on the human multidrug resistance 1 (MDR1) gene promoter. Cancer Letters 1996;98:199–205.

188. Mitenberger RJ, Farham PJ, Smith DE, Stommell JM, Cornwell MM. v-Raf activates transcription of growth-responsive promoters via GC-rich sequences that bind the transcription factor Sp1. Cell Growth Diff 1995;6:549–556.

189. Basu A, Haldar S. Microtubule-damaging drugs triggered bcl2 phosphorylation-requirement of phosphorylation on both serine-70 and serine-87 residues of bcl2 protein. Int J Oncol 1998;13:659–664.

190. Haldar S, Jena N, Croce CM. Antiapoptosis potential of bcl-2 oncogene by dephosphorylation. Biochem cell Biol 1994;72:455–462.

191. Friesen C, Fulda S, Debatin KM. Induction of CD95 ligand and apoptosis by doxorubicin is modulated by the redox state in chemosensitive- and drug-resistant tumor cells. Cell Death Differ 1999;6:471–480.

192. Kobayashi D, Tokino T, Watanabe N. Contribution of caspase-3 differs by p53 status in apoptosis induced by X-irradiation. Jpn J Cancer Res 2001;92:475–481.

193. Yoshikawa T, Kokura S, Oyamada H, et al. Antitumor effect of ischemia-reperfusion injury induced by transient embolization. Cancer Res 1994;54:5033–5035.

194. Oleinick NL, Evans HH. The photobiology of photodynamic therapy: cellular targets and mechanisms. Radiat Res 1998;150:S146–S156.

195. Pervaiz S. Reactive oxygen-dependent production of novel photochemotherapeutic agents. Faseb J 2001;15:612–617.

196. Dvorakova K, Payne CM, Tome ME, Briehl MM, McClure T, Dorr RT. Induction of oxidative stress and apoptosis in myeloma cells by the aziridine-containing agent imexon. Biochem Pharmacol 2000;60:749–758.

197. McLaughlin KA, Osborne BA, Goldsby RA. The role of oxygen in thymocyte apoptosis. Eur J Immunol 1996;26:1170–1174.

198. Matroule JY, Carthy CM, Granville DJ, Jolois O, Hunt DW, Piette J. Mechanism of colon cancer cell apoptosis mediated by pyropheophorbide-a methylester photosensitization. Oncogene 2001;20:4070–4084.

199. Robinson MJ, Cobb MH. Mitogen-activated protein kinase pathways. Curr Opin Cell Biol 1997;9:180–186.

200. Yuan J. Transducing signals of life and death. Curr Opin Cell Biol 1997;9:247–251.

201. Tong Z, Singh G, Rainbow AJ. Sustained activation of the extracellular signal-regulated kinase pathway protects cells from photofrin-mediated photodynamic therapy. Cancer Res 2002;62:5528–5535.

202. Lee K, Esselman WJ. cAMP potentiates $H(2)O(2)$-induced ERK1/2 phosphorylation without the requirement for MEK1/2 phosphorylation. Cell Signal 2001;13:645–652.

203. Zhougang S, Schnellmann RG. H2O2-induced transactivation of EGF receptor requires Src and mediates ERK1/2, but not Akt, activation in renal cells. Am J Physiol Renal Physiol 2004;286:F858–F865.

204. Blanc A, Pandey NR, Srivastava AK. Synchronous activation of ERK 1/2, p38mapk and PKB/Akt signaling by H2O2 in vascular smooth muscle cells: potential involvement in vascular disease (review). Int J Mol Med 2003;11:229–234.

205. Lander H, Jacovina A, Davis R, Tauras J. Differential activation of mitogen-activated protein kinases by nitric oxide-related species. J Biol Chem 1996;271:19,705–19,709.

206. Hoyos B, Imam A, Korichneva I, Levi E, Chua R, Hammerling U. Activation of c-Raf kinase by ultraviolet light. Regulation by retinoids. J Biol Chem 2002;277:23,949–23,957.

207. Abe J, Berk BC. Fyn and JAK2 mediate Ras activation by reactive oxygen species. J Biol Chem 1999;274:21,003–21,010.

208. Abe J, Okuda M, Huang Q, Yoshizumi M, Berk BC. Reactive oxygen species activate p90 ribosomal S6 kinase via Fyn and Ras. J Biol Chem 2000;275:1739–1748.

209. Accorsi K, Giglione C, Vanoni M, Parmeggiani A. The Ras GDP/GTP cycle is regulated by oxidizing agents at the level of Ras regulators and effectors. FEBS Lett 2001;492: 139–145.

210. Ye J, Ding M, Zhang X, Rojanasakul Y, Shi X. On the role of hydroxyl radical and the effect of tetrandrine on nuclear factor—kappaB activation by phorbol 12-myristate 13-acetate. Ann Clin Lab Sci 2000;30:65–71.

211. Korichneva I, Hoyos B, Chua R, Levi E, Hammerling U. Zinc release from protein kinase C as the common event during activation by lipid second messenger or reactive oxygen. J Biol Chem 2002;277:44,327–44,331.

212. Ellis CA, Glark G. The importance of being K-Ras. Cellular Signalling 2000;12:425–434.

213. Heimbrook DC, Huber HE, Stirdivant SM, et al. Identification of potent, selective kinase inhibitors of Raf. Proc Amer Assoc Cancer Res 1998;39:558.

214. Hall-Jackson CA, Eyers PA, Cohen P, et al. Paradoxical activation of Raf by a novel Raf inhibitor. Cehm Biol 1999;6:559–568.

215. Lyons JF, Wilhelm S, Hibner B, Bollag G. Discovery of a novel Raf kinase inhibitor. Endocrine-Related Cancer 2001;8:219–225.

216. Lee JT, Jr, and McCubrey JA. BAY 43-9006 Bayer Current Opinion in Investigational Drugs 2003;4:757–763.

217. Shelton JG, Moye PW, Steelman LS, et al. Differential effects of kinase cascade inhibitors on neoplastic and cytokine-mediated cell proliferation. Leukemia 2003;17:1765–1782.

218. Blagosklonney MV. Hsp-90-associated oncoproteins: multiple targets of geldanamycin and its analogs. Leukemia 2002;16:455–462.

219. Strumberg D, Richly H, Hilger RA, et al. Phase I clinical and pharmacokinetic study of the Novel Raf kinase and vascular endothelial growth factor receptor inhibitor BAY 43-9006 in patients with advanced refractory solid tumors. J Clin Oncol 2005;23: 965–972.

220. Workman P. Altered states:selectively drugging the Hsp90 cancer chaperone. Trends in Molecular Medicine 2004;10:47–51.

221. Neckers L. Hsp90 inhibitors as novel cancer chemotherapeutic agents. Trends in Molecular Medicine 2002;8:S55–S61.

222. Marcu MG, Schulte TW, Neckers L. Novobiocin and related coumarins and depletion of heat shock protein 90-dependent signaling proteins. JNCI 2000;92:242–248.

223. Strumberg D, Voliotis D, Moeller JG, et al. Results of phase I pharmacokinetic and pharmacodynamic studies of the Raf kinase inhibitor BAY 43-9006 in patients with solid tumors. Int J Clin Pharma Thera 2002;40:580–581.

224. Wilhelm SM, Carter C, Tang L, et al. BAY 43-9006 exhibits broad spectrum oral antitumor activity and targets the RAF/MEK/ERK pathway and receptor tyrosine kinases involved in tumor progression and angiogenesis. Cancer Res 2004;64:7099–7109.

225. Richly H, Kupsch P, Passaga K, et al. A phase I clinical and pharmacokinetic study of the Raf kinase inhibitor (RKI) BAY 43-9006 administered in combination with doxorubicin in patients with solid tumors. Int J Clin Pharma Thera 2003;41:620–621.

226. Mross K, Steinbild S, Baas F, et al. Drug-drug interaction pharacokinetic study with the Raf kinase inhibitor (RKI) BAY 43-9006 administered in combination with irinotecan (CPT-11) in patients with solid tumors. Int J Clin Pharma Thera 2003;41:618–619.

227. Heim M, Sharifi M, Hilger RA, Scheulen ME, Seeber S, Strumberg D. Antitumor effect and potentiation or reduction in cytotoxic drug activity in human colon carcinoma cells by Raf kinase inhibitor (RKI) BAY 43-9006. Int J Clin Pharma Thera 2003;41:616–617.

228. Yu C, Dai Y, Dent P, Grant S. Coadministration of UCN-01 with MEK1/2 inhibitors potently induces apoptosis in BCR/ABL+ leukemia cells sensitive and resistant to STI571. Cancer Biol Ther 2002;1:674–682.

229. Allen LF, Sebolt-Leopold J, Meyer MB. CI-1040 (PD184352), a targeted signal transduction inhibitor of MEK (MAPKK). Sem Oncol 2003;30:105–116.

230. Milella M, Estrov Z, Kornblau SM, et al. Synergistic induction of apoptosis by simultaneous disruption of the Bcl-2 and MEK/MAP pathways in acute myelogenous leukemia. Blood 2002;99:3461–3464.

231. Milella M, Kornblau SM, Estrov Z, et al. Therapeutic targeting of the MEK/MAPK signal transduction module in acute myeloid leukemia. J Clin Inv 108:851–859.

232. Mazzucchelli C, Vantaggiato C, Ciamei A, et al. Knockout of ERK1 MAP kinase enhances synaptic plasticity in the striatum and facilitates striatal-mediated learning and memory. Neuron 2002;34:807–820.

233. Pages G, Guerin S, Grall D, et al. Defective thymocyte maturation in p44 MAP kinase (Erk 1) knockout mice. Science 1999;286:1374–1377.

234. Selcher JC, Nekrasova T, Paylor R, Landreth GE, Sweatt JD. Mice lacking the ERK1 isoform of MAP kinase are unimpaired in emotional learning. Learning & Memory 2001;8:11–19.

235. Loda M, Capodieci P, Mishra R, et al. Expression of mitogen-activated protein kinase phosphatase-1 in the early phases of human epithelial carcinogenesis. Amer J Pathol 1996;149:1553–1564.

236. Magi-Galluzzi C, Mishra R, Fiorentino M, et al. Mitogen-activated protein kinase phosphatase 1 is overexpressed in prostate cancers and is inversely related to apoptosis. Lab Invest 1997;76:37–51.

6

MAPK Signaling in Human Diseases

Philippe P. Roux and John Blenis

Summary

Conserved signaling pathways that activate the mitogen-activated protein kinases (MAPKs) are involved in a vast array of biological processes, including cell proliferation, differentiation, motility, and survival. Three subfamilies of MAPK have been extensively studied: extracellular signal-regulated kinases (ERK), p38-MAPK, and c-Jun N-terminal kinase (JNK). The ERKs play roles in cell proliferation, survival, and motility, and inhibitors of this pathway are currently being tested as potential anticancer agents. The p38-MAPKs are involved in the immune response and the response to stress, and have been suggested to contribute to diseases such as asthma and autoimmunity. The JNKs are critical regulators of transcription and apoptosis, and JNK pathway inhibitors are currently being tested for the treatment of rheumatoid arthritis and neurodegenerative diseases. This chapter reviews the biological functions of the MAPKs with an emphasis on the regulation of cell survival and proliferation by these enzymes, and will also discuss the development of MAPK pathway inhibitors for the treatment of human diseases.

Key Words: MAPK; ERK; JNK; p38; RSK; signaling; cancer; proliferation; survival.

1. Introduction

Cells respond to extracellular stimuli by engaging specific signaling cascades, such as those leading to activation of the mitogen-activated protein kinases (MAPKs). Five groups of MAPKs have been characterized in mammals: ERK1/2 (extracellular signal-regulated kinases 1/2), JNK1/2/3 (c-Jun amino (N)-terminal kinases 1/2/3), p38-MAPKs (α, β, γ, δ), ERK3/4, and ERK5 *(1,2)*. These MAPKs have been involved in many cellular activities, including gene expression, cell-cycle regulation, metabolism, motility, survival, and apoptosis, but the ERKs, JNKs, and p38-MAPKs represent the most extensively studied groups of vertebrate MAPKs. It is generally thought that ERK1/2 are preferentially activated in response to mitogens such as growth factors and phorbol esters, and that the JNKs and p38-MAPKs are more responsive to stress stimuli, including osmotic shock, ionizing radiation, and cytokine stimulation *(3)* *(see* Fig. 1).

Although each MAPK has unique characteristics a number of features are shared by the MAPK pathways studied to date. First, each pathway is composed of a set of three evolutionarily conserved sequentially acting kinases: a MAPK, a MAPK kinase (MAPKK), and a MAPKK kinase (MAPKKK) *(see* Fig. 1). Second, MAPKK activation leads to the stimulation of MAPK activity through dual phosphorylation on Thr and Tyr residues (Thr-Xxx-Tyr) located in the activation loop of the MAPK. Finally, once activated, all MAPKs phosphorylate target substrates on a Ser or Thr followed by a Pro residue. Substrate selectivity is often conferred by specific interaction motifs located on physiological substrates,

From: *Apoptosis, Cell Signaling, and Human Diseases: Molecular Mechanisms, Volume 2*
Edited by R. Srivastava © Humana Press Inc., Totowa, NJ

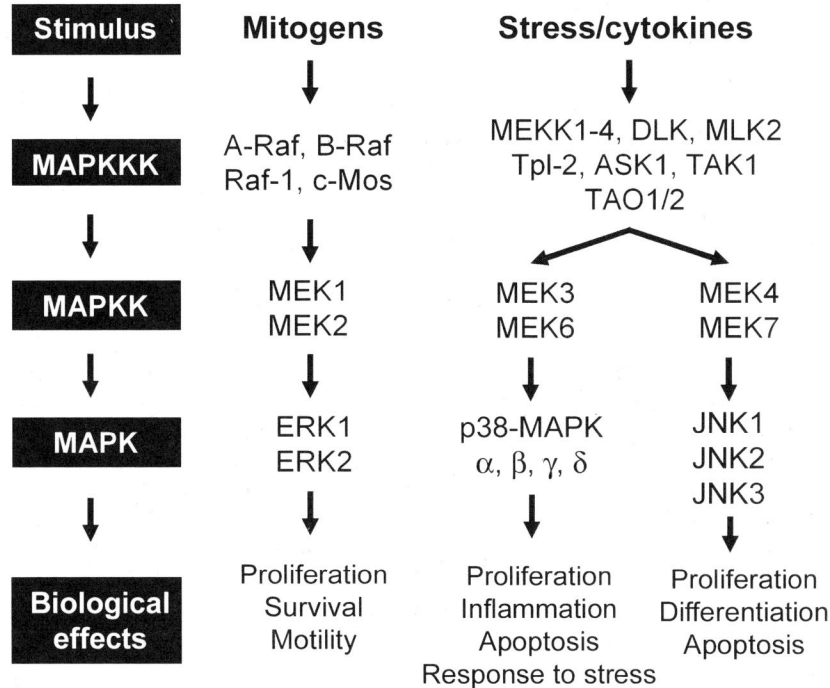

Fig. 1. Signaling cascades leading to activation of the MAPKs. Mitogens and cellular stresses lead to activation of ERK1/2, p38-MAPK and JNK. Activation of these signaling cascades is responsible for the regulation of a vast array of biological processes.

such as D and/or DEF domains *(4)*, but MAPK cascade specificity can also be determined through interaction with scaffolding proteins that organize pathways in specific modules through simultaneous binding of several components.

The wide range of MAPK functions are mediated through phosphorylation of several downstream substrates, including phospholipases, transcription factors, cytoskeletal proteins, and kinases. This chapter will review the biological functions of the MAPKs and their substrates, and will discuss the roles played by the different MAPKs in human diseases.

2. ERK

2.1. Properties

ERK1 was discovered 15 yr ago as a Ser/Thr kinase activated by extracellular stimuli that phosphorylated microtubule-associated protein-2 (MAP2) *(5)*. ERK1 and ERK2 are highly homologous proteins (83% amino acid identity) expressed to varying levels in all tissues *(1)*. ERK1/2 are activated downstream of the *classical* MAPK cascade that comprises MEK1 and MEK2 (MAPKK), as well as A-Raf, B-Raf and Raf-1 (MAPKKK) (*see* Fig. 1). This pathway is activated by growth factors, serum, phorbol esters, ligands of the heterotrimeric G protein-coupled receptors (GPCRs), cytokines and, to a lesser extent, by insulin (under most physiological conditions), osmotic stress, and microtubule disorganization *(6)*.

2.2. Activation Mechanisms

Typically, cell surface receptors such as tyrosine kinases (RTK) and GPCRs transmit activating signals to the Raf/MEK/ERK cascade through different isoforms of the small GTP-binding protein Ras (H-Ras, K-Ras and N-Ras) *(7)*. Activation of membrane-associated Ras isoforms is often achieved through recruitment of son of sevenless (SOS), a Ras-activating guanine nucleotide exchange factor (GEF). SOS stimulates Ras to exchange GDP to GTP, allowing it to interact with a wide range of downstream effector proteins, including isoforms of the Ser/Thr kinase Raf *(8)*. Upon activation at the membrane, Raf stimulates the phosphorylation and activation of the dual-specificity kinases MEK1 and MEK2, which are responsible for the activation of ERK1/2. MEK1/2 phosphorylate ERK1/2 at a conserved Thr-Glu-Tyr (TEY) motif within the activation loop segment, resulting in full kinase activation of ERK1/2. Amplification through the Ras/ERK signaling cascade is such that it is estimated that activation of only 5% of Ras molecules is sufficient to induce full activation of ERK1/2 *(9)*. Two structurally unrelated compounds are commonly used to specifically inhibit the ERK1/2 pathway in cultured cells. Both U0126 and PD98059 are noncompetitive inhibitors of MEK1/2/5 and prevent stimulation-mediated activation of ERK1/2/5 *(10)*.

2.3. Substrates and Functions

Upon stimulation of the Raf/MEK/ERK cascade, ERK1/2 redistribute from the cytosol to the nucleus *(11–13)*. Whereas the mechanisms involved in nuclear accumulation remain elusive, nuclear retention, dimerization, phosphorylation, and release from cytoplasmic anchors are thought to play a role *(14)*. Activated ERK1/2 phosphorylate numerous substrates in different cellular compartments, including various membrane proteins (CD120a, Syk, calnexin), nuclear substrates (SRC-1, Pax6, NF-AT, Elk-1, MEF2, c-Fos, c-Myc, and STAT3), cytoskeletal proteins (neurofilaments and paxillin), and several kinases (RSK, MSK, and MNK) *(see* Fig. 2) *(1,15)*. Both the *erk1* and *erk2* genes have been disrupted in mice by homologous recombination (Table 1). Whereas *erk1*$^{-/-}$ animals are viable and fertile *(16)*, disruption of the *erk2* locus leads to embryonic lethality early in mouse development after the implantation stage *(17)*. ERK2 mutant embryos fail to form the ectoplacental cone and extra-embryonic ectoderm, which give rise to mature trophoblast derivatives in the fetus. ERK1 deficient animals have impaired thymocyte proliferation and maturation in response to T-cell receptor ligation *(16)*.

Studies in several model systems have indicated the involvement of the Ras/ERK cascade in the control of cell survival *(10)*. Members of the Raf family have been described as important antiapoptotic effectors using both MEK-dependent and -independent mechanisms *(18)*. Independent of its ability to promote MEK/ERK signaling, activated Raf directly antagonizes the death-promoting activity of ASK1 (apoptosis signal-regulating kinase 1) *(19)* and promotes the prosurvival function of the transcription factor NF-κB *(20)*. In a MEK-dependent manner, Raf-mediated activation of ERK promotes survival in part through the phosphorylation and activation of members of the RSK (p90 ribosomal S6 kinase) family of kinases. The RSK family contains four human isoforms (RSK1, RSK2, RSK3, and RSK4) that play roles in gene transcription,

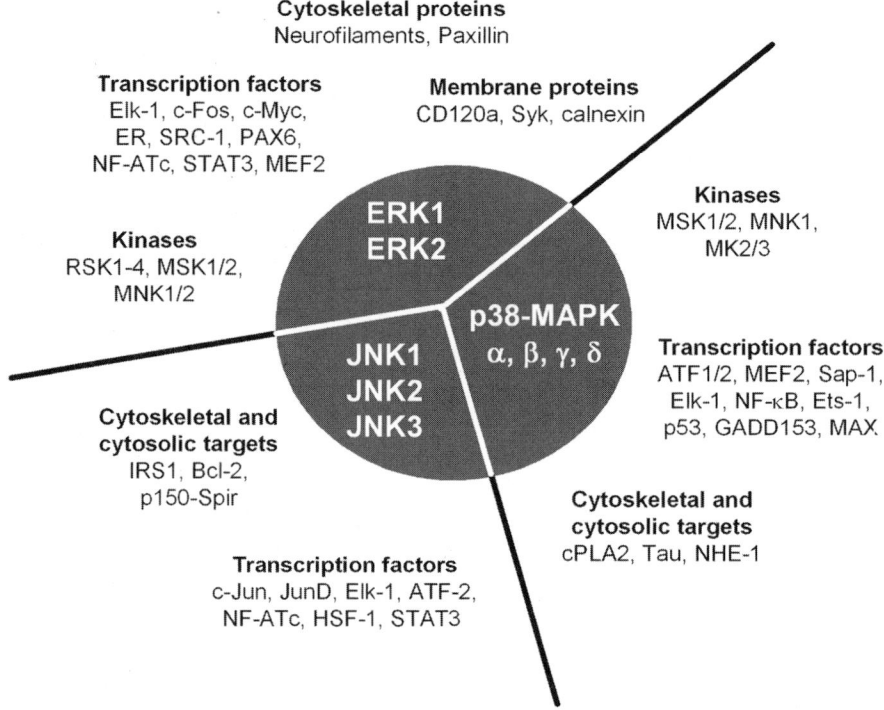

Fig. 2. MAPK substrates. ERK1/2, p38-MAPK and JNK family members regulate many biological processes through the phosphorylation of several downstream substrates. Among these substrates include various transcription factors, cytoskeletal proteins, cytosolic proteins, and kinases.

Table 1
Phenotype of Mice With Disrupted MAPKs

MAPK	Phenotype	Ref.
$erk1^{-/-}$	Viable and fertile. Defective in thymocyte proliferation and maturation.	16
$erk2^{-/-}$	Embryonic lethal at E8.5 from defects in trophoblast development.	17
$erk5^{-/-}$	Embryonic lethal at E10.5 from defects in vascular system development.	107
$p38a^{-/-}$	Embryonic lethal at midgestation due to defects in placental angiogenesis.	65–68
$jnk1^{-/-}$	Viable and fertile. Defective in T cell differentiation.	97
$jnk2^{-/-}$	Viable and fertile. Defective in T cell differentiation and activation.	98,99
$jnk3^{-/-}$	Viable and fertile.	100
$jnk1/jnk2^{-/-}$	Embryonic lethal at E11.5 from defects in neuronal apoptosis.	99,101

cell survival, proliferation, and motility *(15)*. Neurotrophic factor-mediated stimulation of the Ras/ERK cascade leading to RSK2 activation was found to promote survival of primary granule neurons through both transcription-dependent and -independent mechanisms *(21)*. RSK2 phosphorylates the pro-apoptotic protein Bad on Ser112, thereby repressing its death promoting activity *(21)*. A similar regulation of Bad phosphorylation was seen in a haematopoietic cell line, where RSK1-mediated survival required Bad phosphorylation and inactivation *(22)*. Moreover, RSK2-mediated phosphorylation of the transcription factor CREB on Thr133 was found to promote survival of primary neurons through increased transcription of survival-promoting genes *(21,23,24)*. RSK1 was similarly shown to promote survival through the modulation of NF-κB-dependent transcription, by phosphorylating the NF-κB inhibitor IκBα on Ser32 *(25,26)*. RSK1 promotes survival of hepatic stellate cells by phosphorylating C/EBPβ on Thr217 *(27)*, a site also found to promote cell proliferation in response to TGF-α, suggesting that RSK1 may coordinate both cell survival and proliferation.

Growth factors promote proliferation by linking the Ras/ERK signaling pathway with the cell cycle machinery *(29)*. Many groups demonstrated that inhibition of ERK activity using MEK inhibitors reduces proliferation of various cell types *(30–32)*. One link between cell-cycle progression and the Ras/ERK signaling cascade is provided by cyclin D1, whose gene is induced following mitogenic stimulation and regulate cell-cycle entry *(33,34)*. Induction of cyclin D1 is thought to be mediated by activation of the AP-1 transcription factor, which consists of heterodimeric forms of c-Jun and c-Fos proteins *(35)*. The phosphorylation of a network of immediate-early gene products, such as c-Fos, c-Myc and c-Jun, by ERK has been shown to dictate biological outcome *(36)*. For example, sustained ERK signal duration results in c-Fos phosphorylation and stabilization, and promotion of cell-cycle entry *(37,38)*. RSK is another c-Fos kinase that promotes c-Fos protein stability *(39)*, and interestingly, RSK signaling also appears to be required for proper cell proliferation of various cancer cell lines (P. Roux, unpublished results). Moreover, RSK1 was recently shown to phosphorylate and inactivate the TSC2 tumor suppressor protein *(40,41)*, indicating that RSK1 may in fact coordinate cell growth and cell proliferation. Interestingly, a novel RSK-specific inhibitor was recently found to reduce breast and prostate cancer cell proliferation *(42,43)*. Consistent with a role for RSK1 and RSK2 in cell proliferation, these kinases were also found to be amplified in breast and prostate cancer tissue *(42,43)*, suggesting that they may be good molecular targets for anticancer treatments.

Metastasis is the main cause of cancer deaths, and cell motility is a key component of this process. In addition to its role in regulating cell proliferation and survival, ERK signaling has been highlighted in a number of studies of invasive growth and metastasis *(44,45)*. Using effector domain mutants of Ras designed to activate specific downstream effectors (Raf, PI3K, or Ral-GEF), it was shown that activation of the ERK pathway was required to induce lung metastasis in nude mice *(46)*. Although pathways such as PI3K (phosphoinositide 3-kinase) have been shown to play important roles in cell motility, numerous studies have indicated that the ERK pathway can directly promote cell motility and the migration of tumor cells *(44)*. The ERK pathway has been shown to promote cell adhesion turnover and thereby facilitates the process of cell migration through the direct activation of the protease calpain-2 and MLCK *(47–49)*. Cell motility is also controlled by ERK1/2 in the nucleus through the regulation of gene expression. ERK1/2 can regulate the activity of the small GTPases Rac1 and RhoA through the stabilization of the Fra-1 transcription factor, permitting the formation of cellular protrusions and thereby contributing to tumor cell motility and invasion *(50)*. Interestingly, RSK1 has been recently shown to phosphorylate Filamin A *(51)*, suggesting that RSK may also play roles in the regulation of the cytoskeleton downstream of the Ras/ERK pathway. Thus, inhibition of ERK (and possibly RSK) using small molecules may provide a way to block tumor cell invasion and metastasis, as well as survival and proliferation.

2.4. Oncogenic Mutations

Regulation of both Ras and Raf is crucial for the proper maintenance of cell proliferation as activating mutations in these genes lead to oncogenesis *(52)*. Indeed, K-Ras activating mutations have been detected in 15 to 20% of non-small cell lung cancers *(53)*, 40% of colon adenomas *(54)*, and 95% of pancreatic adenocarcinomas *(55)*. Activating mutations in H-Ras and N-Ras have also been observed in some tumors, such as in thyroid cancers *(56)*. B-Raf mutations are most prevalent in melanomas, occurring in nearly 70% of tumor samples analyzed *(57)*. B-Raf mutations are also found at a high frequency in other cancers, such as cancers of colorectal, ovarian, and thyroid origins *(58)*.

In addition to mutational activation, the Ras genes have been shown to be amplified or overexpressed in many types of tumors, including renal, breast, and bladder cancers. Ras signaling can also be stimulated through the increase in coupling to cell surface receptors. Notably, members of the epidermal growth factor (EGF) family of RTK such as EGFR (also called ErbB and HER1) and ErbB2 (also called HER2 and Neu) are commonly amplified in cancers *(59,60)*. Because Ras/ERK signaling has been implicated in cancer cell proliferation, survival and motility, inhibitors of the ERK pathway are entering clinical trials as potential anticancer agents *(58,61)*. ERK inhibitors are currently not available, but the use of the MEK inhibitor PD184352 (CI-1040) gave some promising results in clinical trials with pancreatic cancer patients *(58)*.

3. p38-MAPK

3.1. Properties

The first member of the mammalian p38-MAPK family was simultaneously identified 10 yr ago by two groups, as a kinase that is phosphorylated on Tyr residues upon changes in osmolarity *(62)* and a kinase inhibited by pyridinyl-imidazole that regulates cytokine biosynthesis *(63)*. There are four isoforms of p38-MAPK in mammals, known as p38α, p38β, p38γ, and p38δ, which can all be phosphorylated by the dual-specificity kinases MEK3 and MEK6 (also termed MKK3 and MKK6). In mammalian cells, the p38-MAPK pathway is strongly activated by environmental stresses and inflammatory cytokines, which leads to the activation of several MAPKKKs, including MEKK1-4, MLK2/3, DLK, ASK1, Tpl2, and Tak1 *(2)*. p38α has 50% aa identity with ERK2 and bears significant homology to the product of the budding yeast *hog1* gene which is activated in response to hyper-osmolarity *(62–64)*.

3.2. Activation Mechanisms

Whereas some stress stimuli will also activate the ERK pathway, the p38-MAPK pathway is strongly activated in response to various physical and chemical stresses, such as oxidative stress, ultraviolet (UV) irradiation, hypoxia, ischemia, and various cytokines including interleukin-1 (IL-1) and tumor necrosis factor (TNF-α) *(1)* *(see* Fig. 1). MEK3 and MEK6 show a high degree of specificity towards p38-MAPK as they do not activate other MAPKs. Although MEK6 activates all p38-MAPK isoforms, MEK3 is somewhat selective as it preferentially phosphorylates the p38α and p38β isoforms. Activation of the p38-MAPK isoforms results from the MEK3/6-catalyzed phosphorylation of a conserved Thr-Gly-Tyr (TGY) motif in their activation loop. Most stimuli that activate p38-MAPK also activate JNK, but only p38-MAPK is inhibited by the anti-inflammatory drug SB203580, which has been extremely useful in delineating the biological functions of these stress-activated kinases *(63)*.

3.3. Substrates and Functions

Activated p38-MAPK has been shown to phosphorylate several targets in different cellular compartments, including cytosolic phospholipase A2 (cPLA2), microtubule-associated protein Tau, and transcription factors ATF1/2, MEF2A, Sap-1, Elk-1, NF-κB, Ets-1, and p53 *(2)* *(see* Fig. 2). p38-MAPK also activates several kinases, including MSK1/2, MNK1 and MK2/3, which themselves regulate gene transcription, mRNA translation, and actin reorganization *(15)*. Genetic disruption of p38α has been described by several groups (Table 1), each one concluding that the loss of p38α leads to death during embryogenesis due to defects in the labyrinthine placenta and subsequent poor placental function *(65–68)*.

p38-MAPK appears to play a major role in apoptosis, differentiation, proliferation, inflammation, and other stress responses. p38-MAPK activity was shown to be required for Cdc42-induced cell cycle arrest at G_1/S, and this inhibitory role may be mediated by the inhibition of cyclin D expression *(69)*. Cyclin D was in fact shown to be located at a point of convergence between the ERK1/2 and p38-MAPK pathways *(33)*, and its regulation may play a crucial role in the regulation of the cell cycle. Interestingly, p38-MAPK activity has also been shown to be required for the proliferation of certain cell types, including fibroblasts, T-cells, and breast cancer cells *(70)*. However, the intrinsic oncogenic potential of p38-MAPK, either alone or in combination with other oncogenes has not been reported so far.

A large body of evidence indicates that p38-MAPK activity is critical for normal immune function and inflammatory responses *(71)*. Whereas the exact mechanisms involved in p38-MAPK immune functions are starting to emerge, p38-MAPK activation has been shown to regulate cytokine production through the stabilization of mRNAs involved in this process *(72,73)*. Interestingly, the p38-MAPK-activated kinase MK2 (MAPKAPK2) represents a likely target of p38-MAPK mediating this function *(15,74)*. Recent data using MK2-deficient mice indicated that the catalytic activity of MK2 is necessary for its effects on cytokine production and migration *(75)*, suggesting that MK2 phosphorylates targets involved in mRNA stability. Consistent with this, MK2 has been shown to bind and/or phosphorylate hnRNP A0 (heterogeneous nuclear ribonucleoprotein A0), TTP (tristetraprolin), PABP1 (poly(A)-binding protein), and HuR *(76–80)*. Because the p38-MAPKs are key regulators of inflammatory cytokine expression, they are thought to be involved in human diseases such as asthma and autoimmunity.

The involvement of p38-MAPK in apoptosis is diverse. Although p38-MAPK signaling promotes cell death under certain circumstances *(81)*, it has also been shown to promote survival in others *(81,83)*. Consistent with an antiproliferative role for p38-MAPK, inactivation of p38-MAPK in mice through the disruption of the *mek3* and *mek6* genes, was associated with enhanced proliferation and increased tumorigenesis *(84)*. Induction of p38-MAPK activity in response to UV radiation was shown to initiate cell cycle arrest at the G_2/M *(85,86)* and G_1 checkpoints *(87)*, indicating that p38-MAPK regulates growth arrest, differentiation and apoptosis. Whereas most efforts have been invested in the development of p38-MAPK inhibitors for the treatment of inflammatory diseases, possible applications of the drugs should also be tested for the treatment of cancer *(88)*.

4. JNK

4.1. Properties

The first member of the JNK family was originally isolated from rat livers injected with cycloheximide *(89)*. JNK1, JNK2, and JNK3 (also known as SAPKγ, SAPKα, and SAPKβ) were later cloned by two groups using two different polymerase chain reaction (PCR) based strategies *(90,91)*. The JNKs exist as 10 or more different spliced forms, but aside from having slightly different substrate specificities *(92)*, the significance of these isoforms is not clear *(2)*. All JNK family members are ubiquitously expressed, with the exception of JNK3 which is present primarily in the brain. The JNKs are strongly activated in response to cytokines, UV irradiation, growth factor deprivation, and DNA damaging agents, and to a lesser extent by some GPCRs, serum, and growth factors *(2)*.

4.2. Activation Mechanisms

JNKs are directly activated by the phosphorylation of Thr and Tyr residues on a conserved Thr-Pro-Tyr (TPY) motif located in their activation loop, which is catalyzed by the dual-specificity kinases MEK4 and MEK7. These MAPKKs are themselves phosphorylated and activated by several MAPKKKs, including MEKK1-4, MLK2/3, Tpl-2, DLK, TAO1/2, TAK1, and ASK1/2 *(2)* (*see* Fig. 1). Activation of JNK is also regulated by scaffold proteins such as JIP, β-arrestin, and JSAP1 *(93–95)*.

4.3. Substrates and Functions

The JNK signal transduction pathway is implicated in multiple physiological processes, including cytokine production and other aspects of the inflammatory response, apoptosis, actin reorganization, and cell proliferation *(2)*. Whereas the JNKs have not been shown to activate kinases, a well known substrate for JNK is the transcription factor c-Jun. Phosphorylation of c-Jun on Ser63 and Ser73 by JNK leads to increased c-Jun-dependent transcription *(96)*. Several other transcription factors have been shown to be phosphorylated by the JNKs, such as ATF-2, Elk-1, p53, NF-ATc1, HSF-1, and STAT3 *(1,2)* (*see* Fig. 2).

To elucidate the roles of the JNK isoforms in vivo and in vitro, all JNK alleles have been deleted in mice (Table 1). While the single knockout mice for the *jnk1* or *jnk2* genes are fertile and only display differences in T-cell differentiation *(97–99)*, the single *jnk3* gene knockout mouse has no obvious phenotype *(100)*. Because JNK3 is mostly expressed in the brain, this organ was tested in response to stress such as during kainate-induced seizures. Loss of JNK3 resulted in milder kainate-induced seizures correlating with reduced apoptosis within the hippocampus of knockout animals *(100)*, indicating that JNK3 is influential in neuronal apoptosis following excitotoxic stress. Animals lacking different *jnk* genes were also crossed, and interestingly, the *jnk1/jnk2* double knockout mouse was found to die at E11.5 from defects in programmed neuronal cell death *(101)*. The apoptotic role of JNK has also been demonstrated in response to many types of stress, including UV and γ-irradiation, hyperosmolarity, toxins, heat shock, chemotherapeutic drugs, peroxide, and inflammatory cytokines such as TNF-α*(2,18)*. Furthermore, nerve growth factor (NGF) deprivation of PC12 pheochromocytoma cells leads to sustained JNK activation and c-Jun phosphorylation, which correlates with apoptosis *(83)*. Consistent with this, expression of a dominant-negative MAPKKK or inactive c-Jun has been shown to suppress apoptosis of PC12 cells after NGF withdrawal *(83)*, and other cell types following irradiation or heat shock *(102)*. Because of the potential involvement of the JNK pathway in neurodegenerative diseases, these enzymes represent attractive therapeutic targets, and inhibitors are currently being tested for the treatment of Alzheimer's and Parkinson's diseases *(103)*.

The JNK isoforms have been implicated as tumor suppressors not only in experimental models, but also in human cancer. The *jnk3* gene was found to be mutated in 50% of brain tumors tested *(104)*. The JNK upstream activator MEK4 was also found to be inactivated in pancreatic, breast, colorectal, and prostate cancers, which correlated with decreased JNK activity and increased tumor aggressiveness *(105)*. Because of their ability to enhance chemotherapy-induced inhibition of tumor cell growth and mediate apoptosis following diverse types of stress, the JNK isoforms may be targeted for the treatment of certain human conditions, such as cancer and rheumatoid arthritis *(106)*.

5. Conclusions

In summary, the MAPK pathways play important roles in the regulation of many physiological processes during vertebrate development and homeostasis. Their importance in controlling cellular responses to the environment and in regulating gene expression, cell growth, proliferation, apoptosis, and the immune response has made them a priority for research related to many human illnesses.

There is a high level of interest in targeting the Ras/ERK pathway for the development of improved cancer therapies, which is exemplified by the inhibitors of MEK, Raf, and Ras farnesylation currently in clinical trials *(58)*. Whereas inhibitors against these proteins are currently being tested, a multitude of additional components of this pathway may be more suitable targets, such as the downstream target RSK, which also regulates cell proliferation, survival and motility *(15,42)*. Thus, inhibition of components within the Ras/ERK pathway may provide a way to block tumor cell survival, proliferation and motility. The unequivocal roles played by the p38-MAPKs and JNKs in the regulation of different aspects of the immune response make them ideal targets for the treatment of diseases such as autoimmunity, asthma, rheumatoid arthritis, and haematopoietic malignancies. Similar to the Ras/ERK cascade, generation of inhibitors to more specific components of these pathways, such as MK2, may provide more specific inhibition of certain aspects of the immune response without unwanted side effects.

Finally, the ERK, JNK, and p38-MAPK pathways represent relatively new molecular targets for drug development, and MAPK inhibitors will undoubtedly be an important group of drugs developed for the treatment of various human diseases.

Acknowledgments

We thank Drs. Bryan Ballif and Rana Anjum for critical reading of the manuscript, and members of the Blenis laboratory for helpful discussions. Because of space limitations, we regret having referred to review articles rather than original findings. P.P.R. is supported by a fellowship from the International Human Frontier Science Program Organization (HFSPO).

References

1. Chen Z, Gibson TB, Robinson F, et al. MAP kinases. Chem Rev 2001;101:2449–2476.
2. Kyriakis JM, Avruch J. Mammalian mitogen-activated protein kinase signal transduction pathways activated by stress and inflammation. Physiol Rev 2001;81:807–869.
3. Pearson G, Robinson F, Beers Gibson T, et al. Mitogen-activated protein (MAP) kinase pathways: regulation and physiological functions. Endocr Rev 2001;22:153–183.
4. Tanoue T, Nishida E. Molecular recognitions in the MAP kinase cascades. Cell Signal 2003;15:455–462.
5. Boulton TG, Yancopoulos GD, Gregory JS, et al. An insulin-stimulated protein kinase similar to yeast kinases involved in cell cycle control. Science 1990;249:64–67.
6. Lewis TS, Shapiro PS, Ahn NG. Signal transduction through MAP kinase cascades. Adv Cancer Res 1998;74:49–139.
7. Colicelli J. Human RAS superfamily proteins and related GTPases. Sci STKE 2004; 2004:RE13.
8. Wellbrock C, Karasarides M, Marais R. The RAF proteins take centre stage. Nat Rev Mol Cell Biol 2004;5:875–885.
9. Hallberg B, Rayter SI, Downward J. Interaction of Ras and Raf in intact mammalian cells upon extracellular stimulation. J Biol Chem 1994;269:3913–3916.
10. Ballif BA, Blenis J. Molecular mechanisms mediating mammalian mitogen-activated protein kinase (MAPK) kinase (MEK)-MAPK cell survival signals. Cell Growth Differ 2001;12:397–408.
11. Chen R-H, Sarnecki C, Blenis J. Nuclear localization and regulation of the erk- and rsk-encoded protein kinases. Mol Cell Biol 1992;12:915–927.
12. Lenormand P, Sardet C, Pages G, L'Allemain G, Brunet A, Pouyssegur J. Growth factors induce nuclear translocation of MAP kinases (p42mapk and p44mapk) but not of their activator MAP kinase kinase (p45mapkk) in fibroblasts. J Cell Biol 1993;122:1079–1088.
13. Gonzalez FA, Seth A, Raden DL, Bowman DS, Fay FS, Davis RJ. Serum-induced translocation of mitogen-activated protein kinase to the cell surface ruffling membrane and the nucleus. J Cell Biol 1993;122:1089–1101.
14. Pouyssegur J, Volmat V, Lenormand P. Fidelity and spatio-temporal control in MAP kinase (ERKs) signalling. Biochem Pharmacol 2002;64:755–763.
15. Roux PP, Blenis J. ERK and p38 MAPK-Activated Protein Kinases: a Family of Protein Kinases with Diverse Biological Functions. Microbiol Mol Biol Rev 2004;68:320–344.
16. Pages G, Guerin S, Grall D, et al. Defective thymocyte maturation in p44 MAP kinase (Erk 1) knockout mice. Science 1999;286:1374–1377.
17. Saba-El-Leil MK, Vella FD, Vernay B, et al. An essential function of the mitogen-activated protein kinase Erk2 in mouse trophoblast development. EMBO Rep 2003;4:964–968.
18. Wada T, Penninger JM. Mitogen-activated protein kinases in apoptosis regulation. Oncogene 2004;23:2838–2849.
19. Chen J, Fujii K, Zhang L, Roberts T, Fu H. Raf-1 promotes cell survival by antagonizing apoptosis signal-regulating kinase 1 through a MEK-ERK independent mechanism. Proc Natl Acad Sci USA 2001;98:7783–7788.

20. Baumann B, Weber CK, Troppmair J, et al. Raf induces NF-kappaB by membrane shuttle kinase MEKK1, a signaling pathway critical for transformation. Proc Natl Acad Sci USA 2000;97:4615–4620.

21. Bonni A, Brunet A, West AE, Datta SR, Takasu MA, Greenberg ME. Cell survival promoted by the Ras-MAPK signaling pathway by transcription-dependent and -independent mechanisms. Science 1999;286:1358–1362.

22. Shimamura A, Ballif BA, Richards SA, Blenis J. Rsk1 mediates a MEK-MAP kinase cell survival signal. Curr Biol 2000;10:127–135.

23. Ginty DD, Bonni A, Greenberg ME. Nerve growth factor activates a Ras-dependent protein kinase that stimulates c-fos transcription via phosphorylation of CREB. Cell 1994; 77:713–725.

24. Xing J, Ginty DD, Greenberg ME. Coupling of the RAS-MAPK pathway to gene activation by RSK2, a growth factor-regulated CREB kinase. Science 1996;273:959–963.

25. Schouten GJ, Vertegaal ACO, Whiteside ST, et al. IkBa is a target for the mitogen-activated 90 kDa ribosomal S6 kinase. The EMBO Journal 1997;16:3133–3144.

26. Ghoda L, Lin X, Greene WC. The 90-kDa ribosomal S6 kinase (pp90rsk) phosphorylates the N-terminal regulatory domain of IkappaBalpha and stimulates its degradation in vitro. J Biol Chem 1997;272:21,281–21,288.

27. Buck M, Poli V, Hunter T, Chojkier M. C/EBPbeta phosphorylation by RSK creates a functional XEXD caspase inhibitory box critical for cell survival. Mol Cell 2001; 8: 807–816.

28. Buck M, Poli V, van der Geer P, Chojkier M, Hunter T. Phosphorylation of rat serine 105 or mouse threonine 217 in C/EBP beta is required for hepatocyte proliferation induced by TGF alpha. Mol Cell 1999;4:1087–1092.

29. Roovers K, Assoian RK. Integrating the MAP kinase signal into the G1 phase cell cycle machinery. Bioessays 2000;22:818–826.

30. Pages G, Lenormand P, L'Allemain G, Chambard JC, Meloche S, Pouyssegur J. Mitogen-activated protein kinases p42mapk and p44mapk are required for fibroblast proliferation. Proc Natl Acad Sci USA 1993;90:8319–8323.

31. Dudley DT, Pang L, Decker SJ, Bridges AJ, Saltiel AR. A synthetic inhibitor of the mitogen-activated protein kinase cascade. Proc Natl Acad Sci USA 1995;92:7686–7689.

32. Alessi DR, Cuenda A, Cohen P, Dudley DT, Saltiel AR. PD 098059 is a specific inhibitor of the activation of mitogen-activated protein kinase kinase in vitro and in vivo. J Biol Chem 1995;270:27,489–27,494.

33. Lavoie JN, L'Allemain G, Brunet A, Muller R, Pouyssegur J. Cyclin D1 expression is regulated positively by the p42/p44MAPK and negatively by the p38/HOGMAPK pathway. J Biol Chem 1996;271:20,608–20,616.

34. Cheng M, Sexl V, Sherr CJ, Roussel MF. Assembly of cyclin D-dependent kinase and titration of p27Kip1 regulated by mitogen-activated protein kinase kinase (MEK1). Proc Natl Acad Sci USA 1998;95:1091–1096.

35. Cook SJ, Aziz N, McMahon M. The repertoire of fos and jun proteins expressed during the G1 phase of the cell cycle is determined by the duration of mitogen-activated protein kinase activation. Mol Cell Biol 1999;19:330–341.

36. Murphy LO, MacKeigan JP, Blenis J. A network of immediate early gene products propagates subtle differences in mitogen-activated protein kinase signal amplitude and duration. Mol Cell Biol 2004;24:144–153.

37. Murphy LO, Smith S, Chen RH, Fingar DC, Blenis J. Molecular interpretation of ERK signal duration by immediate early gene products. Nat Cell Biol 2002;4:556–564.

38. Okazaki K, Sagata N. The Mos/MAP kinase pathway stabilizes c-Fos by phosphorylation and augments its transforming activity in NIH 3T3 cells. Embo J 1995;14:5048–5059.

39. Chen RH, Juo PC, Curran T, Blenis J. Phosphorylation of c-Fos at the C-terminus enhances its transforming activity. Oncogene 1996;12:1493–1502.
40. Roux PP, Ballif BA, Anjum R, Gygi SP, Blenis J. Tumor-promoting phorbol esters and activated Ras inactivate the tuberous sclerosis tumor suppressor complex via p90 ribosomal S6 kinase. Proc Natl Acad Sci USA 2004;101:13,489–13,494.
41. Ballif BA, Roux PP, Gerber SA, MacKeigan JP, Blenis J, Gygi SP. Quantitative phosphorylation profiling of the ERK/p90 ribosomal S6 kinase-signaling cassette and its targets, the tuberous sclerosis tumor suppressors. Proc Natl Acad Sci USA 2005;102:667–672.
42. Smith JA, Poteet-Smith CE, Xu Y, Errington TM, Hecht SM, Lannigan DA. Identification of the first specific inhibitor of p90 ribosomal S6 kinase (RSK) reveals an unexpected role for RSK in cancer cell proliferation. Cancer Res 2005;65:1027–1034.
43. Clark DE, Errington TM, Smith JA, Frierson HF, Jr, Weber MJ, Lannigan DA. The Serine/Threonine Protein Kinase, p90 Ribosomal S6 Kinase, Is an Important Regulator of Prostate Cancer Cell Proliferation. Cancer Res 2005;65:3108–3116.
44. Vial E, Pouyssegur J. Regulation of Tumor Cell Motility by ERK Mitogen-Activated Protein Kinases. Ann N Y Acad Sci 2004;1030:208–218.
45. Reddy KB, Nabha SM, Atanaskova N. Role of MAP kinase in tumor progression and invasion. Cancer Metastasis Rev 2003;22:395–403.
46. Webb CP, Van Aelst L, Wigler MH, Woude GF. Signaling pathways in Ras-mediated tumorigenicity and metastasis. Proc Natl Acad Sci USA 1998;95:8773–8778.
47. Klemke RL, Cai S, Giannini AL, Gallagher PJ, de Lanerolle P, Cheresh DA. Regulation of cell motility by mitogen-activated protein kinase. J Cell Biol 1997;137:481–492.
48. Carragher NO, Westhoff MA, Fincham VJ, Schaller MD, Frame MC. A novel role for FAK as a protease-targeting adaptor protein: regulation by p42 ERK and Src. Curr Biol 2003; 13:1442–1450.
49. Mansfield PJ, Shayman JA, Boxer LA. Regulation of polymorphonuclear leukocyte phagocytosis by myosin light chain kinase after activation of mitogen-activated protein kinase. Blood 2000;95:2407–2412.
50. Vial E, Sahai E, Marshall CJ. ERK-MAPK signaling coordinately regulates activity of Rac1 and RhoA for tumor cell motility. Cancer Cell 2003;4:67–79.
51. Woo MS, Ohta Y, Rabinovitz I, Stossel TP, Blenis J. Ribosomal S6 Kinase (RSK) Regulates Phosphorylation of Filamin A on an Important Regulatory Site. Mol Cell Biol 2004;24: 3025–3035.
52. Chong H, Vikis HG, Guan KL. Mechanisms of regulating the Raf kinase family. Cell Signal 2003;15:463–469.
53. Mitsuuchi Y, Testa JR. Cytogenetics and molecular genetics of lung cancer. Am J Med Genet 2002;115:183–188.
54. Grady WM, Markowitz SD. Genetic and epigenetic alterations in colon cancer. Annu Rev Genomics Hum Genet 2002;3:101–128.
55. Jaffee EM, Hruban RH, Canto M, Kern SE. Focus on pancreas cancer. Cancer Cell 2002; 2:25–28.
56. Garcia-Rostan G, Zhao H, Camp RL, et al. ras mutations are associated with aggressive tumor phenotypes and poor prognosis in thyroid cancer. J Clin Oncol 2003;21:3226–3235.
57. Davies H, Bignell GR, Cox C, et al. Mutations of the BRAF gene in human cancer. Nature 2002;417:949–954.
58. Sebolt-Leopold JS, Herrera R. Targeting the mitogen-activated protein kinase cascade to treat cancer. Nat Rev Cancer 2004;4:937–947.
59. Janes PW, Daly RJ, deFazio A, Sutherland RL. Activation of the Ras signalling pathway in human breast cancer cells overexpressing erbB-2. Oncogene 1994;9:3601–3608.

60. Clark GJ, Der CJ. Aberrant function of the Ras signal transduction pathway in human breast cancer. Breast Cancer Res Treat 1995;35:133–144.
61. Kohno M, Pouyssegur J. Pharmacological inhibitors of the ERK signaling pathway: application as anticancer drugs. Prog Cell Cycle Res 2003;5:219–224.
62. Han J, Lee JD, Bibbs L, Ulevitch RJ. A MAP kinase targeted by endotoxin and hyperosmolarity in mammalian cells. Science 1994;265:808–811.
63. Lee JC, Laydon JT, McDonnell PC, et al. A protein kinase involved in the regulation of inflammatory cytokine biosynthesis. Nature 1994;372:739–746.
64. Rouse J, Cohen P, Trigon S, et al. A novel kinase cascade triggered by stress and heat shock that stimulates MAPKAP kinase-2 and phosphorylation of the small heat shock proteins. Cell 1994;78:1027–1037.
65. Allen M, Svensson L, Roach M, Hambor J, McNeish J, Gabel CA. Deficiency of the stress kinase p38alpha results in embryonic lethality: characterization of the kinase dependence of stress responses of enzyme-deficient embryonic stem cells. J Exp Med 2000;191:859–870.
66. Adams RH, Porras A, Alonso G, et al. Essential role of p38alpha MAP kinase in placental but not embryonic cardiovascular development. Mol Cell 2000;6:109–116.
67. Tamura K, Sudo T, Senftleben U, Dadak AM, Johnson R, Karin M. Requirement for p38alpha in erythropoietin expression: a role for stress kinases in erythropoiesis. Cell 2000;102:221–231.
68. Mudgett JS, Ding J, Guh-Siesel L, et al. Essential role for p38alpha mitogen-activated protein kinase in placental angiogenesis. Proc Natl Acad Sci USA 2000;397:10,454–10,459.
69. Takenaka K, Moriguchi T, Nishida E. Activation of the protein kinase p38 in the spindle assembly checkpoint and mitotic arrest. Science 1998;280:599–602.
70. Engelberg D. Stress-activated protein kinases-tumor suppressors or tumor initiators? Semin Cancer Biol 2004;14:271–282.
71. Ono K, Han J. The p38 signal transduction pathway: activation and function. Cell Signal 2000;12:1–13.
72. Kontoyiannis D, Kotlyarov A, Carballo E, et al. Interleukin-10 targets p38 MAPK to modulate ARE-dependent TNF mRNA translation and limit intestinal pathology. EMBO J 2001;20:3760–3770.
73. Vasudevan S, Peltz SW. Regulated ARE-mediated mRNA decay in Saccharomyces cerevisiae. Mol Cell 2001;7:1191–1200.
74. Winzen R, Kracht M, Ritter B, et al. The p38 MAP kinase pathway signals for cytokine-induced mRNA stabilization via MAP kinase-activated protein kinase 2 and an AU-rich region-targeted mechanism. Embo J 1999;18:4969–4980.
75. Kotlyarov A, Yannoni Y, Fritz S, et al. Distinct cellular functions of MK2. Mol Cell Biol 2002;22:4827–4835.
76. Rousseau S, Morrice N, Peggie M, Campbell DG, Gaestel M, Cohen P. Inhibition of SAPK2a/p38 prevents hnRNP A0 phosphorylation by MAPKAP-K2 and its interaction with cytokine mRNAs. Embo J 2002;21:6505–6514.
77. Mahtani KR, Brook M, Dean JL, Sully G, Saklatvala J, Clark AR. Mitogen-activated protein kinase p38 controls the expression and posttranslational modification of tristetraprolin, a regulator of tumor necrosis factor alpha mRNA stability. Mol Cell Biol 2001;21:6461–6469.
78. Bollig F, Winzen R, Gaestel M, Kostka S, Resch K, Holtmann H. Affinity purification of ARE-binding proteins identifies polyA-binding protein 1 as a potential substrate in MK2-induced mRNA stabilization. Biochem Biophys Res Commun 2003;301:665–670.

79. Han Q, Leng J, Bian D, et al. Rac1-MKK3-p38-MAPKAPK2 pathway promotes urokinase plasminogen activator mRNA stability in invasive breast cancer cells. J Biol Chem 2002; 277:48,379–48,385.

80. Tran H, Maurer F, Nagamine Y. Stabilization of urokinase and urokinase receptor mRNAs by HuR is linked to its cytoplasmic accumulation induced by activated mitogen-activated protein kinase-activated protein kinase 2. Mol Cell Biol 2003;23:7177–7188.

81. Porras A, Zuluaga S, Black E, et al. P38 alpha mitogen-activated protein kinase sensitizes cells to apoptosis induced by different stimuli. Mol Biol Cell 2004;15:922–933.

82. Park JM, Greten FR, Li ZW, Karin M. Macrophage apoptosis by anthrax lethal factor through p38 MAP kinase inhibition. Science 2002;297:2048–2051.

83. Xia Z, Dickens M, Raingeaud J, Davis RJ, Greenberg ME. Opposing effects of ERK and JNK-p38 MAP kinases on apoptosis. Science 1995;270:1326–1331.

84. Brancho D, Tanaka N, Jaeschke A, et al. Mechanism of p38 MAP kinase activation in vivo. Genes Dev 2003;17:1969–1978.

85. Bulavin DV, Higashimoto Y, Popoff IJ, et al. Initiation of a G2/M checkpoint after ultraviolet radiation requires p38 kinase. Nature 2001;411:102–107.

86. Garner AP, Weston CR, Todd DE, Balmanno K, Cook SJ. Delta MEKK3:ER* activation induces a p38 alpha/beta 2-dependent cell cycle arrest at the G2 checkpoint. Oncogene 2002;21:8089–8104.

87. Todd DE, Densham RM, Molton SA, et al. ERK1/2 and p38 cooperate to induce a p21CIP1-dependent G1 cell cycle arrest. Oncogene 2004;23:3284–3295.

88. Schultz RM. Potential of p38 MAP kinase inhibitors in the treatment of cancer. Prog Drug Res 2003;60:59–92.

89. Kyriakis JM, Avruch J. pp54 microtubule-associated protein 2 kinase. A novel serine/threonine protein kinase regulated by phosphorylation and stimulated by poly-L-lysine. J Biol Chem 1990;265:17,355–17,363.

90. Kyriakis JM, Banarjee P, Nikolakaki E, et al. The stress-activated protein kinase subfamily of c-Jun kinases. Nature 1994;369:156–160.

91. Derijard B, Hibi M, Wu I-H, et al. JNK1: A protein kinase stimulated by UV light and Ha-Ras that binds and phosphorylates the c-Jun activation domain. Cell 1994;76:1025–1037.

92. Gupta S, Barrett T, Whitmarsh AJ, et al. Selective interaction of JNK protein kinase isoforms with transcription factors. Embo J 1996;15:2760–2770.

93. McDonald PH, Chow CW, Miller WE, et al. Beta-arrestin 2: a receptor-regulated MAPK scaffold for the activation of JNK3. Science 2000;290:1574–1577.

94. Whitmarsh AJ, Cavanagh J, Tournier C, Yasuda J, Davis RJ. A mammalian scaffold complex that selectively mediates MAP kinase activation. Science 1998;281:1671–1674.

95. Ito M, Yoshioka K, Akechi M, et al. JSAP1, a novel jun N-terminal protein kinase (JNK)-binding protein that functions as a Scaffold factor in the JNK signaling pathway. Mol Cell Biol 1999;19:7539–7548.

96. Weston CR, Davis RJ. The JNK signal transduction pathway. Curr Opin Genet Dev 2002; 12:14–21.

97. Dong C, Yang DD, Wysk M, Whitmarsh AJ, Davis RJ, Flavell RA. Defective T cell differentiation in the absence of Jnk1. Science 1998;282:2092–2095.

98. Yang DD, Conze D, Whitmarsh AJ, et al. Differentiation of CD4+ T cells to Th1 cells requires MAP kinase JNK2. Immunity 1998;9:575–585.

99. Sabapathy K, Hu Y, Kallunki T, et al. JNK2 is required for efficient T-cell activation and apoptosis but not for normal lymphocyte development. Curr Biol 1999;9:116–125.

100. Yang DD, Kuan CY, Whitmarsh AJ, et al. Absence of excitotoxicity-induced apoptosis in the hippocampus of mice lacking the Jnk3 gene. Nature 1997;389:865–870.

101. Kuan CY, Yang DD, Samanta Roy DR, Davis RJ, Rakic P, Flavell RA. The Jnk1 and Jnk2 protein kinases are required for regional specific apoptosis during early brain development. Neuron 1999;22:667–676.

102. Verheij M, Bose R, Lin XH, et al. Requirement for ceramide-initiated SAPK/JNK signalling in stress-induced apoptosis. Nature 1996;380:75–79.

103. Resnick L, Fennell M. Targeting JNK3 for the treatment of neurodegenerative disorders. Drug Discov Today 2004;9:932–939.

104. Yoshida S, Fukino K, Harada H, et al. The c-Jun NH2-terminal kinase3 (JNK3) gene: genomic structure, chromosomal assignment, and loss of expression in brain tumors. J Hum Genet 2001;46:182–187.

105. Kim HL, Vander Griend DJ, Yang X, et al. Mitogen-activated protein kinase kinase 4 metastasis suppressor gene expression is inversely related to histological pattern in advancing human prostatic cancers. Cancer Res 2001;61:2833–2837.

106. Han Z, Boyle DL, Chang L, et al. c-Jun N-terminal kinase is required for metalloproteinase expression and joint destruction in inflammatory arthritis. J Clin Invest 2001;108:73–81.

107. Yan L, Carr J, Ashby PRE, et al. Knockout of ERK5 causes multiple defects in placental and embryonic development. BMC Dev Biol 2003;3:11.

7

Serine/Threonine Protein Phosphatases in Apoptosis

Gro Gausdal, Camilla Krakstad, Lars Herfindal, and Stein Ove Døskeland

Summary

The Ser/Thr protein phosphatases are of crucial importance for controlling the phosphorylation state of survival and cell death proteins like Bad, Bid, myosin light chain, lamin B, some caspases and their inhibitors, as well as transcriptions factors like CREB and NF-κB/I-κB. Most is known about the protein phosphatases (PPs) PP1, PP2A, PP2B/calcineurin, PP4/PPX, and PP5. With the possible exception of PP5, which so far appears to mainly promote apoptosis, the phosphatases can promote both death and survival signaling. Here, a particular emphasis is put on the natural toxins targeting PP2A and PP1 and their CaMKII dependent mode of action.

Key Words: Cell death; protein phosphorylation; calmodulin-dependent protein kinase; okadaic acid; microcystin; PP2A; myosin light chain.

1. Introduction

The importance of protein phosphorylation in apoptosis was appreciated relatively late, the first reviews appearing in the mid 1990s *(1,2)*. The explosive growth of knowledge since then is reflected by the many excellent recent reviews describing the role in apoptosis of protein kinases like receptor tyrosine kinases *(3)*, the DNA-dependent protein kinase *(4)*, Ras/Raf/MEK/ERK *(5)*, PI3K/Akt and mTOR controlled kinases *(6,7)*, PKC *(8,9)*, focal adhesion kinase (FAK) *(10)* and stress activated kinases like JNK and p38 MAP *(11)*. In general, the Ser/Thr phosphatases have received less attention than the Ser/Thr protein kinases. This is because their catalytic subunits have broad substrate specificity compared with the kinases and because they are hard to study by overexpression because cells tend to keep the level of active phosphatase within strict limits. This in itself is presumably a sign of their vital importance in cell signaling. It appears however, that the lack of variation of catalytic subunits is compensated by an array of regulatory subunits. These subunits are either modulators of the PP activity or scaffolding subunits *(12–15)*.

Previous reviews on protein phosphatases and apoptosis have concentrated mainly on the ability of phosphatases to antagonize mitogenesis and survival signaling *(16–21)*. This is in line with early observations that the PP2A/PP1 inhibitor okadaic acid is a tumor promoter in skin *(22)* and that the growth and survival promoting SV40 t antigens inhibit PP2A *(23–25)*.

The present review will also focus on the mechanisms of the antiapoptotic effects of the major protein phosphatases PP2A and PP1, and in particular on the role of the multifunctional calmodulin-dependent protein kinase II (CaMKII) as mediator of PP-inhibitor induced apoptosis. This is in accordance with the fact that inhibitors of PP2A/PP1, at

From: *Apoptosis, Cell Signaling, and Human Diseases: Molecular Mechanisms, Volume 2*
Edited by R. Srivastava © Humana Press Inc., Totowa, NJ

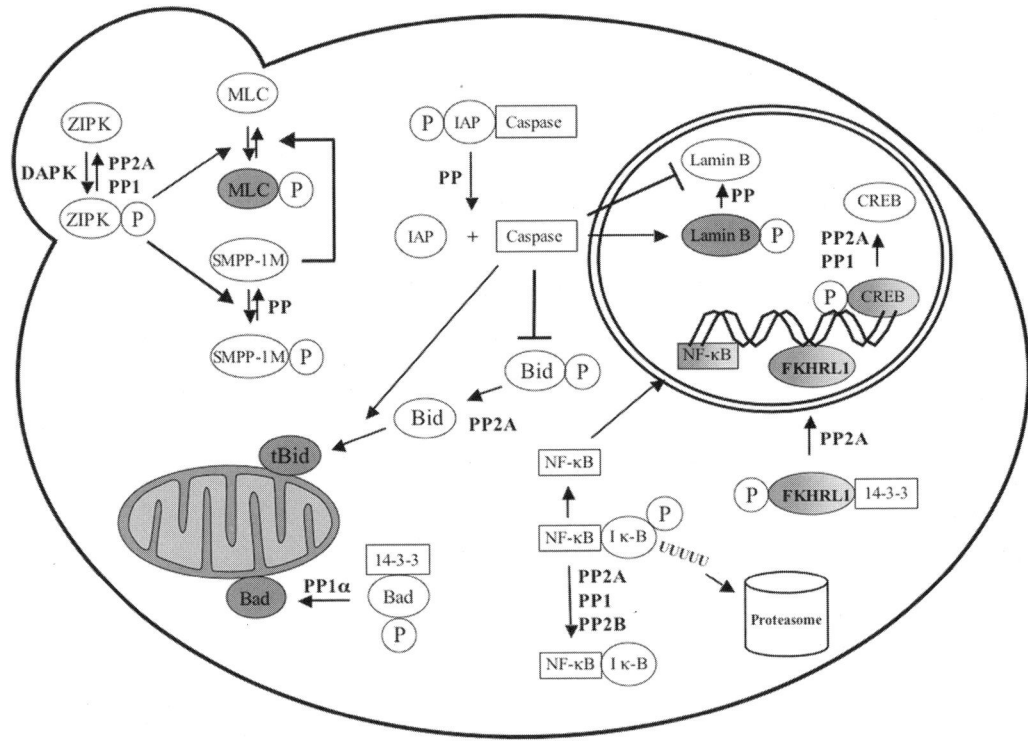

Fig. 1. Key apoptotic regulators controlled by dephosphorylation. The figure shows selected phosphoproteins regulating mitochondrial death pathways, the transcription of anti- or proapoptotic gene products and cytoskeletal integrity. The effect of dephosphorylation is indicated. Proteins believed to be mainly proapoptotic are red, while proteins with pro- and antiapoptotic function also have green color.

concentrations sufficient to inhibit most of the cellular PP2A/PP1 activity, are nearly universal inducers of apoptosis *(1)* and that knock-out of the catalytic subunit of PP2A induces apoptosis *(26,27)*.

First we will briefly give an overview of the mechanisms whereby dephosphorylation of key phosphoproteins can enhance or counteract apoptosis, and then dwell in more depth on selected Ser/Thr phosphatases and their involvement in apoptosis.

2. Ways in Which Protein Dephosphorylation May Affect Cell Viability

Phosphorylation/dephosphorylation can modulate enzyme activity, protein–protein interactions and protein processing that can lead to protein translocation and altered gene transcription. Some examples of apoptosis-relevant phosphoproteins targeted by Ser/Thr protein phosphatases are shown in Fig. 1.

Two examples of phosphorylation-dependent cytoskeletal rearrangements are shown. The nuclear lamina is often disassembled in apoptosis (e.g., like that induced by the PP2A/PP1 inhibitor calyculin A *[28]*). This can be related to the fact that only hyperphosphorylated lamin B is a target for caspase-mediated proteolysis *(29)*. Also, the actin-capping protein alpha-adducin is more prone to caspase-mediated cleavage when phosphorylated *(30)*.

One of the characteristics of apoptosis is rearrangements of the actin cytoskeleton, leading to cell budding and, eventually, the release of apoptotic bodies. Such budding can be induced by inhibitors of PP2A/PP1 and is tightly correlated with hyperphosphorylation of myosin light chain (MLC) *(31)*. Several protein kinases, all incriminated in apoptosis, are known to phosphorylate the myosin light chain: MLC-kinase *(32)*, CaMKII *(33)*, Rho-asssociated protein kinase ROCK *(34)*, PKC *(35)*, MAPKAP2 *(36)*, p21-activated protein kinase *(37)*, and Zipper-interacting protein kinase (ZIPK) *(38)*. Several of the mentioned kinases are themselves phosphoproteins subject to dephosphorylation by PPs and therefore prone to activation by inhibitors of PP2A/PP1. Another mechanism of MLC hyperphosphorylation involves the negative regulation of the specific MLC phosphatase SMPP-1M by ZIPK *(39,40)*, which itself is activated by phosphorylation by the death associated protein kinase, DAPK *(41)* (*see* Fig. 1).

Also the caspases can have their activity modulated by direct PP phosphorylation *(42–44)*. In addition, their activity can be modulated indirectly through phosphorylation. Firstly, the caspase inhibitor proteins (IAP) survivin and XIAP are active only when phosphorylated. Upon dephosphorylation, survivin dissociates from the caspase *(43)*, whereas XIAP becomes subject to proteasomal degradation *(46)*. Secondly, some caspase substrates, like cytokeratin *(47)* and Bid *(48,49)*, are protected from caspase cleavage when phosphorylated (*see* Fig. 1).

The 14-3-3 proteins sequester phospho-proteins in the cytoplasm *(50)*. When bound to 14-3-3, phospho-Bad cannot compromise the antiapoptotic activity of the mitochondrial Bcl-2 family members. Upon dephosphorylation Bad will translocate to the mitochondria and promote apoptosis *(21,51–55)*. Sequestering by 14-3-3 can also promote apoptosis. The antiapoptotic apoptosis signal-regulating kinase 1 (ASK 1) is able to protect cells against death only when dephosphorylated and released from 14-3-3 *(56)*.

Phosphorylation can modulate the expression of a large number of apoptosis-relevant genes through modulation of key transcription factors. The transcription factor Forkhead can induce pro- as well as antiapoptotic genes. Upon dephosphorylation by PP2A it is released from 14-3-3 and translocates to the nucleus to initiate gene transcription *(57,58)* (*see* Fig. 1).

The transcription factors NF-κB, p53, and CREB can induce gene products associated with both cell survival and death. Their activity is subjected to control by protein phosphorylation at several levels. Typically, multisite phosphorylations catalyzed by several kinases and counteracted by several phosphatases occur not only on the transcription factor itself but also on its protein partners. This leads to altered activity, stability, and subcellular localization, which in some cases probably result in a changed expression of pro- and antiapoptotic genes *(59–67)*.

I-κB, the inhibitor of NF-κB, is targeted to the proteasome for degradation when phosphorylated, allowing NF-κB to translocate to the nucleus and enhance gene transcription *(68)* (*see* Fig. 1). This process can be antagonized upon dephosphorylation of phospho-I-κB by either PP2B *(69)*, PP2A, or PP1 *(70)*.

The cAMP response element-binding factor CREB, when phosphorylated at Ser133, assumes a conformation allowing more efficient binding of transcriptional co-activators (*see* Fig. 1). The duration of CREB activation is limited even at maximal cAMP stimulation, presumably as a result of dephosphorylation of phospho-CREB by PP1 *(71)* and PP2A *(72)*. The genotoxic stress activated ATM kinase can phosphorylate CREB at residue(s) N-terminal to Ser133. This curbs the activity of CREB, leading to cell death *(73)*.

In conclusion, dephosphorylation of proteins can have a bewildering number of apoptosis-relevant effects through modulation of enzyme or transcriptional activity, protein degradation, or protein translocation. Obviously, a general stimulation or inhibition of dephosphorylation would lead to chaotic cell signaling, calling for strict control of phosphoprotein phosphatase activity in the intact cell.

3. The Role of Specific Phosphoprotein Phosphatases in Apoptosis

Protein phosphatases are classified into three families based on primary sequence and crystal structure. The PPPs and PPMs (Mg^{2+}-dependent phosphatases) dephosphorylate phospho-serine or phospho-threonine residues, whereas the PTPs dephosphorylate phospho-tyrosine residues. An interesting subgroup of the PTPs has dual substrate specificity and can dephosphorylate both serine, threonine, and tyrosine residues. Another interesting PTP is PTEN, a phosphatase which prefers the phosphoinositide second messenger PIP_3 as its physiological substrate *(74)*. Additionally, a histidine phosphatase (PHP) has been isolated from rat liver. The peptide sequence of this phosphatase shows no homology with other known protein phosphatases *(75)*. So far, the PHP has not been directly linked to apoptosis.

3.1. An Overview of the PPP Species

The Mg^{2+}-dependent phosphatase PP2C can promote cell survival by dephosphorylating phospho-Bad on Ser155, thereby preventing its dimerization with Bcl-X_L *(54)*. On the other hand, the overexpression of PP2C can cause apoptosis, possibly through the p53 dependent expression of proapoptotic gene products *(76)*.

PP2B (also known as calcineurin) differs from the other members of the PPP family by being stimulated by Ca^{2+}/calmodulin. PP2B is the target of the T-cell immunosuppressing compounds cyclosporine and FK506 *(77)*. PP2B appears to be involved in both pro- and antiapoptotic signaling. One example of this is observed in a myeloid cell line, in which activation of PP2B protects against apoptosis induced by enforced expression of wild type p53, but enhances apoptosis in the absence of p53 *(78)*. Another example is found in the heart. PP2B A$\beta^{-/-}$ mice are predisposed to ischemia-induced apoptosis suggesting an overall protective role of PP2B *(79)*. On the other hand, the anticancer drug doxorubicin, feared for its cardiotoxic side effects, appears to activate PP2B, its downstream transcription factor NFAT and activate caspases *(80)*. Because PP2B can be activated through caspase-mediated truncation *(81)*, there is a potential for an autoactivatory apoptotic loop once caspases have been activated in a PP2B-dependent manner.

As indicated in Fig. 1, the dephosphorylation of phospho-Bad is an important proapoptotic effect, not only of PP2B, but also PP2C, PP2A, and PP1. This effect of PP2B is presumably more important in tissues where its relative expression is high, for example, the brain.

PP6/Sit4 and PP7 are the only members of the PPP family that appear not to have been implicated in vertebrate cell apoptosis. However, in yeast, Sit4 is suggested to have a death facilitating role since *Saccharomyces cerevisiae* devoid of Sit4 is resistant to ceramide-induced death *(82)*.

PP5 is ubiquitously expressed, but at a lower level than most other Ser/Thr protein phosphatases. Unlike the other phosphatases, for which the catalytic subunit is complexed with structural or regulatory subunits, PP5 exists mainly as a single chain. However, this chain both has catalytic, regulatory and targeting functions, in part conveyed through the N-terminal TPR domain. This domain serves to link PP5 to the Hsp90 chaperone *(83,84)*, which harbors a number of important prosurvival proteins whose dephosphorylation certainly can affect apoptosis. PP5 is a p53 phosphatase, and cells depleted of PP5 show p53-dependent growth arrest *(85)*. Cells that overexpress PP5 have increased resistance toward either oxidative stress *(86)* or hypoxic stress *(87)*, presumably through antagonism of the ASK 1/MKK-4/JNK signaling cascade. Because mammary carcinoma cells with increased PP5 expression grow better in a hypoxic in vivo model *(87)*, PP5 may be an interesting target in cancer therapy.

The natural compound and drug rapamycin inhibits the kinase mTOR, resulting in apoptosis in some cells. One mechanism for this effect is through dissociation of the PP2A-B subunit from its complex with PP5 and ASK 1. This in turn leads to downregulation of PP5 activity and activation of ASK 1, resulting in apoptosis. The effect of PP5 is specific, because overexpression of PP5, but not of PP2A catalytic subunit, protects cells against rapamycin-induced apoptosis *(88)*. This demonstrates an interesting and non-redundant interaction between PP5 and PP2A signaling. The DNA-dependent protein kinase is related to mTOR, and autophosphorylation of this kinase can alter the susceptibility of the host cell to radiation-induced death. PP5 appears to interact with and dephosphorylate the DNA-kinase *(89)*. Knock-down of PP5 in *Drosophila* cells did not induce cell death *(27)*. In conclusion, although PP5 has been studied for only a short period of time and much remains to be known, it appears to have a prosurvival function at least in mammalian cells under hypoxic or oxidative stress.

PP4/PPX is found associated with the centrosome and is composed of a catalytic, a structural and a regulatory subunit *(90)*. At least three regulatory subunits exist. PP4/PPX can dephosphorylate and associate with a number of signaling proteins known to be important in apoptosis *(91)*. Downregulation of PP4/PPX can confer resistance to irradiation induced death, suggesting that PP4 may have a proapoptotic role *(92)*. In line with this, knock-down of PP4 in *Drosophila* cells did not induce cell death *(27)*.

Tumor necrosis factor (TNF)-α activates PP4 *(93)* which again binds to and activates NF-κB, presumably by dephosphorylation of Thr435 of NF-κB p65 *(59)*. This effect of PP4 is opposite to that of the closely related PP2A. PP2A-catalyzed dephosphorylation of NF-κB p65 inhibits transcription *(94)*. On the other hand, PP2A (and PP1, PP2B) can activate NF-κB indirectly through dephosphorylation of Iκ-B *(see* Fig. 1). TNF-α has a dual effect on cell viability, the prosurvival effect being mediated in part through NF-κB-induced transcription *(95)*.

PP2A accounts for as much as 1% of the total cellular protein content and for the major portion of the Ser/Thr phosphatase activity in most cells *(96)*. PP2A is a heterotrimer consisting of a catalytic subunit (C), a scaffolding subunit (A) and a regulatory subunit (B), where the C-subunit always associates with an A-subunit. The C-subunit itself is affected by post-translational modifications like phosphorylation and methylation *(97)*, but most of its regulation is supposed to be mediated through the B subunits.

Distinct classes of regulatory B-subunits, termed B, B', and B'' exist, all encoded by separate genes *(98)*. It has been estimated that more than 50 subunits of PP2A exist when counting the many splice variants of the B-subunits *(14)*. In addition, PP2A forms stable complexes with an impressive array of other proteins, including tightly binding substrates, caspase 3, viral proteins including SV40 small t antigens *(14)* and the endogenous inhibitor SET known to be an oncogene in leukaemia *(99)*. Some of the subunits and associated proteins act to direct PP2A to specific subcellular locations. Mitochondrially located PP2A is thought to be particularly important in the dephosphorylation of phospho-Bcl-2 *(100)*. The mature neutrophil is prone to spontaneous apoptosis, presumably in part because NF-κB is inhibited by a nuclear form of I-κB. Nuclear located PP2A can dephosphorylate nuclear I-κB and thereby protect the neutrophil against apoptosis *(70)*.

Traditionally, PP2A has been considered a purely proapoptotic enzyme. This results partly from the fact that it counteracts a number of protein kinases important in survival signaling (e.g., Akt *[14,21]*). Another reason is that the PP2A inhibiting/dislocating SV40 small t antigens and SET are oncogens rather than tumor supressors. A third reason is that PP2A is activated through cleavage by caspase 3 *(101)*, suggesting that active PP2A may be a feature of advanced caspase-dependent cell death. A fourth reason is that at low concentrations, the PP2A inhibitor okadaic acid is a tumor promoter *(22)* and can protect, at least transiently, against certain apoptotic stimuli *(16,18,20,102)*. A fifth reason is the role of PP2A to promote dephosphorylation of phospho-Bad and sequester NF-κB in an inactive complex with its inhibitor (*see* Fig. 1). The recent finding that PP2A is activated in the early stage of Fas-induced neutrophil apoptosis, further supports a proapoptotic role for PP2A. In these cells, PP2A can reverse the p38-MAPK catalyzed activatory phosphorylation of caspase 3 and 8 *(44,103)*.

Beginning with the realization that the PP2A inhibitor okadaic acid at high concentrations was a near universal apoptosis inducer *(104)*, evidence has slowly emerged that PP2A may also be involved in survival signaling. This prosurvival function of PP2A has gained additional credibility by the demonstration that CaMKII, whose activation is induced by PP2A, can be a cell death inducer *(105)*. An additional argument is the near perfect correlation between the ability of the adenovirus coded E4orf4 protein to suppress PP2A activity (through binding to the B55 subunit) and its ability to induce apoptosis *(106)*. The most direct evidence for a survival function of PP2A is that elimination of the catalytic subunit of PP2A using RNA interference technology (RNAi), causes cell death *(26,27)*. In line with this, RNAi-mediated knock-down of the scaffolding alpha subunit of PP2A, which also decreased the expression of the catalytic subunit of PP2A, induced cell death and was associated with decreased survival signaling through Akt *(107)*.

In view of the complexity of the PP2A signaling, it should not come as a surprise that PP2A can mediate both pro- and antiapoptotic signals. Knock-down of the B56 subunit of PP2A promotes apoptosis, presumably by allowing a proapoptotic protein upstream of caspase activation to become active *(26,27)*. On the other hand, knock-down of the R2/B subunit enhanced ERK-mediated signaling *(27)*.

PP1 is a heterodimer composed of a catalytic and a regulatory subunit. About 50 tightly binding endogenous peptide or protein partners (potential regulatory subunits) are known *(12)*. PP1 binds to the prosurvival proteins Bcl-2 and Bcl-X_L through a

RVXF motif, rendering the phosphatase in a particular favorable position to dephosphorylate phospho-Bad upon binding of the latter to Bcl-X_L *(108)*.

An important feature in apoptosis is the expression of alternatively spliced mRNA species coding for proapoptotic versions of gene products. PP1 stimulation in ceramide-induced apoptosis acts on phosphorylated splicing factors promoting the expression of the short proapoptotic Bcl-x(s) *(109)*.

3.2. PP1, PP2A, PP4, PP5: Targets for a Number of Nonvertebrate Toxins

A number of toxins from a variety of nonvertebrate species targeting Ser/Thr phosphatases have evolved (*see* Fig. 2A). The reason for this unusual convergence must be that the phosphatases are well conserved in evolution *(19,96)* and essential for vital functions in a number of organisms. It is noteworthy that one single binding site (shown for PP1 in Fig. 2C), conserved in PP1, PP2A, PP4, PP5, and PP7, is targeted by toxins as disparate as the microcystin family of cyclic cyanobacterial peptides, nodularin, the dinoflagellate fatty acid derivative okadaic acid, tautomycins and the blister beetle compound cantharidin *(110)*. The toxin binding site overlaps with the recognition area for the PP1-specific endogenous inhibitors DARPP32/inhibitor 1 *(111–113)* (*see* Fig. 2C). There is a relative PP isozyme preference for some of the toxins. Okadaic acid binds with about 100-fold higher affinity to PP2A than to PP1 *(113)*, whereas tautomycetin binds with 50-fold higher affinity to PP1 than to PP2A *(114)*. The PP2A is targeted specifically by the bacterial toxin fostriecin, which binds to a site that does not overlap with the okadaic acid binding site *(115)*. The toxins are presumably synthesized for defense against predators. Obviously, one efficient way of doing this is by inducing cell death of the predator. There are more than 100 reports of apoptotic cell death induced by okadaic acid or other toxins targeting the same site on the protein phosphatases. As shown in Fig. 3, different types and concentrations of phosphatase inhibitors can give different apoptotic morphologies in the same cell type, even if they target the same phosphatase binding site. The morphologically distinct phenotypes are accompanied by differences in biochemical parameters like DNA and RNA fragmentation *(28)*.

3.3. Phosphatase Inhibitors Like Microcystin and Okadaic Acid Can Induce Apoptosis Via a CaMKII-Dependent Pathway Shared With γ-Irradiation-Induced Apoptosis

In spite of the variation in death type induced by phosphatase inhibitors (*see* Fig. 3), a death type induced very rapidly by phosphatase inhibitors appears to be distinct. In hepatocytes, which have an efficient uptake system for the cyclic peptide phosphatase inhibitors microcystin and nodularin *(116,117)*, cell death with the features of apoptosis is induced in less than 2 min *(31)*. In cells lacking an efficient uptake mechanism, rapid apoptosis could be induced by the same toxins when microinjected *(105)*. A rapid hyperphosphorylation of proteins precedes the first signs of apoptosis in cells treated with microcystin. Basically, for a phosphatase inhibitor to act rapidly it must counteract a fast dephosphorylation process. The ubiquitously expressed and multifunctional calmodulin-dependent protein kinase II (CaMKII) is known to be tightly controlled by activatory autophosphorylation, and is subjected to rapid cycles of phosphorylation and dephosphorylation *(118,119)*. Its dephosphorylation is catalyzed by PP2A and PP1,

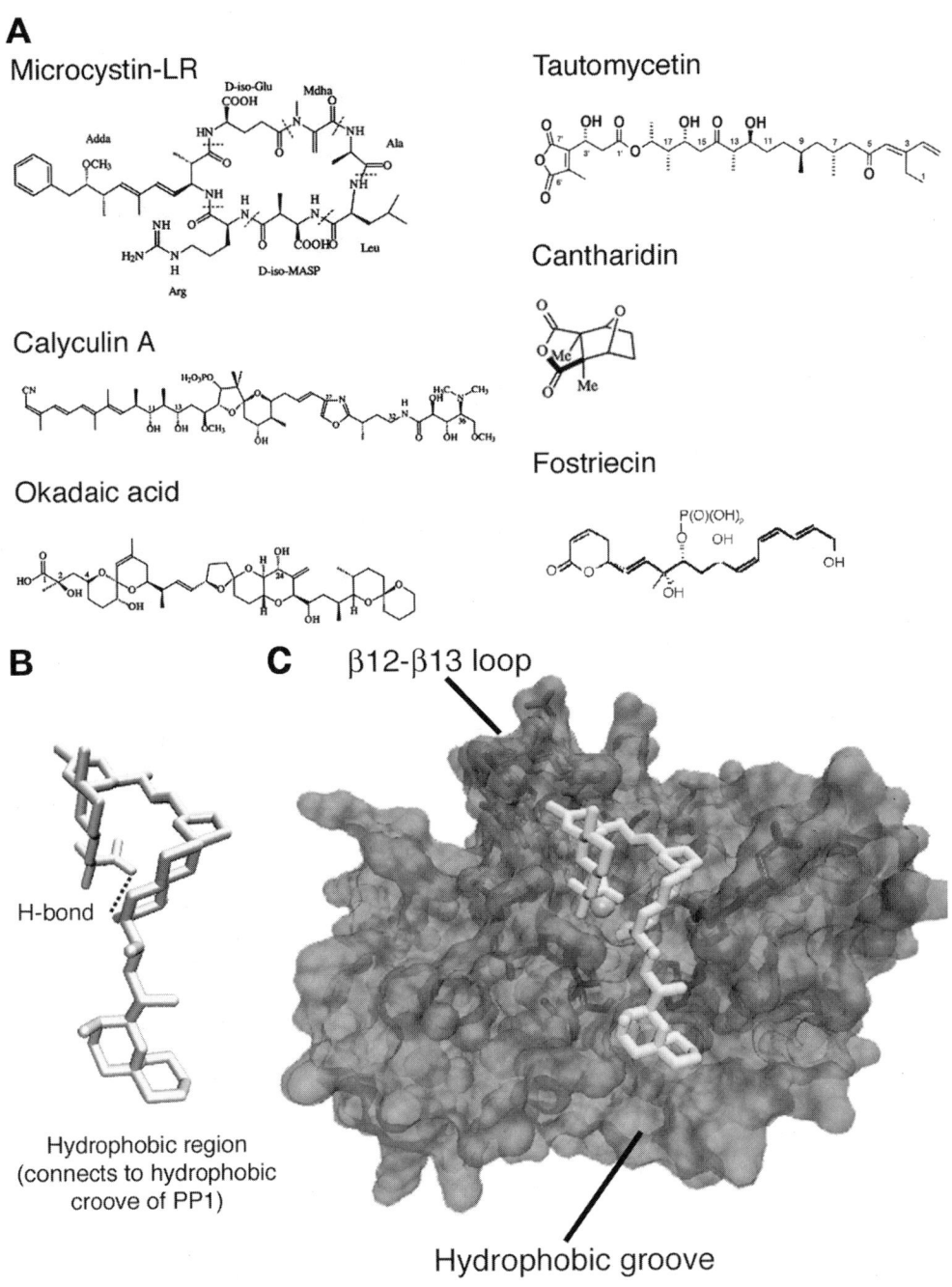

Fig. 2. Structures of Ser/Thr protein phosphatase inhibitors. Panel **A** shows the chemical structure of some protein phosphatase inhibitors. Note the structural diversity of the different compounds. The three-dimensional structure of okadaic acid (OA) is shown in panel **B**. Panel **C** shows the highly conserved hydrophobic region of PP1C with bound OA (yellow). The two manganese ions in the catalytic core are in green. Amino acid residues in close proximity to OA are in blue. Residues needed for recognition of PP1C by DARPP32/Inhibitor 1 are in red, and residues that interact with both OA and DARPP32/Inhibitor 1 are in purple. For further details see refs. *111* and *113*. The images in B and C were created in the VMD software system *(129)*.

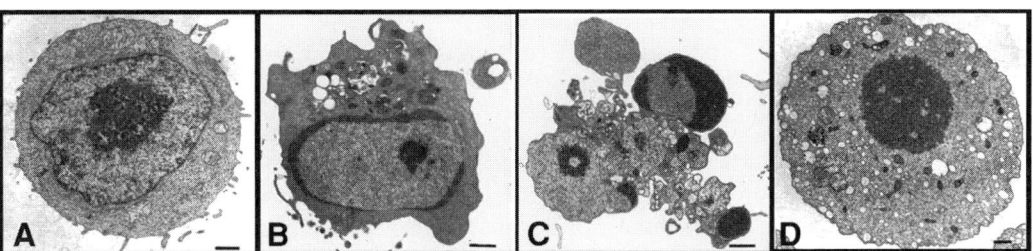

Fig. 3. Distinct IPC-cell death types induced by protein phosphatase inhibitors. Electromicrographs showing typical ultrastructure of IPC-81 myelogenic leukaemia cells undergoing apoptosis in response to protein phosphatase inhibitors. The cells were exposed to vehicle (**A**), to 1 μ*M* okadaic acid for 6 h (**B**), to 3 n*M* calyculin A for 24 h (**C**), or to 100 n*M* calyculin A for 6 h (**D**). See also ref. *28*.

both of which are targeted by microcystin, nodularin and okadaic acid. The autophosphorylated kinase has such a high affinity for Ca^{2+}/calmodulin that no increase of intracellular Ca^{2+} is required for its activation *(120,121)*. It appears that most of the microcystin-induced hyperphosphorylation events as well as the induction of apoptosis itself, was prevented by CaMKII inhibitors, suggesting that CaMKII is essential for at least some forms of phosphatase inhibitor induced apoptosis *(105)*. The death induction can occur in the absence and presence of Bcl-2 overexpression and is enhanced, but not dependent on, caspase activation *(31,122)*.

In order to elucidate the pathways between protein hyperphosphorylation and cell death, a cDNA library was introduced into fibroblasts, and the cells were subsequently treated with a lethal concentration of okadaic acid. A surviving clone expressed mRNA coding for the coiled-coil domain of the protein Irod/Gimap5/Ian5 *(122,123)*. This protein is mutated and inactive in the T-cell lymphopenic BB-rat *(124,125)*, indicating that it has a prosurvival role in the intact animal. Overexpression of the protein protected Jurkat T-cells selectively against phosphatase inhibitors and γ-irradiation. Intriguingly, also CaMKII inhibitors protected against these death stimuli. This suggested that the phosphatase inhibitors might act by short-circuiting a phylogenetically ancient cell death pathway originally activated by damage similar to that induced by γ-irradiation. One component common to cells exposed to γ-irradiation and phosphatase inhibitors is the increased formation of reactive oxygen species (ROS). The formation of ROS has been demonstrated during microcystin-induced cell death *(126,127)*. Because ROS may inhibit PP2A and thereby activate CaMKII *(128)*, a positive feed-back loop involving CaMKII and ROS may be formed.

4. Conclusion

It appears that Ser/Thr protein phosphatases compensate for their low number of catalytic subunits by creating protein–protein complexes to control key proteins, such as protein kinases. In principle, death signals decide an irreversible event, and is therefore bound to be controlled in a stringent manner. As described in this chapter, several protein phosphatases are involved in both pro- and antiapoptotic signaling. Certainly, we will learn more about the involvement of protein phosphatases in death signaling during the next few years. This will contribute to a better understanding of the complex regulation of apoptosis, and aid the development of new drugs targeting apoptosis linked diseases.

References

1. Gjertsen BT, Doskeland SO. Protein phosphorylation in apoptosis. Biochim Biophys Acta 1995;1269:187–199.
2. Lavin MF, Watters D, Song Q. Role of protein kinase activity in apoptosis. Experientia 1996;52:979–994.
3. Zwick E, Bange J, Ullrich A. Receptor tyrosine kinase signalling as a target for cancer intervention strategies. Endocr Relat Cancer 2001;8:161–173.
4. Burma S, Chen DJ. Role of DNA-PK in the cellular response to DNA double-strand breaks. DNA Repair (Amst) 2004;3:909–918.
5. Chang F, Steelman LS, Shelton JG, et al. Regulation of cell cycle progression and apoptosis by the Ras/Raf/MEK/ERK pathway (Review). Int J Oncol 2003;22:469–480.
6. Franke TF, Hornik CP, Segev L, Shostak GA, Sugimoto C. PI3K/Akt and apoptosis: size matters. Oncogene 2003;22:8983–8998.
7. Bjornsti MA, Houghton PJ. Lost in translation: dysregulation of cap-dependent translation and cancer. Cancer Cell 2004;5:519–523.
8. Brodie C, Blumberg PM. Regulation of cell apoptosis by protein kinase c delta. Apoptosis 2003;8:19–27.
9. Webb PR, Wang KQ, Scheel-Toellner D, Pongracz J, Salmon M, Lord JM. Regulation of neutrophil apoptosis: a role for protein kinase C and phosphatidylinositol-3-kinase. Apoptosis 2000;5:451–458.
10. Hanks SK, Ryzhova L, Shin NY, Brabek J. Focal adhesion kinase signaling activities and their implications in the control of cell survival and motility. Front Biosci 2003; 8:d982–d996.
11. Wada T, Penninger JM. Mitogen-activated protein kinases in apoptosis regulation. Oncogene 2004;23:2838–2849.
12. Cohen PT. Protein phosphatase 1—targeted in many directions. J Cell Sci 2002;115: 241–256.
13. Cohen P. The structure and regulation of protein phosphatases. Annu Rev Biochem 1989; 58:453–508.
14. Millward TA, Zolnierowicz S, Hemmings BA. Regulation of protein kinase cascades by protein phosphatase 2A. Trends Biochem Sci 1999;24:186–191.
15. Ceulemans H, Bollen M. Functional diversity of protein phosphatase-1, a cellular economizer and reset button. Physiol Rev 2004;84:1–39.
16. Van Hoof C, Goris J. Phosphatases in apoptosis: to be or not to be, PP2A is in the heart of the question. Biochim Biophys Acta 2003;1640:97–104.
17. Van Hoof C, Goris J. PP2A fulfills its promises as tumor suppressor: which subunits are important? Cancer Cell 2004;5:105–106.
18. Gehringer MM. Microcystin-LR and okadaic acid-induced cellular effects: a dualistic response. FEBS Lett 2004;557:1–8.
19. Mumby MC, Walter G. Protein serine/threonine phosphatases: structure, regulation, and functions in cell growth. Physiol Rev 1993;73:673–699.
20. Klumpp S, Krieglstein J. Serine/threonine protein phosphatases in apoptosis. Curr Opin Pharmacol 2002;2:458–462.
21. Garcia A, Cayla X, Guergnon J, et al. Serine/threonine protein phosphatases PP1 and PP2A are key players in apoptosis. Biochimie 2003;85:721–726.

22. Suganuma M, Fujiki H, Suguri H, et al. Okadaic acid: an additional non-phorbol-12-tetra-decanoate-13-acetate-type tumor promoter. Proc Natl Acad Sci USA 1988;85:1768–1771.

23. Pallas DC, Shahrik LK, Martin BL, et al. Polyoma small and middle T antigens and SV40 small t antigen form stable complexes with protein phosphatase 2A. Cell 1990;60:167–176.

24. Yang SI, Lickteig RL, Estes R, Rundell K, Walter G, Mumby MC. Control of protein phosphatase 2A by simian virus 40 small-t antigen. Mol Cell Biol 1991;11:1988–1995.

25. Scheidtmann KH, Mumby MC, Rundell K, Walter G. Dephosphorylation of simian virus 40 large-T antigen and p53 protein by protein phosphatase 2A: inhibition by small-t antigen. Mol Cell Biol 1991;11:1996–2003.

26. Li X, Scuderi A, Letsou A, Virshup DM. B56-associated protein phosphatase 2A is required for survival and protects from apoptosis in Drosophila melanogaster. Mol Cell Biol 2002;22:3674–3684.

27. Silverstein AM, Barrow CA, Davis AJ, Mumby MC. Actions of PP2A on the MAP kinase pathway and apoptosis are mediated by distinct regulatory subunits. Proc Natl Acad Sci USA 2002;99:4221–4226.

28. Gjertsen BT, Cressey LI, Ruchaud S, Houge G, Lanotte M, Døskeland SO. Multiple apoptotic death types triggered through activation of separate pathways by cAMP and inhibitors of protein phosphatases in one (IPC leukemia) cell line. J Cell Sci 1994;107:3363–3377.

29. Cross T, Griffiths G, Deacon E, et al. PKC-delta is an apoptotic lamin kinase. Oncogene 2000;19:2331–2337.

30. van de Water B, Tijdens IB, Verbrugge A, et al. Cleavage of the actin-capping protein alpha-adducin at Asp-Asp-Ser-Asp633-Ala by caspase-3 is preceded by its phosphorylation on serine 726 in cisplatin-induced apoptosis of renal epithelial cells. J Biol Chem 2000;275:25,805–25,813.

31. Fladmark KE, Brustugun OT, Hovland R, et al. Ultrarapid caspase-3 dependent apoptosis induction by serine/threonine phosphatase inhibitors. Cell Death Differ 1999;6:1099–1108.

32. Mills JC, Stone NL, Erhardt J, Pittman RN. Apoptotic membrane blebbing is regulated by myosin light chain phosphorylation. J Cell Biol 1998;140:627–636.

33. Edelman AM, Lin WH, Osterhout DJ, Bennett MK, Kennedy MB, Krebs EG. Phosphorylation of smooth muscle myosin by type II Ca2+/calmodulin-dependent protein kinase. Mol Cell Biochem 1990;97:87–98.

34. Amano M, Ito M, Kimura K, et al. Phosphorylation and activation of myosin by Rho-associated kinase (Rho-kinase). J Biol Chem 1996;271:20,246–20,249.

35. Ikebe M, Stepinska M, Kemp BE, Means AR, Hartshorne DJ. Proteolysis of smooth muscle myosin light chain kinase. Formation of inactive and calmodulin-independent fragments. J Biol Chem 1987;262:13,828–13,834.

36. Komatsu S, Hosoya H. Phosphorylation by MAPKAP kinase 2 activates Mg(2+)-ATPase activity of myosin II. Biochem Biophys Res Commun 1996;223:741–745.

37. Van Eyk JE, Arrell DK, Foster DB, et al. Different molecular mechanisms for Rho family GTPase-dependent, Ca2+-independent contraction of smooth muscle. J Biol Chem 1998; 273:23,433–23,439.

38. Niiro N, Ikebe M. Zipper-interacting protein kinase induces Ca(2+)-free smooth muscle contraction via myosin light chain phosphorylation. J Biol Chem 2001;276:29,567–29,574.

39. Endo A, Surks HK, Mochizuki S, Mochizuki N, Mendelsohn ME. Identification and characterization of zipper-interacting protein kinase as the unique vascular smooth muscle myosin phosphatase-associated kinase. J Biol Chem 2004;279:42,055–42,061.

40. Graves PR, Winkfield KM, Haystead TA. Regulation of zipper-interacting protein kinase activity in vitro and in vivo by multisite phosphorylation. J Biol Chem 2005;280:9363–9374.

41. Shani G, Marash L, Gozuacik D, et al. Death-associated protein kinase phosphorylates ZIP kinase, forming a unique kinase hierarchy to activate its cell death functions. Mol Cell Biol 2004;24:8611–8626.

42. Cardone MH, Roy N, Stennicke HR, et al. Regulation of cell death protease caspase-9 by phosphorylation. Science 1998;282:1318–1321.

43. Martins LM, Kottke TJ, Kaufmann SH, Earnshaw WC. Phosphorylated forms of activated caspases are present in cytosol from HL-60 cells during etoposide-induced apoptosis. Blood 1998;92:3042–3049.

44. Alvarado-Kristensson M, Melander F, Leandersson K, Ronnstrand L, Wernstedt C, Andersson T. p38-MAPK signals survival by phosphorylation of caspase-8 and caspase-3 in human neutrophils. J Exp Med 2004;199:449–458.

45. O'Connor DS, Grossman D, Plescia J, et al. Regulation of apoptosis at cell division by p34cdc2 phosphorylation of survivin. Proc Natl Acad Sci USA 2000;97:13,103–13,107.

46. Dan HC, Sun M, Kaneko S, et al. Akt phosphorylation and stabilization of X-linked inhibitor of apoptosis protein (XIAP). J Biol Chem 2004;279:5405–5412.

47. Ku NO, Omary MB. Effect of mutation and phosphorylation of type I keratins on their caspase-mediated degradation. J Biol Chem 2001;276:26,792–26,798.

48. Desagher S, Osen-Sand A, Montessuit S, et al. Phosphorylation of bid by casein kinases I and II regulates its cleavage by caspase 8. Mol Cell 2001;8:601–611.

49. Pinna LA, Donella-Deana A. Phosphorylated synthetic peptides as tools for studying protein phosphatases. Biochim Biophys Acta 1994;1222:415–431.

50. Tzivion G, Shen YH, Zhu J. 14-3-3 proteins; bringing new definitions to scaffolding. Oncogene 2001;20:6331–6338.

51. Wang HG, Reed JC. Bcl-2, Raf-1 and mitochondrial regulation of apoptosis. Biofactors 1998;8:13–16.

52. Soane L, Cho HJ, Niculescu F, Rus H, Shin ML. C5b-9 terminal complement complex protects oligodendrocytes from death by regulating Bad through phosphatidylinositol 3-kinase/Akt pathway. J Immunol 2001;167:2305–2311.

53. Wang HG, Pathan N, Ethell IM, et al. Ca2+-induced apoptosis through calcineurin dephosphorylation of BAD. Science 1999;284:339–343.

54. Klumpp S, Selke D, Krieglstein J. Protein phosphatase type 2C dephosphorylates BAD. Neurochem Int 2003;42:555–560.

55. Chiang CW, Kanies C, Kim KW, et al. Protein phosphatase 2A dephosphorylation of phosphoserine 112 plays the gatekeeper role for BAD-mediated apoptosis. Mol Cell Biol 2003; 23:6350–6362.

56. Subramanian RR, Zhang H, Wang H, Ichijo H, Miyashita T, Fu H. Interaction of apoptosis signal-regulating kinase 1 with isoforms of 14-3-3 proteins. Exp Cell Res 2004;294:581–591.

57. Yellaturu CR, Bhanoori M, Neeli I, Rao GN. N-Ethylmaleimide inhibits platelet-derived growth factor BB-stimulated Akt phosphorylation via activation of protein phosphatase 2A. J Biol Chem 2002;277:40,148–40,155.

58. Burgering BM, Medema RH. Decisions on life and death: FOXO Forkhead transcription factors are in command when PKB/Akt is off duty. J Leukoc Biol 2003;73:689–701.

59. Yeh PY, Yeh KH, Chuang SE, Song YC, Cheng AL. Suppression of MEK/ERK signaling pathway enhances cisplatin-induced NF-kappaB activation by protein phosphatase 4-mediated NF-kappaB p65 Thr dephosphorylation. J Biol Chem 2004;279:26,143–26,148.

60. Appella E, Anderson CW. Post-translational modifications and activation of p53 by genotoxic stresses. Eur J Biochem 2001;268:2764–2772.

61. Ou YH, Chung PH, Sun TP, Shieh SY. p53 C-Terminal Phosphorylation by CHK1 and CHK2 Participates in the Regulation of DNA-Damage-induced C-Terminal Acetylation. Mol Biol Cell 2005;16:1684–1695.

62. Tao X, Finkbeiner S, Arnold DB, Shaywitz AJ, Greenberg ME. Ca2+ influx regulates BDNF transcription by a CREB family transcription factor-dependent mechanism. Neuron 1998;20:709–726.

63. Riccio A, Ahn S, Davenport CM, Blendy JA, Ginty DD. Mediation by a CREB family transcription factor of NGF-dependent survival of sympathetic neurons. Science 1999;286: 2358–2361.

64. Shieh PB, Hu SC, Bobb K, Timmusk T, Ghosh A. Identification of a signaling pathway involved in calcium regulation of BDNF expression. Neuron 1998;20:727–740.

65. Nishihara H, Hwang M, Kizaka-Kondoh S, Eckmann L, Insel PA. Cyclic AMP promotes cAMP-responsive element-binding protein-dependent induction of cellular inhibitor of apoptosis protein-2 and suppresses apoptosis of colon cancer cells through ERK1/2 and p38 MAPK. J Biol Chem 2004;279:26,176–26,183.

66. Ruchaud S, Seite P, Foulkes NS, Sassone-Corsi P, Lanotte M. The transcriptional repressor ICER and cAMP-induced programmed cell death. Oncogene 1997;15:827–836.

67. Lanotte M, Riviere JB, Hermouet S, et al. Programmed cell death (apoptosis) is induced rapidly and with positive cooperativity by activation of cyclic adenosine monophosphate-kinase I in a myeloid leukemia cell line. J Cell Physiol 1991;146:73–80.

68. Zandi E, Karin M. Bridging the gap: composition, regulation, and physiological function of the IkappaB kinase complex. Mol Cell Biol 1999;19:4547–4551.

69. Carballo M, Marquez G, Conde M, et al. Characterization of calcineurin in human neutrophils. Inhibitory effect of hydrogen peroxide on its enzyme activity and on NF-kappaB DNA binding. J Biol Chem 1999;274:93–100.

70. Miskolci V, Castro-Alcaraz S, Nguyen P, Vancura A, Davidson D, Vancurova I. Okadaic acid induces sustained activation of NFkappaB and degradation of the nuclear IkappaBalpha in human neutrophils. Arch Biochem Biophys 2003;417:44–52.

71. Alberts AS, Montminy M, Shenolikar S, Feramisco JR. Expression of a peptide inhibitor of protein phosphatase 1 increases phosphorylation and activity of CREB in NIH 3T3 fibroblasts. Mol Cell Biol 1994;14:4398–4407.

72. Wadzinski BE, Wheat WH, Jaspers S, et al. Nuclear protein phosphatase 2A dephosphory-lates protein kinase A-phosphorylated CREB and regulates CREB transcriptional stimulation. Mol Cell Biol 1993;13:2822–2834.

73. Shi Y, Venkataraman SL, Dodson GE, Mabb AM, LeBlanc S, Tibbetts RS. Direct regulation of CREB transcriptional activity by ATM in response to genotoxic stress. Proc Natl Acad Sci USA 2004;101:5898–5903.

74. Leslie NR, Downes CP. PTEN function: how normal cells control it and tumour cells lose it. Biochem J 2004;382:1–11.

75. Zolnierowicz S, Bollen M. Protein phosphorylation and protein phosphatases. De Panne, Belgium, September 19-24, 1999. Embo J 2000;19:483–488.

76. Ofek P, Ben-Meir D, Kariv-Inbal Z, Oren M, Lavi S. Cell cycle regulation and p53 activation by protein phosphatase 2C alpha. J Biol Chem 2003;278:14,299–14,305.

77. Cardenas ME, Sanfridson A, Cutler NS, Heitman J. Signal-transduction cascades as targets for therapeutic intervention by natural products. Trends Biotechnol 1998;16:427–433.

78. Lotem J, Kama R, Sachs L. Suppression or induction of apoptosis by opposing pathways downstream from calcium-activated calcineurin. Proc Natl Acad Sci USA 1999; 96:12,016–12,020.

79. Bueno OF, Lips DJ, Kaiser RA, et al. Calcineurin Abeta gene targeting predisposes the myocardium to acute ischemia-induced apoptosis and dysfunction. Circ Res 2004;94:91–99.

80. Kalivendi SV, Konorev EA, Cunningham S, Joseph J, Kalyanaraman B. Doxorubicin activates nuclear factor of activated T-lymphocytes and Fas ligand transcription: role of mitochondrial reactive oxygen species and calcium. Biochem J 2005;389:527–539.

81. Mukerjee N, McGinnis KM, Gnegy ME, Wang KK. Caspase-mediated calcineurin activation contributes to IL-2 release during T cell activation. Biochem Biophys Res Commun 2001; 285:1192–1199.

82. Nickels JT, Broach JR. A ceramide-activated protein phosphatase mediates ceramide-induced G1 arrest of Saccharomyces cerevisiae. Genes Dev 1996;10:382–394.

83. Silverstein AM, Galigniana MD, Chen MS, Owens-Grillo JK, Chinkers M, Pratt WB. Protein phosphatase 5 is a major component of glucocorticoid receptor.hsp90 complexes with properties of an FK506-binding immunophilin. J Biol Chem 1997;272:16,224–16,230.

84. Chen MX, McPartlin AE, Brown L, Chen YH, Barker HM, Cohen PT. A novel human protein serine/threonine phosphatase, which possesses four tetratricopeptide repeat motifs and localizes to the nucleus. Embo J 1994;13:4278–4290.

85. Chinkers M. Protein phosphatase 5 in signal transduction. Trends Endocrinol Metab 2001; 12:28–32.

86. Morita K, Saitoh M, Tobiume K, et al. Negative feedback regulation of ASK1 by protein phosphatase 5 (PP5) in response to oxidative stress. Embo J 2001;20:6028–6036.

87. Zhou G, Golden T, Aragon IV, Honkanen RE. Ser/Thr protein phosphatase 5 inactivates hypoxia-induced activation of an apoptosis signal-regulating kinase 1/MKK-4/JNK signaling cascade. J Biol Chem 2004;279:46,595–46,605.

88. Huang S, Shu L, Easton J, et al. Inhibition of mammalian target of rapamycin activates apoptosis signal-regulating kinase 1 signaling by suppressing protein phosphatase 5 activity. J Biol Chem 2004;279:36,490–36,496.

89. Wechsler T, Chen BP, Harper R, et al. DNA-PKcs function regulated specifically by protein phosphatase 5. Proc Natl Acad Sci USA 2004;101:1247–1252.

90. Brewis ND, Cohen PT. Protein phosphatase X has been highly conserved during mammalian evolution. Biochim Biophys Acta 1992;1171:231–233.

91. Zhou G, Boomer JS, Tan TH. Protein phosphatase 4 is a positive regulator of hematopoietic progenitor kinase 1. J Biol Chem 2004;279:49,551–46,561.

92. Mourtada-Maarabouni M, Kirkham L, Jenkins B, et al. Functional expression cloning reveals proapoptotic role for protein phosphatase 4. Cell Death Differ 2003;10:1016–1024.

93. Zhou G, Mihindukulasuriya KA, MacCorkle-Chosnek RA, et al. Protein phosphatase 4 is involved in tumor necrosis factor-alpha-induced activation of c-Jun N-terminal kinase. J Biol Chem 2002;277:6391–6398.

94. Yang J, Fan GH, Wadzinski BE, Sakurai H, Richmond A. Protein phosphatase 2A interacts with and directly dephosphorylates RelA. J Biol Chem 2001;276:47,828–47,833.

95. Ward C, Walker A, Dransfield I, Haslett C, Rossi AG. Regulation of granulocyte apoptosis by NF-kappaB. Biochem Soc Trans 2004;32:465–467.

96. Cohen PT. Novel protein serine/threonine phosphatases: variety is the spice of life. Trends Biochem Sci 1997;22:245–251.

97. Goldberg Y. Protein phosphatase 2A: who shall regulate the regulator? Biochem Pharmacol 1999;57:321–328.

98. Sontag E. Protein phosphatase 2A: the Trojan Horse of cellular signaling. Cell Signal 2001; 13:7–16.

99. Li M, Makkinje A, Damuni Z. The myeloid leukemia-associated protein SET is a potent inhibitor of protein phosphatase 2A. J Biol Chem 1996;271:11,059–11,062.

100. Tamura Y, Simizu S, Osada H. The phosphorylation status and anti-apoptotic activity of Bcl-2 are regulated by ERK and protein phosphatase 2A on the mitochondria. FEBS Lett 2004;569:249–255.

101. Santoro MF, Annand RR, Robertson MM, et al. Regulation of protein phosphatase 2A activity by caspase-3 during apoptosis. J Biol Chem 1998;273:13,119–13,128.

102. Harmala-Brasken AS, Mikhailov A, Soderstrom TS, et al. Type-2A protein phosphatase activity is required to maintain death receptor responsiveness. Oncogene 2003;22:7677–7686.

103. Alvarado-Kristensson M, Andersson T. Protein phosphatase 2A regulates apoptosis in neutrophils by dephosphorylating both p38 MAPK and its substrate caspase 3. J Biol Chem 2005;280:6238–6244.

104. Bøe R, Gjertsen BT, Vintermyr OK, Houge G, Lanotte M, Døskeland SO. The protein phosphatase inhibitor okadaic acid induces morphological changes typical of apoptosis in mammalian cells. Exp Cell Res 1991;195:237–246.

105. Fladmark KE, Brustugun OT, Mellgren G, et al. Ca2+/calmodulin-dependent protein kinase II is required for microcystin-induced apoptosis. J Biol Chem 2002;277:2804–2811.

106. Shtrichman R, Sharf R, Barr H, Dobner T, Kleinberger T. Induction of apoptosis by adenovirus E4orf4 protein is specific to transformed cells and requires an interaction with protein phosphatase 2A. Proc Natl Acad Sci USA 1999;96:10,080–10,085.

107. Strack S, Cribbs JT, Gomez L. Critical role for protein phosphatase 2A heterotrimers in mammalian cell survival. J Biol Chem 2004;279:47,732–47,739.

108. Ayllon V, Cayla X, Garcia A, Fleischer A, Rebollo A. The anti-apoptotic molecules Bcl-xL and Bcl-w target protein phosphatase 1alpha to Bad. Eur J Immunol 2002;32:1847–1855.

109. Chalfant CE, Rathman K, Pinkerman RL, et al. De novo ceramide regulates the alternative splicing of caspase 9 and Bcl-x in A549 lung adenocarcinoma cells. Dependence on protein phosphatase-1. J Biol Chem 2002;277:12,587–12,595.

110. Fujiki H, Suganuma M. Unique features of the okadaic acid activity class of tumor promoters. J Cancer Res Clin Oncol 1999;125:150–155.

111. Connor JH, Kleeman T, Barik S, Honkanen RE, Shenolikar S. Importance of the beta12-beta13 loop in protein phosphatase-1 catalytic subunit for inhibition by toxins and mammalian protein inhibitors. The Journal of Biological Chemistry 1999;274:22,366–22,372.

112. Dawson JF, Holmes CF. Molecular mechanisms underlying inhibition of protein phosphatases by marine toxins. Front Biosci 1999;4:646–658.

113. Holmes CF, Maynes JT, Perreault KR, Dawson JF, James MN. Molecular enzymology underlying regulation of protein phosphatase-1 by natural toxins. Current Medicinal Chemistry 2002;9:1981–1989.

114. Mitsuhashi S, Matsuura N, Ubukata M, Oikawa H, Shima H, Kikuchi K. Tautomycetin is a novel and specific inhibitor of serine/threonine protein phosphatase type 1, PP1. Biochem Biophys Res Commun 2001;287:328–331.

115. Walsh AH, Cheng A, Honkanen RE. Fostriecin, an antitumor antibiotic with inhibitory activity against serine/threonine protein phosphatases types 1 (PP1) and 2A (PP2A), is highly selective for PP2A. FEBS Letters 1997;416:230–234.

116. Runnegar MT, Gerdes RG, Falconer IR. The uptake of the cyanobacterial hepatotoxin microcystin by isolated rat hepatocytes. Toxicon 1991;29:43–51.

117. Eriksson JE, Gronberg L, Nygard S, Slotte JP, Meriluoto JA. Hepatocellular uptake of 3H-dihydromicrocystin-LR, a cyclic peptide toxin. Biochim Biophys Acta 1990;1025:60–66.

118. Schulman H, Hanson PI. Multifunctional Ca2+/calmodulin-dependent protein kinase. Neurochem Res 1993;18:65–77.

119. Colbran RJ. Targeting of calcium/calmodulin-dependent protein kinase II. Biochem J 2004;378:1–16.

120. Schworer CM, Colbran RJ, Soderling TR. Reversible generation of a Ca2+-independent form of Ca2+(calmodulin)-dependent protein kinase II by an autophosphorylation mechanism. J Biol Chem 1986;261:8581–584.

121. Lai Y, Nairn AC, Greengard P. Autophosphorylation reversibly regulates the Ca2+/calmodulin-dependence of Ca2+/calmodulin-dependent protein kinase II. Proc Natl Acad Sci USA 1986;83:4253–4257.

122. Sandal T, Ahlgren R, Lillehaug J, Doskeland SO. Establishment of okadaic acid resistant cell clones using a cDNA expression library. Cell Death Differ 2001;8:754–766.

123. Sandal T, Aumo L, Hedin L, Gjertsen BT, Doskeland SO. Irod/Ian5: an inhibitor of gamma-radiation- and okadaic acid-induced apoptosis. Mol Biol Cell 2003;14:3292–3304.

124. Hornum L, Romer J, Markholst H. The diabetes-prone BB rat carries a frameshift mutation in Ian4, a positional candidate of Iddm1. Diabetes 2002;51:1972–1979.

125. MacMurray AJ, Moralejo DH, Kwitek AE, et al. Lymphopenia in the BB rat model of type 1 diabetes is due to a mutation in a novel immune-associated nucleotide (Ian)-related gene. Genome Res 2002;12:1029–1039.

126. Ding WX, Nam Ong C. Role of oxidative stress and mitochondrial changes in cyanobacteria-induced apoptosis and hepatotoxicity. FEMS Microbiol Lett 2003;220:1–7.

127. Ding WX, Shen HM, Ong CN. Critical role of reactive oxygen species and mitochondrial permeability transition in microcystin-induced rapid apoptosis in rat hepatocytes. Hepatology 2000;32:547–555.

128. Howe CJ, Lahair MM, McCubrey JA, Franklin RA. Redox Regulation of the Calcium/Calmodulin-dependent Protein Kinases. J Biol Chem 2004;279:44,573–44,581.

129. Humphrey W, Dalke A, Schulten K. VMD: visual molecular dynamics. J Mol Graph 1996; 14:33–38, 27–28.

8

Urokinase/Urokinase Receptor-Mediated Signaling in Cancer

Sreerama Shetty and Steven Idell

Summary

Experimental oncogenic transformation or in spontaneous human cancers, mitogenesis and expression of fibrinolytic components such as urokinase (uPA), its receptor (uPAR), and its major inhibitor plasminogen activator inhibtor-1 (PAI-1) or -2 (PAI-2) are activated by common signaling mechanisms. In tumor cells of mesenchymal or epithelial origin uPA, uPAR, and PAIs or metallo-proteinases (MMPs) are overexpressed and these molecules are implicated in tumor invasion or metastasis. Oncogenic stimuli constitutively activate extracellular-regulated kinases (Erk1/2) and NH2-Jun-kinase (Jnk). Tumor or transformed cells typically overexpress uPA and uPAR and show increased activation of the above signaling modules. These signaling intermediaries are involved in the expression of uPA and uPAR as well as mitogenesis, neoplastic growth and metaststic spread and tissue remodeling as occurs in the pathogenesis of neoplasia.

Key Words: Urokinase; urokinase receptor; plasminogen activator inhibitor; cellular proliferation; signaling; apoptosis.

1. Disordered Fibrinolysis in Neoplasia: A Brief Overview

The pathogenesis of neoplasia includes key components including cellular transformation, invasion, and metastasis. These are all multifaceted processes that involve several tumor-derived and host-derived factors. Tumor cells exhibit anchorage-independent growth, an increased propensity to migrate, the capacity to produce large quantities of tumor proteases, and enhanced proliferation. The ability of tumor cells to assume an invasive phenotype depends in part on the balance between proteases and their inhibitors. This balance generally favors increased activity of proteases vs their inhibitors in animal and human tumors. Among the tumor-related proteases, procoagulants favor extravascular fibrin deposition in solid neoplasms (1). Increased extravascular fibrin deposition promotes the desmoplastic response as well as the growth and spread of solid tumors (1).

Abnormalities of the plasminogen activator system have likewise been implicated in the pathogenesis of tumor invasiveness and metastatic spread. Urokinase plasminogen activator (uPA) and plasmin are overexpressed in wide range of solid tumors. These include breast, lung, bladder, kidney, colorectal, stomach, brain, ovarian, endometrial cancers, and malignant melanoma (2). Several reports link the expression of uPA, uPAR, PAI-1, and PAI-2 to both tumor cell invasiveness and to prognosis in cancer patients. For instance, high levels of uPA have been correlated with poor outcome or shorter survival in a variety of tumors, including lung and breast cancers (3). Increased expression of PAI-1 has likewise been linked to virulent tumor behavior (3). Increased

From: *Apoptosis, Cell Signaling, and Human Diseases: Molecular Mechanisms, Volume 2*
Edited by R. Srivastava © Humana Press Inc., Totowa, NJ

uPAR expression has also been associated with poor prognosis in some forms of malignancy, including breast or colorectal cancer *(3,4)*. Different histological forms of tumors and those linked to poor prognosis have been associated with increased expression of uPA, uPAR, and PAI-1. These observations suggest that clinical outcome extends beyond predictable alterations in fibrinolytic capacity. It is conceivable that outcomes may be broadly related to pathophysiologic responses initiated by these proteins through cellular signaling. Mechanisms by which the uPA–uPAR system contributes to the pathogenesis of tumor invasiveness include proteolytic degradation of basement membrane and extracellular matrix constituents and intracellular signaling that initiates mitogenic or altered adhesion responses in tumor cells.

The serine protease uPA has an approximate molecular weight of 50 kDa and is secreted in proenzyme form known as single chain urokinase or pro-uPA. Pro-uPA is converted into active two-chain uPA. Besides plasminogen activation and degradation of several extracellular matrix proteins, uPA also activates matrix metalloproteinases (MMPs). The MMPS are responsible for collagen degradation as well as remodeling of the extracellular matrix. Most of the biological activity of uPA depends on its association with its receptor uPAR. This interaction is an extremely important determinant of the invasive ability of tumor cells.

uPAR is highly glycosylated membrane proteins with five potential glycosylation sites. The aminoterminal portion of this receptor provides the uPA binding site. Because uPAR is a receptor that lacks both cytoplasmic and transmembrane domains, it was not immediately evident that uPAR could participate in signal transduction. The mitogenic activity of uPA is either dependent on receptor interaction *(5–8)* or uPA catalytic activity *(9)*.

Plasminogen activator inhibitors (PAI) belong to the serpin protease inhibitor superfamily. These serpins rapidly inactivate the plasminogen activators. PAI are present in most body fluids and tissues *(3)*. PAI-1 is a 50 kDa glycoprotein serpin secreted by several cell types, including many types of neoplastic cells. PAI-1 can bind to free and receptor bound uPA, thereby inhibiting uPA-mediated degradation of the extracellular matrix. Receptor bound PAI-1 also mediates internalization of trimeric cell surface uPA-PAI-1-uPAR complexes and the subsequent recycling of uPAR to the cell surface *(5)*. PAI-1 is also involved in the regulation of cell adhesion and migration.

PAI-2 is a 47-kDa protein and is less potent in inhibiting receptor bound uPA than is PAI-1. PAI-2 exists both in intracellular and secreted form and the latter participates in the control of tissue remodeling and fibrinolysis. PAI-2 is usually undetectable in the circulation, except in pregnancy and in association with selected neoplasms.

2. Regulation of the Plasminogen Activator System by Cytokines and Growth Factors

The expression of uPA, uPAR or PAI-1 in tumor tissues is regulated by the products of several different cell types, including tumor cells, those in surrounding tissues and those that infiltrate these tissues. Tumor associated macrophages secrete a wide range of cytokines and growth factors such as interleukin-1 (IL-1β), ineterleukin-2 (IL-2), tumor necrosis factor (TNF-α), vascular endothelial growth factor (VEGF), fibroblast growth factor (FGF), and granulocyte-macrophage colony stimulating factor (GM-CSF). Expression of several of these mediators is increased in neoplastic cells or those that occur

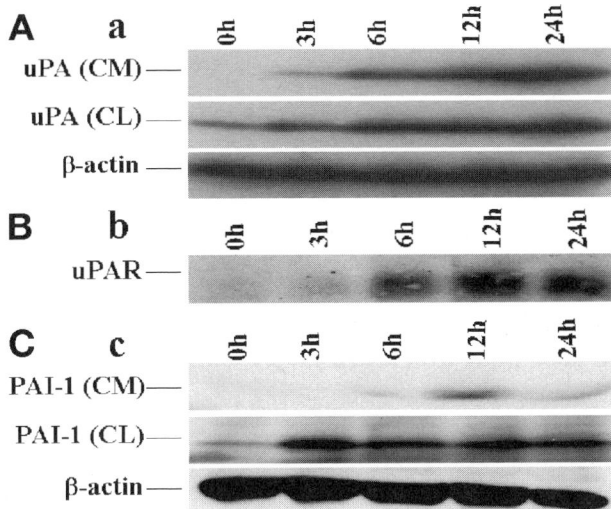

Fig. 1. Regulation of uPA, uPAR and PAI-1 expression by uPA. Beas2B cells grown to confluence were treated with uPA (1000 ng/mL) for 0–24 h at 37°C in basal medium containing 0.5% bovine serum albumin (BSA). Total proteins from the cell lysates (CL) and conditioned media (CM) were separated on 8% sodium dodecyl sulfate (SDS)-polyacrylamide gels, then transferred to a nitrocellulose membrane. The membrane was immunoblotted with anti-uPA (**A**) and –PAI-1 (**C**) antibodies. In the case of uPAR (**B**) the membrane proteins were subjected to Western blotting using anti-uPAR antibody. (Reprinted from refs. *13–15*, with permission.)

in surrounding tissues. These mediators generally up-regulate the expression of either uPA or uPAR or both and these effects may be observed in various cell types *(3)*. These cytokines and growth factors may also regulate the initiation and extant of tumor angiogenesis.

The propagation of solid tumors and tissue inflammation share common features such as vascular leakage that are likewise regulated by common mediators *(1,10)*. Increased vascular permeability occurs in both settings as a result of permeability factors, activation of the proximate endothelium, and the invasion of leukocytes. Macrophage infiltration was also been reported to correlate with the neovascularization of several tumor types *(3)*. Proangiogenic mediators and tumor promoting agents like hepatocyte growth factor (HGF), epidermal growth factor (EGF), and insulin-like growth factor (IGF) also induce uPA–uPAR expression. Similarly, hypoxia, which stimulates tumor neovascularization, also induces uPAR expression *(11)*. Interventions that alter expression of components of the uPA–uPAR system and target proinflammatory mediators can alter neoplastic growth. For example, nonsteroidal anti-inflammatory drugs (NSAIDs) down regulate uPAR expression in monocytes and suppress metastasis in animal models *(3)*.

3. Regulation of the Plasminogen Activator System by Urokinase

Recent reports confirm that uPA itself can stimulate its own expression or that of uPAR or PAI-1 in lung epithelial cells in culture (*see* Fig. 1) and other tumor cells *(12–15)*. Whereas clear evidence of the role of these pathways in cancer has yet to be determined, they may contribute to the derangements of fibrin turnover that have been reported to occur in neoplasia. Tumor cell invasion is facilitated by saturation of uPAR with endogenous uPA originating from neoplastic or surrounding nonneoplastic cells.

uPA-mediated induction of components of the fibrinolytic system could enhance this process, as well as cellular migration or proliferation. As it now appears that uPA, uPAR, and PAI-1 are integrally involved in the pathogenesis of carcinomas, it may be that uPA-mediated autoinduction or induction of PAI-1 and uPAR could potentiate over-expression of these molecules by neoplastic cells. Although the clinical relevance of these newly described pathways remains to be established, it now appears that clinical prognosis may importantly rely upon autocrine or paracrine induction by uPA.

4. Transcriptional Regulation of the Plasminogen Activator System

Expression of uPA, uPAR, PAI-1, and PAI-2 genes are regulated at both transcriptional and posttranscriptional levels *(16–30)*. uPA gene expression is highly responsive to phorbol esters, growth factors, steroid hormones, and cytoskeletal reorganization, ultraviolet (UV) light, changes in cell morphology and contact and, as noted above, by uPA itself. Under many of these conditions, uPA gene transcription is upregulated by selected transcription factors including AP1, Ets-1 and Ets-2. The best characterized regulatory regions of the uPA promoter are the Ets/AP1a composite site, the AP1b site and the connecting 74 bp cooperation mediator region. Ets1 and Ets2 have been shown to activate the uPA promoter. Activation of uPA gene expression through these pathways likely contributes to cellular proliferation, invasion and metastasis in several human cancers as well as inflammation and injury.

uPAR gene expression is induced by PMA, TGF-β, TNF-α, cAMP, EGF, PDGF, VEGF, HGF, and hypoxia. The uPAR promoter contains AP1, SP1, AP2, and SP1/3 binding sites that are involved in activation of uPAR gene activation. PAI-1 gene expression is likewise regulated by many growth factors and cytokines, including TGF-β, EGF, PDGF, FGF, IL-1, and TNF-α and hormones. The PAI-1 promoter contains SP1 and AP2 binding sites and also binds a 72-kDa unknown component. These interactions all facilitate PAI-1 gene activation. Transcription factor nuclear factor I and the ubiquitous factor (USF) are also responsive to TGF-β induction in the PAI-1 promoter. Like uPA, uPAR, PAI-1, and PAI-2 gene expression is regulated by multiple growth factors, hormones, cytokines, and vasoactive peptides. By site-directed mutagenesis studies, basal and PMA-induced PAI-2 transcription appears to depend on AP1a and CRE-like sites.

5. Post-Transcriptional Regulation of the Plasminogen Activator System

uPA, uPAR, PAI-1, and PAI-2 mRNAs are all regulated by posttranscriptional mechanisms that operate at the level of mRNA stability. The uPA mRNA 3′ untranslated region (3′UTR) contains multiple instability determinants and specific mRNA binding protein recognition sequences *(18,22,23)*. Interaction of two specific mRNA binding proteins with uPA mRNA have been reported *(18,22)*. uPAR mRNA is also induced by several agents including PMA, LPS, or TGF-β through regulation at the posttranscriptional level *(21)*. Stability of uPAR mRNA is regulated by determinants present both in the coding and 3′untranslated regions. A 50-kDa binding protein interacts with the uPAR mRNA coding region *(16)* and a 40-kDa protein likewise interacts with 3′UTR determinants to regulate uPAR mRNA stability *(19)*. The posttranscriptional regulation of uPA and uPAR mRNA stability is regulated through cellular phosphorylation *(18,22)*.

PAI-1 mRNA stability is altered via TGF-β, cAMP, and insulin-like growth factor. At least two stability determinants have been identified within the PAI-1 3′UTR *(20,25–27)*. The PAI mRNA 3′UTR interacts with multiple PAI-1 mRNA binding proteins *(25,26)*. Recently, we found that uPA regulates uPA, uPAR and PAI-1 expression through posttranscriptional stabilization of each of these transcripts *(13–15)*. PAI-2 mRNA is also induced by PMA and TNF-α via posttranscriptional regulatory mechanisms. Post-transcriptional regulation of PAI-2 mRNA is regulated by stability determinants present in both the coding region and 3′UTR *(29,30)*.

6. Signal Transduction of the uPA/uPAR System and Cell Proliferation

The balance between fibrinolysis and coagulation is aberrant with fibrin deposition observed in many human tumors *(1)*. Inhibition of fibrinolytic activity resulting from excess PAI-1 expression leads to extravascular fibrin deposition in various solid tumors and in inflammatory conditions *(1,10)*. Extravascular fibrin provides a neomatrix which can undergo remodeling with ultimate fibrotic repair. This process involves organization with migration of endothelial cells and neovascularization. The fibrin gel undergoes remodeling through the action of uPA and other proteases. uPA expression is also upregulated in tumor cells by growth factors such as HGF/SF, VEGF, EGF, IGF-I and -II, bFGF, LPA, CSF-1, vasopressin, and α-thrombin, agonists that signal through PLC, PKC, PLD, Ral, Ras, Raf, Mek-1, and Erk1/2. Rho, Rac, and Cdc42 represent other signaling intermediates activated by uPA. Activation of EGFR, in turn, may be one of the key mechanisms by which tumor cells up regulate uPA though activation of appropriate signaling mechanisms.

Induction of uPA by neoplastic or other cells may involve selective cellular signaling. Proto-oncogene HER2/neu increases uPA expression and utilizes sequential activation of a tyrosine kinase transmembrane receptor, c-Ras, c-Raf, Mek-1, and Erk which leads activation and/or expression of transcription factors c-Jun, c-fos, Ha-Ras, and c-Ets *(2)*. These observations suggest that growth factors and oncogenes use common signaling pathways to induce uPA expression and influence processes such as mitogenesis.

The biological effect of the uPA-uPAR interaction are not only restricted to mitogenesis, but include chemotaxis, cellular adhesion, cytoskeletal reorganization and migration *(31)*. uPA when bound to its receptor stimulates growth in an autocrine fashion. Although the mechanism is still unclear, one can speculate that uPA can support cell proliferation and matrix remodeling by virtue of signaling for plasmin-mediated growth factor and pro-MMP activation as well as uPAR. The ability of uPAR, a GPI-anchored protein, to transduce signals suggests the association of uPAR with transmembrane proteins is capable of coupling ligand binding through uPAR to the cytoplasm. Localization of uPAR to caveolae suggests that these structures are related to signaling through uPAR *(32)*. The association of uPAR with β1, β2, and β3-integrins *(33)* suggests that these integrins may likewise be related to uPAR-mediated cellular signaling. β1 and β2 integrins are involved in the uPAR dependent adhesion and migration in tumor cells *(33)*. Receptor bound uPA activates several tyrosines kinases from Src family (Fyn, Lck, Hck, Yes) which are involved in monocyte chemotaxis *(34)*. uPA-uPAR communicates with extracellular matrix via binding to vitronectin, where it is immobilized in extracellular matrices in cancer and inflammation.

uPA binding to uPAR has also been shown to stimulate kinase activity and phophory-lation of proteins in JAK/STAT pathway *(34,35)*, translocation of glucose transporter to the plasma membrane, and downstream up-regulation of various early response genes such as *c-myc*, *c-fos*, and *c-jun (7,36)* associated with cell growth. uPAR dependent JAK/STAT signaling may be involved in the migration of vascular smooth muscle cells *(35)*. The JAK/STAT pathway is activated via clustering of the uPA-uPAR complex, rep-resenting a sequence that promotes cellular migration *(34,35)*. In breast cancer cells, the uPA–uPAR interaction results in cellular migration through the activation of *Erk1/Erk2 (33)*. An inhibitor targeting MAPK kinase, a member of the JAK family of kinases, sup-presses uPA induced uPAR-dependent activation of *Erk1/Erk2* in these cells *(3)*. The MAPK pathway is likewise activated in cytokine-mediated signaling and has been implicated as a major signal-transduction mechanism in angiogenesis. The uPA-uPAR interaction activates and releases growth and angiogenic factors such as HGF, TGF-β, and VEGF, responses that in turn activate the MAPK pathway.

uPAR is further associated with tyrosine kinases JAK1 and Tyk2 *(34)*. Tyk2 interacts with downstream signaling cascade involving PI3 kinase in vascular smooth muscle cells (VSMC). uPA activation of PI3 kinase is abolished in VSMC expressing dominant negative Tyk2 and uPA increases PI3 kinase activity in Tyk2 immunoprecipitates. Inhibition of PI3 kinase inhibits uPA-induced VSMC migration *(38)*. The growth factor activity of uPA is associated with a rapid transient activation of early response genes (*c-fos*, *c-jun*, and *c-myc*) and the subsequent down regulation of p53 *(39)* and p21CIPI with constant expression of MEK1. Mitogenic activity of uPA involves PTK and PKC *(3)*. Expression of Wt. p53 but not mutant p53 diminished phosphorylation of Stat3 and reduced Stat3 DNA binding activity *(39)*. Src kinases activate Stat3 dependent tran-scription in mammary epithelial cells and EGFR activation can lead to activation of Stat1 and Stat3 *(40)*. uPA also interacts with EGFR and the aminoterminal fragment (ATF) of uPA activates EGFR through Src and MMP. This interaction of uPAR with EGFR activates *Erk*. Similarly, the interaction of EGF with EGFR also activates *Erk*. The promigratory effect of uPA and EGF are mediated though MEK1 and Rho-Rho kinases *(41)*. The growth factor domain of uPA (residues 13–19) and EGF (residues 14–20 show considerable homology and induces proliferation in variety of cell types. However, EGF does not interfere with ATF receptor binding.

The uPA-uPAR interaction is physically associated with αvβ1 *(32)*. In cells express-ing low levels of uPAR, the frequency of the αvβ1 association was significantly reduced leading to lower adhesion of cells to fibronectin *(42)*. Adhesion to fibronectin results in robust *Erk1/2* activation, and inhibition of the uPAR-αvβ1 integrin interaction reduced fibronectin-dependent *Erk1/2* activation *(42)*. Cells overexpressing uPAR by activation of α5β1 integrin initiates intracellular signal through FAK and Src leading to Erk acti-vation and tumorigenicity in vivo. FAK, a nonreceptor tyrosine kinase is overexpressed in several human cancers and activation of FAK induces survival, proliferation and motility of cells in culture *(31)*. uPAR also activates αvβ1 integrins and Erk1 signaling *(33)*. This effect requires the uPA-binding domain I of uPAR and FAK linking αvβ1 integrin with EGFR signaling. In addition, FAK signaling can activate the PI3-Akt/PKB pathways *(11)*, which are important for growth factor and matrix dependent induction of cell survival and migration in vitro. Loss of cell–cell interaction (E-cadherin-based) and increased uPAR–integrin association resulting from fibronectin-matrix interactions constitutively activates the Mek-Erk MAPK (mitogenic) pathway and suppress p38 SAPK (growth suppressive) pathways *(43)*. Downregulation of uPAR induces dormancy *(44)*; lack of cellular proliferation by reversing Erk signaling.

Higher uPAR expression and its interaction with active $\alpha5\beta1$ integrin is required for phosphorylation of FAK and Src in tumor cells *(11)*. The association of uPAR-$\alpha5\beta1$ integrin activates FAK and Src which in turn activates Ras *(41)*. This uPAR-integrin interaction is the best-characterized activator of the Raf-Mek-Erk signaling cascade. Src family protein tyrosine kinases p60 fyn, p53/56 lyn, p58/64hck, and p59fgr are also associated with uPAR *(45)*. Therefore activation of FAK and Src signaling is necessary for uPAR-integrin-mediated tumorigenicity and inhibition leads to arrest of tumor growth *(44)*.

In addition to remodeling the basement membrane, receptor bound uPA mediates phosphorylation of focal adhesion proteins and the activation of MAP kinases *(11)*. uPA catalytic activity also induces the activation of PKC whereas receptor occupancy effects MAP kinase activation *(2)*. These two pathways could operate simultaneously during cell migration or invasion. Therefore, the design of approaches to interrupt these signaling pathways should be to block both uPA activity and its receptor interaction *(3)*.

PAI-1 through inhibition of uPA-mediated angiostatin production induces endothelial cell apoptosis and migration indicating the importance of the contribution of the uPA system to angiogenesis *(3)*. High uPA and PAI-1 levels results in worst prognosis in a wide variety of solid tumors *(3)*, suggesting that their effects on neovascularization could contribute to clinical outcome. These observations also underscore the intricate relationship between uPA, uPAR, and PAI-1 to the pathogenesis of tumor angiogenesis and metastasis.

The signaling pathways activated by the uPA/uPAR interaction seem to utilize the same pathway to induce their own expression. Interaction of uPA with uPAR results in activation of Erk *(33)*. Activated Erk in turn controls many physiological processes such as cell growth, differentiation, apoptosis and migration as well as cancer invasion and metastasis *(2)*. uPA-mediated Erk activation occurs via a growth factor coupled Ras, Raf and Mek- dependent signaling pathway. In aggressive cancer cells, endogenous uPA serves as a major determinant of the basal level of Erk activation, in the absence of stimulants *(3)*. Cytoskeletal reorganization induces uPA expression by a mechanism independent of PKC and PKA *(2)*. Cytoskeletal reorganization induces Erk and Jnk, Erk and Jnk in turn phosphorylate and activate c-Jun *(3)*.

7. Signal Transduction of the uPA/uPAR System and Apoptosis

Recent evidence suggests the involvement of the uPA/uPAR system in programmed cell death. Failure of tumor cells implanted in uPA$^{-/-}$ mice to proliferate suggests that alteration in host expression of uPA affects the balance between tumor cell death and proliferation *(11,36)*. Regulation of the tumor suppressor protein p53 level by uPA occurs in a concentration-dependent manner through mdm2 mediated degradation *(see* Fig. 2) *(38)*. Increased responsiveness of cells bearing reduced levels of uPAR to TNF-α mediates apoptosis-inducing ligand-induced apoptosis, suggesting the involvement of the uPA/uPAR system in cell death *(11)*. The antiapoptotic ability of the uPA/uPAR system may be caused by its ability to activate Ras-Erk signaling pathway in diverse cell types as observed in MDA-MB-231 breast cancer cells, in which inhibition of the

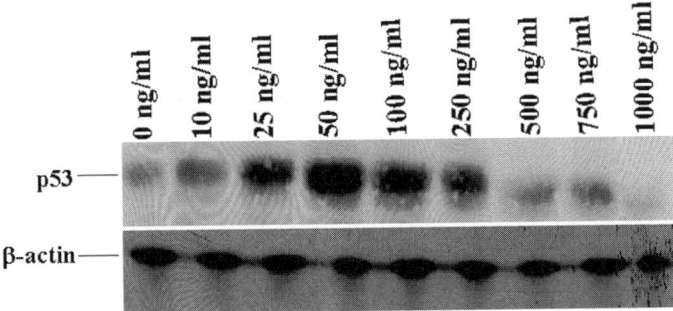

Fig. 2. Effect of uPA concentration on p53 expression. Beas2B cells grown to confluence were treated with varying amounts of uPA (0–1000 ng/mL) for 24 h at 37°C in basal medium containing 0.5% BSA. Total proteins were separated on 8% SDS-polyacrylamide gels, then transferred to a nitrocellulose membrane. The membrane was immunoblotted with anti-p53 antibody. (Reprinted from ref. *38*, with permission.)

uPA/uPAR interaction decreased phopshorylated Erk followed by apoptosis *(11)*. These observations suggest that the uPA/uPAR system may be involved in the regulation of cell death through Erk signaling. Similarly, overexpression of uPA or uPAR and activation of the PI3K-Akt-dependent antiapoptotic pathway has been reported in globlastoma cells *(46)*.

8. Conclusions

During tumor progression, cells acquire a variety of capabilities that promote tumor growth and spread. Acquisition of these capabilities involves signaling interactions, prominent among which are those that involve the uPA–uPAR system. Neoplastic cells acquire sustained, effective growth signals, insensitivity to antigrowth signals, the ability to evade apoptosis and increased replicative potential. Cellular signaling also initiates angiogenesis, which facilitates tissue invasion and metastasis. The uPA/uPAR system otherwise supports the malignant phenotype through several mechanisms including remodeling of the extracellular matrix, increased cell motility through cytoskelatal and focal adhesion formation, and cellular proliferation. Collectively, these findings support the concept that the uPA/uPAR system and signaling pathways that involve the system are appropriate targets for development of novel therapeutic interventions for various forms of neoplasia.

References

1. Dvorak HF. Tumors: wounds that do not heal. Similarities between tumor stroma generation and wound healing. New Eng J Med 1986;315(26):1650–1659.
2. Ghiso JAA, Alonso DF, Farias EF, Gomez DE, Joffe EB. Deregulation of the signaling pathways controlling urokinase production: its relationship with the invasive phenotype. Eur J Biochem 1999;263:295–304.
3. Mazar AP, Henkin J, Goldfarb RH. The urokinase plasminogen activator system in cancer: Implications for tumor angiogenesis and metastasis. Angiogenesis 2000;3:15–32.
4. Dano K, Behrendt N, Brunner N, Ellis V, Ploug M, Pyke C. The urokinase receptor: protein Structure and role in plasminogen activation and cancer Invasion. Fibrinolysis 1994;1:189–203.

5. Shetty S, Kumar A, Johnson A, Idell S. Regulation of mesothelial cell mitogenesis by anti-sense oligonucleotides for the urokinase receptor. Antisense Res Dev 1995;5:307–314.

6. Shetty S, Kumar A, Johnson A, Pueblitz S, Idell S. Urokinase receptor in human malignant mesothelioma cells: role in tumor cell mitogenesis and proteolysis. Am J Physiol 1995;268:L972–L982.

7. Rabbani SA, Gladu J, Mazar AP, Henkin J, Goltzman D. Induction in human osteoblastic cells (SaOS2) of the early response genes fos, jun, and myc by the amino terminal fragment (ATF) of urokinase. J Cell Physiol 1997;172(2):137–145.

8. Xing RH, Mazar A, Henkin J, Rabbani SA. Prevention of breast cancer growth, invasion, and metastasis by antiestrogen tamoxifen alone or in combination with urokinase inhibitor B-428. Cancer Res 1997;57:3585–3593.

9. Bhat GJ, Gunaje JJ, Idell S. Urokinase-type plasminogen activator induces tyrosine phosphorylation of a 78-kDa protein in H-157 cells. Am J Physiol 1998;277:L301–L309.

10. Idell S. Anticoagulants for acute respiratory distress syndrome: Can they work. Am J Resp Crit Care Med 2001;164:517–520.

11. Alfano D, Franco P, Vocca I, et al. The urokinase plasminogen activator and its receptor: Role in cell growth and apoptosis. Thromb Haemostat 2005;93:205–211.

12. Montuori N, Salzano S, Rossi G, Ragno P. Urokinase-type plasminogen activator up-regulates the expression of its cellular receptor FEBS Letters 2000;476:166–170.

13. Shetty S, Idell S. Urokinase induces expression of its own receptor in Beas2B lung epithelial cells. J Biol Chem 2001;276:24,549–24,556.

14. Shetty S, Pendurthi UR, Halady PKS, Azghani AO, Idell S. Urokinase induces its own expression in Beas2B lung epithelial cells. Am J Physiol (Lung Cell Mol Physiol) 2002; 283:L319–L328.

15. Shetty S, Bdeir K, Cines DB, Idell S. Induction of plasminogen activator inhibitor-1 by urokinase in lung epithelial cells. J Biol Chem 2003;278(20):18,124–18,131.

16. Shetty S, Muniyappa H, Halady PK, Idell S. Regulation of urokinase receptor expression by phosphoglycerate kinase. Am J Resp Cell Mol Biol 2004;31(1):100–106.

17. Shetty S, Idell S. Urokinase receptor mRNA stability involves tyrosine phosphorylation in lung epithelial cells. Am J Resp Cell Mol Biol 2004;30(1):69–75.

18. Shetty S, Idell S. Post-transcriptional regulation of urokinase mRNA: Identification of a novel urokinase mRNA-binding protein in human lung epithelial cells in vitro. J Biol Chem 2000;275(18):13,771–13,779.

19. Shetty S. Regulation of Urokinase Receptor mRNA Stability by hnRNPC in Lung Epithelial Cells. Mol Cell Biochem 2005;272:107–118.

20. Shetty S, Idell S. Posttranscriptional regulation of plasminogen activator inhibitor-1 in human lung carcinoma cells in vitro. Am J Physiol- Lung Cell Mol Physiol 2000;278(1):L148–L156.

21. Shetty S, Kumar A, Idell S. Posttranscriptional regulation of urokinase receptor mRNA: identification of a novel urokinase receptor mRNA binding protein in human mesothelioma cells. Mol Cell Biol 1997;17(3):1075–1083.

22. Tran H, Maurer F, Nagamine Y. Stabilization of urokinase and urokinase receptor mRNAs by HuR is linked to its cytoplasmic accumulation induced by activated mitogen-activated protein kinase-activated protein kinase 2. Mol Cell Biol 2003;23(20):7177–7188.

23. Montero L, Nagamine Y. Regulation by p38 mitogen-activated protein kinase of adenylate- and uridylate-rich element-mediated urokinase-type plasminogen activator (uPA) messenger RNA stability and uPA-dependent in vitro cell invasion. Cancer Res 1999;59(20): 5286–5293.

24. Nanbu R, Menoud PA, Nagamine Y. Multiple instability-regulating sites in the 3′ untranslated region of the urokinase-type plasminogen activator mRNA. Mol Cell Biol 1994;14(7): 4920–4928.

25. Heaton JH, Dlakic WM, Gelehrter TD. Posttranscriptional regulation of PAI-1 gene expression. Thromb Haemost 2003;89(6):959–966.

26. Heaton JH, Dlakic WM, Dlakic M, Gelehrter TD. Identification and cDNA cloning of a novel RNA-binding protein that interacts with the cyclic nucleotide-responsive sequence in the Type-1 plasminogen activator inhibitor mRNA. J Biol Chem 2001;276(5):3341–3347.

27. Heaton JH, Tillmann-Bogush M, Leff NS, Gelehrter TD. Cyclic nucleotide regulation of type-1 plasminogen activator-inhibitor mRNA stability in rat hepatoma cells. Identification of cis-acting sequences. J Biol Chem 1998;273(23):14,261–14,268.

28. Yu H, Stasinopoulos S, Leedman P, Medcalf RL. Inherent instability of plasminogen activator inhibitor type 2 mRNA is regulated by tristetraprolin. J Biol Chem 2003; 278(16):13,912–13,918.

29. Tierney MJ, Medcalf RL. Plasminogen activator inhibitor type 2 contains mRNA instability elements within exon 4 of the coding region. Sequence homology to coding region instability determinants in other mRNAs. J Biol Chem 2001;276(17):13,675–13,684.

30. Maurer F, Tierney M, Medcalf RL. An AU-rich sequence in the 3′-UTR of plasminogen activator inhibitor type 2 (PAI-2) mRNA promotes PAI-2 mRNA decay and provides a binding site for nuclear HuR. Nuc Acid Res 1999;27(7):1664–1673.

31. Carlin SM, Resink TJ, Tamm M, Roth M. Urokinase signal transduction and its role in cell migration. FASEB J 2005;19:195–202.

32. Wei Y, Yang X, Liu Q, Wilkins JA, Chapman HA. A role for caveolin and the urokinase receptor in integrin-mediated adhesion and signaling. J Cell Biol 1999;144:1285–1294.

33. Ossowski L, Auguirre-Ghiso JA. Urokinase receptor and integrin partnership: Coordination of signaling for cell adhesion, migration and growth. Curr Opin Cell Biol 2000;12:613–620.

34. Dulmer I, Weis A, Mayboroda OA, et al. The Jak/Stat pathway and urokinase receptor signaling in human aortic vascular smooth muscle cells. J Biol Chem 1998;273:315–321.

35. Koshelnick Y, Ehart M, Hufnag P, Heinrich PC, Binder BR. Urokinase receptor is associated with the components of the JAK1/STAT1 signaling pathway and leads to activation of this pathway upon receptor clustering in the human kidney epithelial tumor cell line TCL-598. J Biol Chem 1997;272:28,563–28,567.

36. Dumler I, Petri T, Schleuning WD. Induction of c-fos gene expression by urokinase-type plasminogen activator in human ovarian cancer cells. FEBS Letter 1994;343:103–106.

37. Kunigal S, Kusch A, Tkachuk N, et al. Monocyte-expressed urokinase inhibits vascular smooth muscle cell growth by activating Stat1. Blood 2003;102:4377–4383.

38. Shetty S, Gyetko MR, Mazar AP. Induction of p53 by urokinase in lung epithelial cells. J Biol Chem 2005;280:28,133–28,141.

39. Lin J, Jin X, Rothman K, Lin H, Tang H, Burke W. Modulation of signal transducer and activator of transcription 3 activities by p53 tumor suppressor in breast cancer cells. Cancer Res 2002;62:376–380.

40. Berclaz G, Altermatt HJ, Rohrbach V, Siragusa A, Dreher E, Smith PD. EGFR dependent expression of Stat3 but not Stat1 in breast cancer. Int J Oncol 2001;19(6):1155–1160.

41. Jo M, Thomas KS, O'Donnell DM, Gonias SL. Epidermal growth factor receptor dependent and—independent cell signaling pathways originating from the urokinase receptor. J Biol Chem 2003;278:1642–1646.

42. Liu D, Auguirre-Ghiso JA, Estrada Y, Ossowski L. EGFR is a tranducer of the urokinase receptor initiated signal that is required for in vivo growth of a human carcinoma. Cancer Cell 2002;1:445–457.

43. Aguirre-Ghiso JA, Liu D, Mignatti A, Kovalski K, Ossowski L. Urokinase receptor and fibronectin regulate the ERK(MAPK) to p38(MAPK) activity ratios that determine carcinoma cell proliferation or dormancy in vivo. Mol Biol Cell 2001;12(4):863–879.

44. Ghiso JAA. Inhibition of FAK signaling activated by urokinase receptor induces dormancy in human carcinoma cells in vivo. Oncogene 2002;21(16):2513–2524.
45. Bohuslav J, Horejsi V, Hansmann C, et al. Urokinase plasminogen activator receptor, beta 2-integrins, and Src-kinases within a single receptor complex of human monocytes. J Exptl Med 1995;181(4):1381–1390.
46. Chandrasekar N, Mohanam S, Gujrati M, Olivero WC, Dinh DH, Rao JS. Down regulation of uPA inhibits migration and PI3k/Akt signaling in glioblastoma cells. Oncogene 2003; 22(3):392–400.

II

MOLECULAR BASIS OF CELL DEATH

9

Signaling Pathways That Protect the Heart Against Apoptosis Induced by Ischemia and Reperfusion

Zheqing Cai and Gregg L. Semenza

Summary

Ischemia and reperfusion injury commonly occurs in ischemic heart disease, resulting in apoptotic or necrotic cell death. Apoptotic cell death is highly regulated. Two mechanisms of apoptosis involve the extrinsic death receptor pathway and the intrinsic mitochondrial pathway. Both pathways lead to the activation of effector caspases, resulting in cell death. The mitochondrial pathway plays a key role in initiating apoptosis after ischemia and reperfusion. The phosphatidylinositol 3-kinase (PI3K), protein kinase C (PKC), and extracellular signal-regulated kinase (ERK) signaling pathways protect the heart against ischemia and reperfusion injury. They inhibit mitochondrial cytochrome c release into the cytosol by regulating the Bcl-2 family proteins and activating the mitoK$_{ATP}$ channel, thereby blocking the process of apoptosis.

Key words: Apoptosis; signal transduction; cardioprotection; ischemia and reperfusion injury; mitochondria; caspases; Bcl-2; PI3K; PKC; ERK.

1. Introduction

Ischemic heart disease is a leading cause of death in the United States. Each year millions of people suffer a heart attack, and hundreds of thousands die of acute myocardial infarction. The survivors from heart attack suffer ischemia and reperfusion injury. Cell death can occur in two different ways, necrosis and apoptosis. Necrosis is an irreversible process, leading to membrane disruption, cell swelling, and cellular debris that stimulate an inflammatory response. In contrast, apoptosis is a highly regulated form of cell death, characterized by cell shrinkage, chromatin condensation, DNA fragmentation, and organelle dismantling without an inflammatory response (1). To maintain normal heart function, cardiomyocytes need be protected against apoptosis. The loss of cardiomyocytes results in greater demands on the remaining myocytes, which have to work harder to compensate. The added stress induces further cardiomyocyte apoptosis resulting in a vicious cycle leading to congestive heart failure. Therefore, controlling the process of apoptosis is a logical strategy to prevent heart failure in patients with ischemic heart disease. This chapter provides a review of the mechanism of apoptosis during ischemia and reperfusion, then discuss signaling pathways that regulate the process of apoptosis to prevent cell death in the heart.

2. Ischemia and Reperfusion Injury

It is well known that ischemia followed by reperfusion induces damage to the heart; however, it is still controversial whether ischemia alone can cause apoptotic

From: *Apoptosis, Cell Signaling, and Human Diseases: Molecular Mechanisms, Volume 2*
Edited by R. Srivastava © Humana Press Inc., Totowa, NJ

cell death. Fliss and Gattinger *(2)* demonstrated apoptosis in ischemic rat myocardium in the absence of reperfusion. Kajstura et al. *(3)* reported both apoptosis and necrosis contribute to infarct size after a prolonged period of ischemia in the rat heart. Several studies *(2,4,5)* demonstrated that reperfusion accelerated the apoptotic cell death initiated by ischemia, but Ohno et al. *(6)* reported that necrotic but not apoptotic cell death was detected in the rabbit heart after coronary artery occlusion by using immunogold electron microscopy and *in situ* nick end labeling. In the rabbit myocardium, Gottlieb et al. *(7)* reported that apoptosis was detectable only after reperfusion. Anversa et al. *(8)* showed the transition from apoptosis to necrosis in the ischemic myocardium. Although apoptosis and necrosis are different mechanisms of cell death, they may share the same early events. Differences in detection methods, time points, and animal species may contribute to these different findings. Further studies are needed to determine the mechanisms of cell death induced by ischemia and ischemia-reperfusion.

3. Mechanism of Apoptosis

Apoptosis can be induced by either the intrinsic mitochondrial pathway or the extrinsic cell death receptor pathway (Fig. 1). Both pathways are activated in the heart subjected to ischemia and reperfusion. Ischemia and reperfusion increase intracellular calcium and free radicals, stimulating the intrinsic mitochondrial pathway, and also increase levels of death receptor ligands such as Fas ligand (FasL) *(9,10)* and tumor necrosis factor-1α (TNF-1α) in the heart *(11)*, leading to activation of the extrinsic pathway. The mitochondria play a key role in initiating the process of apoptosis, which is regulated by Bcl-2 family proteins. Cell death is mediated by caspases *(12)*.

4. Death Receptor Pathway

The death receptors belong to the tumor necrosis factor receptor (TNFR) gene superfamily. They are composed of an N-terminal extracellular region required for ligand binding, a single transmembrane spanning region, and an intracellular C-terminal region containing the death domain. After binding of ligands, death receptors such as Fas and the TNF-1α receptor form a homotrimeric complex, which recruits adaptor proteins that interact with the death domains, leading to the recruitment and subsequent activation of caspase-8 *(13,14)*. The activation of caspase-8 then activates downstream effectors caspase-3 and caspase-7, executing cells by apoptosis *(15,16)*. The death receptors are regulated by ischemia and reperfusion. Increased expression of Fas has been reported in the heart subjected to ischemia and reperfusion *(4)*. The coronary effluent from postischemic isolated and perfused hearts contained significantly increased levels of FasL. Isolated hearts from Fas deficient mice had a marked reduction in cell death and infarct size following ischemia and reperfusion *(17)*. Signaling through the TNF-1α receptor may also play a role in ischemia and reperfusion injury. Over-expression of TNF-1α in transgenic mice increased cardiac apoptosis. Increased levels of TNF-1α were found in the hearts after ischemia and reperfusion *(11)*. The circulating levels of TNF-1α are directly related to disease severity *(18)*, and inhibition of TNF-1α production improved heart function *(19)*. However, several studies reported that TNF-1α may have a beneficial effect in the heart. TNF-1α pretreatment decreased lactate dehydrogenase (LDH)

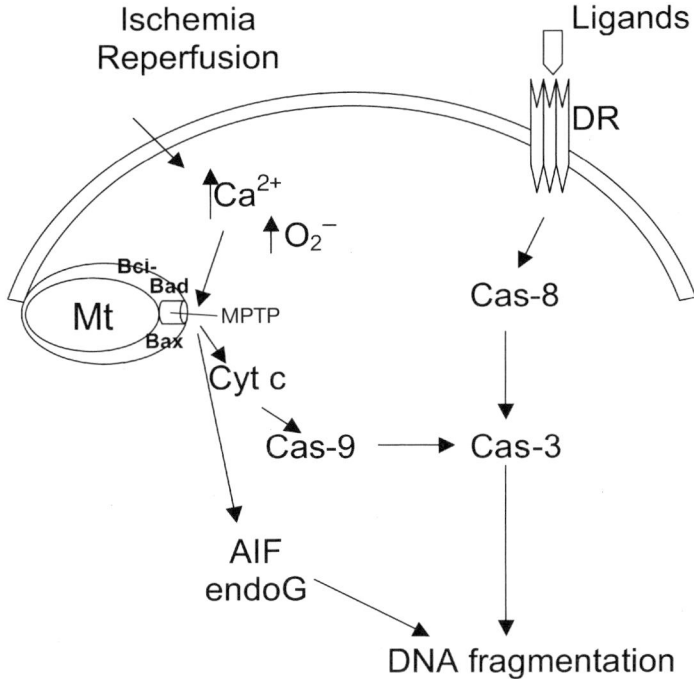

Fig. 1. Mechanisms of apoptosis. Ischemia and reperfusion induce apoptosis in the heart by activating the mitochondrial pathway and the death receptor pathway. The mitochondrial pathway plays a key role in initiating apoptosis in response to ischemia and reperfusion. Increased intracellular calcium and free radicals open MPTP in the mitochondria, leading to the release of cytochrome *c* (Cyt c) into cytosol and then the activation of caspase-9 (Cas-9). Activated cas-9 cleaves pro-caspase-3 and generates active cas-3, which is also activated by the death receptor pathway through caspase-8 (cas-8). Cas-3 causes DNA fragmentation in cardiomyocytes. Mt, mitochondrion; MPTP, mitochondrial permeability transition pore; DR, death receptor; AIF, apoptosis-inducing factor; endoG, endonuclease G.

release after ischemia *(20)*. TNF-receptor deficient mice showed increased infarct size and apoptotic cells after ischemia and reperfusion *(21)*. The beneficial effect is related to NF-κB activation, which is mediated by TNF-1α. NF-κB is a transcription factor which induces the expression of survival genes and inhibits apoptosis *(22)*. In response to TNF-1α, inhibition of NF-κB activation increased apoptosis in cardiomyocytes infected with an adenovirus expressing a dominant-negative form of IκB *(23)*. TNF-1α activates multiple signal pathways in the heart. The predominant effects of TNF-1α may depend on specific physiological conditions.

5. Mitochondrial Pathway

Mitochondria play a dual role within the cell *(24)*. They not only produce ATP to meet the large energy requirement of cardiac myocytes, but also are actively involved in the regulation of cell death. The mitochondrial inner membrane contains the adenine nucleotide transporter (ANT) and components of the electron transport chain, thereby maintaining the proton gradient required for energy production. The mitochondrial

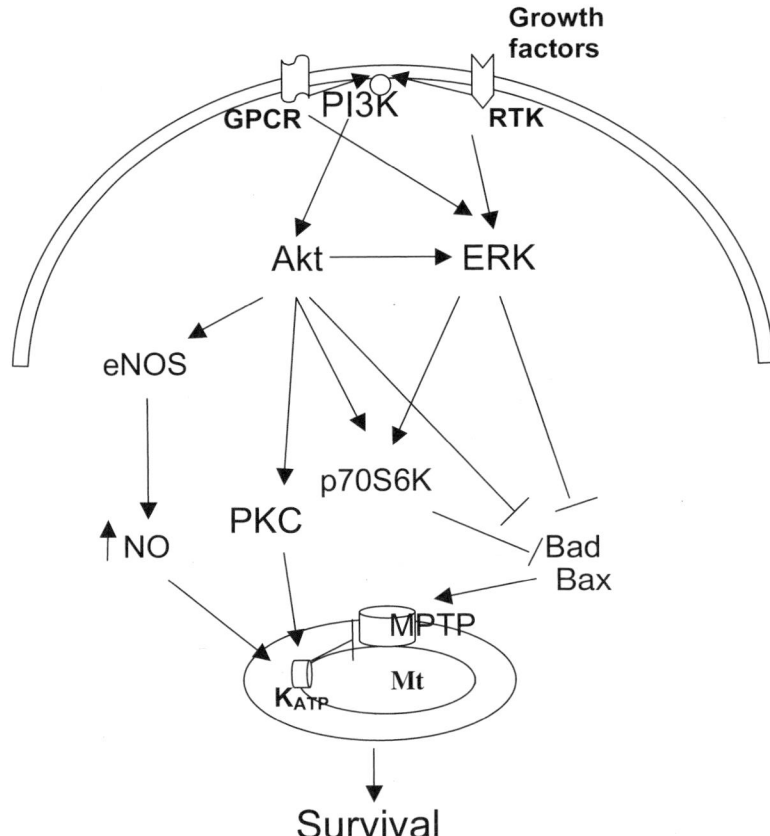

Fig. 2. Mechanism of cell protection mediated by PI3K, PKC, and ERK. The PI3K, PKC, and ERK pathways are activated through RTK or GPCR signaling induced by IPC or binding of growth factors and hormones. PI3K signaling is upstream of PKC activation and crosstalks with the ERK pathway. The PI3K and ERK signal transduction pathways target MPTP through regulating the mitoK$_{ATP}$ channel and Bcl-2 family proteins, thereby inhibiting the release of cytochrome *c* and promoting cell survival in response to apoptotic stimuli. IPC, ischemic preconditioning; PI3K, phosphatidylinositol 3-kinase; PKC, protein kinase C; ERK, extracellular signal-regulated kinase; RTK, receptor protein tyrosine kinase; GPCR, G protein-coupled receptor; K$_{ATP}$, mitochondrial ATP dependent potassium channel; MPTP, mitochodrial permeability transition pore; Mt, mitochondrion.

outer membrane contains a voltage-dependent anion channel (VDAC), which associates with ANT to form a large nonspecific pore, the mitochondrial permeability transition pore (MPTP). In response to apoptotic stimuli, cyclophilin D binds to ANT. Under physiological conditions, ANT transfers ATP and ADP across the inner mitochondrial membrane. After cyclophilin-D binding, ANT undergoes a conformational change that converts a specific transporter into a nonspecific pore. The reaction is regulated by intracellular calcium, free radicals, ATP, and inorganic phosphate *(25)*. MPTP opening is also dependent on VDAC which allows translocation of low-molecular-weight solutes across the outer membrane *(26)*. The VDAC pore can be modulated by Bcl-2 family proteins *(27)*. Once MPTP is activated, cytochrome *c* is released into the cytosol. The release of cytochrome c is a critical step in the initiation of apoptosis. When cytochrome c enters the cytosol, it interacts with apoptosis-activating factor-1 (Apaf-1) along with caspase 9 and dATP/ATP to form the apoptosome, leading to caspase 3 activation and apoptosis *(28)*. MPTP also releases other proapoptotic proteins such as Smac/Diablo protein, apoptosis-inducing factor (AIF), and endonuclease G (endoG) from the inter-membrane space into the cytosol. Smac/Diablo protein promotes caspase activation by sequestering the inhibitor of apoptosis protein (IAP) *(29,30)*. AIF and endoG translocate from the mitochondria to the nucleus where they cause chromatin condensation and DNA fragmentation *(31–33)*. The mitochondrial pathway appears to play a major role in cardiomyocyte death in ischemia and reperfusion injury. Increased cytochrome c release from the mitochondria has been demonstrated in cardiomyocytes, intact hearts, and isolated cardiac mitochondria in response to various stimuli, including hypoxia, serum, and glucose deprivation. Several studies *(34,35)* reported that ischemia and reperfusion caused the release of cytochrome c and the activation of caspase-9 in isolated perfused hearts.

6. Bcl-2 Family Proteins

Bcl-2 family proteins play a key role in regulating apoptosis. Some of them are anti-apoptotic, such as Bcl-2 and Bcl-xL, whereas others are proapoptotic, such as Bax, Bad, and Bak *(36)*. The antiapoptotic Bcl-2 family members are mainly localized to the cyto-plasmic side of the mitochondrial outer membrane, where they inhibit the release of cytochrome c in part by sequestering proapoptotic Bax *(37)*. The antiapoptotic effects of Bcl-2 have been reported in several studies. Overexpression of Bcl-2 in the heart sig-nificantly reduced infarct size and myocyte apoptosis after ischemia and reperfusion *(38)*. Adult cardiomyocytes overexpressing Bcl-2 showed less apoptosis, decreased cytochrome c release, and decreased activation of caspases 3 and 9 after hypoxia and reoxgenation *(39)*. Proapoptotic proteins of the Bcl-2 family promote release of cytochrome c from the mitochondrial membrane into the cytosol *(40)*. In response to apoptotic stimuli, the proapoptotic proteins are regulated by phosphorylation or proteo-lytic cleavage, then they translocate from the cytosol to the mitochondria, where they undergo a conformational change and integrate into the mitochondrial outer membrane. This increases the release of cytochrome c from the intermembrane space into the cytosol *(12)*.

7. Caspases

At least 14 different mammalian caspases have been identified *(41)*. These caspases mediate the cleavage of survival signaling molecules and structural proteins *(42,43)*. In response to apoptotic stimuli, caspases can be activated by the death receptor pathway or the mitochondrial pathway. Caspase-8 is the initial caspase that is activated through the death receptor pathway. Activated caspase-8 cleaves effector caspases such as cas-pase-3, and -7. In contrast, the mitochondrial pathway first activates caspase-9, which also cleaves caspase-3, and -7, promoting cell death. It has been reported that there is crosstalk between these two pathways. Caspase-8 cleaves the proapoptotic Bcl-2 pro-tein Bid. Activated Bid translocates to the mitochondria and stimulates the release of cytochrome c, leading to the activation of the mitochondrial pathway *(44,45)*. The inhi-bition of caspase-8 activation was shown to decrease both caspase-8 and caspase-9 activity in hearts that were subjected to ischemia and reperfusion. Ischemia has been shown to activate caspase-9 whereas caspase-8 activation is found in the heart only after reperfusion *(46)*. Caspase-3 activation leads to cardiac contractile dysfunction and increased infarct size after ischemia and reperfusion. Mocanu et al. *(47)* reported that hearts treated with an inhibitor of caspase-3 during the early reperfusion phase had reduced infarct size. Cleaved caspase-3 was found in hearts with ischemia-reperfusion injury *(48)*. Furthermore, the overexpression of caspase-3 in mice increased infarct size and depressed cardiac function *(49)*. Active caspase-3 was shown to cleave myosin light chain and damage sarcomeres *(50)*, thereby leading to contractile dysfunction.

8. Signaling Pathways in Cardioprotection

According to the mechanism of apoptosis after ischemia and reperfusion injury, several approaches can be taken to prevent cell death. First, by inhibiting effector caspases, the cell death process is blocked. For example, inhibition of caspase-3 reduces cell apoptosis and limits infarct size in hearts subjected to ischemia and reperfusion. Second, by sequestering proapoptotic Bcl-2 family members in the cytosol, their translocation to the mitochondria is inhibited, thereby blocking apoptosis. It has been reported that some Bcl-2 family proteins such as Bad and Bax are inactivated by Akt, a protective signaling protein. Third, by blocking formation of the MPTP, cytochrome *c* release is inhibited. Calcium overload, ATP deletion, and free radicals all contribute to MPTP opening. Cyclosporine A, an immunosuppressive agent, exerts its actions by inhibiting the MTPT opening *(51)*. Over the past two decades, signaling pathways that protect against cell death have been intensely investigated. Advances in this field have greatly contributed to the study of ischemic preconditioning (IPC), a phenomenon in which brief periods of ischemia and reperfusion generate profound protection *(52)*. Several pathways have been implicated in protection against apoptosis associated with IPC (Fig. 2). Some signaling proteins have been found to target Bcl-2 family proteins, caspases, mitochondrial K_{ATP} channels, and the MPTP, whereas other signal transduction pathways regulate gene transcription to promote cell survival. Here, we focus on the signaling pathways that induce early cardiac protection during ischemia and reperfusion.

9. PI3K

The phosphatidylinositol 3-kinase (PI3K) pathway has been shown to be protective in numerous studies. Several groups *(52–56)* independently reported that erythropoietin (EPO) protected the heart against ischemia-reperfusion injury. Our group *(53,57)* demonstrated that pretreatment with recombinant human EPO before ischemia reduced the number of TUNEL positive cells and decreased caspase-3 activity in isolated rat hearts exposed to 30 min of ischemia and 45 min of reperfusion. The protective effects of EPO were blocked by the PI3K inhibitor wortmannin. We also showed that EPO administration increased the association of its receptor with the regulatory subunit of PI3K, leading to increased Akt phosphorylation in the heart. Calvillo et al. *(52)* reported EPO-mediated PI3K activation and decreased cardiomyocyte death after hypoxia and re-oxygenation. Parsa et al. *(54)* demonstrated increased Akt phosphorylation in the myocardium subjected to ischemia and reperfusion injury in vivo.

The protective effects of insulin are also mediated by the PI3K pathway. Jonassen et al. *(58)* reported that infusion of insulin during early reperfusion limited infarct size in isolated rat hearts, an effect that was blocked by wortmannin; furthermore, increased levels of phosphorylated Akt were found in the insulin-treated heart. Growth factors such as insulin-like growth factor (IGF-1) *(59)*, transforming growth factor-β1 (TGF-β1) *(12)*, and vascular endothelial growth factor (VEGF) *(60)* are well known as survival factors. Buerke et al. *(59)* reported a cardioprotective effect of IGF-1 in myocardial ischemia-reperfusion. Over-expression of IGF-1 in transgenic mouse hearts attenuated apoptosis after ischemia and reperfusion. Increased levels of phosphorylated Akt were demonstrated in the hearts and were sensitive to wortmannin treatment *(61)*.

Aside from growth factors, atorvastatin, a hydroxyl-3-methyglutaryl (HMG)-co-enzyme A reductase inhibitor, has been reported to limit myocardial infarct size through PI3K-Akt signaling in hearts subjected to ischemia and reperfusion *(62,63)*. IPC-mediated protection has been shown to be mediated by the PI3K pathway. Tong et al. *(64)* reported that both wortmannin and LY294002, another PI3K inhibitor, blocked the PC-induced improvement in recovery of postischemic function. Mocanu et al. *(65)* showed that inhibition of wortmannin and LY294002 significantly decreased the IPC-induced reduction in infarct size. Bradykinin and adenosine were implicated in triggering IPC. Bell et al. *(66)* reported that bradykinin limits the infarction in the mouse heart by stimulating the PI3K-Akt pathway. Activation of PI3K and Akt may be sufficient to mediate protection against apoptosis. Matsui et al. *(67)* demonstrated that adenovirus gene transfer of activated PI3K and Akt inhibited apoptosis of hypoxic cardiomyocytes in vitro.

Although experimental evidence clearly indicates that activation of the PI3K-Akt pathway leads to protective effects in vitro and in vivo, the upstream signals and the downstream targets of PI3K have not been fully delineated. There are three classes of PI3K. Only class I is well characterized. Class Ia is composed of a 110-kDa catalytic sub-unit and an 85-kDa regulatory subunit. Class Ib is a heterodimer of a 110-kDa catalytic subunit and a 101-kDa regulatory subunit. PI3K is a lipid kinase that phosphorylates the inositol ring at D3 position, converting phosphatidylinositol, phosphatidylinositol-4-phosphate, and phosphatidylinositol-4,5-bisphosphate to phosphatidylinositol-3-phosphate, phosphatidylinositol-3,4-phosphate, and phosphatidylinositol-3,4,5-phosphate *(68)*. These lipid products interact with the pleckstrin homology domains and src homology domains of protein kinases, including 3-phosphoinositide-dependent kinase 1 (PDK-1) and Akt, thereby regulating their activity. PI3K can be activated through receptor tyrosine kinase (RTK) and G protein-coupled receptor (GPCR) signaling. RTKs can phosphorylate the 85-kDa regulatory subunit of PI3K. βγ subunits of heterotrimeric G proteins in GPCR interact with the 110-kDa catalytic subunit of PI3K, leading to its activation *(69)*. Sequestration of the βγ subunits of G proteins blocked activation of PI3K *(70)*. There may be crosstalk between GPCR and RTK pathways. Acetylcholine activates PI3K through GPCRs *(71)*, but activation of PI3K is blocked by the RTK inhibitor AG-1478. The PI3K-Akt pathway can be activated by down-regulation of PTEN, a phosphatase and tensin homolog on chromosome ten. Cardiomyocytes expressing a dominant-negative form of PTEN had an elevated level of phosphorylated Akt, less caspase activity, and less cell apoptosis *(72)*.

PI3K downstream targets include PDK1 and Akt. Akt has been shown to modulate several protective pathways. Akt substrates include glycogen synthase kinase-3 (GSK-3β), endothelial nitric oxide (NO) synthase (eNOS), Bad, Bax, caspase-9, p70 S6 kinase (p70S6K), and protein kinase C (PKC). Akt phosphorylates and inactivates GSK-3β. Tong et al. *(73)* reported that GSK-3β inhibition reduced infarct size and improved recovery of cardiac function after ischemia and reperfusion. Akt phosphorylates and activates eNOS, leading to NO production *(74,75)*. NO has been shown to activate the mitochondrial ATP dependent potassium (mitoK$_{ATP}$) channel *(76)*. Phosphorylation of Bad by Akt results in its sequestration by 14-3-3 proteins in the cytosol *(77)*. Akt also regulates cell survival and apoptosis by inhibiting the conformational change of Bax *(78)* and by maintaining Bcl-2 levels in the mitochondrial membrane. P70S6K activation was shown to promote cell survival through phosphorylation of Bad *(79)*. PKC can be activated by PI3K. PI3K inhibition blocks PKC translocation and IPC-mediated heart protection *(80)*.

10. PKC

Many studies have shown that activation of PKC mediates heart protection *(81–88)*. Speechly-Dick et al. *(81)* reported that IPC was blocked by the inhibitor of PKC, chelerythrine, and that 1,2-dioctanoyl-sn-glycerol (DOG), a diacylglycerol analogue and specific antagonist of PKC, mimics the protective effect of IPC. Phamacological treatment with catecholamines *(82)*, bradykinin *(83)*, endocannabinoids *(84)*, ethanol *(85)*, and TGF-1 *(86)* have been shown to mediate cardiac protection against ischemia-reperfusion injury through the activation of PKC. PKC overexpression in transgenic animals or selective PKCε activation in the isolated perfused heart prevents ischemic damage *(87,88)*.

PKC is a serine/threonine kinase. There are several PKC isoforms. The classical isoforms α, β, γ are dependent on the lipid cofactor diacylglycerol (DAG) and calcium for their activation. The novel isoforms δ, η, ε require only DAG to be activated. The atypical isoform ζ is calcium- and DAG-independent in activity. Each activated isoform binds to a docking protein known as a receptor for activated C kinase (RACK), which brings the isoform to a specific substrate *(89)*. Although the roles of each isoform are still largely unknown, PKCε may mediate protective effects *(88)*. In contrast, the activation of PKCδ may increase cardiac injury after ischemia and reperfusion *(90,91)*.

PKC activity is regulated through RTK *(86)* and GPCR *(82,83)* signaling as observed in the PI3K pathway. PI3K may be an upstream activation of PKC. Ping et al. *(75)* showed that NO stimulated PKCε translocation that was blocked by the NOS inhibitor, L-NAME. Tong et al. *(80)* reported that inhibition of PI3K abolished PKC activation. Because PI3K can activate eNOS, PI3K may activate PKC through NO.

Several downstream targets of PKCε have been identified. PKCε activates mitoK$_{ATP}$ channels. The activation of mitoK$_{ATP}$ channels leads to cell protection either by depolarizing the inner membrane or by increasing the matrix volume *(92)*. PKC has been shown to associate with the MPTP *(93)*. Hu et al. *(94)* demonstrated that PKCε inhibited the voltage-dependent calcium channel, preventing calcium overload. PKC overexpression in cardiomyocytes selectively activates p42/p44 ERK, another protective signaling molecule *(95)*; moreover, ERK activation is PKC-dependent during ischemia and reperfusion in the rabbit heart *(96)*.

11. ERK

Activation of the serine-threonine kinase ERK promotes cell survival, proliferation, and differentiation. It is a member of the mitogen-activated protein kinase (MAPK) family *(97)*. Lefer et al. *(12)* reported that TGF-β1 protected against ischemia-reperfusion injury in the rat heart. Baxter et al. *(98)* demonstrated that TGF-β1 limited infarct size in the isolated perfused rat heart and reduced cell apoptosis in cardiomyocytes in a ERK dependent manner. Other growth factors such as IGF-1 *(99)*, VEGF *(60)* also have been shown to be cardioprotective. Their effects are mediated by activation of ERK. When RTKs bind growth factor ligands, ERK is activated. Like the PI3K pathway, ERK also can be activated by stimulation of GPCRs *(97)*. Administration of AMP579, an adenosine A1/A2a receptor agonist, reduced infarct size after ischemia and reperfusion in the rabbit heart in vivo. When the ERK inhibitor PD098059 was given at reperfusion, AMP579-induced protection was abrogated *(100)*. In many circumstances, both ERK and PI3K are found to be activated when protection is induced. Liao et al. *(101)* reported that cardiotrophin-1 (CT-1), a member of the interleukin-6 family of cytokines, induced cardioprotection when added both prior to ischemia and at reperfusion via activation of the ERK pathway. Brar et al. *(102)* showed that CT-1-mediated heart protection is mediated by the PI3K-Akt and ERK pathways. Hausenloy et al. *(103)* reported IPC increased phosphorylation of ERK and Akt at early reperfusion. The presence of either PD098059 or LY294002 blocked the IPC-induced phosphorylation of ERK and Akt, respectively, and cardioprotective effects. Strohm et al. *(104)* also reported that inhibition of the ERK pathway with PD98059 blocked IPC in the pig myocardium. However, Mocanu et al. *(65)* reported IPC-mediated protection was not inhibited by PD98059. ERK shares some downstream targets with Akt, thereby preventing apoptosis. Activated ERK phoshorylates Bad directly or indirectly through p70S6K activation *(79,105)*, resulting in its sequestration in the cytosol; moreover, the kinase inhibits the conformational change in BAX, blocking its translocation to the mitochondrial membrane *(106,107)*. Crosstalk may exist between the PI3K, PKC, and ERK pathways. ERK may be connected to the PI3K pathway through PKC signaling. It has been reported that PKC activation increases ERK phosphorylation in IPC *(96,108)*. Tong et al. *(70)* showed that ERK activation was blocked by inhibition of PI3K.

12. Conclusions

In summary, ischemia-reperfusion injury is a common phenomenon following heart attack. Apoptosis and necrosis are usually seen in the myocardium. The mechanisms of cell death are not fully understood. Both the death receptor pathway and the mitochondrial pathway play a role in mediating apoptosis in ischemia-reperfusion injury, but the mitochondrial pathway appears to be a more important contributor to cell death. Ischemia and reperfusion cause elevated intracellular calcium and free radicals, which increase the release of cytochrome *c* into the cytosol, leading to caspase activation and apoptosis. The release of cytochrome *c* is dependent on the MPTP, which is regulated by Bcl-2 family proteins. PI3K, PKC, and ERK are three signaling pathways that mediate heart protection against ischemia and reperfusion injury in numerous studies. They promote cell survival by inhibiting proapoptotic proteins and activating the mitoK$_{ATP}$ channel. There is extensive crosstalk between the PI3K, PKC and ERK pathways. In many circumstances, they are activated in the myocardium. They may have synergistic effects leading to cardioprotection. Further studies are warranted to increase understanding of the mechanisms of apoptosis so that we can block the process of cell death by activating specific survival signaling pathways or by blocking apoptotic pathways that lead to ischemia and reperfusion injury.

References

1. Searle J, Kerr JF, Bishop CJ. Necrosis and apoptosis: Distinct modes of cell death with fundamentally different significance. Pathol Annu 1982;17(Pt2):229–259.
2. Fliss H, Gattinger D. Apoptosis in ischemic and reperfused rat myocardium. Circ Res 1996;79(5):949–956.
3. Kajstura J, Cheng W, Reiss K, et al. Apoptotic and necrotic myocyte cell deaths are independent contributing variables of infarct size in rats. Lab Invest 1996;74(1):86–107.
4. Scarabelli TM, Knight RA, Rayment NB, et al. Quantitative assessment of cardiac myocyte apoptosis in tissue sections using the fluorescence-based tunel technique enhanced with counterstains. J Immunol Methods 1999;228(1–2):23–28.
5. Freude B, Masters TN, Robicsek F, et al. Apoptosis is initiated by myocardial ischemia and executed during reperfusion. J Mol Cell Cardiol 2000;32(2):197–208.
6. Ohno M, Takemura G, Ohno A, et al. ''Apoptotic'' myocytes in infarct area in rabbit hearts may be oncotic myocytes with DNA fragmentation: analysis by immunogold electron microscopy combined with In situ nick end-labeling. Circulation 1998;98(14):1422–1430.
7. Gottlieb RA, Burleson KO, Kloner RA, Babior BM, Engler RL. Reperfusion injury induces apoptosis in rabbit cardiomyocytes. J Clin Invest 1994;94(4):1621–1628.
8. Anversa P, Cheng W, liu Y, Ledaelli G, Kajstura J. Apoptosis and myocardial infarction. Basic Res Cardio 1998;93(suppl 3):8–12.
9. Lee P, Sata M, Lefer DJ, Factor SM,Walsh K, Kitsis RN. Fas pathway is a critical mediator of cardiac myocyte death and MI during ischemia-reperfusion in vivo. Am J Physiol Heart Circ Physiol 2003;284:H456–H463.
10. Jeremias I, Kupatt C, Martin-Villalba A, et al. Involvement of CD95/Apol/Fas in cell death after myocardial ischemia. Circulation 2000;102:915–920.
11. Lefer AM, Tsao P, Aoki N, Palladino MA, Jr. Mediation of cardioprotection by transforming growth factor-beta. Science 1990;249:61–64.
12. Desagher S, Martinou JC. Mitochondria as the central control point of apoptosis. Trends Cell Biol 2000;10:369–377.
13. Boldin MP, Goncharov TM, Goltsev YV, Wallach D. Involvement of MACH, a novel MORTI/FADD-interacting protease, in Fas/APO-1- and TNF receptor-induced cell death. Cell 1996;85:803–815.
14. Muzio M, Chinnaiyan AM, Kischkel FC, et al. FLICE, a novel FADD-homologous ICE/CED-3–like protease, is recruited to the CD95 (Fas/APO-1) death-inducing signaling complex. Cell 1996;85:817–827.
15. Hirata H, Takahashi A, Kobayashi et al. Caspases are activated in a branched protease cascade and control distinct downstream processes in Fas-induced apoptosis. J Exp Med 1998; 187:587–600.
16. Slee EA, Harte MT, Kluck RM, et al. Ordering the cytochrome c-initiated caspase cascade: Hierarchical activation of caspases-2, -3, -6, -7, -8, and -10 in a caspase-9-dependent manner. J Cell Biol 1999;144:281–292.
17. Lee P, Sata M, Lefer DJ, Factor SM,Walsh K, Kitsis RN. Fas pathway is a critical mediator of cardiac myocyte death and MI during ischemia-reperfusion in vivo. Am J Physiol Heart Circ Physiol 2003;284:H456–H463.
18. Ferrari R, Bachetti T, Confortini R, et al. Tumor necrosis factor soluble receptors in patients with various degrees of congestive heart failure. Circulation 1995;92:1479–1486.
19. Sliwa K, Skudicky D, Candy G, Wisenbaugh T, Sareli P. Randomised investigation of effects of pentoxifylline on leftventricular performance in idiopathic dilated cardiomyopathy. Lancet 1998;351:1091–1093.

20. Eddy LJ, Goeddel DV, Wong GH. Tumor necrosis factor-alpha pretreatment is protective in a rat model of myocardial ischemia reperfusion injury. Biochem Biophys Res Commun 1992;184:1056–1059.

21. Kurrelmeyer KM, Michael LH, Baumgarten G, et al. Endogenous tumor necrosis factor protects the adult cardiac myocyte against ischemic induced apoptosis in a murine model of acute myocardial infarction. Proc Natl Acad Sci USA 2000;97:5456–5461.

22. Liu ZG, Hsu H, Goeddel DV, Karin M. Dissection of TNF receptor 1 effector functions: JNK activation is not linked to apoptosis while NF-kappaB activation prevents cell death. Cell 1996;87:565–576.

23. Bergmann MW, Loser P, Dietz R, Von Harsdorf R. Effect of NFκB inhibition on TNF-α-induced apoptosis and downstream pathways in cardiomyocytes. J Mol Cell Cardiol 2001;33:1223–1232.

24. McFalls EO, Liem D, Schoonderwoere K, Lamers J, Sluiter W, and Duncker D. Mitochondrial function: the heart of myocardial preservation. J Lab Clin Med 2003;142:141–149.

25. Halestrap A, Davidson A. Inhibition of Ca2+-induced large amplitude swelling of liver and heart mitochondria by cyclosporin A is probably caused by the inhibitor binding to mitochondrial matrix peptidyl-propyl cis-trans isomerase and preventing it interacting within the adenine nucleotide translocase. Biochem J 1990;268:153–160.

26. Shimizu S, Matsaoka Y, Shinohara Y, Yomeda Y, Tsujimoto Y. Essential role of voltage-dependent anion channel in various forms of apoptosis in mammalian cells. J Cell Biol 2001;152:237–250.

27. Antosson B, Conti F, Ciavatta A, et al. Inhibition of Bax channel-forming activity by Bcl-2. Science 1997;277:370–372

28. Zou H, Li Y, Liu X, Wang X. An APAF-1 cytochrome c multimeric complex is a functional apoptosome that activates procaspase-9. J Biol Chem 1999;274:11,549–11,556

29. Du C, Fang M, Li Y, Li L, Wang X. Smac, a mitochondrial protein that promotes cytochrome c-dependent caspase activation by eliminating IAP inhibition. Cell 2000;102:33–42.

30. Verhagen AM, Ekert PG, Pakusch M, et al. of DIABLO, a mammalian protein that promotes apoptosis by binding to and antagonizing IAP proteins. Cell 2000;102:43–53.

31. Susin SA, Lorenzo HK, Zamzami N, et al. Molecular characterization of mitochondrial apoptosis-inducing factor. Nature 1999;397:441–446.

32. Li LY, Luo X, Wang X. Endonuclease G is an apoptotic DNase when released from mitochondria. Nature 2001;412:95–99

33. Parrish J, Li L, Klotz K, Ledwich D, Wang X, Xue D. Mitochondrial endonuclease G is important for apoptosis in C elegans. Nature 2001;412:90–94.

34. Chen M, He H, Zhan S, Krajewski S, Reed JC, Gottlieb RA. Bid is cleaved by calpain to an active fragment in vitro and during myocardial ischemia/reperfusion. J Biol Chem 2001;276:30,724–30,728.

35. Scheubel RJ, Bartling B, Simm A, et al. Apoptotic pathway activation from mitochondria and death receptors without caspase-3 cleavage in failing human myocardium: Fragile balance of myocyte survival? J Am Coll Cardiol 2002;39:481–488.

36. Adams JM, Cory S. The Bcl-2 protein family: Arbiters of cell survival. Science 1998;281:1322–1326.

37. Cheng EH, Wei MC, Weiler S, et al. BCL-2, BCL-X(L) sequester BH3 domain-only molecules preventing BAX- and BAK-mediated mitochondrial apoptosis. Mol Cell 2001;8:705–711.

38. Chen Z, Chua CC, Ho YS, Hamdy RC, Chua BH. Overexpression of Bcl-2 attenuates apoptosis and protects against myocardial I/R injury in transgenic mice. Am J Physiol Heart Circ Physiol 2001;280:H2313–H2320.

39. Kang PM, Haunstetter A, Aoki H, Usheva A, Izumo S. Morphological and molecular characterization of adult cardiomyocyte apoptosis during hypoxia and reoxygenation. Circ Res 2000;87:118–125.

40. Zha J, Harada H, Yang E, Jocket J, Korsmeyer SJ. Serine phosphorylation of death agonist BAD in response to survival factor results in binding to 14–3–3 not BCL-X(L). Cell 1996;87:619–628.

41. Troy CM, Salvesen GS. Caspases on the brain. J Neurosci Res 2002;69:145–150.

42. Cohen GM. Caspases: the executioners of apoptosis. Biochem J 1997;326:1–16.

43. Thornberry NA, Lazebnik Y. Caspases: Enemies within. Science 1998;281:1312–1316.

44. Li H, Zhu H, Xu CJ, Yuan J. Cleavage of BID by caspase 8 mediates the mitochondrial damage in the Fas pathway of apoptosis. Cell 1998;94:491–501.

45. Luo X, Budihardjo I, Zou H, Slaughter C, Wang, X. Bid, a Bcl2 interacting protein, mediates cytochrome c release from mitochondria in response to activation of cell surface death receptors. Cell 1998;94:481–490.

46. Stephanou A, Brar B, Liao Z, Scarabelli T, Knight RA, Latchman DS. Distinct initiator caspases are required for the induction of apoptosis in cardiac myocytes during ischaemia versus reperfusion injury. Cell Death Differ 2001;8:434–435.

47. Scarabelli TM, Stephanou A, Pasini E, et al. Different signaling pathways induce apoptosis in endothelial cells and cardiac myocytes during ischemia/ reperfusion injury. Circ Res 2002;90:745–748.

47. Mocanu MM, Baxter GF, Yellon DM. Caspase inhibition and limitation of myocardial infarct size: Protection against lethal reperfusion injury. Br J Pharmacol 2000;130:197–200.

48. Holly TA, Drincic A, Byun Y, et al. Caspase inhibition reduces myocyte cell death induced by myocardial ischemia and reperfusion in vivo. J Mol Cell Cardiol 1999;31:1709–1715.

49. Condorelli G, Roncarati R, Ross J, et al. Heart-targeted overexpression of caspase3 in mice increases infarct size and depresses cardiac function. Proc Natl Acad Sci USA 2001;98: 9977–9982.

50. Laugwitz KL, Moretti A, et al. Blocking caspase activated apoptosis improves contractility in failing myocardium. Hum Gene Ther 2001;12:2051–2063.

51. Crompton M, Ellinger H, Costi A. Inhibition by cyclosporin A of a Ca2+ -dependent pore in heart mitochondria activated by inorganic phosphate and oxidative stress. Biochem J 1988;255:357–360.

52. Calvillo L, Latini R, Kajstura J, et al. Recombinant human erythropoietin protects the myocardium from ischemia – reperfusion injury and promotes beneficial remodeling. Proc Natl Acad Sci USA 2003;100:4802–4806.

53. Cai Z, Manalo DJ, Wei G, et al. Hearts from rodents exposed to intermittent hypoxia or erythropoietin are protected against ischemia–reperfusion injury. Circulation 2003;108:79–85.

54. Parsa CJ, Matsumoto A, Kim J, et al. A novel protective effect of erythropoietin in the infarcted heart. J Clin Invest 2003;112:999–1007.

55. Tramontano AF, Muniyappa R, Black AD, et al. Erythropoietin protects cardiac myocytes from hypoxia-induced apoptosis through an Akt-dependent pathway. Biochem Biophys Res Commun 2003;112:990–994.

56. Moon C, Krawczyk M, Ahn D, et al. Erythropoietin reduces myocardial infarction and left ventricular functional decline after coronary artery ligation in rats. Proc Natl Acad Sci USA 2003;100:11,612–11,617.

57. Cai Z, Semenza GL. Phosphatidylinositol-3-kinase signaling is required for erythropoietin-mediated acute protection against myocardial ischemia/reperfusion injury. Circulation 2004;109:2050–2053.

58. Jonassen AK, Sack MN, Mjos OD, Yellon DM. Myocardial protection by insulin at reperfusion requires early administration and is mediated via Akt and p70s6 kinase cell-survival signaling. Circ Res 2001;89(12):1191–1198.

59. Buerke M, Murohara T, Skurk C, Nuss C, Tomaselli K, Lefer AM. Cardioprotective effect of insulin-like growth factor I in myocardial ischemia followed by reperfusion. Proc Natl Acad Sci USA 1995;92(17):8031–8035.

60. Luo Z, Diaco M, Murohara T, Ferrara N, Isner JM, Symes JF. Vascular endothelial growth factor attenuates myocardial ischemia–reperfusion injury. Ann Thorac Surg 1997;64(4):993–998.

61. Yamashita K, Kajstura J, Discher DJ, et al. Reperfusion-activated Akt kinase prevents apoptosis in transgenic mouse hearts overexpressing insulin-like growth factor-1. Circ Res 2001;88(6):609–614.

62. Kureishi Y, Luo Z, Shiojima I, et al. The HMG-CoA reductase inhibitor simvastatin activates the protein kinase Akt and promotes angiogenesis in normocholesterolemic animals. Nat Med 2000;6(9):1004–1010.

63. Bell RM, Yellon DM. Atorvastatin, administered at the onset of reperfusion, and independent of lipid lowering, protects the myocardium by up-regulating a pro-survival pathway. J Am Coll Cardiol 2003;41:508–515.

64. Tong H, Chen W, Steenbergen C, Murphy E. Ischemic preconditioning activates phosphatidylinositol-3-kinase upstream of protein kinase C. Circ Res 2000;87:309–315.

65. Mocanu M, Bell R, Yellon D. PI3 kinase and not p42/p44 appear to be implicated in the protection conferred by ischemic preconditioning. J Mol Cell Cardiol 2002;34:661–668.

66. Bell RM, Yellon DM. Bradykinin limits infarction when administered as an adjunct to reperfusion in mouse heart: the role of PI3K, Akt and eNOS. J Mol Cell Cardiol 2003;35(2):185–193.

67. Matsui T, Li L, del Monte F, et al. Adenovirus gene transfer of activated phosphatidylinositol-3′-kinase and Akt inhibits apoptosis of hypoxic cardiomyocytes in vitro. Circulation 1999;100:2373–2379.

68. Chan T, Rittenhouse S, Tsichlis P. Akt/PKB and other D3 phosphoinositide-regulated kinases: kinase activation by phosphoinositide-dependent phosphorylation. Ann Rev Biochem 1999;68:965–1014.

69. Cantley L. The phosphoinositide 3–kinase pathway. Science 2002;296:1655–1657.

70. Tong H, Steenbergen C, Koch WJ, Murphy E. G protein-dependent signaling pathway in ischemic preconditioning: a role for endosomal signaling. Circulation 2003;108(suppl IV): IV-37. Abstract.

71. Oldenburg O, Qin Q, Sharma AR, Cohen MV, Downey JM, Benoit JN. Acetylcholine leads to free radical production dependent on KATP channels, Gi proteins, phosphatidylinositol 3-kinase and tyrosine kinase. Cardiovasc Res 2002;55:544–552.

72. Schwartzbauer G, Robbins J. The tumor suppressor gene PTEN can regulate cardiac hypertrophy and survival. J Biol Chem 2001;276:35,786–35,793.

73. Tong H, Imahashi K, Steenbergen C, Murphy E. Phosphorylation of glycogen synthase kinase 3b during preconditioning through a phosphatidylinositol-3-kinase-dependent pathway is cardioprotective. Circ Res 2002;90:377–379.

74. Bell RM, Yellon DM. The contribution of endothelial nitric oxide synthase to early ischaemic preconditioning: the lowering of the preconditioning threshold: an investigation in eNOS knockout mice. Cardiovasc Res 2001;52:274–280.

75. Ping P, Takano H, Zhang J, et al. Isoform-selective activation of protein kinase C by nitric oxide in the heart of conscious rabbits: a signaling mechanism for both nitric oxide-induced and ischemia-induced preconditioning. Circ Res 1999;84:587–604.

76. Sasaki N, Sato T, Ohler A, O'Rourke B, Marbán E. Activation of mitochondrial ATP-dependent potassium channels by nitric oxide. Circulation 2000;101:439–445.

77. Scheid MP, Woodgett JR. PKB/Akt: functional insights from genetic models. Nat Rev Mol Cell Biol 2001;2:760–768.

78. Yamaguchi H, Wang HG. The protein kinase PKB/Akt regulates cell survival and apoptosis by inhibiting Bax conformational change. Oncogene 2001;20(53):7779–7786.

79. Harada H, Andersen JS, Mann M, Terada N, Korsmeyer SJ. p70S6 kinase signals cell survival as well as growth, inactivating the proapoptotic molecule BAD. Proc Natl Acad Sci USA 2001;98(17):9666–9670.

80. Tong H, Chen W, Steenbergen C, Murphy E. Ischemic preconditioning activates phosphatidylinositol-3-kinase upstream of protein kinase C. Circ Res 2000;87:309–315.

81. Speechly-Dick ME, Mocanu MM, Yellon DM. Protein kinase C: its role in ischemic preconditioning in the rat. Circ Res 1994;75:586–590.

82. Mitchell MB, Meng X, Ao L, Brown JM, Harken AH, Banerjee A. Preconditioning of isolated rat heart is mediated by protein kinase C. Circ Res 1995;76:73–81.

83. Goto M, Liu Y, Yang XM, Ardell JL, Cohen MV, Downey JM. Role of bradykinin in protection of ischemic preconditioning in rabbit hearts. Circ Res 1995;77:611–621.

84. Lepicier P, Bouchard JF, Lagneux C, Lamontagne D. Endocannabinoids protect the rat isolated heart against ischaemia. Br J Pharmacol 2003;139:805–815.

85. Miyamae M, Rodriguez MM, Camacho SA, Diamond I, Mochly-Rosen D, Figueredo VM. Activation of epsilon protein kinase C correlates with a cardioprotective effect of regular ethanol consumption. Proc Natl Acad Sci USA 1998;95:8262–8267.

86. Palmen M, Daemen MJ, De Windt LJ, et al. Fibroblast growth factor-1 improves cardiac functional recovery and enhances cell survival after ischemia and reperfusion: a fibroblast growth factor receptor, protein kinase C, and tyrosine kinase-dependent mechanism. J Am Coll Cardiol 2004;44:1113–1123.

87. Tian R, Miao W, Spindler M, et al. Long-term expression of protein kinase C in adult mouse hearts improves postischemic recovery. Proc Natl Acad Sci USA 1999;96:13,536–13,541.

88. Dorn GW 2nd, Souroujon MC, Liron T, et al. Sustained in vivo cardiac protection by a rationally designed peptide that causes protein kinase C translocation. Proc Natl Acad Sci USA 1999;96:12,798–12,803.

89. Mackay K, Mochly-Rosen D. Location, anchoring, and functions of protein kinase C isozymes in the heart. J Mol Cell Cardiol 2001;33:1301–1307.

90. Chen L, Hahn H, Wu G, et al. Opposing cardioprotective action and parallel hypertrophic effects of δPKC and εPKC. Proc Natl Acad Sci USA 2001;98:11,114–11,119.

91. Mayr M, Metzler B, Chung YL, et al. Ischemic preconditioning exaggerates cardiac damage in PKC-delta null mice. Am J Physiol Heart Circ Physiol 2004;287:H946–H956.

92. Sato T, O'Rourke B, Marbán E. Modulation of mitochondrial ATP dependent K channels by protein kinase C. Circ Res 1998;83:110–114.

93. Baines CP, Song CX, Zheng YT, et al. Protein kinase C interacts with and inhibits the permeability transition pore in cardiac mitochondria. Circ Res 2003;92:873–880.

94. Hu K, Mochly-Rosen D, Boutjdir M. Evidence for functional role of εPKC isozyme in the regulation of cardiac Ca^{2+} channels. Am J Physiol 2000;279:H2658–H2664.

95. Heidkamp MC, Bayer AL, Martin JL, Smarel AM. Differential activation of mitogen-activated protein kinase cascades and apoptosis by protein kinase C-ε and δ in neonatal rat ventricular myocytes. Circ Res 2001;89:882–890.

96. Ping P, Zhang J, Cao X, et al. PKC-dependent activation of p44/p42 MAPKs during myocardial ischemia-reperfusion in conscious rabbits. Am J Physiol 1999;276:H1468–H1481.

97. Widmann C, Gibson S, Jarpe MB, Johnson GL. Mitogen-activated protein kinase: conservation of a three-kinase module from yeast to human. Physiol Rev 1999;79:143–180.

98. Baxter GF, Mocanu MM, Brar BK, Latchman DS, Yellon DM. Cardioprotective effects of transforming growth factor-beta1 during early reoxygenation or reperfusion are mediated by p42/p44 MAPK. J Cardiovasc Pharmacol 2001;38:930–939.

99. Parrizas M, Saltiel AR, LeRoith D. Insulin-like growth factor 1 inhibits apoptosis using the phosphatidylinositol 3V-kinase and mitogenactivated protein kinase pathways. J Biol Chem 1997;272:154–161.

100. Xu Z, Yang XM, Cohen MV, Neumann T, Heusch G, Downey JM. Limitation of infarct size in rabbit hearts by the novel adenosine receptor agonist AMP 579 administered at reperfusion. J Mol Cell Cardiol 2000;32:2339–2347.

101. Liao Z, Brar BK, Cai Q, et al. Cardiotrophin-1 (CT-1) can protect the adult heart from injury when added both prior to ischaemia and at reperfusion. Cardiovasc Res 2002;53:902–910.

102. Brar BK, Stephanou A, Liao Z, et al. Cardiotrophin-1 can protect cardiac myocytes from injury when added both prior to simulated ischaemia and at reoxygenation. Cardiovasc Res 2001;51:265–274.

103. Hausenloy DJ, Tsang A, Mocanu MM, Yellon DM. Ischemic preconditioning protects by activating prosurvival kinases at reperfusion. Am J Physiol Heart Circ Physiol 2005 Feb; 288:H971–H976.

104. Strohm C, Barancik T, Bruhl ML, Kilian SA, Schaper W. Inhibition of the ER-kinase cascade by PD98059 and UO126 counteracts ischemic preconditioning in pig myocardium. J Cardiovasc Pharmacol 2000;36:218–229.

105. Datta SR, Dudek H, Tao X, et al. Akt phosphorylation of BAD couples survival signals to the cell-intrinsic death machinery. Cell 1997;91(2):231–241.

106. Yamaguchi H, Wang HG. The protein kinase PKB/Akt regulates cell survival and apoptosis by inhibiting Bax conformational change. Oncogene 2001;20(53):7779–7786.

107. Tsuruta F, Masuyama N, Gotoh Y. The phosphatidylinositol 3-kinase (PI3K)–Akt pathway suppresses Bax translocation to mitochondria. J Biol Chem 2002;277:14,040–14,047.

108. Baines CP, Zhang J, Wang GW, et al. Mitochondrial PKCε and MAPK form signaling modules in the murine heart: enhanced mitochondrial PKC-MAPK interactions and differential MAPK activation in PKCε-induced cardioprotection. Circ Res 2002;90:390–397.

10

Cyclooxygenase-2 Gene Expression

Transcriptional and Posttranscriptional Controls in Intestinal Tumorigenesis

Shrikant Anant and Sripathi M. Sureban

Summary

Cyclooxygenases (COX) also known as prostaglandin (PG) synthases, are present in two forms, COX-1 and COX-2. Whereas COX-1 is responsible for cytoprotective functions in a number of organs, COX-2, which is normally absent at basal levels, is induced under certain conditions including pathophysiological states like acute inflammation, arthritis, as well as in cancer and cancer-related angiogenesis. Overexpression of COX-2 enhances PGE_2 synthesis, thereby resulting in increased cellular proliferation, which is an important role for the molecule in cancer progression. In addition, increased PGE_2 levels protect the cancer cells from the deleterious effects of ionization radiation (IR). COX-2 expression is tightly controlled under normal conditions. At the transcriptional level, COX-2 gene expression is controlled by multiple elements in a 800-bp region proximal to the transcription start site. Many cellular transcription factors bind these elements to regulate the COX-2 gene transcription. Among them, the key factors are NF-κB and β-catenin. Of these, β-catenin is especially interesting because it can egulate COX-2 both directly by binding to the promoter elements as a comples with either TCF-4/LEF or p300, or indirectly by inducing expression of PEA-3, which subsequently binds to its cognate element in the COX-2 promoter to induce transcription. COX-2 mRNA is also tightly regulated at the post-transcriptional levels of mRNA stability and translation. This is mediated by AU-rich sequence elements located in the 3'UTR of the COX-2 mRNA. Multiple RNA binding proteins have been identified that bind to the AU-rich sequences in the COX-2 3'-UTR to mediate this process. HuR, a ubiquitously expressed protein, is overexpressed in colon and other cancer cells, and it increases the stability and translation of COX-2 mRNA. In contrast, CUGBP2 is induced in cells undergoing apoptosis and it inhibits translation of COX-2 mRNA. Another protein that have been identified induce the translation is hnRNPA1, whereas those that inhibit the translation are TIA1, TTP, TIAR, and AUF1.

Key Words: Posttranscriptional; gene regulation; cyclooxygenase-2; AU-rich sequences; prostaglandins; transcription; mRNA translation; RNA binding protein; 3'-untranslated region; HuR; CUGBP2; tristetraprolin.

1. Introduction

Cancer is a hyperproliferative disorder in which invasion and angiogenesis lead to tumor metastasis. The World Health Organization (WHO) has estimated that 1,300,000 new cases of cancer occur each year and 55,000 people will die in the United States alone. Colorectal cancer is the second leading cause of cancer-related deaths in the western world. The estimated number of new cases of colorectal cancer in the United States for 2002 was 107,000, and approx 48,000 people dies from the cancer or its complications *(1)*. To prevent the onset of cancer, the National Institute of Health (NIH) in

From: *Apoptosis, Cell Signaling, and Human Diseases: Molecular Mechanisms, Volume 2*
Edited by R. Srivastava © Humana Press Inc., Totowa, NJ

the United States has recommended a high-fiber, low-fat diet, consisting of more fruits and vegetables. The incidence in many countries, such as India, had been very low in the past, but more recently a rapid increase was projected resulting from adaptation of more Western-style diets *(2)*. This is because the diet from these regions was comprised primarily of fruit and vegetables, as well as spices such as turmeric that contain the active anticancer ingredient, curcumin *(3–5)*. However, Western diets are high in fats and red meat and low in fiber, a major risk factor for colon carcinogenesis *(2)*. Additional risk factors for colon carcinogenesis include age (>50 yr), gender (women > men) and lifestyle factors such as alcohol abuse, smoking, and sedentary habits *(6–12)*.

Colon cancer, if identified early, can be treated. Screening of colon cancer can be done by colonoscopy to find polyps; removing these polyps at an early stage can prevent cancer progression. Long-term polyps may develop into cancer but screening tests can find these lesions early and they can be treated. Hence, regular colonoscopy is recommended in the United States for those over 50 yr of age *(13,14)*. The reoccurrences of colon cancer is common and it is estimated in about 40% of the cases; cancer will return after 3 to 5 yr of treatment *(15)*. Cancer may recur in the colon or rectum, or in another part of the body.

Colorectal cancers result from the progression of a normal colonic mucosa to an adenomatous polyp, and eventually to a malignant cancer (*see* Fig. 1). There are at least five to seven major events that occur in the cancer progression. Two major pathways may lead to cancer—chromosomal instability and microsatellite instability *(16–18)*. Because of the acquired or inherited mutations in DNA repair-related proteins, a defect in the DNA repair is observed, resulting in the microsatellite instability-related cancers *(19–21)*. About 85% of colorectal cancers are due to chromosomal instability. Many genes are involved in the pathway to tumorigenesis, including the adenomatous polyposis coli (APC, chromosome 5q), deleted in colon cancer (DCC, 18q) and p53 (17p) *(22,23)*.

The loss of the APC gene is a primary event in cancer progression. APC, located in chromosome 5q21, is considered a gatekeeper and mutations in the gene result in loss of signal transduction and cell adhesion *(24–28)*. A major function of APC is to control β-Catenin levels in the cell. β-Catenin is an important transcription factor that controls the expression of many genes involved in cancer progression. It is also a member of the Wnt signaling pathways and plays a role in development *(29–32)*. β-Catenin is also a component of the cell–cell adhesion machinery. It binds to cytosolic tail of E-cadherin and connects actin filaments through β-catenin to form the cytoskeleton *(33,34)*. APC binds to β-catenin and targets itself for ubiquitination-mediated degradation in the cytoplasm of the epithelial cell. Loss of APC results in loss of this degradation event thereby increasing the accumulation of β-catenin, which in turn translocates to the nucleus and transcriptionally activates expression of genes essential for tumor progression. A key target of β-catenin is the cyclooxygenase (COX)-2 gene, which is discussed below *(35–38)*. Other genes that are consistently observed to have mutations are p53 and K-Ras, and both also regulate COX-2 expression *(38–40)*. These mutations are not the only cause of cancer, but overexpression and loss of other gene expression by other mechanisms can also lead to the cancer phenotype. An example of decreased expression in colon carcinogenesis is that of transforming growth factor (TGF)-β, which occurs late in the cancer progression, whereas induction of COX-2 occurs early and is sustained throughout the cancer progression *(41–44)*. In addition, COX-2 levels may

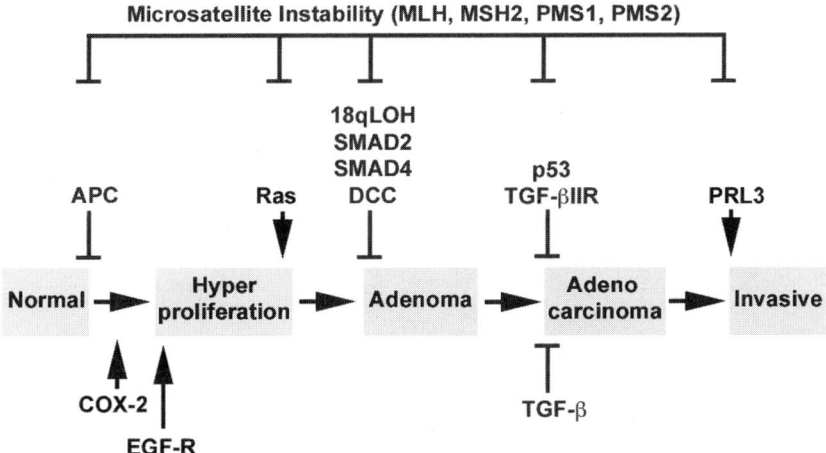

Fig. 1. Progression of a normal cell to an invasive carcinoma. Genes whose mutation or change in expression levels result in loss (inhibitory arrows), and gain (arrows) of function are shown along the left-to-right corresponding to the progression of a normal cell to a cancer cell. Genes whose mutation led to change in function are shown above, whereas those whose expression is modulated are shown below. Mutation in one of the four mismatch-repair genes (MLH, MSH2, PMS1 and PMS2) would lead to a deficiency of mismatch-repair proteins, resulting in microsatellite instability. This in turn could result in mutations of genes such as the APC and K-ras. However, mutation in the microsatellite instability related genes could occur at anytime during the progression of cancer phenotype. COX-2 is one of the first genes that is over-expressed, and this can be observed even before any gross phenotypical changes are observed.

increase at a very early stage in colon carcinogenesis, even before the appearance of adenomas, implying that COX-2 may be a marker for colon cancer progression *(45,46)*.

Prostaglandins are the products of the cycoloxygenase-mediated metabolism of arachidonic acid. This is a multistep process with the first step being the liberation of arachidonic acid, a polyunsaturated fatty acid formed from the membrane phospholipids by phospholipase A2 *(47–49)*. The phospholipids are degraded to diacylglycerol by phosphoinositol-specific phopholipase C. The diacylglycerol that is formed is catabolized to arachidonic acid by diacylglycerol lipase. Once released, free intracellular arachidonic acid is oxidized via three major metabolic pathways: the prostaglandin G/H synthase (PGHS) (COX), the lipoxygenase and the cytochrome *P-450* monooxygenase pathways. In PGHS pathway, the arachidonic acid is converted to PGH_2 by the COX enzymes. Arachidonic acid is also converted to 5-L-hydroperoxy-5,8,10,14-eicosate-traenoic acid (5-HPETE) by lipoxygenase. COX introduces two molecules of O_2 into arachidonic acid to form prostaglandin (PG) endoperoxides, from which the eicosanoids PGE_2, PGD_2, $PGF_{2\alpha}$, PGI_2, PGJ_2, and thromboxanes (Tx) A_2 are formed *(see* Fig, 2) *(50)*. The types and amounts of PGs and thromboxanes are highly variable in different cell types because of differences in the distal synthases and have different functions. The most important PG in the gastrointestinal tract is PGE_2 *(51–53)*.

PGE_2 effects are mediated by a family of G protein coupled receptors, called EP_1, to EP_4 *(see* Fig. 3) *(54–58)*. There are four PGE_2 receptors (EP receptors) and are categorised into three groups basis of their signal transduction and action: the relaxant

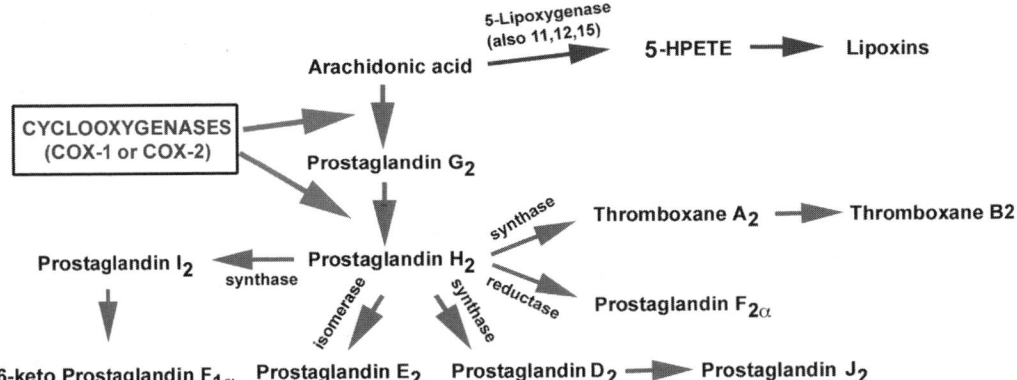

Fig. 2. Cyclooxygenases are the rate-limiting enzymes in prostaglandin synthesis pathway. Arachidonic acid, the precursor for eicosinoid, is catalytically converted by one of the two cyclooxygenases, COX-1 and COX-2 to Prostalandin H_2. Depending on the enzyme, namely synthase, reductase or isomerase, the different PGs are subsequently produced. The most important PG in the intestinal tract is PGE_2, which is generated by PGH_2 isomerase. PGE_2 is important enzyme in colon cancer development because it enhances proliferation and inhibits apoptosis of the cancer cells. However, in other tissues PGJ_2 also have antimicrobial and antiproliferative effects. Alternatively, arachidonic acid metabolism is catalyzed by 5-lipoxygenase resulting in the eventual generation of lipoxins via lipoxygenase pathway and the other is via cytochrome *P-450* monooxygenase pathway.

receptors, the contractile receptors and the inhibitory receptors. The relaxant receptor, which mediates increase in cAMP and induces smooth muscle relaxation, consists of EP_2 and EP_4 receptors. The contractile receptor is the EP1 receptor, which mediates Ca^{2+} mobilization and induces smooth muscle contraction. The EP3 receptor is an inhibitory receptor that mediates decreases in camp and inhibits smooth muscle relaxation. EP_1, EP_2, and EP_4 have been demonstrated to play a major role in colon carcinogenesis *(59–66)*. Furthermore, PGE_2 was observed to promote cell growth and motility through EP_2 and EP_4 receptors by activating the T-cell factor (TCF)/lymphoid-enhancer factor (LEF)-mediated transcription activation of the protein kinase A and phosphotidylinositol 3-kinase (PI3K)-protein kinase B (AKT/PKB) dependent pathways, respectively *(67)*. PGE_2 results in a decrease in the cells undergoing apoptosis, this occurs via induction of Bcl-2 expression. Increased Bcl-2 expression results in reduced caspase-3 and -9 activation, thereby inhibiting the apoptosis *(68,69)*. Bcl-2 is also induced by PGE_2 by activating the MAPK *(70,71)*. These result in the cell proliferation and maintenance of tumor integrity.

1.1. Cyclooxygenases

COXs are the key enzymes involved in the arachidonate metabolism. They catalyze the conversion of arachidonic acid to PGH_2, the precursor for PGs and thromboxanes *(see* Fig. 2) *(45–49)*. Two isoforms of this enzyme exist; COX-1 and COX-2. Both COX enzymes are membrane bound and are present on the luminal surfaces of the endoplasmic reticulum, and the inner and outer membranes of the nuclear envelope *(72)*. COX-1 and COX-2 have structural similarities but they differ in their role in tissue biology and

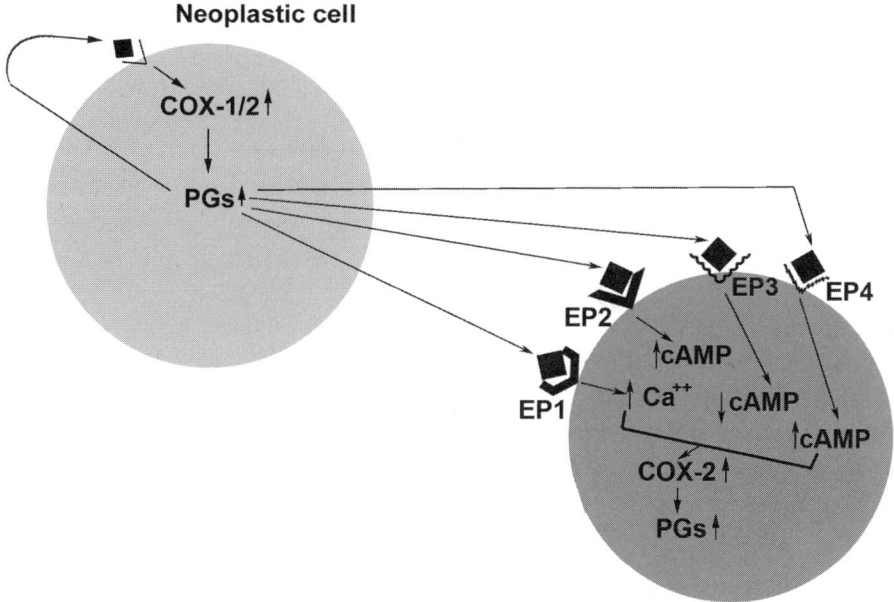

Fig. 3. Prostaglandin signals through the EP receptors. PGE_2, which is formed from the catalytic action of COX-2, is secreted out of the cell, where it can act in an autocrine or paracrine fashion. PGE_2 binds to one of the EP receptors (EP1 to EP4) on the cell surface, resulting in either increased calcium (EP1- contractile receptor), or changes in cAMP (EP2 to EP4). Specifically, signaling through either EP2 or EP4 results in increased cAMP (involved in the smooth muscle relaxation) whereas that with EP3 (inhibitory receptor) results in decreases in cellular cAMP levels and inhibition of smooth muscle relaxation. These cellular changes caused by the EP receptors eventually results in a cellular response that may either be pro- or antiproliferative depending on the status of the cell. Among other things, PGE2 binding to its cognate EP receptors activates transcription of COX-2, suggesting a positive feedback induction.

disease progression. COX-1 is a housekeeping gene that is constitutively expressed in many tissues and plays a major role in tissue homeostasis. On the other hand, COX-2, which was discovered in 1991, is an early response gene and is expressed at low levels in some tissues such as the stomach, intestine, and kidney, but is highly induced by growth factors, cytokines, and inflammatory agents *(45)*. As mentioned above, overexpression of COX-2 occurs at the early stage in epithelial malignancy and colon cancer, showing its importance in tumorogenesis. COX-2 expression levels may vary based on the type of cancer. It is not only overexpressed in colon tumors but also in other cancer such as mammary tumors, neuroblastoma, prostate cancer, ovarian cancer, and melanoma *(73–80)*. One study determined that COX-2 levels were lower in hereditary nonpolyposis colorectal cancer (HNPCC) as compared with familial adenomatous polyposis, and sporadic colorectal cancers *(81)*. Nonsteroidal antiinflammatory drugs (NSAIDs) have reduced the incidence of colon carcinoma by 40 to 50% *(82–85)*. Deletion of the COX-2 gene in mice suggests that it plays a major role in the development of intestinal tumors *(86,87)*.

COX-2 expression is increased in the early stages of tumorigenesis, including the early stages of adenoma formation *(88–90)*. An example is the formation of the intestinal tumors in APC^min/+ mice, the murine model for familial polyposis syndrome, which demonstrate high levels of COX-2. Furthermore, treatment of these mice with a non-specific cyclooxygenase inhibitor, sulindac also demonstrated a reduction in intestinal polyps *(91–94)*. In chemically induced tumors in rodents, where the animals were administered azoxymethane (AOM), the tumors demonstrated high levels of COX-2 *(93)*. Furthermore, dietary fat induces COX-2 expression and greater numbers of aberrant crypts, whereas inhibition of COX-2 by chemical inhibitors results in decreased aberrant crypts in rats following AOM treatment *(95–100)*. In mice that are genetically modified to lack COX-2 and have a deletion of one allele of the APC gene (APC^min/+), there was a significant reduction in intestinal polyps, suggesting that overexpression of COX-2 is a critical step in the tumorigeneisis process *(87,101,102)*. However, the mechanisms responsible for the increased level of COX-2 in the adenoma are not entirely understood. Many studies have determined the presence of both transcriptional and post-transcriptional mechanisms in inducing COX-2 gene expression. The rest of this article will focus on our current knowledge of the mechanisms that regulate COX-2 gene expression and describe the role of several cellular factors that are involved in this process.

2. Regulation of COX-2 Gene Expression

2.1. Transcription

The human COX-2 gene is localized to the long arm of chromosome 1 at position q25.2-25.3 *(103)*. The gene spans approx 8 kb and consists of 10 exons *(see* Fig. 4) *(103,104)*. Transcription initiation occurs 134 nt upstream of the ATG translation initiation site *(103)*. Out of the 1.69-kb region of nucleotides upstream of the transcriptional start site (the 5′-flanking region), the first 800 bp contain a canonical TATA box located 31 nt upstream of the transcription start site, as well as response elements for transcription factors CArG box, NF-IL6, PEA-1, PEA-3, myb, xenobiotic response elements, three SP1 sites, C/EBP motif, Est-1, AP2, two NF-κB sites, GATA-1, 12-*O*-tetrade-canoyl-phorbol-13-acetate-response elements, and cAMP-response element (CRE) binding protein. This suggests that expression of this gene may involve the complex interaction of various transcription enhancing factors *(105)*. Furthermore, the first 275 nt of the human COX-2 promoter demonstrated approx 65% homology with both the mouse and rat COX-2 promoters but no homology to the COX-1 promoter, suggesting that this region is critical in regulating COX-2 expression *(103)*. Additional promoter analyses demonstrated that a 460-nt sequence upstream of the transcription start site is sufficient to drive the expression of a luciferase gene in a human vascular endothelial cell line *(105)*.

TNF-α, a cytokline whose expression is elevated in inflammation potently induces COX-2 gene transcription through transcription factors NF-κB and NF-IL6. In contrast, dexamethasone completely suppressed the interleukin (IL)-1-mediated induction of COX-2 expression ø. The 5′-flanking region of COX-2 gene encodes two NF-κB binding sites, and a strong correlation exists between COX-2 and NF-κB expression in colorectal tumors *(109)*. In normal cells, NF-κB activity is mainly controlled at the protein level and is silenced by sequestration and degradation in the cytoplasm by the inhibitory protein IκBα *(110)*. However, following activation, IκB is phosphorylated, releasing NF-κB from the complex, thereby allowing NF-κB to migrate to the nucleus and inducing transcription of its target genes, including COX-2. Treatment of neuronal cells with

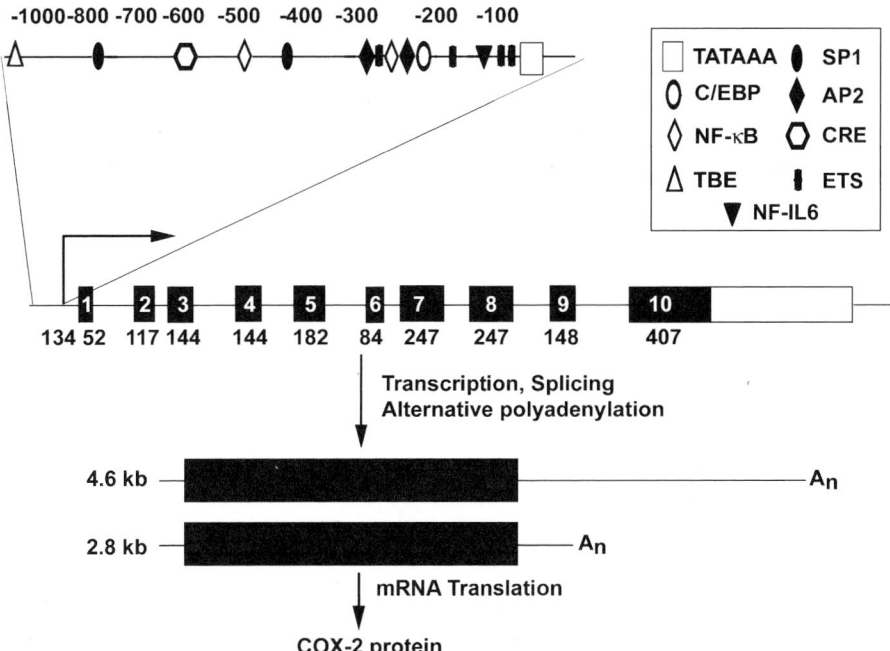

Fig. 4. Schematic representation of the human COX-2 gene with the promoter. The exons are shown by a black box and their sizes are indicated below. Location of the transcription start site is depicted by an arrow. The 1050-nt region upstream of the transcription start site has been expanded and the location of the TATAA box and the various transcription factor binding sites including c/EBP, NF-κB, SP1, AP2, CRE, and TBE are shown. Use of alternative polyadenylation sites in exon 10 results in the generation of two distinct transcripts differing in the length of the 3′-UTR. Accordingly, this results in two distinct isoforms of COX-2 mRNA of 2.8 kb and 4.6 kb in length. Translation of either COX-2 mRNA results in COX-2 protein production.

the phorbol ester TPA induced COX-2 promoter activity, which was a result of NF-κB activity *(111)*. The role of NF-κB in the TPA-mediated induction of COX-2 promoter was further confirmed in a murine neuroblastoma cell line that had constitutively NF-κB activity, where COX-2 expression was not further induced by TPA *(111)*. Bacterial lipopolysaccharide (LPS), interferon (IFN)-γ and TNF-α also induce COX-2 expression by inducing IκB degradation resulting in activation of NF-κB as well as by turning on interferon-regulatory-factor (IRF)-1 *(112–114)*. Furthermore, LPS was unable to induce COX-2 in IRF-1 deficient mice suggesting that LPS mediated induction of COX-2 is mediated by IRF-1 *(113)*. Collectively these data provide a role for NF-κB and IRF-1 in regulating COX-2 gene transcription. One of the other mechanisms of COX-2 activation may be by inhibiting NF-κB binding, by IκBα–mediated shuttling p65 back to cytosol. IκBα is known to enter the nucleus under certain conditions *(115)*. It is possible that IκBα enters the nucleus and brings p65 back to cytosol, thereby reducing the p65 levels in the nucleus.

TPA transactivates COX-2 in a time dependent manner. COX-2 protein levels remain high for 6 h and decrease by 12 h post administration. TPA treatment. In addition to activation of NF-κB, TPA mediates the binding of several other transcription factors to the COX-2 promoter *(116)*. TPA enhances binding of the cyclic AMP response element binding protein, CREB binding, and c-Jun/c-Fos binding to CRE region in COX-2 promoter *(117–122)*. TPA stimulates phosphorylation of C/EBPβ via ERK1/2 pathway and phosphorylated C/EBPβ exhibits active binding to the CREB site *(123)*. TPA also stimulates AP-1 and CREB/ATF to bind to the COX-2 promoter region and recruits the transcription coactivator p300, together they interact with the TFII-B in the transcription machinery to trigger the polymerase activity. In addition, following TPA mediated enhanced binding of p300 to the COX-2 promoter, p300 interacts with C/EBPβ, CREB, and c-Jun *(124)*. p300 is essential for TNF-α, IL-1β, and LPS mediated induction of COX-2 transcription *(124)*. However, while TNFα also induces the binding of p50/p65 to the COX-2 promoter, it does not affect the CREB, c-Jun/c-Fos, or c/EBPβ binding *(124)*. In addition, p65 NFκB recruits p300 to the COX-2 promoter region for the transactivation *(124)*. Collectively, these data suggest that C/EBPβ and NFκB are dynamically regulated, and may account for a majority of the COX-2 transcriptional program in the cell.

Tumor suppressor protein p53 is a transcription factor that is important in the suppression of cellular growth and transformation *(125–128)*. p53 can either increase or suppress the expression of a number of target genes, and it was observed to inhibit COX-2 gene transcription *(129–133)*. p53 suppresses transcription from TATA containing promoters probably by interacting with components of the basal transcriptional machinery *(134–138)*. This may explain why COX-2 mRNA is not detected in normal intestinal cells whereas it is overexpressed in colon cancer cells, suggesting an important link between p53 and cancer. However, no p53 binding sites are known to be present in the COX-2 promoter, suggesting that p53 mediated suppression of COX-2 transcription likely occurs indirectly, either by inducing a transcriptional repressor, which in turn represses COX-2 gene transcription, or by interacting with an inhibitor complex that binds to the COX-2 promoter to directly repress the transcription.

More recently, COX-2 has been shown to be regulated by the Wnt and ras pathways. Loss of functional APC in APC$^{min/+}$ mice activates Wnt signaling pathways, resulting in the accumulation of β-catenin which then binds to TCF-4/LEF and turns on COX-2 (*see* Fig. 5) *(38)*. Given that PGE$_2$ induces TCF-4/LEF by signaling through EP2 or EP4 receptors, a closed positive feedback occurs between PGE$_2$ and COX-2 is implied *(38)*. Furthermore, PEA-3, a transcription factor of the Ets family is upregulated by β-catenin and PEA-3 induces COX-2 expression (*see* Fig. 5) *(36,139)*. Taken together, these data suggest that multiple mechanisms may be in place for inducing COX-2 expression, and this is dependent on the inducer that is present near the target cell.

2.2. RNA Stability and Translation: cis-Acting Elements

In addition to the regulation at the transcriptional level, COX-2 gene expression is tightly controlled at the post-transcriptional levels of mRNA stability and translation. Regulation of mRNA stability is often mediated by the sequences that are located within the 3′-UTR. COX-2 mRNA is present in two forms, 2.8 and 4.6 kb in length. The only difference between the two transcripts is the length of the 3′-UTR, which is believed to arise as a result of alternative polyadenylation site usage (*see* Fig. 5) *(140–143)*. An important difference between the two transcripts is that the 4.6-kb COX-2 mRNA is degraded at a faster rate as compared with the 2.8-kb mRNA in HCA-7cells, when the cells were treated with dexamethasone *(140)*. This implies the presence of one or more sequence elements within the COX-2 3′-UTR that is present in the 4.6-kb but not in the

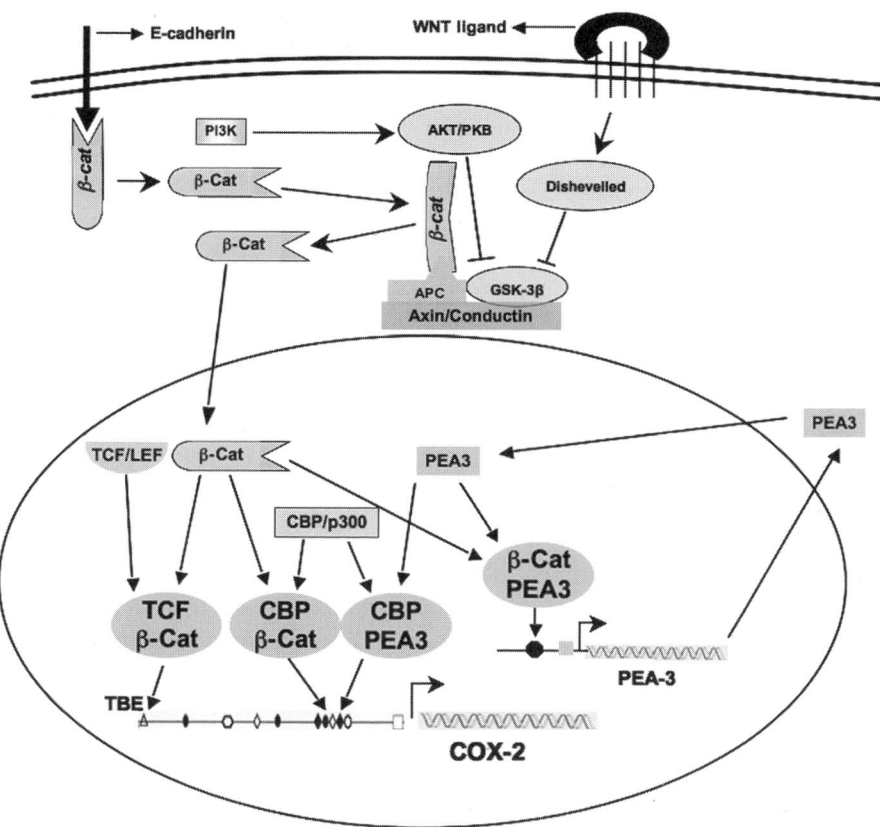

Fig. 5. β-catenin and COX-2 transcription. β-catenin is a part of the cytoplasmic protein complex consisting of the adenomatous polyposis coli (APC) protein, glycogen synthase kinase-3β (GSK-3β), axin and conductin. β-catenin can also be found bound to the membrane where it acts in the cell–cell adhesion machinery. Either inactivation of GSK-3β by WNT ligand or growth factor signaling, or mutation of APC or conductin results in β-catenin stabilization. β-catenin translocates to the nucleus and binds to T-cell factor (TCF)/lymphoid enhancing factor (LEF) or with the CBP/p300 and act as a transcription factor. The β-catenin-TCF complex binds to the TCF4-binding element (TBE) and the β-catenin-CBP/p300 complex binds to c/EBP site to activate COX-2 gene transcription. In addition, the β-catenin-CBP/p300 complex can activate transcription of the PEA3 gene. The PEA3 subsequently binds either alone or in a complex with p300 to the ETS (−75/−72) and the NF-IL6 sites in the COX-2 promoter to activate COX-2 gene transcription.

2.8-kb isoform that mediates the degradation activity. IL1-β has been shown to increase the half-life of COX-2 mRNA in rat mesengial cells through activation of the JNK and MAPK pathways *(144–147)*. Furthermore, many cellular proteins were identified to bind the first 150 nt of the 3′-UTR, suggesting a major role for this region in modulating COX-2 mRNA stability *(148,149)*. An interesting recent observation is that only the 2.8-kb COX-2 mRNA is present in spermatogonial cells of mature rat testis, and that testosterone and follicle stimulating hormones increased expression of COX-2 protein *(150)*. However, it remains to be seen whether the hormones affected COX-2 mRNA stability and/or translation.

An important mechanism for post-transcriptional gene regulation in mammalian cells is rapid degradation of mRNAs mediated by AU rich elements (AREs) in their 3′-UTR *(151)*. The enhanced mRNA stability in tumor cells suggests that altered recognition of AU rich sequences in neoplasia may lead to improper function of AREs. Inspection of the COX-2 3′-UTR revealed the presence of multiple AUUUA sequence motifs (*see* Fig. 6) *(104,152)*. There are 22 copies of AUUUA throughout human COX-2 3′-UTR *(104,152)*. Of these, many are located as tandem repeats in the first sixty nucleotides of the COX-2 3′-UTR and form the minimal ARE sequence *(104,152,153)*. In addition, the nonamer UUAUUUAU/AU/A has been identified as a key AU-rich sequence motif that mediates mRNA degradation *(154)*. Within the first sixty nucleotides of COX-2 3′-UTR, three such elements were identified as tandem repeats (*see* Fig. 6) *(155)*. Deletion analysis has identified the first 60 nucleotides as a major mRNA stability and translational control element (*see* Fig. 6) *(152)*. In addition, other downstream regions of the COX-2 3′-UTR are involved in mRNA stability and translational control (*see* Fig. 6) *(152)*. These regions also contain scattered AUUUA sequences but do not conform to the canonical nonamer ARE that may be essential for ARE-mediated degradation activity. This suggests that a complex system of regulation exists for modulating COX-2 expression at the post-transcriptional level, which would include the *cis*-acting elements in the 3′-UTR as well as cellular *trans*-acting factors that may bind to these elements in order to modulate activity. These COX-2 AREs play a major role in regulating the COX-2 expression and in cancer cells, and cellular defects in the regulation of mRNA stability can contribute to the elevated COX-2 protein expression, which thereby promotes the cell growth *(156)*. Thus, the tight control of the stability of COX-2 mRNA (and other angiogenic factor mRNAs) rests primarily on the RNA-binding proteins to properly interact with each other and with the AREs.

2.3. RNA Stability and Translation: trans-Acting Factors

Stability and translation of COX-2 mRNA is likely controlled through a complex network of RNA/protein interactions involving the recognition of specific targets in the mRNA. Currently, many RNA binding proteins (RNABPs) have been identified to bind the AREs in a variety of transcripts and some have been demonstrated to alter the stability and translation. These include HuR, HuB/Hel-N1, HuC, HuD, T-cell internal antigen-1 (TIA-1), TIA 1-related protein (TIAR), CUG binding protein 2 (CUGBP2), tristetraprolin (TTP), RBM3, AUF1 (also known as heterogeneous nuclear ribonucleoproteins [hnRNP D]), and other hnRNPs including hnRNP A1, CPF-A (hnRNP A/B), hnRNP A0, hnRNP A2/B1, hnRNP A3, and hnRNP U *(149–163)*. Adding to the complexity of these interactions is the fact that these proteins have alternative spliced variants, which may have differential roles following binding to the ARE. As mentioned before, the first 60 nucleotides in COX-2 3′-UTR are rich in ARE sequence elements, and many of the above mentioned proteins bind to this region, including AUF1, HuR, TIA1, TIAR, CPF-A and CUGBP2 *(149,153,156,158)*. In addition, RBM3 was identified in a complex that bound to this region, but its direct binding is not known *(149)*. In contrast, TTP and hnRNP A0 bind to the distal part of the 3′-UTR *(159,164)*.

2.4. HuR

HuR is a ubiquitously expressed mammalian protein that is an ortholog of the *Drosophila melanogaster* protein, embryonic lethal abnormal vision (Antic and Keene,

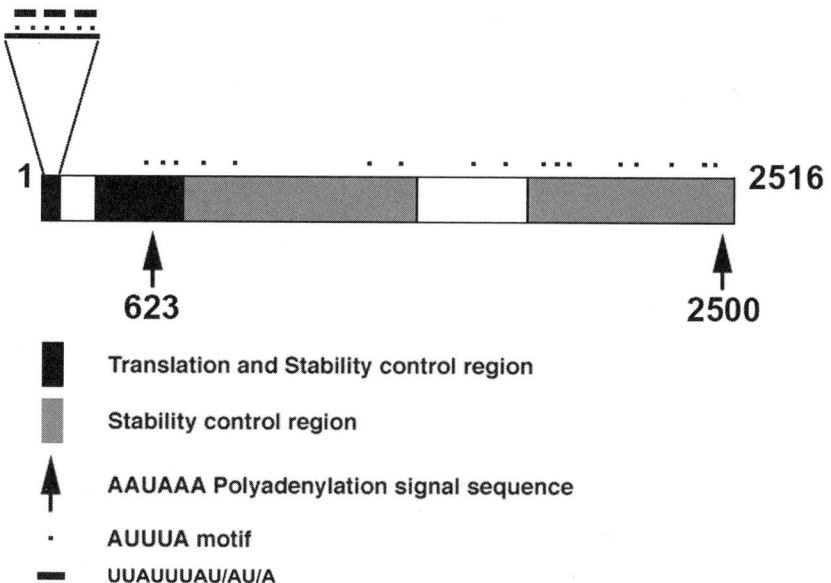

Fig. 6. COX-2 3′-UTR and its role in posttranscriptional control of COX-2 mRNA. Multiple control elements are present in the COX-2 3′-UTR that regulate COX-2 gene expression at the posttranscriptional level. AUUUA motifs present throughout the COX-2 3′-UTR is shown in the form of "dots." The nanomer sequence motif (UUAUUUAU/AU/A), which is a tandem repeat of AUUUA is present in the first sixty nucleotides and is shown by a "horizontal line." There are three such motifs present in the first sixty nucleotides and is the site for specific interactions of RNA binding proteins (Cok et al. 2001, Mukhopadhyay et al. 2003). Regions of COX-2 3′-UTR that regulate both mRNA stability and translation are depicted in black, while those that regulate mRNA stability only are shown in grey. Regions that neither affects stability nor translation are shown in white. The two polyadenylation signals located at nt positions 623 and 2500 of the 3′-UTR (arrows) are also shown.

1997). HuR was originally identified to bind and stabilize AU-rich containing mRNAs such as c-myc, c-fos, and TNF-α *(165–169)*. Recently, overexpression of HuR in HT-29 cells, a colon cancer cell line was shown to increase COX-2 expression, lengthen mRNA half-lives and allows for translation *(156)*. Furthermore, it was observed HuR in lysates of these cells can bind to COX-2 ARE sequences, and upon binding were able to increase the stability of a chimeric mRNA containing the coding region of firefly luciferase and the 3′-UTR of COX-2 mRNA. There was an additional inference from this manuscript that the limited ARE-binding of the higher molecular weight complex (60–90 kDa) reduced competition for COX-2 ARE-binding sites by other proteins *(156)*. HuR binding to the ARE containing messages and subsequent stabilization suggests that a decrease in HuR levels should result in a decline in COX-2 levels. Indeed, reduction of HuR levels using antisense oligonucleotide or silencer RNA resulted in a significant decrease in COX-2 protein levels *(170)*. These data suggest that a critical amount of HuR is essential for optimal expression of COX-2. Therefore, regulation of COX-2 by HuR is an important event in colorectal carcinogenesis.

HuR has been shown to bind to three high affinity binding sites in COX-2 3′-UTR located between nt 48–54, 1155–1187, and 1249–1256 *(170)*. The authors, as well as other investigators, have previously demonstrated that the first 60 nucleotides are very AU-rich and contain class II type ARE sequences *(152,153,171)*. Furthermore, the first 60 nucleotides nt of the COX-2 3′-UTR are highly conserved during evolution, suggesting a very important role for this region in regulating COX-2 gene expression *(170)*. Moreover, this region was identified as the minimal element required for COX-2 mRNA stability in response to p38 MAPK *(171)*. On the other hand, the nt site between 1249–1256, which also interacts with HuR, is essentially made up of U-rich sequences *(170)*. We have now determined that HuR binds to this region with a very high affinity, much higher than that observed with HuR binding to either the 1155–1187 or the 1249–1256 sequences (Sureban and Anant, unpublished observations). The significance of this binding is currently unknown. However, as alluded to above, the first 60 nucleotides regulate both stability and translation of COX-2 mRNA *(152,153)*. It has also been reported that, following IL-1β stimulation, there is an increase in cytosolic HuR protein levels and HuR-COX-2 mRNA complexes *(152)*. Given that IL-1β increases COX-2 levels, it is suggested that IL-1β regulates the process by increasing the binding of HuR to the COX-2 AREs and increase the transport of COX-2 mRNA from nucleus to cytoplasm to facilitate the translation of the mRNA.

2.5. Tristetraprolin

TTP, also known as TIS11 and Nup475, is an immediate early gene and a prototype for a family of zinc-binding Cys(3) His motif proteins that is required for regulating TNF-α mRNA stability in macrophages *(172)*. TTP is observed in the intestinal mucosa, with increased expression observed to precede the adaptive hyperplastic response after small bowel resection *(173,174)*. Like HuR, TTP was also found to bind to AREs of RNAs from immediate early genes such as c-fos, IL-8, and TNF-α. However, in contrast with HuR, TTP binding to the ARE results in rapid degradation of the mRNA *(172,175–177)*. TTP was found to be upregulated in HCA-7, a colon cancer cell line in a confluence-dependent fashion, and upon induction was observed to bind to COX-2 3′-UTR *(178)*. Furthermore, TTP was found to bind to a region between nucleotide positions 3125 and 3432, which contain a UAUUUA sequence at nucleotide position 3369. This site is present only in the 4.6-kb COX-2 mRNA, but is lacking in the 2.8-kb mRNA *(178)*. Indeed, TTP affected the degradation of only 4.6-kb isoform, whereas the 2.8-kb isoform was unaffected *(178)*. Consistent with this increased degradation of COX-2 mRNA, there was decreased COX-2 protein expression in the cells *(178)*.

2.6. CUGBP2

CUGBP2 , also known as ETR-3, BRUNOL, and NAPOR1, is a prototype of the CELF (for CUG-BP- and ETR-3-like factors) family of RNA binding proteins that was originally identified to bind to expanded CUG triplet repeats in the 3′-UTR of an mRNA encoding a protein kinase involved in myotonic dystrophy *(179–181)*. CUGBP2 is ubiquitously expressed protein, with significantly higher levels observed in skeletal and cardiac muscle. In neuroblastoma cells, CUGBP2 expression is not observed at base line, but is significantly induced when the cells are stimulated to undergo apoptosis *(180)*. It binds to the AREs present in the COX-2 3′-UTR and inhibits the mRNA translation. Recently, we identified that CUGBP2 expression is induced in intestinal epithelial cells when the cells were subjected to radiation treatment. COX-2 mRNA was also induced in the intestinal epithelial cells following radiation, but translation of the transcript was inhibited *(153)*. Further studies determined that the block in COX-2 translation resulted from CUGBP2 binding to the ARE sequences located in the first 60 nucleotides of COX-2 3'-UTR. Surprisingly, CUGBP2 binding to the ARE sequences increased the stability of COX-2 mRNA but inhibited mRNA translation. In contrast, when CUGBP2 expression was suppressed using antisense methods, there was increased COX-2 mRNA translation and PGE_2 synthesis *(153)*. The *D. ortholog* of CUGBP2 is Bruno, which binds to the Bruno-response element in the 3'-UTR of Oskar mRNA *(182–184)*. The coupled regulation of oskar mRNA localization and translation in time and space is critical for correct anteroposterior patterning of the Drosophila embryo. Oskar mRNA is located throughout the developing embryo, but the RNA localized at the posterior of the oocyte is selectively translated and those localized in the other regions remains in a translationally repressed state as a result of the binding of Bruno to the BRE in the oskar 3'-UTR *(182)*. CUGBP2 mimics Bruno function in the vertebrates by regulating COX-2 mRNA translation. It is not known whether CUGBP2 binding to the COX-2 mRNA results in polysome disassembly, but it would not be too far fetched to speculate that CUGBP2 bound COX-2 mRNA is transported to stress granules when the cells are subjected to high levels of radiation stress. This has been demonstrated with TIAR, which translocates to cytoplasm under heat stress, and accumulates into the stress granules along with the untranslated mRNA, resulting in translation inhibition of granule-associated RNA *(185,186)*. This increased levels of COX-2 mRNA, but inhibition of its translation comes as a surprise and raises questions as to whether the traditional concept that more mRNA means more protein is correct under all conditions. It remains to be seen, however, whether this is a global phenomenon or if it is unique to this situation. In any case, the inhibition of mRNA translation without enhancing degradation makes the mechanism of CUGBP2 action rather unique.

2.7. Other RNA Binding Proteins

TIA-1 and TIAR have been shown to bind to transcripts and promote polysome disassembly, thereby allowing the complex to be transported into stress granules *(187–189)*. Overexpression of TIA-1 causes the silencing of the COX-2 *(158)*. Furthermore, TIA-1, TIAR, and their splice variants have also been shown to bind to ARE sequences in the COX-2 and TNF-α 3'-UTRs, resulting in the silencing of the message translation *(149)*. hnRNP U also binds to the AREs present in the first 60 nt of the COX-2 3'-UTR and may act as a scaffolding protein and mediate interactions between target mRNA ands and proteins regulating the mRNA expression *(190)*. HnRNP A1, a predominantly nuclear protein, relocalizes to the cytoplasm where it binds to the AREs and is sequestered in stress granules *(191)*. This is reminiscent of the mechanism by which CUGBP2 regulates COX-2 mRNA translation, suggesting a common pathway for the two proteins.

Additional proteins that are thought to bind COX-2 3'-UTR and regulate translation are AUF-1 and CPF-A *(149)*. In addition, splice variants exist for CPF-A and AUF-1, which also bind to the proximal part of the COX-2 3'-UTR. Many hnRNPs, such as hnRNP A3 and A2/B1, are implicated in cytoplasmic trafficking of RNAs and these too bind to the COX-2 3'-UTR. It is possible that these proteins may either assist in the transportation of the message from the cytoplasm to the nucleus, may regulate COX-2 mRNA stability and translation, or both. Studies on demonstrating the effects of these hnRNPs on the posttranscriptional regulation of COX-2 and other mRNAs will no doubt yield some very interesting results.

Acknowledgments

Studies from the authors laboratory are supported by NIH grants DK-62265 and CA-109269.

References

1. Jemal A, Clegg LX, Ward E, et al. Annual report to the nation on the status of cancer, 1975–2001, with a special feature regarding survival. Cancer 2004;101:3–27.
2. Mohandas KM. Dietary fiber and colorectal cancer. N Engl J Med 1999;340:1925–1926.
3. Rao CV, Rivenson A, Simi B, Reddy BS. Chemoprevention of colon cancer by dietary curcumin. Ann N Y Acad Sci 1995;768:201–204.
4. Rao CV, Rivenson A, Simi B, Reddy BS. Chemoprevention of colon carcinogenesis by dietary curcumin, a naturally occurring plant phenolic compound. Cancer Res 1995;55:259–266.
5. Reddy BS, Rao CV. Novel approaches for colon cancer prevention by cyclooxygenase-2 inhibitors. J Environ Pathol Toxicol Oncol 2002;21:155–164.
6. Slattery ML, Berry TD, Kerber RA. Is survival among women diagnosed with breast cancer influenced by family history of breast cancer? Epidemiology 1993;4:543–548.
7. Slattery ML, Kerber RA. A comprehensive evaluation of family history and breast cancer risk. The Utah Population Database. JAMA 1993;270:1563–1568.
8. Slattery ML, West DW. Smoking, alcohol, coffee, tea, caffeine, and theobromine: risk of prostate cancer in Utah (United States). Cancer Causes Control 1993;4:559–563.
9. Reddy BS. Overview of diet and colon cancer. Prog Clin Biol Res 1998;279:111–121.
10. Potter JD. Colon cancer—do the nutritional epidemiology, the gut physiology and the molecular biology tell the same story? J Nutr 1993;123:418–423.
11. Potter JD, Slattery ML, Bostick RM, Gapstur SM. Colon cancer: a review of the epidemiology. Epidemiol Rev 1993;15:499–545.
12. DeCosse JJ, Ngoi SS, Jacobson JS, Cennerazzo WJ. Gender and colorectal cancer. Eur J Cancer Prev 1993;2:105–115.
13. Hough DM, Malone DE, Rawlinson J, et al. Colon cancer detection: an algorithm using endoscopy and barium enema. Clin Radiol 1994;49:170–175.
14. Mainguet P, Jouret A. Colon cancer prevention: role of the endoscopy. Review of the new histopathological techniques. Eur J Cancer Prev 1993;2:261–262.
15. Topal B, Basha G, Penninckx F. Mechanisms and prevention of recurrent colorectal cancer. Hepatogastroenterology 1999;46:701–708.
16. Peltomaki P. Role of DNA mismatch repair defects in the pathogenesis of human cancer. J Clin Oncol 2003;21:1174–1179.
17. Muller A, Fishel R. Mismatch repair and the hereditary non-polyposis colorectal cancer syndrome (HNPCC). Cancer Invest 2002;20:102–109.
18. Jiricny J, Nystrom-Lahti M. Mismatch repair defects in cancer. Curr Opin Genet Dev 2000; 10:157–161.
19. Fedier A, Fink D. Mutations in DNA mismatch repair genes: implications for DNA damage signaling and drug sensitivity (review). Int J Oncol 2004;24:1039–1047.
20. De Jong AE, Morreau H, Van Puijenbroek M, et al. The role of mismatch repair gene defects in the development of adenomas in patients with HNPCC. Gastroenterology 2004;126:42–48.
21. Carethers JM, Chauhan DP, Fink D, et al. Mismatch repair proficiency and in vitro response to 5-fluorouracil. Gastroenterology 1999;117:123–131.
22. Vogelstein B, Kinzler KW. The multistep nature of cancer. Trends Genet 1993;9:138–141.

23. Vogelstein B, Kinzler KW. Cancer genes and the pathways they control. Nat Med 2004;10: 789–799.
24. Spirio L, Olschwang S, Groden J, et al. Alleles of the APC gene: an attenuated form of familial polyposis. Cell 1993;75:951–957.
25. Brensinger JD, Laken SJ, Luce MC, et al. Variable phenotype of familial adenomatous polyposis in pedigrees with 3′ mutation in the APC gene. Gut 1998;43:548–552.
26. Pedemonte S, Sciallero S, Gismondi V, et al. Novel germline APC variants in patients with multiple adenomas. Genes Chromosomes Cancer 1998;22:257–267.
27. Soravia C, Berk T, Madlensky L, et al. Genotype-phenotype correlations in attenuated adenomatous polyposis coli. Am J Hum Genet 1998;62:1290–1301.
28. Laken SJ, Papadopoulos N, Petersen GM, et al. Analysis of masked mutations in familial adenomatous polyposis. Proc Natl Acad Sci USA 1999;96:2322–2326.
29. Luu HH, Zhang R, Haydon RC, et al. Wnt/beta-catenin signaling pathway as a novel cancer drug target. Curr Cancer Drug Targets 2004;4:653–671.
30. Bienz M, Hamada F. Adenomatous polyposis coli proteins and cell adhesion. Curr Opin Cell Biol 2004;16:528–535.
31. Behrens J, Lustig B. The Wnt connection to tumorigenesis. Int J Dev Biol 2004;48:477–487.
32. Clevers H. Wnt breakers in colon cancer. Cancer Cell 2004;5:5–6.
33. Barth AI, Nathke IS, Nelson WJ. Cadherins, catenins and APC protein: interplay between cytoskeletal complexes and signaling pathways. Curr Opin Cell Biol 1997;9:683–690.
34. Behrens J. Cadherins and catenins: role in signal transduction and tumor progression. Cancer Metastasis Rev 1999;18:15–30.
35. Wong NA, Pignatelli M. Beta-catenin—a linchpin in colorectal carcinogenesis? Am J Pathol 2002;160:389–401.
36. Howe LR, Crawford HC, Subbaramaiah K, Hassell JA, Dannenberg AJ, Brown AM. PEA3 is up-regulated in response to Wnt1 and activates the expression of cyclooxygenase-2. J Biol Chem 2001;276:20,108–20,115.
37. Dimberg J, Hugander A, Sirsjo A, Soderkvist P. Enhanced expression of cyclooxygenase-2 and nuclear beta-catenin are related to mutations in the APC gene in human colorectal cancer. Anticancer Res 2001;21:911–915.
38. Araki Y, Okamura S, Hussain SP, et al. Regulation of cyclooxygenase-2 expression by the Wnt and ras pathways. Cancer Res 2003;63:728–734.
39. Iacopetta B. TP53 mutation in colorectal cancer. Hum Mutat 2003;21:271–276.
40. Iacopetta B. Aberrant DNA methylation: have we entered the era of more than one type of colorectal cancer? Am J Pathol 2003;162:1043–1045.
41. Kishimoto Y, Morisawa T, Hosoda A, Shiota G, Kawasaki H, Hasegawa J. Molecular changes in the early stage of colon carcinogenesis in rats treated with azoxymethane. J Exp Clin Cancer Res 2002;21:203–211.
42. Mascaux C, Martin B, Verdebout JM, Ninane V, Sculier JP. COX-2 expression during early lung squamous cell carcinoma oncogenesis. Eur Respir J 2005;26:198–203.
43. Yan M, Rerko RM, Platzer P, et al. 15-Hydroxyprostaglandin dehydrogenase, a COX-2 oncogene antagonist, is a TGF-beta-induced suppressor of human gastrointestinal cancers. Proc Natl Acad Sci USA 2004;101:17,468–17,473.
44. Morita T, Tomita N, Ohue M, et al. Molecular analysis of diminutive, flat, depressed colorectal lesions: are they precursors of polypoid adenoma or early stage carcinoma? Gastrointest Endosc 2002;56:663–671.
45. Williams CS, Mann M, DuBois RN. The role of cyclooxygenases in inflammation, cancer, and development. Oncogene 1999;18:7908–7916.
46. Williams CS, Sheng H, Brockman JA, et al. A cyclooxygenase-2 inhibitor (SC-58125) blocks growth of established human colon cancer xenografts. Neoplasia 2001;3:428–436.

47. Dubois RN, Abramson SB, Crofford L, et al. Cyclooxygenase in biology and disease. Faseb J 1998;12:1063–1073.

48. Williams CS, DuBois RN. Prostaglandin endoperoxide synthase: why two isoforms? Am J Physiol 1996;270:G393–G400.

49. Vane JR, Bakhle YS, Botting RM. Cyclooxygenases 1 and 2. Annu Rev Pharmacol Toxicol 1998;38:97–120.

50. Cook JA, Geisel J, Halushka PV, Reines HD. Prostaglandins, thromboxanes, leukotrienes, and cytochrome P-450 metabolites of arachidonic acid. New Horiz 1993;1:60–69.

51. Johansson C, Bergstrom S. Prostaglandin and protection of the gastroduodenal mucosa. Scand J Gastroenterol Suppl 1982;77:21–46.

52. Johansson C, Kollberg B. Effects of E2 prostaglandins on gastric secretion and gastrointestinal mucosa. Scand J Gastroenterol Suppl 1980;58:93–97.

53. Eberhart CE, Dubois RN. Eicosanoids and the gastrointestinal tract. Gastroenterology 1995;109:285–301.

54. Hull MA, Ko SC, Hawcroft G. Prostaglandin EP receptors: targets for treatment and prevention of colorectal cancer? Mol Cancer Ther 2004;3:1031–1039.

55. Regan JW. EP2 and EP4 prostanoid receptor signaling. Life Sci 2003;74:143–153.

56. Breyer RM, Bagdassarian CK, Myers SA, Breyer MD. Prostanoid receptors: subtypes and signaling. Annu Rev Pharmacol Toxicol 2001;41:661–690.

57. Breyer MD, Breyer RM. Prostaglandin E receptors and the kidney. Am J Physiol Renal Physiol 2000;279:F12–F23.

58. Chiarugi V, Magnelli L, Gallo O. Cox-2, iNOS and p53 as play-makers of tumor angiogenesis (review). Int J Mol Med 1998;2:715–719.

59. Mutoh M, Watanabe K, Kitamura T, et al. Involvement of prostaglandin E receptor subtype EP(4) in colon carcinogenesis. Cancer Res 2002;62:28–32.

60. Bamba H, Ota S, Kato A, Kawamoto C, Matsuzaki F. Effect of prostaglandin E1 on vascular endothelial growth factor production by human macrophages and colon cancer cells. J Exp Clin Cancer Res 2000;19:219–223.

61. Ushikubi F, Sugimoto Y, Ichikawa A, Narumiya S. Roles of prostanoids revealed from studies using mice lacking specific prostanoid receptors. Jpn J Pharmacol 2000;83:279–285.

62. Watanabe K, Kawamori T, Nakatsugi S, et al. Inhibitory effect of a prostaglandin E receptor subtype EP(1) selective antagonist, ONO-8713, on development of azoxymethane-induced aberrant crypt foci in mice. Cancer Lett 2000;156:57–61.

63. Sonoshita M, Takaku K, Sasaki N, et al. Acceleration of intestinal polyposis through prostaglandin receptor EP2 in Apc(Delta 716) knockout mice. Nat Med 2002;7:1048–1051.

64. Takafuji V, Lublin D, Lynch K, Roche JK. Mucosal prostanoid receptors and synthesis in familial adenomatous polyposis. Histochem Cell Biol 2001;116:171–181.

65. Takafuji V, Cosme R, Lublin D, Lynch K, Roche JK. Prostanoid receptors in intestinal epithelium: selective expression, function, and change with inflammation. Prostaglandins Leukot Essent Fatty Acids 2000;63:223–235.

66. Kawamori T, Uchiya N, Kitamura T, et al. Evaluation of a selective prostaglandin E receptor EP1 antagonist for potential properties in colon carcinogenesis. Anticancer Res 2001;21:3865–3869.

67. Fujino H, West KA, Regan JW. Phosphorylation of glycogen synthase kinase-3 and stimulation of T-cell factor signaling following activation of EP2 and EP4 prostanoid receptors by prostaglandin E2. J Biol Chem 2002;277:2614–2619.

68. Tinhofer I, Bernhard D, Senfter M, et al. Resveratrol, a tumor-suppressive compound from grapes, induces apoptosis via a novel mitochondrial pathway controlled by Bcl-2. Faseb J 2001;15:1613–1615.

69. Susin SA, Lorenzo HK, Zamzami N, et al. Mitochondrial release of caspase-2 and -9 during the apoptotic process. J Exp Med 1999;189:381–394.

70. Wikstrom K, Ohd JF, Sjolander A. Regulation of leukotriene-dependent induction of cyclooxygenase-2 and Bcl-2. Biochem Biophys Res Commun 2003;302:330–335.

71. Sheng H, Shao J, Morrow JD, Beauchamp RD, DuBois RN. Modulation of apoptosis and Bcl-2 expression by prostaglandin E2 in human colon cancer cells. Cancer Res 1998;58:362–366.

72. Spencer AG, Woods JW, Arakawa T, Singer II, Smith WL. Subcellular localization of prostaglandin endoperoxide H synthases-1 and -2 by immunoelectron microscopy. J Biol Chem 1998;273:9886–9893.

73. Johnsen JI, Lindskog M, Ponthan F, et al. Cyclooxygenase-2 is expressed in neuroblastoma, and nonsteroidal anti-inflammatory drugs induce apoptosis and inhibit tumor growth in vivo. Cancer Res 2004;64:7210–7215.

74. Munkarah A, Ali-Fehmi R. COX-2: a protein with an active role in gynecological cancers. Curr Opin Obstet Gynecol 2005;17:49–53.

75. Hussain T, Gupta S, Mukhtar H. Cyclooxygenase-2 and prostate carcinogenesis. Cancer Lett 2003;191:125–135.

76. Singh B, Lucci A. Role of cyclooxygenase-2 in breast cancer. J Surg Res 2002;108:173–179.

77. Lee JL, Kim A, Kopelovich L, Bickers DR, Athar M. Differential expression of E prostanoid receptors in murine and human non-melanoma skin cancer. J Invest Dermatol 2005;125:818–825.

78. Karim A, Fowler M, Jones L, et al. Cyclooxygenase-2 expression in brain metastases. Anticancer Res 2005;25:2969–2971.

79. Figueiredo A, Caissie AL, Callejo SA, et al. Cyclooxygenase-2 expression in uveal melanoma: novel classification of mixed-cell-type tumours. Can J Ophthalmol 2003;38: 352–356.

80. Denkert C, Kobel M, Berger S, et al. Expression of cyclooxygenase 2 in human malignant melanoma. Cancer Res 2000;61:303–308.

81. Sinicrope FA, Lemoine M, Xi L, et al. Reduced expression of cyclooxygenase 2 proteins in hereditary nonpolyposis colorectal cancers relative to sporadic cancers. Gastroenterology 1999;117:350–358.

82. Muir KR, Logan RF. Aspirin, NSAIDs and colorectal cancer—what do the epidemiological studies show and what do they tell us about the modus operandi? Apoptosis 1999;4: 389–396.

83. Herendeen JM, Lindley C. Use of NSAIDs for the chemoprevention of colorectal cancer. Ann Pharmacother 2003;37:1664–1674.

84. Morgan G. Beneficial effects of NSAIDs in the gastrointestinal tract. Eur J Gastroenterol Hepatol 1999;11:393–400.

85. Jolly K, Cheng KK, Langman MJ. NSAIDs and gastrointestinal cancer prevention. Drugs 2002;62:945–956.

86. Sasai H, Masaki M, Wakitani K. Suppression of polypogenesis in a new mouse strain with a truncated Apc(Delta474) by a novel COX-2 inhibitor, JTE-522. Carcinogenesis 2000; 21:953–958.

87. Chulada PC, Thompson MB, Mahler JF, et al. Genetic disruption of Ptgs-1, as well as Ptgs-2, reduces intestinal tumorigenesis in Min mice. Cancer Res 2000;60:4705–4708.

88. Fujita M, Fukui H, Kusaka T, et al. Relationship between cyclooxygenase-2 expression and K-ras gene mutation in colorectal adenomas. J Gastroenterol Hepatol 2000;15:1277–1281.

89. Fujita M, Fukui H, Kusaka T, Ueda Y, Fujimori T. Immunohistochemical expression of cyclooxygenase (COX)-2 in colorectal adenomas. J Gastroenterol 2000;35:488–490.

90. Hao X, Bishop AE, Wallace M, et al. Early expression of cyclo-oxygenase-2 during sporadic colorectal carcinogenesis. J Pathol 1999;187:295–301.

91. Beazer-Barclay Y, Levy DB, Moser AR, et al. Sulindac suppresses tumorigenesis in the Min mouse. Carcinogenesis 1996;17:1757–1760.

92. Boolbol SK, Dannenberg AJ, Chadburn A, et al. Cyclooxygenase-2 overexpression and tumor formation are blocked by sulindac in a murine model of familial adenomatous polyposis. Cancer Res 1996;56:2556–2560.

93. DuBois RN, Giardiello FM, Smalley WE. Nonsteroidal anti-inflammatory drugs, eicosanoids, and colorectal cancer prevention. Gastroenterol Clin North Am 1996;25:773–791.

94. Chiu CH, McEntee MF, Whelan J. Sulindac causes rapid regression of preexisting tumors in Min/+ mice independent of prostaglandin biosynthesis. Cancer Res 1997;57:4267–4273.

95. Reddy BS, Furuya K, Lowenfels A. Effect of neomycin on azoxymethane-induced colon carcinogenesis in F344 rats. J Natl Cancer Inst 1984;73:275–279.

96. Reddy BS, Tanaka T, El-Bayoumy K. Inhibitory effect of dietary p-methoxybenzeneselenol on azoxymethane-induced colon and kidney carcinogenesis in female F344 rats. J Natl Cancer Inst 1985;74:1325–1328.

97. Reddy BS, Maruyama H, Kelloff G. Dose-related inhibition of colon carcinogenesis by dietary piroxicam, a nonsteroidal antiinflammatory drug, during different stages of rat colon tumor development. Cancer Res 1987;47:5340–5346.

98. Reddy BS, Tanaka T. Interactions of selenium deficiency, vitamin E, polyunsaturated fat, and saturated fat on azoxymethane-induced colon carcinogenesis in male F344 rats. J Natl Cancer Inst 1986;76:1157–1162.

99. Singh J, Hamid R, Reddy BS. Dietary fat and colon cancer: modulation of cyclooxygenase-2 by types and amount of dietary fat during the postinitiation stage of colon carcinogenesis. Cancer Res 1997;57:3465–3470.

100. Singh J, Hamid R, Reddy BS. Dietary fat and colon cancer: modulating effect of types and amount of dietary fat on ras-p21 function during promotion and progression stages of colon cancer. Cancer Res 1997;57:253–258.

101. Oshima M, Dinchuk JE, Kargman SL, et al. Suppression of intestinal polyposis in Apc delta716 knockout mice by inhibition of cyclooxygenase 2 (COX-2). Cell 1996;87:803–809.

102. Vane J. Suppression of intestinal polyposis by inhibition of COX-2 in Apc knockout mice. Jpn J Cancer Res 1997;88:inside front cover.

103. Kosaka T, Miyata A, Ihara H, et al. Characterization of the human gene (PTGS2) encoding prostaglandin-endoperoxide synthase 2. Eur J Biochem 1994;221:889–897.

104. Appleby SB, Ristimaki A, Neilson K, Narko K, Hla T. Structure of the human cyclooxygenase-2 gene. Biochem J 1994;302 (Pt 3):723–727.

105. Tazawa R, Xu XM, Wu KK, Wang LH. Characterization of the genomic structure, chromosomal location and promoter of human prostaglandin H synthase-2 gene. Biochem Biophys Res Commun 1994;203:190–199.

106. Geng Y, Blanco FJ, Cornelisson M, Lotz M. Regulation of cyclooxygenase-2 expression in normal human articular chondrocytes. J Immunol 1995;155:796–801.

107. Crofford LJ. COX-1 and COX-2 tissue expression: implications and predictions. J Rheumatol Suppl 1994;49:15–19.

108. Berg J, Stocher M, Bogner S, Wolfl S, Pichler R, Stekel H. Inducible cyclooxygenase-2 gene expression in the human thyroid epithelial cell line Nthy-ori3-1. Inflamm Res 2000;49:139–143.

109. Charalambous MP, Maihofner C, Bhambra U, Lightfoot T, Gooderham NJ. Upregulation of cyclooxygenase-2 is accompanied by increased expression of nuclear factor-kappa B and I kappa B kinase-alpha in human colorectal cancer epithelial cells. Br J Cancer 2003; 88:1598–1604.

110. Stancovski I, Baltimore D. NF-kappaB activation: the I kappaB kinase revealed? Cell 1994;91:299–302.

111. Kaltschmidt B, Linker RA, Deng J, Kaltschmidt C. Cyclooxygenase-2 is a neuronal target gene of NF-kappaB. BMC Mol Biol 2002;3:16.

112. Blanco JC, Contursi C, Salkowski CA, et al. Interferon regulatory factor (IRF)-1 and IRF-2 regulate interferon gamma-dependent cyclooxygenase 2 expression. J Exp Med 2000; 191:2131–2144.

113. Zhang S, Thomas K, Blanco JC, Salkowski CA, Vogel SN. The role of the interferon regulatory factors, IRF-1 and IRF-2, in LPS-induced cyclooxygenase-2 (COX-2) expression in vivo and in vitro. J Endotoxin Res 2002;8:379–388.

114. Upreti M, Kumar S, Rath PC. Replacement of 198MQMDII203 of mouse IRF-1 by 197IPVEVV202 of human IRF-1 abrogates induction of IFN-beta, iNOS, and COX-2 gene expression by IRF-1. Biochem Biophys Res Commun 2004;314:737–744.

115. Crepieux P, Kwon H, Leclerc N, et al. I kappaB alpha physically interacts with a cytoskeleton-associated protein through its signal response domain. Mol Cell Biol 1997;17:7375–7385.

116. Chen CC, Sun YT, Chen JJ,. Chiu KT. TNF-alpha-induced cyclooxygenase-2 expression in human lung epithelial cells: involvement of the phospholipase C-gamma 2, protein kinase C-alpha, tyrosine kinase, NF-kappa B-inducing kinase, and I-kappa B kinase 1/2 pathway. J Immunol 2000;165:2719–2728.

117. Deng WG, Zhu Y, Wu KK. Role of p300 and PCAF in regulating cyclooxygenase-2 promoter activation by inflammatory mediators. Blood 2004;103:2135–2142.

118. Duque J, Fresno M, Iniguez MA. Expression and function of the nuclear factor of activated T cells in colon carcinoma cells: involvement in the regulation of cyclooxygenase-2. J Biol Chem 2005;280:8686–8693.

119. Kim SP, Park JW, Lee SH, et al. Homeodomain protein CDX2 regulates COX-2 expression in colorectal cancer. Biochem Biophys Res Commun 2004;315:93–99.

120. Subbaramaiah K, Cole PA, Dannenberg AJ. Retinoids and carnosol suppress cyclooxygenase-2 transcription by CREB-binding protein/p300-dependent and -independent mechanisms. Cancer Res 2002;62:2522–2530.

121. Saunders MA, Sansores-Garcia L, Gilroy DW, Wu KK. Selective suppression of CCAAT/enhancer-binding protein beta binding and cyclooxygenase-2 promoter activity by sodium salicylate in quiescent human fibroblasts. J Biol Chem 2001;276:18,897–18,904.

122. Wong BC, Jiang XH, Lin MC, et al. Cyclooxygenase-2 inhibitor (SC-236) suppresses activator protein-1 through c-Jun NH2-terminal kinase. Gastroenterology 2004;126:136–147.

123. Wang HQ, Kim MP, Tiano HF, Langenbach R, Smart RC. Protein kinase C-alpha coordinately regulates cytosolic phospholipase A(2) activity and the expression of cyclooxygenase-2 through different mechanisms in mouse keratinocytes. Mol Pharmacol 2001;59: 860–866.

124. Deng WG, Zhu Y, Montero A, Wu KK. Quantitative analysis of binding of transcription factor complex to biotinylated DNA probe by a streptavidin-agarose pulldown assay. Anal Biochem 2003;323:12–18.

125. Prives C, Hall PA. The p53 pathway. J Pathol 1999;187:112–126.

126. Eliyahu D, Michalovitz D, Eliyahu S, Pinhasi-Kimhi O, Oren M. Wild-type p53 can inhibit oncogene-mediated focus formation. Proc Natl Acad Sci USA 1989;86:8763–8767.

127. Levine AJ, Finlay CA, Hinds PW. P53 is a tumor suppressor gene. Cell 2004;116:S67-69, 61 p following S69.

128. Finlay CA, Hinds PW, Levine AJ. The p53 proto-oncogene can act as a suppressor of transformation. Cell 1989;57:1083–1093.

129. Oliner JD. Discerning the function of p53 by examining its molecular interactions. Bioessays 1993;15:703–707.

130. Nakamura Y. Isolation of p53-target genes and their functional analysis. Cancer Sci 2004;95:7–11.

131. Subbaramaiah K, Altorki N, Chung WJ, et al. Inhibition of cyclooxygenase-2 gene expression by p53. J Biol Chem 1999;274:10,911–10,915.

132. Pesch J, Brehm U, Staib C, Grummt F. Repression of interleukin-2 and interleukin-4 promoters by tumor suppressor protein p53. J Interferon Cytokine Res 1996;16:595–600.

133. Lee Y, Chen Y, Chang LS, Johnson LF. Inhibition of mouse thymidylate synthase promoter activity by the wild-type p53 tumor suppressor protein. Exp Cell Res 1997;234:270–276.

134. Mack DH, Vartikar J, Pipas JM, Laimins LA. Specific repression of TATA-mediated but not initiator-mediated transcription by wild-type p53. Nature 1993;363:281–283.

135. Seto E, Usheva A, Zambetti GP, et al. Wild-type p53 binds to the TATA-binding protein and represses transcription. Proc Natl Acad Sci USA 1992;89:12,028–12,032.

136. Ginsberg D, Mechta F, Yaniv M, Oren M. Wild-type p53 can down-modulate the activity of various promoters. Proc Natl Acad Sci USA 1991;88:9979–9983.

137. Crighton D, Woiwode A, Zhang C, et al. p53 represses RNA polymerase III transcription by targeting TBP and inhibiting promoter occupancy by TFIIIB. Embo J 2003;22:2810–2820.

138. Zhai W, Comai L. Repression of RNA polymerase I transcription by the tumor suppressor p53. Mol Cell Biol 2000;20:5930–5938.

139. Liu Y, Borchert GL, Phang JM. Polyoma enhancer activator 3, an ets transcription factor, mediates the induction of cyclooxygenase-2 by nitric oxide in colorectal cancer cells. J Biol Chem 2004;279:18,694–18,700.

140. Newton R, Seybold J, Liu SF, Barnes PJ. Alternate COX-2 transcripts are differentially regulated: implications for post-transcriptional control. Biochem Biophys Res Commun 1997; 234:85–89.

141. Ristimaki A, Narko K, Hla T. Down-regulation of cytokine-induced cyclo-oxygenase-2 transcript isoforms by dexamethasone: evidence for post-transcriptional regulation. Biochem J 1996;318(Pt 1):325–331.

142. Liu SF, Newton R, Evans TW, Barnes PJ. Differential regulation of cyclo-oxygenase-1 and cyclo-oxygenase-2 gene expression by lipopolysaccharide treatment in vivo in the rat. Clin Sci (Lond) 1996;90:301–306.

143. Hla T. Molecular characterization of the 5.2 KB isoform of the human cyclooxygenase-1 transcript. Prostaglandins 1996;51:81–85.

144. Diaz-Cazorla M, Perez-Sala D, Lamas S. Dual effect of nitric oxide donors on cyclooxygenase-2 expression in human mesangial cells. J Am Soc Nephrol 1999;10:943–952.

145. Guan Z, Buckman SY, Miller BW, Springer LD, Morrison AR. Interleukin-1beta-induced cyclooxygenase-2 expression requires activation of both c-Jun NH2-terminal kinase and p38 MAPK signal pathways in rat renal mesangial cells. J Biol Chem 1998;273:28,670–28,676.

146. Guan Z, Buckman SY, Baier LD, Morrison AR. IGF-I and insulin amplify IL-1 beta-induced nitric oxide and prostaglandin biosynthesis. Am J Physiol 1998;274:F673–F679.

147. Guan Z, Baier LD, Morrison AR. p38 mitogen-activated protein kinase down-regulates nitric oxide and up-regulates prostaglandin E2 biosynthesis stimulated by interleukin-1beta. J Biol Chem 1997;272:8083–8089.

148. Srivastava SK, Tetsuka T, Daphna-Iken D, Morrison AR. IL-1 beta stabilizes COX II mRNA in renal mesangial cells: role of 3'-untranslated region. Am J Physiol 1994;267:F504–F508.

149. Cok SJ, Acton SJ, Sexton AE, Morrison AR. Identification of RNA-binding proteins in RAW 264.7 cells that recognize a lipopolysaccharide-responsive element in the 3-untranslated region of the murine cyclooxygenase-2 mRNA. J Biol Chem 2004;279:8196–8205.

150. Neeraja S, Sreenath AS, Reddy PR, Reddanna P. Expression of cyclooxygenase-2 in rat testis. Reprod Biomed Online 2003;6:302–309.

151. Chen CY, Shyu AB. AU-rich elements: characterization and importance in mRNA degradation. Trends Biochem Sci 1995;20:465–470.

152. Cok SJ, Morrison AR. The 3'-untranslated region of murine cyclooxygenase-2 contains multiple regulatory elements that alter message stability and translational efficiency. J Biol Chem 2001;276:23,179–23,185.

153. Mukhopadhyay D, Houchen CW, Kennedy S, Dieckgraefe BK, Anant S. Coupled mRNA stabilization and translational silencing of cyclooxygenase-2 by a novel RNA binding protein, CUGBP2. Mol Cell 2003;11:113–126.

154. Zubiaga AM, Belasco JG, Greenberg ME. The nonamer UUAUUUAUU is the key AU-rich sequence motif that mediates mRNA degradation. Mol Cell Biol 1995;15:2219–2230.

155. Anant S, Murmu N, Houchen CW, et al. Apobec-1 protects intestine from radiation injury through posttranscriptional regulation of cyclooxygenase-2 expression. Gastroenterology 2004;127:1139–1149.

156. Dixon DA, Tolley ND, King PH, et al. Altered expression of the mRNA stability factor HuR promotes cyclooxygenase-2 expression in colon cancer cells. J Clin Invest 2001;108: 1657–1665.

157. Yeo SJ, Yoon JG, Yi AK. Myeloid differentiation factor 88-dependent post-transcriptional regulation of cyclooxygenase-2 expression by CpG DNA: tumor necrosis factor-alpha receptor-associated factor 6, a diverging point in the Toll-like receptor 9-signaling. J Biol Chem 2003;278:40,590–40,600.

158. Dixon DA, Balch GC, Kedersha N, et al. Regulation of cyclooxygenase-2 expression by the translational silencer TIA-1. J Exp Med 2003;198:475–481.

159. Boutaud O, Dixon DA, Oates JA, Sawaoka H. Tristetraprolin binds to the COX-2 mRNA 3′ untranslated region in cancer cells. Adv Exp Med Biol 2003;525:157–160.

160. Dean JL, Sully G, Wait R, Rawlinson L, Clark AR, Saklatvala J. Identification of a novel AU-rich-element-binding protein which is related to AUF1. Biochem J 2002;366:709–719.

161. Nabors LB, Gillespie GY, Harkins L, King PH. HuR, a RNA stability factor, is expressed in malignant brain tumors and binds to adenine- and uridine-rich elements within the 3′ untranslated regions of cytokine and angiogenic factor mRNAs. Cancer Res 2001;61: 2154–2161.

162. Thiele BJ, Doller A, Kahne T, et al. RNA-binding proteins heterogeneous nuclear ribo-nucleoprotein A1, E1, and K are involved in post-transcriptional control of collagen I and III synthesis. Circ Res 2004;95:1058–1066.

163. Hamilton BJ, Nichols RC, Tsukamoto H, et al. hnRNP A2 and hnRNP L bind the 3′UTR of glucose transporter 1 mRNA and exist as a complex in vivo. Biochem Biophys Res Commun 1999;261:646–651.

164. Rousseau S, Morrice N, Peggie M, Campbell DG, Gaestel M, Cohen P. of SAPK2a/p38 prevents hnRNP A0 phosphorylation by MAPKAP-K2 and its interaction with cytokine mRNAs. Embo J 2002;21:6505–6514.

165. Ma WJ, Chung S, Furneaux H. The Elav-like proteins bind to AU-rich elements and to the poly(A) tail of mRNA. Nucleic Acids Res 1997;25:3564–3569.

166. Fan XC, Steitz JA. Overexpression of HuR, a nuclear-cytoplasmic shuttling protein, increases the in vivo stability of ARE-containing mRNAs. Embo J 1998;17:3448–3460.

167. Levy NS, Chung S, Furneaux H, Levy AP. Hypoxic stabilization of vascular endothelial growth factor mRNA by the RNA-binding protein HuR. J Biol Chem 1998;273: 6417–6423.

168. Peng SS, Chen CY, Xu N, Shyu AB. RNA stabilization by the AU-rich element binding protein, HuR, an ELAV protein. Embo J 1998;17:3461–3470.

169. Xu N, Loflin P, Chen CY, Shyu AB. A broader role for AU-rich element-mediated mRNA turnover revealed by a new transcriptional pulse strategy. Nucleic Acids Res 1998;26:558–565.

170. Sengupta S, Jang BC, Wu MT, Paik JH, Furneaux H, Hla T. The RNA-binding protein HuR regulates the expression of cyclooxygenase-2. J Biol Chem 2003;278:25,227–25,233.

171. Gou Q, Liu CH, Ben-Av P, Hla T. Dissociation of basal turnover and cytokine-induced transcript stabilization of the human cyclooxygenase-2 mRNA by mutagenesis of the 3′-untranslated region. Biochem Biophys Res Commun 1998;242:508–512.

172. Carballo E, Gilkeson GS, Blackshear PJ. Bone marrow transplantation reproduces the tristetraprolin-deficiency syndrome in recombination activating gene-2 (-/-) mice. Evidence that monocyte/macrophage progenitors may be responsible for TNFalpha over-production. J Clin Invest 1997;100:986–995.

173. Ehrenfried JA, Townsend CM, Jr, Thompson JC, Evers BM. Increases in nup475 and c-jun are early molecular events that precede the adaptive hyperplastic response after small bowel resection. Ann Surg 1995;222:51–56.

174. Sacks AI, Warwick GJ, Barnard JA. Early proliferative events following intestinal resection in the rat. J Pediatr Gastroenterol Nutr 1995;21:158–164.

175. Mahtani KR, Brook M, Dean JL, Sully G, Saklatvala J, Clark A R. Mitogen-activated protein kinase p38 controls the expression and posttranslational modification of tristetraprolin, a regulator of tumor necrosis factor alpha mRNA stability. Mol Cell Biol 2001;21:6461–6469.

176. Raghavan A, Robison RL, McNabb J, et al. HuA and tristetraprolin are induced following T cell activation and display distinct but overlapping RNA binding specificities. J Biol Chem 2001;276:47,958–47,965.

177. Worthington MT, Pelo JW, Sachedina MA, et al. RNA binding properties of the AU-rich element-binding recombinant Nup475/TIS11/tristetraprolin protein. J Biol Chem 2002; 277:48,558–48,564.

178. Sawaoka H, Dixon DA, Oates JA, Boutaud O. Tristetraprolin binds to the 3′-untranslated region of cyclooxygenase-2 mRNA. A polyadenylation variant in a cancer cell line lacks the binding site. J Biol Chem 2003;278:13,928–13,935.

179. Good PJ, Chen Q, Warner SJ, Herring DC. A family of human RNA-binding proteins related to the Drosophila Bruno translational regulator. J Biol Chem 2000;275:28,583–28,592.

180. Choi DK, Ito T, Mitsui Y, Sakaki Y. Fluorescent differential display analysis of gene expression in apoptotic neuroblastoma cells. Gene 1998;223:21–31.

181. Ladd AN, Charlet N, Cooper TA. The CELF family of RNA binding proteins is implicated in cell-specific and developmentally regulated alternative splicing. Mol Cell Biol 2001;21: 1285–1296.

182. Kim-Ha J, Kerr K, Macdonald PM. Translational regulation of oskar mRNA by bruno, an ovarian RNA-binding protein, is essential. Cell 1995;81:403–412.

183. Webster PJ, Liang L, Berg CA, Lasko P, Macdonald PM. Translational repressor bruno plays multiple roles in development and is widely conserved. Genes Dev 1997;11:2510–2521.

184. Lie YS, Macdonald PM. Translational regulation of oskar mRNA occurs independent of the cap and poly(A) tail in Drosophila ovarian extracts. Development 1999;126:4989–4996.

185. Piecyk M, Wax S, Beck AR, et al. TIA-1 is a translational silencer that selectively regulates the expression of TNF-alpha. Embo J 2000;19:4154–4163.

186. Kedersha N, Cho MR, Li W, et al. Dynamic shuttling of TIA-1 accompanies the recruitment of mRNA to mammalian stress granules. J Cell Biol 2000;151:1257–1268.

187. Kedersha NL, Gupta M, Li W, Miller I, Anderson P. RNA-binding proteins TIA-1 and TIAR link the phosphorylation of eIF-2 alpha to the assembly of mammalian stress granules. J Cell Biol 1999;147:1431–1442.

188. Anderson P, Kedersha N. Visibly stressed: the role of eIF2, TIA-1, and stress granules in protein translation. Cell Stress Chaperones 2002;7:213–221.

189. Kedersha N, Anderson P. Stress granules: sites of mRNA triage that regulate mRNA stability and translatability. Biochem Soc Trans 2002;30:963–969.

190. Cok SJ, Acton SJ, Morrison AR. The proximal region of the 3′-untranslated region of cyclooxygenase-2 is recognized by a multimeric protein complex containing HuR, TIA-1, TIAR, and the heterogeneous nuclear ribonucleoprotein U. J Biol Chem 2003;278:36,157–36,162.

191. Denegri M, Chiodi I, Corioni M, Cobianchi F, Riva S, Biamonti G. Stress-induced nuclear bodies are sites of accumulation of pre-mRNA processing factors. Mol Biol Cell 2001;12: 3502–3514.

11

Death Receptors

Mechanisms, Biology, and Therapeutic Potential

Sharmila Shankar and Rakesh K. Srivastava

Summary

Apoptosis is a genetically controlled process that plays important roles in embryogenesis, meta-morphosis, cellular homeostasis, and as a defensive mechanism to remove infected, damaged or mutated cells. Although a number of stimuli trigger apoptosis, it is mainly mediated through at least three major pathways that are regulated by (1) the death receptors, (2) the mitochondria, and (3) the ER (endoplasmic reticulum). Under certain conditions, these pathways may crosstalk to enhance apoptosis. Death receptor pathways are involved in immune-mediated neutralization of activated or autoreactive lymphocytes, virus-infected cells, and tumor cells. Consequently, dysregulation of death receptor pathway has been implicated in the development of autoimmune diseases, immunodeficiency, and cancer. Increasing evidence indicates that mitochondrial and ER pathways of apoptosis play a critical role in cytokine receptor-mediated apoptosis. Considerable evidence has accrued about the effects of dysregulation of these pathways on drug resistance. Recent data indicate that BH3-only proteins act as mediators that link various upstream signals, including death receptors and DNA damage signaling, to the mitochondrial and the ER pathway. Evidence suggests that these proteins may function as integrators of damage signals, and may be the final decision point as to whether a cell lives or dies. This chapter discusses the molecular mechanisms of apoptotic pathways regulated by the death receptors, mitochondria and endoplasmic reticulum and their potential applications to cancer therapy.

Key Words: TRAIL; Fas; Fas ligand; death receptor; TNF; TNFR; mitochondria; apoptosis; IAP; caspase.

1. Introduction

Apoptosis, programmed cell death, is a genetically controlled process that plays important roles in embryogenesis, metamorphosis, cellular homeostasis, and as a defensive mechanism to remove infected, damaged or mutated cells. Apoptosis is characterized by loss of cellular contact with the matrix, chromatin condensation, cytoplasmic contraction, plasma membrane blebbing, and DNA fragmentation into oligosomes. Failure to undergo apoptosis has been implicated in tumor development and resistance to cancer therapy. Dysregulation of the apoptotic machinery plays a role in the pathogenesis of various diseases and molecules involved in cell death pathways are potential therapeutic targets in immunological, neurological, cancer, infectious, and inflammatory diseases. Strategies for overcoming resistance to apoptosis include: direct targeting of antiapoptotic molecules expressed in tumors, resensitization of previously resistant tumor cells by counteracting survival pathways, and inducing expression or activity of proapoptotic molecules. Molecular insights into the regulation of apoptosis and defects

From: *Apoptosis, Cell Signaling, and Human Diseases: Molecular Mechanisms, Volume 2*
Edited by R. Srivastava © Humana Press Inc., Totowa, NJ

in apoptosis signaling in tumor cells will help define resistance or sensitivity of tumor cells toward antitumor therapy and will provide new targets and approaches for rational chemotherapeutic intervention.

There are three major pathways of apoptosis: the death receptor pathway, the endoplasmic reticulum (ER) pathway, and the mitochondrial pathway. These pathways are linked in certain cell types and appear to enhance apoptosis in a cooperative manner.

2. Death Receptor Pathway

Molecules belonging to the tumor necrosis factor (TNF) and TNF receptor (TNF-R) superfamilies have explosively expanded through the era of genomics and bioinformatics. A growing appreciation of the molecular basis of signalling pathways transduced by TNF-R has provided a framework for better understanding the biology of this expanding family. Death receptors belong to the TNF superfamily and are involved in proliferation, cell metabolism, cytokine production, differentiation, and apoptosis *(1–3)*. They are primarily type I integral receptors with a conserved extracellular domain containing two to four cysteine-rich pseudo-repeats, a single transmembrane region and a conserved intracellular death domain about 80 amino acids in length that binds to adaptor proteins and initiates apoptosis *(4–6)*. Some TNF receptors exist as type II transmembrane proteins (lacking a signal peptide); for example, TRAIL-R3 is anchored by a covalently linked C-terminal glycolipid, and OPG and DcR3 lack a membrane-interacting domain and are secreted as soluble proteins. The members of the TNF family comprise of several genes which encode type II transmembrane proteins (Table 1). They are characterized by conserved C-terminal domain termed (TNF homology domain, THD) which is required for receptor binding. Most of the ligands within this family are synthesized as membrane-bound proteins; soluble form can be generated from alternatively splicing of the primary mRNA transcripts, or by specific proteolysis of the cell surface molecule. The proapoptotic members of TNF ligand family are TNF-α, CD95L/ FasL/ APO-1L, and TRAIL/APO-2L. They bind to their specific receptors and activate signaling pathways to exert their biological activity.

2.1. TNF/TNFR System

TNF-α plays important roles in regulating cell proliferation and differentiation, inflammatoty responses, and immune functions. TNF-α functions through two distinct surface receptors, a 55-kDa receptor 1 (TNF-R1) and a 75-kDa receptor 2 (TNF-R2). TNF-R1 plays the predominant role in induction of cellular responses by soluble TNF-α *(7)*. The binding of TNF-α to the TNFR1 (TNF-α receptor 1) leads to activation of prosurvival pathways followed by proapoptosis pathways. It has been shown that the prosurvival pathways are activated by a rapid recruitment of a protein complex, known as complex I, to the cytosolic portion of the activated TNFR1. The binding of TNF-α to TNF-R1 leads to the recruitment of TNF-R1-associated death domain (TRADD), and TRADD further recruits TNF-receptor-associated factor 2 (TRAF2) *(8)* and receptor-interacting protein (RIP) *(9,10)*. RIP interacts directly with TRADD via its death domain *(9)*. Formation of complex I, including TNFR1, TRADD, RIP, and TRAF2 proteins, leads to activation of the NF-κB pathway, as well as mitogen activated protein kinase (MAPK) pathways such as the extracellular signal-regulated kinases (ERK), Jun N-terminal kinase (JNK), and p38 pathways. The NF-κB pathway is considered to be a

Table 1

Ligand	Receptor	Functions
TNFα/LT/LTβ	LTBR/TNFR/TNFR2-RP/CD18/TNFR-RP/TNFCR/TNF-R-III	Immune response, inflammation, lymphoid neogenesis
TNF/TNFα/DIF	TNFR-1/CD120A/TNF-R1p55/TNFAR/TNF-R55, TNFR-II/CD120B/P75	Immune response, inflammation, proapoptotic, bone resorption, lymphoid neogenesis
TNFC/p33/LTβ	LTβR	Immune response, inflammation, lymphoid neogenesis
OX-40L/gp34/TXGP1	OX40/ACT35/TXGP1L	Cell development, deliver costimulatory signals to augment immune response
CD40LG/IMD3/HIGM1/ CD40L/TRAP/CD154/gp39	CD40/p50/Bp50	Cell growth and development, apoptosis
FasL/CD95L/APO-1L	Fas/CD95/APO-1/APT1	Immune response, proapoptotic
CD70/CD27L/CD27LG	CD27/Tp55/S152	Cell survival, apoptosis, deliver costimulatory signals to augment immune response
CD30LG	CD30/Ki-1/D1S166E	Deliver costimulatory signals to augment immune response
4-1BBL	4-1BB/CD137/ILA	Cell proliferation, antitumor immunity, deliver costimulatory signals to augment immune responses

(Continued)

221

Table 1 (*Continued*)

Ligand	Receptor	Functions
TRAIL/Apo-2L/TL2	TRAIL-R1/DR4/Apo2, TRAIL-R2/DR5/KILLER/TRICK2A, TRAIL-R3/DcR1/LIT/TRID, TRAIL-R4/DcR2/TRUNDD, OPG/OCIF/TR1	Immune response, proapoptotic, lumen formation of mammary epithelial cells
TRANCE/OPGL/RANKL/ ODF	RANK	Bone resorption, lymphoid neogenesis, osteoclastogenesis and bone remodelling
TWEAK	Fn14	Antiangiognic
APRIL	TACI, BCM/BCMA, and proteoglycans	T-B costimulation and B-cell homeostasis, autoimmunity
LIGHT/LTg/HVEM-L	ATAR/HVEM/TR2/LIGHTR/HVEA	Apoptosis, cell proliferation, antitumor immunity
TL1/VEGI	DR3/TRAMP/APO-3/TR3	Antiangiognic, anticancer
AITRL/TL6/gGITRL	AITR/GITR	Cell proliferation
BAFF/BLyS	BAFFR, TACI, and BCMA/BCM	Lymphocyte survival and activation

Abbr: TNF, tumor necrosis factor ; DR, death receptor; Fn14, fibroblast growth factor-inducible 14; LTβR, lymphotoxin beta receptor; HVEM, herpes virus entry mediator; DcR, decoy receptor; VEG1, vascular endothelial cell growth inhibitor; CD27L, CD27 ligand; LIGHT, LT-related inducible ligand that competes for glyco-protein D binding to herpesvirus entry mediator on T-cells; BAFF, B-cell activating factor belonging to the tumor necrosis factor family/ TACI, transmembrane activator and calcium-modulator and cyclophilin ligand interactor; BCMA, B-cell maturation antigen.

major prosurvival pathway. PAK4 (p21 activated kinase) may also facilitate TRADD binding to the TNF receptor and activate survival pathway *(11)*. Similarly, another protein TAK1 is recruited to TNF-α receptor 1 in a RIP-dependent manner and cooperates with MEKK3 leading to NFκB activation *(12)*.

The apoptotic pathways are activated by a second complex, known as complex II or the death inducing signaling complex (DISC), which includes TRADD, RIP, and FADD proteins *(13)*. The molecular mechanisms by which complex I transitions to complex II is not well understood, and it is not clear whether TNFR1 is even included in complex II *(13,14)*. The binding of the adaptor protein FADD is necessary for procaspase 8 cleavage, the beginning of the caspase cascade activation. FADD can recruit and activate the initiator caspases 8 and 10, leading to the activation of two different apoptosis pathways *(15)*. In extrinsic death receptor pathway, caspases 8 and 10 directly cleave and activate caspases 3 and 7. Activated caspases 3 and 7 then regulate the activities of target proteins that play important roles in physiology of cells *(16)*. Intrinsic (mitochondrial) pathway of apoptosis is mediated by cleavage and activation of the Bcl-2 family protein BID by activated caspase-8. The resulting cleaved BID translocates to mitochondria, where it interacts with other Bcl-2 family members to promote release of cytochrome *c* *(17)*. Released cytochrome *c* leads to activation of caspase-9, followed by cleavage and activation of caspase-3 and apoptosis *(18)*.

The mechanisms of TNF-α-mediated apoptosis in liver failure have been studied in several animal models *(19–23)*. Mutaions in TNF ligands and/or receptors have been described in many hereditary diseases. Most of the physiological activities of TNF-α including apoptosis, antiviral activity, and NF-κB activation are mediated by TNFR1. However, in contrast to Fas (CD95), TNFR1 is a strong inducer of NF-κB. NFκB activation through the TNFR1 is, at least in part, mediated by TRAF2. Signals from TRAF2 result in phosphorylation and degradation of IκB and subsequent translocation of NF-κB into the nucleus. The activation NF-κB appears to be a very strong antiapoptotic signal in the TNFR1 pathway *(24–27)*. In order for the cell to survive, full activation of the survival pathways triggered by complex I is critical. Activation of the NF-κB pathway leads to increased expression of several antiapoptotic proteins such as FLIP and c-IAP, which can bind to complex II. If the NF-κB is fully activated and the sufficient amounts of FLIP and c-IAP are presented in complex II, the activation of caspase-8 will be blocked and the cells will survive *(13)*. Because the NF-κB mediated survival depends on production of new proteins it is disrupted by drugs such as protein sysnthesis inhibitor cycloheximide (CHX). The combination of TNF-α and CHX therefore favors activation of the apoptosis pathway leading to cell death. It has been demonstrated that the activity of NF-κB prevents TNF-α-mediated apoptosis in hepatocytes during normal liver regeneration *(28)* and during embryonic liver development *(29,30)*.

The binding of TNFR1 receptor to TNF leads to activation of the IKK/NF-κB, JNK, and p38 mitogen-activated protein kinase (MAPK) signaling pathways *(7,31)*. Upon ligation, TNFR1 binds through its death domain to the adaptor molecule TNFR-associated death domain (TRADD). TRADD in turn binds two other adaptors: RIP1, which also contains a death domain and mediates activation of the IKK and p38 pathways, and TRAF2, which supports activation of IKK and JNK. TNFR1 activates the NF-κB pathway by inducing TRAF2-dependent formation of the signalosome complex, in which scaffold protein NEMO/IKKβ—and two kinases - IKKκ and IKKβ—are important catalytic components *(32)*. Activation of the IKK signalosome causes phosphorylation of the inhibitor of NF-κB (IκB), which in resting cells binds to NF-κB subunits and sequester them in the cytoplasm. Phosphorylation triggers proteosomal degradation of IκB, liberating NF-κB to translocate to the nucleus, where it promotes gene transcription. TRAF2 recruits two additional molecules, cellular inhibitor of apoptosis 1 (cIAP1) and cIAP2, to the TNFR1 complex, but the role of these cIAPs in the TNFR1 signaling complex remains unclear.

In general, TNF activates the IKK, JNK, and p38 kinase pathways, but not apoptotic caspase pathways. Activation of NF-κB by TNF causes induction of anti-apoptotic genes, including FLIP, cIAP1, cIAP2, XIAP, Bcl-X$_L$, and A1, which prevent apoptosis. However, under certain conditions, such as general inhibition of protein synthesis or specific blockade of NF-κB activation, TNF can stimulate a strong proapoptotic signal *(33,34)*. TNF induced apoptosis relies on the formation of a secondary intracellular signaling complex composing of TRADD, RIP1, TRAF2, FADD, and caspase-8 *(13)*. Thus, TNFR1-mediated-signal transduction includes a checkpoint, resulting in cell death (via complex II) in instances where the initial signal (via complex I, NF-κB) fails to be activated.

2.2. Fas/FasL System

FAS receptor (FAS, CD95) and FAS ligand (FAS-L, CD95-L) are complementary members of a particular apoptotic pathway that plays a major role in immune regulation *(1,35–38)*. Some of the functions of this system include selection of T-cell repertoire and deletion of self-reactive T-cells in the thymus; killing of target cells (e.g., virally-infected cells and tumor cells) by cytotoxic T-lymphocytes and natural killer (NK) cells; peripheral deletion of activated lymphocytes at the end of an immune response; killing of inflammatory/immune cells at the "immune privileged" sites (e.g., eyes and testes). Furthermore, the importance of CD95/CD95L pathway in the immune system is supported by its role in the development of autoimmunity in mice (mlr/lpr and gld/gld; lpr and gld are mutations in Fas and FasL, respectively) and humans (autoimmune lymphoproliferative syndrome) with mutation in the Fas or FasL gene *(39)*. We have shown that the chemotherapeutic drugs induce the expression of FasL in tumor cells *(40)*. The upregulated FasL would indicate an increase ability of tumor cells to induce apoptosis in TIL and in the normal tissues invaded *(41)*. However, it is understood that the Fas/FasL system, although essential for apoptosis, is only a contributing factor to the complex process of tumor invasion and antitumor defense.

Fas receptor is expressed at a single locus on chromosome 10 in human cells and chromosome 19 in mouse cells. Several distinct isoforms and splice variants of Fas receptor have been identified in humans. The mechanisms by which cells express several splice variants and thus alter the sensitivity to Fas mediated apoptosis appears to be an important process in vivo but remain to be characterized. In addition to modulating the transcription and alternative splicing of Fas receptor mRNA, the expression of cell surface Fas receptor can be regulated by altering intracellular trafficking of Fas *(42–48)*. Overexpression of soluble Fas receptor has been implicated with the progression of melanoma, and prostate and bladder cancers *(49–51)* and is found to antagonize Fas-receptor-mediated apoptosis in vitro *(52)*.

In contrast to Fas receptors, soluble Fas ligand is not generated by alternative splicing but is instead generated by post-translation modification of membrane bounds Fas ligand at the cell surface. Matrix metalloproteinases (serine proteases) cleave a wide range of extracellular substrates including FasL. Membrane-bound Fas ligand can be cleaved at a conserved cleavage site by matrix metalloproteinase-7 (MMP-7) into a less-active soluble form *(53,54)*. MMP-7 expression plays a role in tumor initiation *(55,56)* and invasion *(57)*, and overexpression of MMP-7 in tumor cells renders them resistant to Fas-mediated apoptosis *(58)*. Furthermore, overexpression of MMP-7 promotes mammary tumor initiation and progression in mice by selecting for tumor cells resistant to Fas-mediated apoptosis *(59)*. Cells expressing a noncleavable variant of Fas ligand or inhibition of MMPs increases the sensitivity to Fas-mediated apoptosis *(60)*. This appears to be a major mechanism by which MMP-7 regulates the sensitivity of cells to Fas-mediated apoptosis.

The single FasL gene is located on human chromosome 1 *(39,61,62)*. FasL is a homotrimeric type II membrane protein of 40 kDa that binds to Fas. In contrast to Fas, which is constitutively expressed in a large number of cell types, FasL expression is restricted. Resting T cells do not express FasL, but its expression can be induced upon activation. Ligation of Fas by FasL leads to receptor timerization through clustering of DD of Fas, setting up a chain of events that culminates in apoptosis. Aggregation of DD of Fas leads to the recruitment of a cytoplasmic adapter protein, FADD through its DD domain. The dead effector domain (DED) of FADD binds to an analogous domain of procaspase-8 (also known as FLICE) which is present in the cytoplasm in the zymogen or precursor form. The assembly of Fas, FADD and procaspase-8 forms a complex, the DISC, which is essential for the subsequent signaling pathway of apoptosis. Formation of DISC leads to the autoprocessing of procaspase-8 into active caspase-8. The activated caspase-8, in turn, activates downstream effector caspases (caspase-3, 6, and 7) which are also present in zymogen precursor form. The active effector caspases cleave a number of cytoplasmic and nuclear substrates, leading to apoptosis.

Primary murine and human hepatocytes have been demonstrated to be very sensitive for Fas (CD95)-mediated apoptosis *(63,64)*. Mice injected intraperitoneally with agonistic Fas (CD95) antibody rapidly developed acute liver failure and died within hours *(65)*. Several studies in humans showed that apoptosis of hepatocytes in Wilson's disease, toxic liver damage, and viral hepatitis is triggered through the Fas (CD95) pathway *(64,66)*. Because of these toxicity problems, the use of Fas to treat human cancers was abandoned.

2.3. TRAIL (APO-2L)/DR4/DR5 System

TNF-related apoptosis-inducing ligand (TRAIL), also designated as Apo2L, was originally identified on the basis of sequence homology to Fas ligand (FasL) and TNF *(2,3)*. TRAIL exerts its biological activity by binding to cell surface receptors on target cells. So far, four human receptors specific for TRAIL have been identified: (1) TRAIL-R1 (DR4) *(67,68)*; (2) TRAIL-R2 (DR5/TRICK2/KILLER) *(69–72)*; (3) TRAIL-R3 (DcR1/TRID/LIT) *(69,71,73)*; and (4) TRAIL-R4 (DcR2/TRUNDD) *(74–76)* (*see* Fig. 1). These receptors have high sequence homology in their extracellular domains. The cytoplasmic region of both DR4 and DR5 contain a death domain homologous to Fas and TNF-R1. A fifth receptor is known as osteoprotegerin, which is a soluble protein and act as a decoy receptor *(77)*. TRAIL can bind to all the receptors, but can induce biological activity only by binding to TRAIL-R1/DR4 and TRAIL-R2/DR5 because these

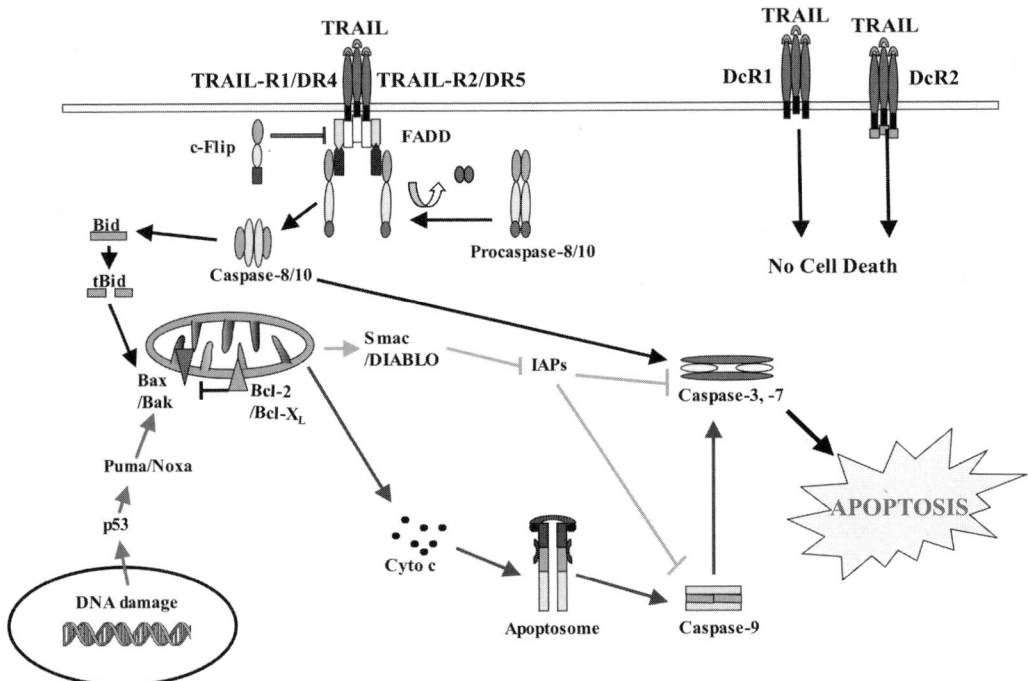

Fig. 1. Intracellular mechanism of TRAIL and its receptors. Trimerization of TRAIL receptors (TRAIL-R1/DR4 and TRAIL-R2/DR5) initiates recruitment of adaptor protein FADD. Decoy receptors DcR1 and DcR2 do not induce apoptosis because of complete or partial loss of cytoplasmic death domain, respectively. Apoptosis pathways activated by TRAIL and mitochondria are depicted. Ligation of death receptors by TRAIL results in the formation of DISC which in turn initiates the activation of caspase-8 and caspase-10 causes activation of effector caspases such as caspase-3 and caspase-7. Active caspase-8 cleaves Bid to truncated Bid which engages and disrupts mitochondria. Cytochrome *c*, along with Apaf-1 and dATP, forms apoptosomes which activate caspase-9. Active forms of caspase-8, -9, and -10 initiate a cascade of effector caspases such as caspase-3 and caspase-7. These caspases then cleave several substrates leading to apoptosis. Bcl-2 and Bcl-X_L antagonize the effects of Bax and Bak at the level of mitochondria. IAPs inhibit caspase-3, caspase-7 and caspase-9 activity. Smac/DIABLO induces apoptosis by inhibiting IAPs functions. In response to DNA damage, activated p53 can directly or indirectly modulate the expression of PUMA, Noxa and other related proteins that control mitochondrial membrane permeability and, therefore, can modulate the release of mitochondrial proteins during apoptosis.

receptors possess intact intracellular cytoplasmic domain *(68,72,78)*. On the other hand, TRAIL-R3/DcR1 and TRAIL-R4/DcR2/TRUNDD lack a functional cytoplasmic domain and do not induce apoptosis *(69,71,73–76)*. DcR1 and DcR2 serve as "decoys" that compete with TRAIL-R1/DR4 and TRAIL-R2/DR5 for binding to the TRAIL. As expected, overexpression of either DcR1 or DcR2 receptors confers protection against TRAIL-induced apoptosis *(67,71)*. The soluble receptor, osteoprotegerin (OPG) can bind to TRAIL with a low affinity at physiological temperature *(77)*.

Like other TNF family members, TRAIL forms a homotrimer and crosslinks three receptor molecules on the surface of target cells. TRAIL specifically kills transformed and cancer cells via binding with specific cell-surface death receptors (TRAIL-R1/DR4 and TRAIL-R2/DR5). Most normal cells appear to be resistant to TRAIL activation *(79–82)*, suggesting a higher activity of TRAIL with its receptors on tumor cells. Binding of TRAIL-R1/DR4 or TRAIL-R2/DR5 with TRAIL results in a caspase-activating signal leading to apoptosis *(6,83,84)*. TRAIL exhibits antitumor activity in a variety of human cancer cell lines in vitro and also in vivo with minimal or no toxicity to nonmalignant human cells *(80,82,85–92)*. Although the physiological role of TRAIL is not fully understood, TRAIL knock-out mice do not show gross abnormality, except for impaired tumor immunosurveillace and higher sensitivity to experimental autoimmune disease, such as collagen-induced arthritis (CIA), streptozotocin-induced diabetes, and experimental autoimmune encephalomyelitis (EAE), suggesting that the main physiological role of TRAIL is played in the immune system *(93–95)*. TRAIL is constitutively expressed in many tissues. It is not expressed in unstimulated peripheral blood T-cells and NK cells, or blood dendritic cells. TRAIL expression can be induced upon activation by interferon (IFN)-γ in peripheral blood T-cells, monocytes, and dendritic cells *(96,97)*. Interestingly, TRAIL is implicated in cytotoxicity mediated by monocytes *(97)*, in activation-induced T-cell death *(98,99)*, NK-mediated cytolysis *(100)*, and T-cell-mediated cytolysis of tumor cells *(98,99)*. Furthermore, TRAIL contributes to IFN-γ-dependent NK cell protection from tumor metastasis. Thus, TRAIL may play a role in the cytolytic effector function of monocytes, dendritic cells, activated T-lymphocytes, and NK cells. These physiological processes will be important in viral clearance, and suppression of autoimmunity and tumor immunity.

The extracellular domains of TNF receptor family members contain multiple disulfide bridges, which stabilize their configuration *(101,102)*. On the other hand, antiparallel β-sheet configuration of the receptor facilitates binding of the ligand *(103)*. Unlike other TNF family members, TRAIL contains an unpaired cysteine residue (Cys230) in its receptor-binding domain, and mutation of Cys230 to alanine or serine strongly affects its ability to kill target cells, suggesting that Cys residue is essential for its proapoptotic activity *(104)*. TRAIL homotrimer induces trimerization of TRAIL-R1/DR4 or TRAIL-R2/DR5, each at the interface between two of its subunits *(105,106)*. In recombinant TRAIL, Cys230 is engaged either in interchain disulfide bridge formation, resulting in poorly active TRAIL, or in the chelation of one zinc atom per TRAIL trimer in the active, proapoptotic form of TRAIL *(107,108)*. Thus, a Zn atom bound by cysteins in the trimeric ligand is essential for optimal biological activity.

Binding of death receptors (TRAIL-R1/DR4 and TRAIL-R2/DR5) with their cognate ligands result in receptor trimerization, and recruitment of adaptor protein called FADD *(38)*. FADD consists of two interaction domains: a DD and DED *(38,78,109)*. FADD binds to the receptor through interactions between DDs and to pro-caspase-8 or pro-caspase-10 through DED interactions to form a complex at the DISC. Recruitment of caspase-8 or caspase-10 through FADD leads to its auto-cleavage and activation. Active caspase-8 or caspase-10, in turn, activates effector caspases such as caspase-3 causing the cell to undergo apoptosis by cleaving of several protein substrates *(38,78,109,110)*. This ultimately results in an irreversible commitment of cells to undergo apoptosis. An endogenous inhibitor, c-FLIP, competes with caspase-8 or caspase-10 for binding to the DISC. The active caspase-8 or caspase-10 cleaves Bid which translocates to mitochondria to activate the intrinsic pathway, thus connecting the two caspase activation pathways and amplifying the death receptor apoptotic signal.

3. Endoplasmic Reticulum Pathway

In addition to nuclear DNA damage and death receptor activation, recent studies suggest that other organelles, including the ER, lysosomes, and the Golgi apparatus, are also major points of integration of proapoptotic signaling or damage sensing *(111)*. Each organelle possesses sensors that detect specific alterations, locally activates signal transduction pathways, and emits signals that ensure interorganellar crosstalk. The ER senses local stress through chaperones, Ca^{2+}-binding proteins, and Ca^{2+} release channels, which might transmit ER Ca^{2+} responses to mitochondria. The ER also contains several Bcl-2-binding proteins, and Bcl-2 has been reported to exert part of its cytoprotective effect within the ER. Upon membrane destabilization, lysosomes release cathepsins that are endowed with the capacity of triggering mitochondrial membrane permeabilization (MMP). The Golgi apparatus constitutes a privileged site for the generation of the proapoptotic mediator ganglioside GD3, facilitates local caspase-2 activation, and might serve as a storage organelle for latent death receptors. Intriguingly, most organelle-specific death responses finally lead to either MMP or caspase activation, both of which might function as central integrators of the death pathway, thereby streamlining lysosome-, Golgi-, or ER-elicited responses into a common pathway. Recent studies have shown the involvement of endoplasmic reticulum in TRAIL signaling *(112–117)*. TRAIL-resistant cells were resensitized to TRAIL by tunicamycin pretreatment, which increased cell surface expression of TRAIL-R1/DR4 and TRAIL-R2/DR5 *(112)*. Some tumor cells may become resistant to TRAIL through regulation of the death receptor cell surface transport and that resistance to TRAIL may be overcome by the glycosylation inhibitor/endoplasmic reticulum stress-inducing agent tunicamycin *(112)*.

4. Mitochondrial Pathway

Mitochondria play a crucial role in regulating cell death, which is mediated by outer membrane permeabilization in response to stress stimuli. There are two major pathways of apoptosis, cell-intrinsic (mitochondria-dependent) and cell-extrinsic (death receptor) pathways *(118–121)*. Activation of death receptor pathway also links cell-intrinsic pathway through Bid. Mitochondrial membrane permeabilization induces the release of cytochrome *c*, Smac/DIABLO, and AIF, which are regulated by proapoptotic and antiapoptotic proteins such as Bax/Bak and Bcl-2/X_L in caspase-dependent and caspase-independent apoptosis pathways *(122,123)*. The antiapoptotic members such as Bcl-2 or Bcl-X_L, inhibit the release of mitochondrial apoptogenic factors (AIF, Smac/DIABLO, Omi/HtrA2, cytochrome *c*, and endonuclease G) whereas the proapoptotic members (e.g., Bax, and Bak) trigger the release *(120,124–129)*. Overexpression of Bcl-2 and Bcl-X_L inhibits TRAIL-induced loss in $\Delta\psi_m$ and apoptosis in several human cancer cell lines *(81,120,130–133)*. Bid promotes apoptosis by activating Bax and Bak, and it might also inactivate pro-survival relatives *(134)*. Bid induces Bax and Bak to oligomerize and form pores in the mitochondrial membrane, but the oligomers do not contain Bid *(135)*, which seems to form homotrimers in the membrane *(136)*. Bid activation indicates the requirement for caspase amplification in the TRAIL-DR pathway by a mechanism reported to take place through translocation of activated Bid to mitochondria, oligomerization of Bax and Bak, facilitating the release of cytochrome *c* release and/or other proteins, and activation of caspase-9 and then caspase-3 and/or caspase-7 leading to apoptosis *(17,78,120)*. Translocation of Bid to mitochondria results in activation of Bax and Bak, providing a mechanism for crosstalk between the DRs and the intrinsic pathway *(17,120,137)*. Synergistic activation of caspase-3 by TRAIL and drugs that damage mitochondrial pathway appears to be one of the mechanisms of inducing apoptosis in cancer cells.

Proapoptotic Bax and Bak are widely distributed, and function mainly at the mitochondria *(120,135,138)*. Inactivation of Bax affect apoptosis only slightly and disruption of Bak has no discernible effect, but inactivation of both genes completely abrogated apoptosis *(120,135,138)*. It appears that the presence of either Bax or Bak is essential for apoptosis in many cell types. Bax is a cytosolic monomer in healthy cells, but it changes conformation during apoptosis, integrates into the outer mitochondrial membrane and oligomerizes *(139–141)*. In healthy cells, Bak is an oligomeric integral mitochondrial membrane protein, but it changes conformation during apoptosis and forms larger aggregates *(135,139,140,142)*. Bax and Bak oligomers are believed to provoke or contribute to the permeabilization of the outer mitochondrial membrane, allowing efflux of apoptogenic proteins *(143)*. Bax and Bak are counteracted by the antiapoptotic family members Bcl2 or Bcl-X_L *(144)*. The "BH-3 only proteins" of the Bcl-2 family (Bid, Bim, PUMA and Noxa) interact with proapoptotic Bcl-2 family members to augment their activity. TRAIL induces activation of Bax and Bak, suggesting that both agents can amplify apoptotic signals at the level of mitochondria. The simultaneous activation of Bax and Bak may have synergistic effects on the activation caspase-3 and apoptosis.

Mitochondrial dysfunction is mediated in two ways. The first is by increased calcium in mitochondria derived from ER; this calcium increase is regulated by Bcl-2 and Bax through the ER-mitochondria connection and the unfolded protein response in the ER. The second is by the lysosomal enzyme cathepsin, which activates Bid through lysosome-mitochondria cross-signaling. The genomic responses in intracellular organelles after DNA damage are controlled and amplified in the cross-signaling via mitochondria; such signals induce apoptosis, autophagy, and other cell death pathways.

5. Role of p53 in Apoptosis

The *p53* gene was first identified in 1979 as a cellular protein that bound to the simian virus (SV40) large T-antigen and accumulated in the nuclei of cancer cells *(145,146)*. The missense mutations in p53 (TP53) gene is the key to understanding the pathological activity of p53. The ability of p53 to form tetramers allows this protein to behave in a dominant negative fashion, whereas the allele-producing mutant p53 suppresses the activity of wild-type p53. Oncogenic human DNA viruses have ability to inactivate p53 functions (e.g., enhanced ubiquitin-dependent proteolysis of p53 by the E6 protein of human papilloma virus (HPV) types 16 and 18) *(147)*.

TP53 was widely recognized as a tumor suppressor gene, mutated or lost in about 50% of all human cancers *(148,149)*. Mice deficient in TP53 are susceptible to spontaneous tumorigenesis and germ-line TP53 mutations occur in individuals with cancer-prone Li-Fraumeni syndrome *(150,151)*. P53 has multiple functions. It acts as a transcription factor, facilitates DNA repair, and is involved in the regulation of cell cycle, apoptosis, development, differentiation, chromosomal segregation, and cellular senescence *(152–156)*. Furthermore, p53 is induced in response to a wide variety of stresses, including DNA damage, hypoxia, and oncogene. Wild-type p53 can react to environmental insult by driving cells into apoptosis. p53 transcriptional activity has been shown to be regulated by a number of posttranslational modifications that affect its confirmation, including phosphorylation, ubiquitination, acetylation, neddylation, and summoylation *(157)*. There is evidence that specific modifications of p53 can affect its binding affinity for different promoters and this may determine whether p53 induces cell-cycle arrest or apoptosis *(158)*.

p53 is involved in both the extrinsic and the intrinsic pathways of apoptosis by initiating apoptosis through mitochondrial depolarization and sensitizing cells to inducers of apoptosis. p53 can induce the apoptotic pathway by transcriptionally activating several genes such as p21 WAF1/CIP1, mdm2, GADD45, cyclin G, Bax, IGF-BP3, CD95/Fas, TRAIL-R2/DR5, Noxa, Bid, Fas/Apo-1, Siva, PTEN, P53AIP, PUMA, APAF-1 PRSS25/HTRA2, and scotin *(158–162)*. p53 has been examined as a marker of proliferation and tumor aggressiveness with respect to prognosis and response to chemotherapy, and only mutated p53 has been associated with a drug-resistant phenotype *(159–162)*. p53 can also promote apoptosis through transcription-independent mechanisms *(163)*. P53 can interact with those proteins that control mitochondrial membrane permeability and therefore can modulate the release of mitochondrial proteins during apoptosis. p53 can also cause permeabilization of the outer mitochondrial membrane directly by forming complexes with, and inhibiting, the protective Bcl-X$_L$ and Bcl-2 proteins, leading to the release of cytochrome *c (164–166)*. These observations suggest that p53 is a cellular gatekeeper.

The authors, along with others, have shown that TRAIL induces apoptosis independently of p53. In some cell types, DR5 expression is regulated in a p53-dependent manner, while in other cells an increase in DR5 expression appears to be p53 independent *(82,89,90)*. By comparison, it is not clear whether p53 regulates the expression of DR4.

6. Caspase Cascade in Apoptosis

Caspases are a group of cysteine proteases which cleave protein substrates at the aspartate cleavage site. The caspase gene family consists of at least 14 mammalian members *(167)*. They are initially expressed as single-chain zymogens, which upon apoptotic signaling are activated by proteolytic processing, either by autoactivation, transactivation, or by cleavage by other caspases *(168,169)*. Once activated, they proteolytically cleave a multitude of cellular proteins, leading to apoptosis. Therefore, the caspase activation is a key regulatory point in the commitment of the cell to apoptosis.

Caspase recruitment and oligomerization mediated by adaptor proteins constitute a basic mechanism of caspase activation. The complex phenotypes of the caspase knockout mice indicate that multiple mechanisms of caspase activation operate in parallel and that death signal transduction pathways are both cell-type and stimulus specific. The BH3-domain-containing proapototic members of Bcl-2 family may be one of the critical links between the initial death signals and the central machinery of apoptosis. TRAIL-mediated signaling involves both death receptor and mitochondrial pathways of apoptosis *(78)*. After initial activation of caspase-8 or caspase-10 by TRAIL-DISC, divergence of signal occurs in two directions: (1) direct activation of caspase-3 without the involvement of mitochondria; and (2) formation of apoptosomes (mitochondrial proteins, dATP and Apaf-1) which lead to activation of caspase-9 *(17,78,137,143)*.

These two pathways appear to converge on caspase-3 *(170)*. In a mechanism not entirely understood, cytochrome *c* and dATP/ATP act as cofactors and stimulate Apaf-1 self-oligomerization. Once activated, caspase-9 can activate effector caspases-3 and -7 that finally dismantle the cell *(119,171)*. The decline in mitochondrial membrane potential ($\Delta\psi_m$) can be blocked by caspase-8 inhibitor, but not by caspase-9 inhibitor *(78,120,172)*. Thus, caspase-8 links the apoptotic signal from the activated TRAIL-DR to mitochondria leading to dissipation of $\Delta\psi_m$ and directly to downstream apoptosis-executing caspases. Cleavage of caspase-3 occurs following activation of either pathway. Knowledge of the relative contributions of these two alternative pathways in TRAIL-induced apoptosis is yet incomplete.

The mechanisms controlling the release of mitochondrial proteins are currently under intensive investigation. They may include opening of a mitochondrial permeability transition pore (PTP), the presence of a specific channel for cytochrome *c* in the outer mitochondrial membrane, or mitochondrial swelling and rupture of the outer membrane without a loss in membrane potential *(168)*. Involvement of the PTP is supported by many reports that the $\Delta\psi_m$ collapses before activation of caspases and apoptosis *(123,173)*. The outer mitochondrial membrane becomes permeable to apoptogenic factors such as cytochrome *c*, Smac/DIABLO (a direct IAP-binding protein with low pI) *(126,129)* and AIF (apoptosis-inducing factor) *(125)*. Furthermore, the proapoptotic activity of Smac/DIABLO is probably independent of binding to IAPs. This can mediate the cleavage of DNA during caspase-independent cell death. TRAIL cleaves several substrates including DNA repair enzyme poly(ADP-ribose) polymerase (PARP) *(78)*.

Treatment of cells with TRAIL results in caspase-8 activation, followed by BID (Bcl-2 inhibitory, BH3-Domain-containing protein) cleavage at its amino terminus *(17,78,120)*. The truncated BID (tBID) translocates and gets inserted into the mitochondrial membrane *(137)*. The presence of tBID in the mitochondria appears to be a stress signal and triggers Bax and/or Bak oligomerization leading to the release of mitochondrial proteins *(168,174)*. Furthermore, BID-deficient mice are resistant to Fas-induced hepatocellular apoptosis *(175)*. Immunodepletion of tBID from subcellular fractions argues that tBID is required for cytochrome *c* release from the mitochondria. Treatment of cells with TRAIL resulted in BID cleavage, suggesting that BID is a substrate for caspase-8 *(78,172)*.

Another form of regulation of caspase activation has recently come to light by the finding that Akt, a serine/threonine kinase involved in some cell-survival pathways, and p21-Ras, an activator of Akt, induce phosphorylation of procaspase-9 *(176)*. In cytosolic extracts prepared from cells expressing either active Ras or Akt, the cytochrome *c*-dependent activation of caspase-9 is abrogated; suggesting that phosphorylation of procaspase-9 inhibits its processing and activation. Although it is unclear how phosphorylation inhibits its processing, it is suggested that it may facilitate enzyme dimerization via an allosteric mechanism.

7. Endogenous Inhibitors of Apoptosis

The inhibitors of apoptosis Proteins (IAPs) have emerged as key regulators of apoptosis by virtue of their ability to directly bind and inhibit distinct caspases. The mammalian IAPs, XIAP (MIHA, hILP), c-IAP1 (MIHB, HIAP2), c-IAP2 (MIHC, HIAP1), NAIP and survivin, can bind to and inhibit caspases. IAPs contain one of three copies of the characteristic baculovirus inhibitor of apoptosis repeat (BIR) domains that are essential for the caspase-inhibitory function of IAP. The BIR domains, in some cases together with the intervening linker regions, directly bind and inhibit caspases. Some of the IAPs also have a C-terminal RING finger domain. Human IAPs—such as XIAP, c-IAP1, and c-IAP2—inhibit both the initiator caspase-9 and the effector caspases-3, and -7 *(177–180)*. In vitro kinetic studies revealed that XIAP is the most potent caspase inhibitor in the IAP family *(177–179)*. New structural data also support the concept that XIAP might inhibit initiator and effector caspases by different means. Thus, a limited number of IAPs can inhibit caspases bound to the apoptosome and consequently might prevent an amplification loop outside the complex. Because TRAIL activates primarily caspases-8, -9, and -3 (and possibly caspase-7), IAP can inhibit TRAIL-induced apoptosis by modulating caspase activity *(78,181)*. Conversely, downregulation of the XIAP or survivin enhances cell death by TRAIL and increases sensitivity against some chemotherapeutic agents in cancer cells *(182)*. These studies suggest that IAPs may be a potential target to increase therapeutic sensitivity.

The Bcl-2 proteins can block only the mitochondrial branch of apoptosis by preventing the release of cytochrome *c*, whereas IAPs block both the mitochondrial- and death receptor-mediated pathways of apoptosis by directly binding to and inhibiting both the initiator and effector caspases. The discovery of cellular proteins that interact with XIAP and modulate its antiapoptotic activity points to the critical role of XIAP in cellular homeostatsis. Overexpression of IAP family proteins inhibits apoptosis induced by Bax and other proapoptotic Bcl-2 family proteins, which are known for their ability to target mitochondria and induce cytochrome *c* release *(179,183,184)*. The IAPs however, do not interfere with Bax-mediated release of cytochrome *c* from mitochondria in vitro, as well as in intact cells *(184)*, an observation that is consistent with other data indicating that the human IAPs block caspase activation and apoptosis downstream of Bax, Bik, Bak, and cytochrome *c* *(177–179,185,186)*. The failure of IAPs to prevent cell death stimuli from triggering cytochrome *c* release has important implications for determining whether cell death will be prevented in the long term or will be merely delayed. These diverse apoptotic inhibitors from mammals, insects, and their associated viral pathogens are providing important insights into the regulatory mechanisms of TRAIL-induced caspase activation and apoptosis.

IAPs are counteracted by mitochondrial proteins Smac/DIABLO, Omi/HtrA2, and GSPT1/eRF3 in mammals *(126–129,187–190)*. Despite the overall sequence differences, these IAP antagonists share a conserved N-terminal IAP-binding motif (IBM). This motif is necessary and sufficient for counteracting IAP's inhibition of caspases. The small subunit of active caspase-9 also binds to XIAP through the same IBM produced by autoprocessing *(189)*. The mechanism allows these IAP antagonists to compete with caspases for IAP-binding and consequently relieves caspases and promotes apoptosis.

Some IAPs also regulate apoptosis through the ubiquitin-protein ligase activity (E3) of their RING domain. These proteins are capable of targeting the polyubiquitinylation of IAP-binding proteins such as caspases and IAP antagonists. An alternative spliced form of Smac, Smac 3, promotes XIAP ubiquitinylation, and degradation. Smac/DIABLO selectively causes the rapid degradation of c-IAP1 and c-IAP2. Thus, IAP antagonists accelerate the disposal of IAPs in addition to releasing captive caspases from

IAPs. Moreover, the mammalian IAP-binding protein Omi/HtrA2 can directly degrade IAP molecules through its serine protease activity. Smac/DIABLO can enhance the therapeutic potential of TRAIL *(120,191)*. These studies suggest that Smac/DIABLO may serve as a key molecule in vivo to selectively reduce the protein level of IAPs through the ubiquitin/proteasomal pathway.

Therapeutic modulation of IAPs could target a key control point in deciding cell fate and TRAIL resistance. There are several outstanding issues that need to be addressed experimentally and will probably provide vital clues about the regulation of IAPs. Understanding the biology of IAPs not only provides the intellectual satisfaction of untangling the complex regulatory networks that control life and death, but it might also supply us with powerful therapeutic approaches for the treatment of several human diseases.

8. Sensitization of TRAIL-Resistant Cancer Cells

Resistance of tumors to cytotoxic therapy may result from disrupted apoptosis programs and remains a major obstacle in cancer treatment. Targeting death receptor to trigger apoptosis in tumor cells is an attractive concept for cancer therapy. TRAIL appears to be a relatively safe and promising death ligand for clinical application. However, the majority of breast, prostate, ovarian and lung carcinoma, multiple myeloma, and leukemia cells are resistant to apoptosis induced by TRAIL. This resistance may be caused by deregulated expression of antiapoptotic molecules. These data suggest that the use of TRAIL alone may not be viable option to treat these cancers. Fortunately, we can sensitize TRAIL-resistant cancer cells by various means outlined below.

8.1. Histone Deacetylase Inhibitors

Histones are part of the core proteins of nucleosomes. The recruitment of histone acetyltransferase (HATs) and histone deacetylases (HDACs) plays an important role in proliferation, differentiation and apoptosis *(192,193)*. Altered HAT or HDAC activity is associated with the development of cancer by changing the expression of several genes *(194,195)*. Hyperacetylation of the N-terminal tails of histones H3 and H4 correlates with gene activation, whereas deacetylation mediates transcriptional repression *(196)*. Treatment of malignant cells with HDAC inhibitors regulates only a small number (1–2%) of genes, as examined by DNA microarray studies *(197)*. HDAC1 interacts directly with other transcription repressors, including all three of the pocket proteins, Rb, p107 and p130, and YY1. HDAC1 causes transcription repression by locally deacetylating histones, leading to a compact nucleosomal structure that prevents transcription factors from accessing DNA to promote transcription. Inhibition of HDAC activity results in p53-independent transcriptional activation of p21 (a cyclin-dependent kinase [CDK] inhibitor). Furthermore, HDAC1 knockout mice were embryonic lethal, possibly due to a proliferative defect upon unrestricted expressions of the cell cycle inhibitors p21 and p27 *(198)*. The authors, along with others, have shown that inhibition of HDAC activity induces apoptosis in various types of cancer *(87,199–202)*.

HDACs play important roles in modulating chromatin structure. The HDACs can be divided into three classes *(203)*. Class I contains HDAC 1, 2, 3, and 8, which have a single deacetylase domain at the N termini and diversified C-terminal regions. Class II includes HDAC 4, 5, 6, 7, 9, and 10 with a deacetylase domain at a more C-terminal position. In addition, HDAC 6 contains a second N-terminal deacetylase domain, which can function independently of its C-terminal counterpart. Recently, mammalian homologs of the yeast Sir2 protein have been identified forming a third class of deacetylases. All of these HDACs apparently exist in the cell as subunits of multiprotein complexes. Class II HDACs have been shown to translocate from the cytoplasm to the nucleus in response to external stimuli, whereas class I HDACs are constitutively nuclear and play important roles in dynamic gene regulation *(204)*. Recently, a number of HDAC inhibitors have been shown to induce growth arrest, differentiation, and/or apoptosis in caner cells *(87,200,205–207)*, and inhibit tumor growth in various xenograft models *(208–215)*. Therefore, HDAC inhibitors are considered as candidate drugs in cancer therapy *(216,217)*.

The authors, along with others, have shown that several HDAC inhibitors can enhance the apoptosis-inducing potential of TRAIL and sensitize TRAIL-resistant breast, prostate, and lung cancer cells, and malignant mesothelioma, leukemia and myeloma cells *(87,200,218–224)*. The sensitization of TRAIL resistant cells appears to be due to downregulation of the anti-apoptotic protein Bcl-X$_L$, upregulation of proapoptotic genes Bax, Bak, TRAIL, Fas, FasL, DR4 ,and DR5, and activation of caspases. HDAC inhibitors upregulate proapoptotic genes only in cancer cells but not in normal cells *(225,226)*. Furthermore, the sensitization of cancer cells to HDAC inhibitors appears to be p53 independent. These results show that sensitivity to HDAC inhibitors in cancer cells is a property of the fully transformed phenotype and depends on activation of a specific death pathway. Because HDAC inhibitors sensitize TRAIL-resistant cancer cells to undergo apoptosis by TRAIL, they appear to be promising candidates for combination chemotherapy.

8.2. Interferon-γ

IFN-γ often modulates the anticancer activities of TNF family members including TRAIL *(227–233)*. However, little is known about the mechanism. IFN-γ pretreatment augmented TRAIL-induced apoptosis in several cancer cell lines. IFN-γ dramatically increased the protein levels of interferon regulatory factor (IRF)-1, but not TRAIL receptors (DR4 and DR5) and proapoptotic (FADD and Bax) and antiapoptotic factors (Bcl-2, Bcl-X$_L$, cIAP-1, cIAP-2, and XIAP). Overexpression of interferon regulatory factor (IRF)-1 by an adenoviral vector AdIRF-1 minimally increased apoptosis, but significantly enhanced TRAIL-induced apoptosis. Specific repression of IRF-1 expression by antisense oligonucleotide abolished enhancer activity of IFN-γ for TRAIL-induced apoptosis. In another study, IFN-γ induced caspase-8 expression through Stat1/IRF1 pathway *(232)*, suggesting that IFN-γ might be an effective strategy to sensitize various resistant tumor cells with deficient caspase-8 expression for chemotherapy or death receptor-induced apoptosis.

8.3. Retinoids

Retinoic acid (RA) is involved in vertebrate morphogenesis, growth, cellular differentiation, and tissue homeostasis. The use of in vitro systems initially led to the identification of nuclear receptor RXR/RAR heterodimers as possible transducers of the RA signal. Retinoids are natural and synthetic analogues of retinoic acid, an active metabolite of vitamin A, and are specific modulators of cell proliferation, differentiation, and morphogenesis in vertebrates. The term "retinoid" is first defined by the chemical structure of vitamin A but is now recognized as the biological term for the ligands of two classes of nuclear receptors that mediate the biological activities of retinoic acid, that is, retinoic acid receptors (RARs) and retinoid X receptors (RXRs). Despite the beneficial activities of retinoids, the scope of retinoid therapy is still limited owing to high toxicity, and only a few retinoids have been clinically used until recently.

Among various nuclear receptor ligands, retinoids are unique. There are two classes of retinoid nuclear receptors, RARs and RXRs, both having three subtypes (α, β, and γ). Endogenous ligands for RARs and RXRs were identified as ATRA and 9-*cis*-retinoic acid (9cRA), respectively, while 9cRA can bind to RARs with as high affinity as to RXRs. Thus, 9cRA is a pan-agonist for all six retinoid nuclear receptors. Most of the retinoidal activities are elicited by the binding of retinoids to the RAR site of RXR-RAR heterodimers. RXRs are the silent partners of RARs, and RXR agonists alone cannot activate the RXR-RAR heterodimers, though RXR agonists allosterically increase the potencies of RAR ligands (retinoid synergism). Besides so-called retinoidal activities, RXRs have significant roles in nuclear receptor actions by heterodimerizing with various nuclear receptors. The heterodimeric partners of RXRs include endocrine nuclear receptors, such as RARs, vitamin D_3 receptor (VDR), and thyroid hormone receptors (TRs), and some orphan nuclear receptors, such as peroxisome proliferator-activated receptors (PPARs), liver X receptors (LXRs), and farnesoid X receptors (FXRs). Nuclear receptors of the latter class have proven to be key regulators of carbohydrate, lipid, and cholesterol metabolism. Because RXR agonists alone can activate these heterodimers, RXR agonists act just like the partner receptor's agonists and have potential as drugs for these metabolic diseases. With this background, compounds with various selectivity among RARs and RXRs have been developed in order to separate the pleiotropic retinoidal activities. In this paper, we discuss the structural evolution of RAR- and RXR-specific ligands and their potential utility in clinical applications.

The therapeutic and preventive activities of retinoids in cancer result partly from their ability to modulate the growth, differentiation, and survival or apoptosis of cancer cells *(234)*. Vitamin A deficiency in experimental animals has been associated with a higher incidence of cancer and with increased susceptibility to chemical carcinogens. This is in agreement with the epidemiological studies indicating that individuals with a lower dietary vitamin A intake are at a higher risk to develop cancer. At the molecular level, aberrant expression and function of nuclear retinoid receptors have been found in various types of cancer including premalignant lesions. Thus, aberrations in retinoid signaling are early events in carcinogenesis. Retinoids at pharmacological doses exhibit a variety of effects associated with cancer prevention. They suppress transformation of cells in vitro, inhibit carcinogenesis in various organs in animal models, reduce premalignant human epithelial lesions, and prevent second primary tumors following curative therapy for epithelial malignancies such as head and neck, lung, liver, and breast cancer.

The majority of ovarian cancer cells are resistant to apoptosis induced by TRAIL. Subtoxic concentrations of the semisynthetic retinoid *N*-(4-hydroxyphenyl)retinamide (4HPR) enhanced TRAIL-mediated apoptosis in ovarian cancer cell lines but not in immortalized nontumorigenic ovarian epithelial cells *(235)*. The enhancement of TRAIL-mediated apoptosis by 4HPR was not caused by changes in the levels of proteins known to modulate TRAIL sensitivity. The combination of 4HPR and TRAIL enhanced cleavage of multiple caspases in the death receptor pathway, including the two initiator caspases, caspase-8 and caspase-9. The 4HPR and TRAIL combination leads to mitochondrial permeability transition, significant increase in cytochrome *c* release, and increased caspase-9 activation. Caspase-9 may further activate caspase-8, generating an amplification loop. Stable overexpression of Bcl-X$_L$ abrogates the interaction between 4HPR and TRAIL at the mitochondrial level by blocking cytochrome *c* release. The synergistic interaction between TRAIL and a synthetic retinoid (CD437) on apoptosis of human lung and prostate cancer cells was observed due to upregulation of death receptors and activation of caspase-3 *(236,237)*. In NB4 acute promyelocytic leukemia cells, retinoids selective for RARα induced an autoregulatory circuitry of survival programs followed by expression of TRAIL *(238)*. In a paracrine mode of action, TRAIL killed NB4 as well as heterologous and retinoic-acid-resistant cells. In the leukemic blasts of freshly diagnosed acute promyelocytic leukemia patients, retinoic acid-induced expression of TRAIL most likely caused blast apoptosis. Thus, induction of TRAIL-mediated death signaling appears to contribute to the therapeutic value of retinoids.

8.4. Chemotherapeutic Drugs

Although TRAIL is capable of inducing apoptosis in tumor cells of diverse origin, recent studies have shown that majority of tumor cells are resistant to TRAIL *(6,239)*, suggesting that cancer treatment by TRAIL may not be useful option. We and others have shown that chemotherapeutic drugs (e.g., cisplatin, carboplatin, etoposide, camptothecin, paclitaxel, vincristine, and vinblastine, doxorubicin, gemcitabine and 5-fluorouracil) can sensitize TRAIL-resistant breast, prostate, colon, bladder, and pancreatic cancer cells to TRAIL *(82,85,240–244)* in vitro and in vivo, suggesting a possibility of combining TRAIL with other commonly used anticancer drugs. We have recently shown that chemotherapeutic drugs not only induce death receptors in vitro, but also in tumor xenografts in nude mice *(82,244)*. Some of the mechanisms by which chemotherapeutic drugs sensitize TRAIL-resistant cells are through upregulation of DR4 and/or DR5, and activation of caspases. We have recently shown that the chemotherapeutic drugs synergize with TRAIL in reducing tumor growth through apoptosis and enhancing survival of xenografted nude mice *(82,244)*. Thus, a combination of chemotherapeutic drug and TRAIL is capable of killing both TRAIL-sensitive and -resistant cancer cells.

8.5. Irradiation

Radiation therapy has been shown to substantially reduce the risk of local recurrence after surgery. It has also been used in combination therapy with chemotherapeutic drugs. It has been shown that the ionizing radiation enhances the therapeutic potential of TRAIL in TRAIL-sensitive cells, and sensitizes TRAIL-resistant cancer cells *(86,89,90,245–251)*. Ionizing radiation induces expression of TRAIL-R2/DR5 which may enhance the apoptosis inducing potential of TRAIL. Furthermore, we have shown that sequential treatment of mice with irradiation followed by TRAIL resulted in enhanced caspase-3 activity and apoptosis in tumor cells, which was accompanied by a regression of tumor growth and an enhancement of survival of xenografted mice *(89,90)*. The synergistic interaction between irradiation and TRAIL on apoptosis was due to upregulation of DR5, Bax and Bak, activation of caspase-8, caspase-3 and caspase-7, and downregulation of Bcl-2 and Mcl-1 *(89,90,247)*. Activation of these caspases causes proteolytic cleavage of cellular proteins such as poly(ADP-ribose) polymerase *(252)*. TRAIL-induced apoptosis required caspase-8, whereas it was not essential for irradiation-induced apoptosis. These data suggest that irradiation may mediate its apoptotic effects not only through increasing expression of DR5 receptors, but also through caspase cascade by directly engaging mitochondria.

The role of p53 in radiation-induced apoptosis is well established. Cells harboring both wild-type and mutant p53 *(89,90)* can be killed by irradiation and TRAIL, suggesting the involvement of p53-independent pathways in regulation of apoptosis. The upregulation of DR5 mRNA expression by p53 has been demonstrated, and the inhibition of p53 by dominant negative or siRNA approach has been shown to block irradiation-induced DR5 expression *(86,89,90)*. In certain cell types, irradiation induces apoptosis through p53 directed *de novo* synthesis of the death agonist Bax *(252)*, indicating the involvement of mitochondrial caspase pathway. p53-independent regulation of DR5 gene has also been demonstrated *(89,90)*. The increase in the induction of apoptosis by combined treatment with irradiation and TRAIL occurred through a change of the intracellular redox state independent of Tp53 status in human carcinoma cell lines *(249)*. Thus, the combination of irradiation and TRAIL can be used to target cancer cells harboring both wild-type and mutant p53. Other transcription factors such as NF-κB, AP-1 and SP-1 may also regulate the expression of DR5 gene *(253,254)*.

The synergistic interaction between TRAIL and irradiation could be the basis for developing novel strategies for pharmacological intervention, with potential for clinical application. As discussed previously, the sequential treatment of cancer patients with irradiation followed by TRAIL could be a better option than single agent alone.

8.6. Chemopreventive Drugs

Chemoprevention of carcinogenesis by using nontoxic chemical substances is regarded as a promising alternative strategy to cancer therapy. Recently, many naturally occurring substances have been shown to prevent multistep process of carcinogenesis. Epidemiological studies have provided convincing evidence that diet, genetic factors, and lifestyle are major causes of cancer *(255)*. Nutritional supplements such as soybean, garlic, grapes, turmeric, and green tea have been shown to possess chemopreventive properties as shown in various tumor models.

8.6.1. Resveratrol

The polyphenolic compound resveratrol (*trans*-3,5,4′-trihydroxystilbene) is a naturally occurring phytochemical found in many plant species, including grapes, peanuts, and various herbs *(256)*. Resveratrol has been shown to have antiinflammatory, antioxidant, antitumor, neuroprotective, cardioprotective, chemopreventive, and immunomodulatory activities *(256–261)*. It also has activity in the regulation of multiple cellular events associated with carcinogenesis *(256–258)*. Resveratrol has also been examined in several model systems for its potential effect against cancer *(262–264)*. Its anticancer effects include its role as a chemopreventive agent, its ability to inhibit cell proliferation, its direct effect in cytotoxicity by induction of apoptosis, and on its potential therapeutic effect in preclinical studies *(265,266)*. Treatment of androgen-sensitive prostate cancer cells (LNCaP) with resveratrol caused downregulation of prostate-specific antigen and p65/NFκB, and induced expressions of p53, p21$^{WAF1/ CIP1}$, p300/CBP ,and apoptotic protease-activating factor (APAF)-1 *(267)*. In addition, resveratrol has been shown to exert sensitization effects on cancer cells that will result in a synergistic cytotoxic activity when resveratrol is used in combination with cytotoxic drugs in drug-resistant tumor cells *(259)*. Resveratrol is a potent sensitizer of tumor cells for TRAIL-induced apoptosis through p53-independent induction of p21 and p21-mediated cell-cycle arrest associated with survivin depletion. Clearly, the studies with resveratrol provide support for its use in human cancer chemoprevention and combination with chemotherapeutic drugs or cytotoxic factors in the treatment of drug or TRAIL refractory tumor cells.

The potential of resveratrol for anticancer therapy may largely reside in its ability to sensitize tumor cells to apoptosis. Resveratrol sensitizes cancer cells to TRAIL, CD95, and several anticancer drugs *(258)*. Because resveratrol significantly potentiates the cytotoxic activity of TRAIL even at relatively low TRAIL concentrations, resveratrol may be used in TRAIL-based therapies to reduce the doses of TRAIL required for inhibition of tumor growth. Also, resveratrol may be particularly useful to sensitize resistant tumor cells to TRAIL-induced apoptosis, without affecting primary nonmalignant human cells (e.g., fibroblasts). It may serve as a novel therapeutic to target survivin expression in cancers through p21-mediated cell-cycle arrest. Resveratrol sensitizes a variety of human cancer cell lines to TRAIL-induced apoptosis independent of wild-type *p53* status through cell-cycle arrest-mediated survivin depletion *(258,268)*. The potential clinical implications of these studies will also depend on whether or not resveratrol can be given safely to humans at doses high enough to achieve pharmacologically active levels.

8.6.2. Curcumin

Curcumin the active component of turmeric, and resveratrol (3,5,4′-trihydroxystilbene), found in grapes, are two such dietary constituents that have received a great deal of attention recently as chemoprotective agents. Phytochemicals inhibit proliferation and induce apoptosis in human and mouse leukemia cell lines. Curcumin augments TRAIL-mediated apoptosis in several types of cancer cell lines *(269–274)*. The induction of apoptosis by combined curcumin and TRAIL treatment involves the activation of initiator/effector caspases (caspase-8, caspase-9, caspase-3), cleavage of proapoptotic Bid, and the release of cytochrome *c* from mitochondria. Thus, combination of TRAIL with curcumin, a pharmacologically safe compound, may provide a more effective adjuvant treatment for prostate cancer.

8.6.3. Selenium

Epidemiological studies, preclinical investigations and clinical intervention trials support the role of selenium (an essential micronutrient and a constituent of antioxidant enzymes) as potent cancer chemopreventive agent for prostate and other cancers *(275,276)*. Induction of apoptosis and inhibition of cell proliferation are considered important cellular events that can account for the cancer preventive effects of selenium. Differential inhibition of PKCδ, NF-κB, and cIAP-2 by selenium may represent important intracellular signaling processes through which selenium induces apoptosis and subsequently exerts its anticarcinogenic effect *(277)*. Selenium prevents clonal expansion of nascent tumors by causing cell-cycle arrest, promoting apoptosis, and modulating p53-dependent DNA repair mechanisms *(278–280)*. Selenium induces apoptosis by upregulating TRAIL-R2/DR5, activating caspase-8 and cleaving Bid in human cancer cells *(281–283)*, thereby suggesting the existence of a potential cross-talk between the death receptor and the mitochondrial pathways. Thus, DR5 is specifically regulated by selenium and its activation may play an important role in selenium-mediated chemoprevention. Furthermore, selenium may sensitize cancer cells to TRAIL treatment by upregulating DR5 receptors.

8.6.4. Luteolin

Luteolin is a dietary flavonoid commonly found in some medicinal plants. Pretreatment with a noncytotoxic concentration of luteolin significantly sensitized TRAIL-induced apoptosis in both TRAIL-sensitive (HeLa) and TRAIL-resistant cancer cells (CNE1, HT29, and HepG2) *(284)*. Such sensitization is achieved through enhanced caspase-8 activation and caspase-3 maturation. Further, the protein level of X-linked inhibitor of apoptosis protein (XIAP) was markedly reduced in cells treated with luteolin and TRAIL, and ectopic expression of XIAP protected against cell death induced by luteolin and TRAIL, showing that luteolin sensitizes TRAIL-induced apoptosis through downregulation of XIAP. Luteolin and TRAIL promoted XIAP ubiquitination and proteasomal degradation. Interestingly, protein kinase C (PKC) activation prevented cell death induced by luteolin and TRAIL via suppression of XIAP downregulation. Moreover, luteolin inhibited PKC activity, and bisindolylmaleimide I, a general PKC inhibitor, simulated luteolin in sensitizing TRAIL-induced apoptosis. PKC activation stabilizes XIAP and thus suppresses TRAIL-induced apoptosis. These data suggest a novel anticancer effect of luteolin and support its potential application in cancer therapy in combination with TRAIL.

Thus, in terms of a clinical perspective, the combination of resveratrol, selenium, Luteolin, or curcumin with TRAIL may be a novel strategy for the treatment of a variety of human cancers that warrants further investigation.

9. Factors Influencing TRAIL Sensitivity in Normal and Malignant Cells

A critical feature of any approach to cancer treatment is the ability to selectively interfere with growth or viability of cancer cells while avoiding or minimizing toxicity to normal, noncancer cells. Several factors appear to be involved in TRAIL sensitivity, such as the relative numbers of death and decoy receptors, expressions of FLICE-inhibitory protein (FLIP), caspase-8 and caspase-10, and the constitutively active AKT/PKB and NF-κB. Delineation of the factors involved in the apoptotic response of TRAIL is complex. The response also varies with the species.

9.1. Expressions of Cell Surface Death and Decoy Receptors

The relative expressions of death (TRAIL-R1/DR4 and TRAIL-R2/DR5) and decoy (DcR1 and DcR2) receptors can be responsible for TRAIL sensitivity. TRAIL induces apoptosis mainly in cancer cells because of the absence of or lower expression of decoy receptors and higher expression of death receptors compared to normal cells *(38,82,86,285)*. In contrast, data revealed that the expressions of death receptors did not correlate with TRAIL sensitivity *(286–288)*. In addition to the two decoy receptors already discussed, a third decoy receptor for TRAIL is a soluble osteoprotegerin (OPG). A role for OPG as a decoy receptor for TRAIL under physiological conditions remains unclear, because the affinity of OPG for TRAIL is rather week compared with the other TRAIL receptors. Recent studies suggest that OPG can act in a paracrine or autocrine manner by binding TRAIL and promoting the survival of breast and prostate carcinoma *(289,290)* and multiple myeloma *(291)*. It is likely that the role of OPG as a survival factor will depend on the relative concentrations, the timing, and the location and expression patterns of OPG and TRAIL in the local microenvironment. Furthermore, TRAIL sensitivity may also be regulated at the intracellular level rather than at the receptor level *(81,120,254,286,292)*. Several alternate mechanisms of TRAIL sensitivity or resistance have been discussed in this chapter.

9.2. Mutations and Stability of Caspase-8 and Caspase-10

Although TRAIL is a new and promising candidate for cancer therapy, resistance of cancer cells to TRAIL posses a challenge in anticancer therapy. Therefore, characterizing the mechanisms of resistance and developing strategies to overcome the resistance are important steps toward successful TRAIL-mediated cancer therapy. Analysis of the DR-DISC components provides evidence for the importance of caspase-8 and caspase-10 in TRAIL-induced apoptosis. The dendritic cells expressing a mutant caspase-10 are resistant to TRAIL signaling *(293)*. Alternatively, a catalytically inactive caspase-10 mutant may act as a nonreleasable substrate trap for caspase-8, thereby inactivating the caspase *(294)*. Therefore, TRAIL sensitivity may differ in different cell types. In another study, instability of caspase-8 protein in a colon cancer DLD1 cell line was found to be a major mechanism of TRAIL resistance *(295)*. Compared with the TRAIL-susceptible DLD1 cell line, TRAIL-resistant DLD1/TRAIL-R cells have a low level of caspase-8 protein, but not its mRNA. Suppression of caspase-8 expression by siRNA in parental DLD1 cells led to TRAIL resistance. Restoration of caspase-8 protein expression by stable transfection rendered the DLD1/TRAIL-R cell line fully sensitive to TRAIL protein, suggesting that the low level of caspase-8 protein expression might be the responsible in TRAIL resistance in DLD1/TRAIL-R cells. A missense mutation in caspase-8 coding region was found in both TRAIL-sensitive and TRAIL-resistant DLD1 cells. The accelerated degradation of caspase-8 protein is one of the mechanisms that lead to TRAIL resistance. Thus, mutations or inappropriate expressions of proteins involved in DISC formation may impose serious risk factors for TRAIL-based therapy.

9.3. FLICE-Inhibitory Protein

FLICE-like inhibitory protein (FLIP) is an antiapoptotic protein that exists in two isoforms, $FLIP_L$ (55 kDa) and $FLIP_S$ (28 kDa) *(294)*. The short form, $FLIP_S$, contains two DEDs and is structurally related to the viral FLIP inhibitors of apoptosis, whereas the long form, $FLIP_L$, contains in addition a caspase-like domain in which the active-center cysteine residue is substituted by a tyrosine residue. $FLIP_S$ and $FLIP_L$ interact with FADD, and caspase-8 and 10, and inhibit apoptosis induced by the activation of death receptors. Similar to procaspases 8 and 10, the FLIP proteins contain a tandem pair of DEDs, but they lack a catalytically active protease domain and thus can operate as transdominant inhibitors of caspase 8 and 10. $FLIP_L$ is expressed during the early stage of T-cell activation, but disappears when T-cells become susceptible to Fas ligand-mediated apoptosis. High levels of $FLIP_L$ protein are also detectable in melanoma cell lines and malignant melanoma tumors. Thus, FLIP may be implicated in tissue homeostasis as an important regulator of apoptosis.

Overexpression of FLIP has been shown to be responsible of TRAIL resistance *(296)*. FLIP plays a significant role in negatively regulating receptor-induced apoptosis *(297,298)*. Addition of protein synthesis inhibitors to TRAIL-resistant melanomas renders them sensitive to TRAIL, indicating that the presence of intracellular apoptosis inhibitors may mediate resistance to TRAIL-mediated apoptosis *(286)*. Expression of FLIP is highest in the TRAIL-resistant melanomas, being low or undetectable in the TRAIL-sensitive melanomas, suggesting the involvement of FLIP in TRAIL resistance *(286)*. Furthermore, addition of actinomycin D to TRAIL-resistant melanomas resulted in decreased intracellular concentrations of FLIP, which correlated with their effectiveness of TRAIL sensitivity *(286)*. Benzyloxycarbonyl-Asp-Glu-Val-Asp-fluoromethyl-ketone, a caspase-3 inhibitor, blocked the cleavage of FLIP(s) and caspase-3 activation in human head and neck squamous cell carcinoma (299). Overexpression of FLIP(s) protected cells from apoptotic death and FLIP(s) cleavage during treatment with TRAIL in combination with cisplatin *(299)*. Caspase-3 is responsible for FLIP(s) cleavage, and the cleavage of FLIP(s) is one of facilitating factors for TRAIL-induced apoptotic death. In HeLa cells, apoptosis induction by TRAIL is dependent on the presence of cycloheximide (CHX), a protein synthesis inhibitor *(300)*. Interestingly, CHX downregulates cFLIP, and overexpression of cFLIP inhibited death receptor-induced NF-κB activation *(300)*. This suggests a novel functional role of cFLIP as a negative regulator of gene induction and apoptosis. Pharmacological downregulation of FLIP might serve as a therapeutic means to sensitize tumor cells to apoptosis induction by TRAIL.

In another study, pretreatment with the proteasome inhibitor MG132 and PS-341 rendered TRAIL-resistant hepatocellular carcinoma (HCC) cell lines but not primary human hepatocytes sensitive for TRAIL-induced apoptosis *(301)*. Downregulation of cFLIP by short interference RNA (siRNA) further sensitized the HCC cell lines. These data have clinical significance because proteasome inhibitors sensitize HCC cells but not primary human hepatocytes for TRAIL-induced apoptosis.

9.4. Akt Activity

The lipid kinase phosphoinositide 3-OH kinase (PI3K) and its downstream target Akt, also known as protein kinase B (PKB), are crucial effectors in oncogenic signaling induced by various receptor-tyrosine kinases. In recent years, data are accumulating that PI3K/Akt signaling components are frequently altered in a variety of human malignancies. PI3K/Akt signaling pathway is an important regulator of a wide spectrum of tumor-related biological processes, including cell proliferation, survival, and motility, as well as neovascularization. Some cancer cells express constitutively AKT/PKB as a result of a complete loss of lipid phosphatase PTEN gene *(302–306)*, a negative regulator of PI-3 kinase pathway. Constitutively active Akt/PKB promotes cellular survival and resistance to chemotherapy and radiation *(307–312)*. We have shown a negative correlation between constitutively active Akt and TRAIL sensitivity *(81,312–316)*. Downregulation of constitutively active Akt by pharmacological (PI-3 kinase inhibitors wortmannin and LY294002) or genetic means (dominant negative Akt) reversed cellular resistance to

TRAIL *(81,312,313,317)*. Conversely, transfecting constitutively active Akt into cells with low Akt activity increased Akt activity and attenuated TRAIL-induced apoptosis. Inhibition of TRAIL sensitivity occurs at the level of BID cleavage, as caspase-8 activity was not affected. Akt activity promotes human gastric cancer cell survival against TRAIL-induced apoptosis via upregulation of FLIP(S), and that the cytotoxic effect of TRAIL can be enhanced by modulating the Akt/FLIP(S) pathway in human gastric cancers *(298)*. TRAIL-induced activation of the intrinsic pathway proceeded by release of mitochondrial factors Smac/DIABLO and cytochrome *c*, and caspase-9 cleavage *(318)*. LY 294002 reduced phosphorylated Akt (p-Akt), with early loss of the short form of cellular FLIP (c-FLIP(S)) and concurrent reduction of Bcl-2. Treatment with small interfering RNA (siRNA) against PI3K also reduced c-FLIP(S) and Bcl-2, and cotreatment with TRAIL triggered caspase-3 cleavage. Combination treatment with p85α or Akt1 siRNA and TRAIL increased apoptosis in TRAIL-resistant colon cancer KM20 and KM12C cells; and these effects were abrogated by caspase-3 inhibitor Z-acetyl-Asp-Glu-Val-Asp-(DEVD)-fmk *(319)*. Furthermore, siRNA-mediated PI3K pathway inhibition resulted in increased expression of the TRAIL-R1/DR4 and TRAIL-R2/DR5 receptors. Inhibition of PI3K/Akt by RNA interference sensitizes resistant colon cancer cells to TRAIL-induced cell death through the induction of TRAIL receptors and activation of caspase-3 and caspase-8.

PTEN-mediated apoptosis involves a FADD-dependent pathway for both death receptor-mediated and drug-induced apoptosis as coexpression of a dominant negative FADD mutant blocked PTEN-mediated apoptosis. PTEN facilitated BID cleavage after treatment with low doses of staurosporine and mitoxantrone *(320)*. BID cleavage was inhibited by dominant negative FADD. It appears that that PTEN promotes drug-induced apoptosis by facilitating caspase-8 activation and BID cleavage through a FADD-dependent pathway. These data clearly demonstrate that Akt promotes survival of cancer cells and modulation of Akt activity, by pharmacological or genetic approaches, alters the cellular responsiveness to TRAIL. Thus, agents that selectively target the PI3K/Akt pathway may enhance the effects of chemotherapeutic agents and provide novel adjuvant treatment for cancers.

9.5. NF-κB Activity

NF-κB plays important roles in immune responses, inflammation, cell division and apoptosis *(321,322)*. NF-κB is a family of dimeric transcription factors formed by the hetero- or homodimerization of proteins from the rel family. There are five rel proteins: RelA (p65), RelB, and cRel, which contain transactivation domains, and p50 and p52, which are expressed as precursor proteins p105 (NF-κB1) and p100 (NF-κB2), respectively, and do not contain transactivation domains *(321,323)*. NF-κB p65/p50 heterodimers remain inactive in the cytoplasm through association with endogenous inhibitor proteins of the IκB family. Activation of NF-κB is achieved by the phosphorylation and activation of the IκB kinase (IKK) complex. The activated IKK complex specifically phosphorylates the IκBs, which are then rapidly polyubiquitinated, targeting them for degradation by the proteosome. Phosphorylation of IκBs in response to proinflammatory stimuli targets them for degradation by the 26S proteosome, allowing the liberated dimers to translocate to the nucleus *(324,325)*. Subsequently, the dimers bind to decameric κB motifs of a large set of genes and regulate their expression *(326)*. NF-κB activates a variety of target genes relevant to the human diseases such as ischemic stroke, Alzheimer's disease, and Parkinson's disease, as well as genes that regulate cell proliferation and mediate cell survival. NF-κB also activates the IκBα gene, thus replenishing the cytoplasmic pool of its own inhibitor. Genes regulated by NF-κB encode defense and signaling proteins including cell surface molecules involved in immune function such as the Ig κL chain, class I and II MHC, and cytokines such as IL-1β, IL-2, IL-6, IL-8, IFN-β, and TNF-α *(323)*. NF-κB regulates the synthesis of a number of MMPs, including MMP-1 (collagenase-1) and MMP-13 (collagenase-3) *(327,328)*. NF-κB dimers also bind to promoters of inducible NO synthase and cyclooxygenase-2 (COX-2), leading to inducible NO synthase and COX-2 expression and the synthesis of NO and PGs *(329–332)*. Whereas p65/p50 heterodimers stimulate these responses, binding of p50/p50 homodimers blocks the transcription of these inflammatory/immune genes.

We have shown that TRAIL activates NF-κB in vitro *(254,333,334)*. Constitutively active NF-κB prevents TRAIL-induced apoptosis in cancer cells *(335,336)*. On the other hand, NF-κB activation is not sufficient for protecting cells from TRAIL-induced apoptosis *(337)*. TRAIL-mediated NF-κB activation increases DR5 expression thereby amplifying the apoptotic response of TRAIL in epithelial-derived cancer cells *(338)*. Based on published data, it appears that various subunits of NF-κB play different roles in regulating cellular response of TRAIL *(254,333)*. We have shown that the RelA/p65 subunit of NF-κB acts as a survival factor by inhibiting expression of DR4/DR5 and caspase-8 and upregulating cIAP1 and cIAP2, whereas c-Rel acts as a proapoptotic factor by enhancing DR4, DR5, and Bcl-Xs, and inhibiting cIAP1, cIAP2 and survivin after TRAIL treatment *(254)*. Thus, the dual function of NF-κB, as an inhibitor or activator of apoptosis, may be cell-type specific where relative levels of RelA and c-Rel subunits determine the ultimate cell fate. Because our study is based on one system, further studies are needed to confirm the role of specific subunits of NF-κB in TRAIL signaling.

9.6. Anti-TRAIL-R1/DR4 and TRAIL-R2/DR5 MAbs as Cancer Therapeutics

Recent studies have shown that monoclonal antibody directed against either TRAIL-R1/DR4 or TRAIL-R2/DR5 can effectively inhibit breast, prostate, lung, glioma and colon tumor cell growth in vitro and in vivo *(339–343)*. TRAIL-R2 monoclonal antibody (TRA-8) induces apoptosis of most TRAIL-sensitive tumor cells by specific binding to DR5 receptors *(340)*. The efficiency of TRA-8-induced apoptosis was variable in different glioma cell lines. Griffith et al. have demonstrated that monoclonal antibodies (MAbs) targeting either DR4 or DR5 could induce apoptosis in melanoma cell lines irrespective of the expression of DcR1 and/or DcR2 *(287)*, suggesting a minor contribution of decoy receptors to resistance of melanoma cells to TRAIL. Furthermore, the apoptosis-inducing potential of TRAIL-R1 or TRAIL-R2 MAb is enhanced by pretreatment of cancer cells with chemotherapeutic agents such as camptothecin or topotecan. Similarly, the combination of TRA-8 treatment and overexpression of Bax overcame TRA-8 resistance of glioma cells in vitro *(340)*. The combination of TRA-8 treatment with specific overexpression of Bax using AdVEGFBax may be an effective approach for the treatment of human malignant gliomas. In another study, agonistic MAbs to the TRAIL death receptors TRAIL-R1 (HGS-ETR1) and TRAIL-R2 (HGS-ETR2) activated caspase-8 and induced cell death in human lymphoma cell lines and primary lymphoma cells *(344)*. HGS-ETR1 and HGS-ETR2 demonstrated comparable activity in the fresh tumor samples, which was independent of TRAIL receptor surface expression, Bax, cFLIP, or procaspase-8 expression, or exposure to prior therapy. Furthermore, both antibodies enhanced the killing effect of doxorubicin and bortezomib. These data collectively suggest that these MAbs can induce cell death in a variety of cultured and primary cancer cells, and thus may have significant therapeutic value.

10. Conclusions and Future Directions

The discovery that death receptor pathway induce apoptosis mainly in cancer cells has opened a path for developing new drugs for the treatment of cancer. Proapoptotic and antiapoptotic proteins are now being targeted to enhance the effect(s) of chemotherapy and radiotherapy. In addition to apoptosis, other types of cell death, such as autophagy and necrosis, may also be responsible for antitumor activity of chemotherapeutic drugs. However, the relationship between apoptosis and autophagy is not clear and needs to be examined because several autophagy-related proteins, including Beclin 1, BNIP3, DAPK, and HSpin 1, may influence apoptosis. Apoptosis and autophagy appear to be mutually exclusive in some cell lines, suggesting an existence of molecular switch between these two processes. Furthermore, alterations of one or both pathways (autophagic or apoptotic) might determine susceptibility to cell death after chemotherapy and radiotherapy, because these agents might trigger both apoptotic caspase-dependent and autophagic caspase-independent pathways at the same time. Studies into the molecular mechanism(s) behind autophagic cell death as a result of chemotherapy might provide a new strategy for overcoming drug resistance by exploring this alternative pathway of cell death.

Strategies for treating cancer by targeting death receptor pathway are likely to provide useful insights into the mechanisms of action of death receptor activation. Current approaches for the development of TRAIL-R1 and/or TRAIL-R2 agonists as anticancer agents include MAbs, soluble TRAIL, and small molecular weight compounds. Based on preclinical data, TRAIL, either alone or in combination with conventional approaches such as chemotherapy or radiation therapy, is very effective as a cancer agent. Furthermore, TRAIL-based gene therapy could be another viable option against cancer. Testing the therapeutic potential of these strategies in treating human cancers in the clinic is eagerly awaited.

Acknowledgments

This work was supported by grants from the Susan G. Komen Breast Cancer Foundation, the Department of Defense, and the National Institutes of Health.

References

1. Krammer PH. CD95(APO-1/Fas)-mediated apoptosis: live and let die. Adv Immunol 1999;71:163–210.
2. Pitti RM, Marsters SA, Ruppert S, Donahue CJ, Moore A, Ashkenazi A. Induction of apoptosis by Apo-2 ligand, a new member of the tumor necrosis factor cytokine family. J Biol Chem 1996;271:12,687–12,690.
3. Wiley SR, Schooley K, Smolak PJ, et al. Identification and characterization of a new member of the TNF family that induces apoptosis. Immunity 1995;3:673–682.

4. Golstein P. Cell death: TRAIL and its receptors. Curr Biol 1997;7:R750–R753.

5. Griffith TS, Lynch DH. TRAIL: a molecule with multiple receptors and control mechanisms. Curr Opin Immunol 1998;10:559–563.

6. Srivastava RK. TRAIL/Apo-2L: mechanisms and clinical applications in cancer. Neoplasia 2001;3:535–546.

7. Chen G, Goeddel DV. TNF-R1 signaling: a beautiful pathway. Science 2002;296: 1634–1635.

8. Hsu H, Shu HB, Pan MG, Goeddel DV. TRADD-TRAF2 and TRADD-FADD interactions define two distinct TNF receptor 1 signal transduction pathways. Cell 1996;84:299–308.

9. Hsu H, Huang J, Shu HB, Baichwal V, Goeddel DV. TNF-dependent recruitment of the protein kinase RIP to the TNF receptor-1 signaling complex. Immunity 1996;4:387–396.

10. Ting AT, Pimentel-Muinos FX, Seed B. RIP mediates tumor necrosis factor receptor 1 activation of NF-kappaB but not Fas/APO-1-initiated apoptosis. EMBO J 1996;15:6189–6196.

11. Li X, Minden A. Pak4 functions in TNFalpha induced survival pathways by facilitating TRADD binding to the TNF receptor. J Biol Chem 2005;280:41,192–41,200.

12. Blonska M, Shambharkar PB, Kobayashi M, et al. TAK1 is recruited to TNF-alpha receptor 1 in a rip-dependent manner and cooperates with MEKK3 leading to NF-kappa B activation. J Biol Chem 2005;280:43,056–43,063.

13. Micheau O, Tschopp J. Induction of TNF receptor I-mediated apoptosis via two sequential signaling complexes. Cell 2003;114:181–190.

14. Schneider-Brachert W, Tchikov V, Neumeyer J, et al. Compartmentalization of TNF receptor 1 signaling: internalized TNF receptosomes as death signaling vesicles. Immunity 2004;21:415–428.

15. Danial NN, Korsmeyer SJ. Cell death: critical control points. Cell 2004;116:205–219.

16. Schultz DR, Harrington WJ, Jr. Apoptosis: programmed cell death at a molecular level. Semin Arthritis Rheum 2003;32:345–369.

17. Luo X, Budihardjo I, Zou H, Slaughter C, Wang X. Bid, a Bcl2 interacting protein, mediates cytochrome c release from mitochondria in response to activation of cell surface death receptors. Cell 1998;94:481–490.

18. Ivanov VN, Bhoumik A, Ronai Z. Death receptors and melanoma resistance to apoptosis. Oncogene 2003;22:3152–3161.

19. Jaeschke H, Fisher MA, Lawson JA, Simmons CA, Farhood A, Jones DA. Activation of caspase-3 (CPP32)-like proteases is essential for TNF-alpha-induced hepatic parenchymal cell apoptosis and neutrophil-mediated necrosis in a murine endotoxin shock model. J Immunol 1998;160:3480–3486.

20. Leist M, Gantner F, Jilg S, Wendel A. Activation of the 55 kDa TNF receptor is necessary and sufficient for TNF-induced liver failure, hepatocyte apoptosis, and nitrite release. J Immunol 1995;154:1307–1316.

21. Pfeffer K, Matsuyama T, Kundig TM, et al. Mice deficient for the 55 kd tumor necrosis factor receptor are resistant to endotoxic shock, yet succumb to L. monocytogenes infection. Cell 1993;73:457–467.

22. Tiegs G, Hentschel J, Wendel A. A T cell-dependent experimental liver injury in mice inducible by concanavalin A. J Clin Invest 1992;90:196–203.

23. Trautwein C, Rakemann T, Brenner DA, et al. Concanavalin A-induced liver cell damage: activation of intracellular pathways triggered by tumor necrosis factor in mice. Gastroenterology 1998;114:1035–1045.

24. Van Antwerp DJ, Martin SJ, Kafri T, Green DR, Verma IM. Suppression of TNF-alpha-induced apoptosis by NF-kappaB. Science 1996;274:787–789.

25. Beg AA, Baltimore D. An essential role for NF-kappaB in preventing TNF-alpha-induced cell death. Science 1996;274:782–784.

26. Liu ZG, Hsu H, Goeddel DV, Karin M. Dissection of TNF receptor 1 effector functions: JNK activation is not linked to apoptosis while NF-kappaB activation prevents cell death. Cell 1996;87:565–576.

27. Wang CY, Mayo MW, Baldwin AS, Jr. TNF- and cancer therapy-induced apoptosis: potentiation by inhibition of NF-kappaB. Science 1996;274:784–787.

28. Iimuro Y, Nishiura T, Hellerbrand C, et al. NF-kappaB prevents apoptosis and liver dysfunction during liver regeneration. J Clin Invest 1998;101:802–811.

29. Beg AA, Sha WC, Bronson RT, Ghosh S, Baltimore D. Embryonic lethality and liver degeneration in mice lacking the RelA component of NF-kappa B. Nature 1995;376:167–170.

30. Li Q, Van Antwerp D, Mercurio F, Lee KF, Verma IM. Severe liver degeneration in mice lacking the IkappaB kinase 2 gene. Science 1999;284:321–325.

31. Chen G, Cao P, Goeddel DV. TNF-induced recruitment and activation of the IKK complex require Cdc37 and Hsp90. Mol Cell 2002;9:401–410.

32. Zhang SQ, Kovalenko A, Cantarella G, Wallach D. Recruitment of the IKK signalosome to the p55 TNF receptor: RIP and A20 bind to NEMO (IKKgamma) upon receptor stimulation. Immunity 2000;12:301–311.

33. Kucharczak J, Simmons MJ, Fan Y, Gelinas C. To be, or not to be: NF-kappaB is the answer—role of Rel/NF-kappaB in the regulation of apoptosis. Oncogene 2003;22:8961–8982.

34. Varfolomeev EE, Ashkenazi A. Tumor necrosis factor: an apoptosis JuNKie? Cell 2004; 116:491–497.

35. Curtin JF, Cotter TG. Live and let die: regulatory mechanisms in Fas-mediated apoptosis. Cell Signal 2003;15:983–992.

36. Barnhart BC, Alappat EC, Peter ME. The CD95 type I/type II model. Semin Immunol 2003;15:185–193.

37. Gupta S. Molecular steps of cell suicide: an insight into immune senescence. J Clin Immunol 2000;20:229–239.

38. Ashkenazi A, Dixit VM. Death receptors: signaling and modulation. Science 1998;281: 1305–1308.

39. Takahashi T, Tanaka M, Brannan CI, et al. Generalized lymphoproliferative disease in mice, caused by a point mutation in the Fas ligand. Cell 1994;76:969–976.

40. Srivastava RK, Sasaki CY, Hardwick JM, Longo DL. Bcl-2-mediated drug resistance: inhibition of apoptosis by blocking nuclear factor of activated T lymphocytes (NFAT)-induced Fas ligand transcription. J Exp Med 1999;190:253–265.

41. Ioachim HL, Decuseara R, Giancotti F, Dorsett BH. FAS and FAS-L expression by tumor cells and lymphocytes in breast carcinomas and their lymph node metastases. Pathol Res Pract 2005;200:743–751.

42. Bennett M, Macdonald K, Chan SW, Luzio JP, Simari R, Weissberg P. Cell surface trafficking of Fas: a rapid mechanism of p53-mediated apoptosis. Science 1998;282:290–293.

43. Sodeman T, Bronk SF, Roberts PJ, Miyoshi H, Gores GJ. Bile salts mediate hepatocyte apoptosis by increasing cell surface trafficking of Fas. Am J Physiol Gastrointest Liver Physiol 2000;278:G992–G999.

44. Ivanov VN, Lopez Bergami P, Maulit G, Sato TA, Sassoon D, Ronai Z. FAP-1 association with Fas (Apo-1) inhibits Fas expression on the cell surface. Mol Cell Biol 2003;23:3623–3635.

45. Augstein P, Dunger A, Salzsieder C, et al. Cell surface trafficking of Fas in NIT-1 cells and dissection of surface and total Fas expression. Biochem Biophys Res Commun 2002;290:443–451.

46. Kawasaki Y, Saito T, Shirota-Someya Y, et al. Cell death-associated translocation of plasma membrane components induced by CTL. J Immunol 2000;164:4641–4648.

47. O'Reilly LA, Divisekera U, Newton K, et al. Modifications and intracellular trafficking of FADD/MORT1 and caspase-8 after stimulation of T lymphocytes. Cell Death Differ. 2004;11:724–736.

48. Tsokos GC, Kovacs B, Liossis SN. Lymphocytes, cytokines, inflammation, and immune trafficking. Curr Opin Rheumatol 1997;9:380–386.

49. Furuya Y, Fuse H, Masai M. Serum soluble Fas level for detection and staging of prostate cancer. Anticancer Res 2001;21:3595–3598.

50. Mizutani Y, Yoshida O, Bonavida B. Prognostic significance of soluble Fas in the serum of patients with bladder cancer. J Urol 1998;160:571–576.

51. Ugurel S, Rappl G, Tilgen W, Reinhold U. Increased soluble CD95 (sFas/CD95) serum level correlates with poor prognosis in melanoma patients. Clin Cancer Res 2001;7:1282–1286.

52. Cheng J, Zhou T, Liu C, et al. Protection from Fas-mediated apoptosis by a soluble form of the Fas molecule. Science 1994;263:1759–1762.

53. Tanaka M, Itai T, Adachi M, Nagata S. Downregulation of Fas ligand by shedding. Nat Med 1998;4:31–36.

54. Powell WC, Fingleton B, Wilson CL, Boothby M, Matrisian LM. The metalloproteinase matrilysin proteolytically generates active soluble Fas ligand and potentiates epithelial cell apoptosis. Curr Biol 1999;9:1441–1447.

55. Rudolph-Owen LA, Chan R, Muller WJ, Matrisian LM. The matrix metalloproteinase matrilysin influences early-stage mammary tumorigenesis. Cancer Res 1998;58: 5500–5506.

56. Shigemasa K, Tanimoto H, Sakata K, et al. Induction of matrix metalloprotease-7 is common in mucinous ovarian tumors including early stage disease. Med Oncol 2000; 17:52–58.

57. Yamamoto H, Adachi Y, Itoh F, et al. Association of matrilysin expression with recurrence and poor prognosis in human esophageal squamous cell carcinoma. Cancer Res 1999;59: 3313–3316.

58. Fingleton B, Vargo-Gogola T, Crawford HC, Matrisian LM. Matrilysin [MMP-7] expression selects for cells with reduced sensitivity to apoptosis. Neoplasia 2001;3:459–468.

59. Vargo-Gogola T, Fingleton B, Crawford HC, Matrisian LM. Matrilysin (matrix metalloproteinase-7) selects for apoptosis-resistant mammary cells in vivo. Cancer Res 2002;62: 5559–5563.

60. Knox PG, Milner AE, Green NK, Eliopoulos AG, Young LS. Inhibition of metalloproteinase cleavage enhances the cytotoxicity of Fas ligand. J Immunol 2003;170:677–685.

61. Takahashi T, Tanaka M, Inazawa J, Abe T, Suda T, Nagata S. Human Fas ligand:gene structure, chromosomal location and species specificity. Int Immunol 1994;6:1567–1574.

62. Lynch DH, Watson ML, Alderson MR, et al. The mouse Fas-ligand gene is mutated in gld mice and is part of a TNF family gene cluster. Immunity 1994;1:131–136.

63. Ni R, Tomita Y, Matsuda K, et al. Fas-mediated apoptosis in primary cultured mouse hepatocytes. Exp Cell Res 1994;215:332–337.

64. Galle PR, Hofmann WJ, Walczak H, et al. Involvement of the CD95 (APO-1/Fas) receptor and ligand in liver damage. J Exp Med 1995;182:1223–1230.

65. Ogasawara J, Watanabe-Fukunaga R, Adachi M, et al. Lethal effect of the anti-Fas antibody in mice. Nature 1993;364:806–809.

66. Strand S, Hofmann WJ, Grambihler A, et al. Hepatic failure and liver cell damage in acute Wilson's disease involve CD95 (APO-1/Fas) mediated apoptosis. Nat Med 1998;4: 588–593.

67. Pan G, Ni J, Wei YF, Yu G, Gentz R, Dixit VM. An antagonist decoy receptor and a death domain-containing receptor for TRAIL. Science 1997;277:815–818.

68. Pan G, O'Rourke K, Chinnaiyan AM, et al. The receptor for the cytotoxic ligand TRAIL. Science 1997;276:111–113.

69. Schneider P, Bodmer JL, Thome M, Hofmann K, Holler N, Tschopp J. Characterization of two receptors for TRAIL. FEBS Lett 1997;416:329–334.

70. Screaton GR, Mongkolsapaya J, Xu XN, Cowper AE, McMichael AJ, Bell JI. TRICK2, a new alternatively spliced receptor that transduces the cytotoxic signal from TRAIL. Curr Biol 1997;7:693–696.

71. Sheridan JP, Marsters SA, Pitti RM, et al. Control of TRAIL-induced apoptosis by a family of signaling and decoy receptors. Science 1997;277:818–821.

72. Walczak H, Degli-Esposti MA, Johnson RS, et al. TRAIL-R2: a novel apoptosis-mediating receptor for TRAIL. EMBO J 1997;16:5386–5397.

73. Degli-Esposti MA, Smolak PJ, Walczak H, et al. Cloning and characterization of TRAIL-R3, a novel member of the emerging TRAIL receptor family. J Exp Med 1997;186: 1165–1170.

74. Degli-Esposti MA, Dougall WC, Smolak PJ, et al. The novel receptor TRAIL-R4 induces NF-kappaB and protects against TRAIL-mediated apoptosis, yet retains an incomplete death domain. Immunity 1997;7:813–820.

75. Marsters SA, Sheridan JP, Pitti RM, et al. A novel receptor for Apo2L/TRAIL contains a truncated death domain. Curr Biol 1997;7:1003–1006.

76. Pan G, Ni J, Yu G, Wei YF, Dixit VM. TRUNDD, a new member of the TRAIL receptor family that antagonizes TRAIL signalling. FEBS Lett 1998;424:41–45.

77. Emery JG, McDonnell P, Burke MB, et al. Osteoprotegerin is a receptor for the cytotoxic ligand TRAIL. J Biol Chem 1998;273:14,363–14,367.

78. Suliman A, Lam A, Datta R, Srivastava RK. Intracellular mechanisms of TRAIL: apoptosis through mitochondrial-dependent and -independent pathways. Oncogene 2001;20: 2122–2133.

79. Ashkenazi A, Dixit VM. Apoptosis control by death and decoy receptors. Curr Opin Cell Biol 1999;11:255–260.

80. Walczak H, Miller RE, Ariail K, et al. Tumoricidal activity of tumor necrosis factor-related apoptosis-inducing ligand in vivo. Nat Med 1999;5:157–163.

81. Chen X, Thakkar H, Tyan F, et al. Constitutively active Akt is an important regulator of TRAIL sensitivity in prostate cancer. Oncogene 2001;20:6073–6083.

82. Singh TR, Shankar S, Chen X, Asim M, Srivastava RK. Synergistic interactions of chemotherapeutic drugs and tumor necrosis factor-related apoptosis-inducing ligand/Apo-2 ligand on apoptosis and on regression of breast carcinoma in vivo. Cancer Res 2003;63: 5390–5400.

83. LeBlanc HN, Ashkenazi A. Apo2L/TRAIL and its death and decoy receptors. Cell Death Differ 2003;10:66–75.

84. French LE, Tschopp J. The TRAIL to selective tumor death. Nat Med 1999;5:146–147.

85. Ashkenazi A, Pai RC, Fong S, et al. Safety and antitumor activity of recombinant soluble Apo2 ligand. J Clin Invest 1999;104:155–162.

86. Chinnaiyan AM, Prasad U, Shankar S, et al. Combined effect of tumor necrosis factor-related apoptosis-inducing ligand and ionizing radiation in breast cancer therapy. Proc Natl Acad Sci USA 2000;97:1754–1759.

87. Singh TR, Shankar S, Srivastava RK. HDAC inhibitors enhance the apoptosis-inducing potential of TRAIL in breast carcinoma. Oncogene 2005;24:4609–4623.

88. Shabbeer S, Carducci MA. Focus on deacetylation for therapeutic benefit. IDrugs 2005;8: 144–154.

89. Shankar S, Singh TR, Chen X, Thakkar H, Firnin J, Srivastava RK. The sequential treatment with ionizing radiation followed by TRAIL/Apo-2L reduces tumor growth and induces apoptosis of breast tumor xenografts in nude mice. Int J Oncol 2004;24:1133–1140.

90. Shankar S, Singh TR, Srivastava RK. Ionizing radiation enhances the therapeutic potential of TRAIL in prostate cancer in vitro and in vivo: Intracellular mechanisms. Prostate 2004; 61:35–49.

91. Mitsiades CS, Treon SP, Mitsiades N, et al. TRAIL/Apo2L ligand selectively induces apoptosis and overcomes drug resistance in multiple myeloma: therapeutic applications. Blood 2001;98:795–804.

92. Pollack IF, Erff M, Ashkenazi A. Direct stimulation of apoptotic signaling by soluble Apo2l/tumor necrosis factor-related apoptosis-inducing ligand leads to selective killing of glioma cells. Clin Cancer Res 2001;7:1362–1369.

93. Cretney E, Takeda K, Yagita H, Glaccum M, Peschon JJ, Smyth MJ. Increased susceptibility to tumor initiation and metastasis in TNF-related apoptosis-inducing ligand-deficient mice. J Immunol 2002;168:1356–1361.

94. Sedger LM, Glaccum MB, Schuh JC, et al. Characterization of the in vivo function of TNF-alpha-related apoptosis-inducing ligand, TRAIL/Apo2L, using TRAIL/Apo2L gene-deficient mice. Eur J Immunol 2002;32:2246–2254.

95. Lamhamedi-Cherradi SE, Zheng SJ, Maguschak KA, Peschon J, Chen H. Defective thymocyte apoptosis and accelerated autoimmune diseases in TRAIL-/- mice. Nat Immunol 2003;4: 255–260.

96. Fanger NA, Maliszewski CR, Schooley K, Griffith TS. Human dendritic cells mediate cellular apoptosis via tumor necrosis factor-related apoptosis-inducing ligand (TRAIL). J Exp Med 1999;190:1155–1164.

97. Griffith TS, Wiley SR, Kubin MZ, Sedger LM, Maliszewski CR, Fanger NA. Monocyte-mediated tumoricidal activity via the tumor necrosis factor-related cytokine, TRAIL. J Exp Med 1999;189:1343–1354.

98. Kayagaki N, Yamaguchi N, Nakayama M, et al. Involvement of TNF-related apoptosis-inducing ligand in human CD4+ T cell-mediated cytotoxicity. J Immunol 1999;162: 2639–2647.

99. Thomas WD, Hersey P. CD4 T cells kill melanoma cells by mechanisms that are independent of Fas (CD95). Int J Cancer 1998;75:384–390.

100. Kayagaki N, Yamaguchi N, Nakayama M, et al. Expression and function of TNF-related apoptosis-inducing ligand on murine activated NK cells. J Immunol 1999;163:1906–1913.

101. Banner DW, D'Arcy A, Janes W, et al. Crystal structure of the soluble human 55 kd TNF receptor-human TNF beta complex: implications for TNF receptor activation. Cell 1993;73:431–445.

102. D'Arcy A, Banner DW, Janes W, et al. Crystallization and preliminary crystallographic analysis of a TNF-beta-55 kDa TNF receptor complex. J Mol Biol 1993;229:555–557.

103. Karpusas M, Hsu YM, Wang JH, et al. 2 A crystal structure of an extracellular fragment of human CD40 ligand. Structure 1995;3:1031–1039.

104. Bodmer JL, Meier P, Tschopp J, Schneider P. Cysteine 230 is essential for the structure and activity of the cytotoxic ligand TRAIL. J Biol Chem 2000;275:20,632–20,637.

105. Hymowitz SG, Christinger HW, Fuh G, et al. Triggering cell death: the crystal structure of Apo2L/TRAIL in a complex with death receptor 5. Mol Cell 1999;4:563–571.

106. Mongkolsapaya J, Grimes JM, Chen N, et al. Structure of the TRAIL-DR5 complex reveals mechanisms conferring specificity in apoptotic initiation. Nat Struct Biol 1999;6:1048–1053.

107. Bodmer JL, Holler N, Reynard S, et al. TRAIL receptor-2 signals apoptosis through FADD and caspase-8. Nat Cell Biol 2000;2:241–243.

108. Hymowitz SG, O'Connell MP, Ultsch MH, et al. A unique zinc-binding site revealed by a high-resolution X-ray structure of homotrimeric Apo2L/TRAIL. Biochemistry 2000;39: 633–640.

109. Schulze-Osthoff K, Ferrari D, Los M, Wesselborg S, Peter ME. Apoptosis signaling by death receptors. Eur J Biochem 1998;254:439–459.

110. Kischkel FC, Lawrence DA, Chuntharapai A, Schow P, Kim KJ, Ashkenazi A. Apo2L/TRAIL-dependent recruitment of endogenous FADD and caspase-8 to death receptors 4 and 5. Immunity 2000;12:611–620.

111. Ferri KF, Kroemer G. Organelle-specific initiation of cell death pathways. Nat Cell Biol 2001;3:E255–E263.

112. Jin Z, McDonald ER, 3rd, Dicker DT, El-Deiry WS. Deficient tumor necrosis factor-related apoptosis-inducing ligand (TRAIL) death receptor transport to the cell surface in human colon cancer cells selected for resistance to TRAIL-induced apoptosis. J Biol Chem 2004;279:35,829–35,839.

113. Sheikh MS, Huang Y. TRAIL death receptors, Bcl-2 protein family, and endoplasmic reticulum calcium pool. Vitam Horm 2004;67:169–188.

114. He Q, Lee DI, Rong R, et al. Endoplasmic reticulum calcium pool depletion-induced apoptosis is coupled with activation of the death receptor 5 pathway. Oncogene 2002;21: 2623–2633.

115. Rudner J, Lepple-Wienhues A, Budach W, et al. Wild-type, mitochondrial and ER-restricted Bcl-2 inhibit DNA damage-induced apoptosis but do not affect death receptor-induced apoptosis. J Cell Sci 2001;114:4161–4172.

116. Kim R. Recent advances in understanding the cell death pathways activated by anticancer therapy. Cancer 2005;103:1551–1560.

117. Daniel PT, Schulze-Osthoff K, Belka C, Guner D. Guardians of cell death: the Bcl-2 family proteins. Essays Biochem 2003;39:73–88.

118. Green DR. Apoptotic pathways: the roads to ruin. Cell 1998;94:695–698.

119. Green DR, Amarante-Mendes GP. The point of no return: mitochondria, caspases, and the commitment to cell death. Results Probl Cell Differ 1998;24:45–61.

120. Kandasamy K, Srinivasula SM, Alnemri ES, et al. Involvement of proapoptotic molecules Bax and Bak in tumor necrosis factor-related apoptosis-inducing ligand (TRAIL)-induced mitochondrial disruption and apoptosis: differential regulation of cytochrome c and Smac/DIABLO release. Cancer Res 2003;63:1712–1721.

121. Kroemer G. Mitochondrial control of apoptosis: an overview. Biochem Soc Symp 66:1–15.

122. Desagher S, Martinou JC. Mitochondria as the central control point of apoptosis. Trends Cell Biol 2000;10:369–377.

123. Kroemer G, Reed JC. Mitochondrial control of cell death. Nat Med 2000;6:513–519.

124. Hegde R, Srinivasula SM, Zhang Z, et al. Identification of Omi/HtrA2 as a mitochondrial apoptotic serine protease that disrupts inhibitor of apoptosis protein-caspase interaction. J Biol Chem 2002;277:432–438.

125. Susin SA, Zamzami N, Kroemer G. Mitochondria as regulators of apoptosis: doubt no more. Biochim Biophys Acta 1998;1366:151–165.

126. Du C, Fang M, Li Y, Li L, Wang X. Smac, a mitochondrial protein that promotes cytochrome c-dependent caspase activation by eliminating IAP inhibition. Cell 2000;102: 33–42.

127. Verhagen AM, Coulson EJ, Vaux DL. Inhibitor of apoptosis proteins and their relatives: IAPs and other BIRPs. Genome Biol 2001;2:REVIEWS3009.

128. Verhagen AM, Ekert PG, Pakusch M, et al. Identification of DIABLO, a mammalian protein that promotes apoptosis by binding to and antagonizing IAP proteins. Cell 2000;102: 43–53.

129. Verhagen AM, Silke J, Ekert PG, et al. HtrA2 promotes cell death through its serine protease activity and its ability to antagonize inhibitor of apoptosis proteins. J Biol Chem 2002;277:445–454.

130. Sun SY, Yue P, Zhou JY, et al. Overexpression of BCL2 blocks TNF-related apoptosis-inducing ligand (TRAIL)-induced apoptosis in human lung cancer cells. Biochem Biophys Res Commun 2001;280:788–797.

131. Nimmanapalli R, Perkins CL, Orlando M, O'Bryan E, Nguyen D, Bhalla KN. Pretreatment with paclitaxel enhances apo-2 ligand/tumor necrosis factor-related apoptosis-inducing ligand-induced apoptosis of prostate cancer cells by inducing death receptors 4 and 5 protein levels. Cancer Res 2001;61:759–763.

132. Munshi A, Pappas G, Honda T, et al. TRAIL (APO-2L) induces apoptosis in human prostate cancer cells that is inhibitable by Bcl-2. Oncogene 2001;20:3757–3765.

133. Rokhlin OW, Guseva N, Tagiyev A, Knudson CM, Cohen MB. Bcl-2 oncoprotein protects the human prostatic carcinoma cell line PC3 from TRAIL-mediated apoptosis. Oncogene 2001;20:2836–2843.

134. Wang K, Yin XM, Chao DT, Milliman CL, Korsmeyer SJ. BID: a novel BH3 domain-only death agonist. Genes Dev 1996;10:2859–2869.

135. Wei MC, Zong WX, Cheng EH, et al. Proapoptotic BAX and BAK: a requisite gateway to mitochondrial dysfunction and death. Science 2001;292:727–730.

136. Grinberg M, Sarig R, Zaltsman Y, et al. tBID Homooligomerizes in the mitochondrial membrane to induce apoptosis. J Biol Chem 2002;277:12,237–12,245.

137. Li H, Zhu H, Xu CJ, Yuan J. Cleavage of BID by caspase-8 mediates the mitochondrial damage in the Fas pathway of apoptosis. Cell 1998;94:491–501.

138. Lindsten T, Ross AJ, King A, et al. The combined functions of proapoptotic Bcl-2 family members bak and bax are essential for normal development of multiple tissues. Mol Cell 2000;6:1389–1399.

139. Nechushtan A, Smith CL, Lamensdorf I, Yoon SH, Youle RJ. Bax and Bak coalesce into novel mitochondria-associated clusters during apoptosis. J Cell Biol 2001;153: 1265–1276.

140. Antonsson B. Bax and other pro-apoptotic Bcl-2 family "killer-proteins" and their victim the mitochondrion. Cell Tissue Res 2001;306:347–361.

141. Mikhailov V, Mikhailova M, Pulkrabek DJ, Dong Z, Venkatachalam MA, Saikumar P. Bcl-2 prevents Bax oligomerization in the mitochondrial outer membrane. J Biol Chem 2001; 276:18,361–18,374.

142. Wei MC, Lindsten T, Mootha VK, et al. tBID, a membrane-targeted death ligand, oligomerizes BAK to release cytochrome c. Genes Dev 2000;14:2060–2071.

143. Gross A, McDonnell JM, Korsmeyer SJ. BCL-2 family members and the mitochondria in apoptosis. Genes Dev 1999;13:1899–1911.

144. Bouillet P, Strasser A. BH3-only proteins - evolutionarily conserved proapoptotic Bcl-2 family members essential for initiating programmed cell death. J Cell Sci 115:1567–1574.

145. Lane DP, Crawford LV. T antigen is bound to a host protein in SV40-transformed cells. Nature 1979;278:261–263.

146. Linzer DI, Levine AJ. Characterization of a 54K dalton cellular SV40 tumor antigen present in SV40-transformed cells and uninfected embryonal carcinoma cells. Cell 1979;17: 43–52.

147. Scheffner M, Werness BA, Huibregtse JM, Levine AJ, Howley PM. The E6 oncoprotein encoded by human papillomavirus types 16 and 18 promotes the degradation of p53. Cell 1990;63:1129–1136.

148. Hollstein M, Sidransky D, Vogelstein B, Harris CC. p53 mutations in human cancers. Science 1991;253:49–53.

149. Levine AJ, Momand J, Finlay CA. The p53 tumour suppressor gene. Nature 1991;351:453–456.

150. Donehower LA, Harvey M, Slagle BL, et al. Mice deficient for p53 are developmentally normal but susceptible to spontaneous tumours. Nature 1992;356:215–221.

151. Srivastava S, Zou ZQ, Pirollo K, Blattner W, Chang EH. Germ-line transmission of a mutated p53 gene in a cancer-prone family with Li-Fraumeni syndrome. Nature 1990;348: 747–749.

152. Harris CC. p53 tumor suppressor gene: from the basic research laboratory to the clinic—an abridged historical perspective. Carcinogenesis 1996;17:1187–1198.

153. Oren M, Rotter V. Introduction: p53—the first twenty years. Cell Mol Life Sci 1999;55:9–11.

154. Adimoolam S, Ford JM. p53 and regulation of DNA damage recognition during nucleotide excision repair. DNA Repair (Amst) 2003;2:947–954.

155. Offer H, Wolkowicz R, Matas D, Blumenstein S, Livneh Z, Rotter V. Direct involvement of p53 in the base excision repair pathway of the DNA repair machinery. FEBS Lett 1999;450:197–204.

156. Zhou J, Ahn J, Wilson SH, Prives CA. role for p53 in base excision repair. EMBO J 2001; 20:914–923.

157. Vogelstein B, Lane D, Levine AJ. Surfing the p53 network. Nature 2000;408:307–310.

158. Vousden KH, Lu X. Live or let die: the cell's response to p53. Nat Rev Cancer 2002; 2:594–604.

159. Morris SM. A role for p53 in the frequency and mechanism of mutation. Mutat Res 2002;511:45–62.

160. Nayak BK, Das GM. Stabilization of p53 and transactivation of its target genes in response to replication blockade. Oncogene 2002;21:7226–7229.

161. Nakamura S, Gomyo Y, Roth JA, Mukhopadhyay T. C-terminus of p53 is required for G(2) arrest. Oncogene 2002;21:2102–2107.

162. Mirza A, McGuirk M, Hockenberry TN, et al. Human survivin is negatively regulated by wild-type p53 and participates in p53-dependent apoptotic pathway. Oncogene 2002;21: 2613–2622.

163. Haupt S, Berger M, Goldberg Z, Haupt Y. Apoptosis—the p53 network. J Cell Sci 2003; 116:4077–4085.

164. Mishra NC, Kumar S. Apoptosis: a mitochondrial perspective on cell death. Indian J Exp Biol 2005;43:25–34.

165. Mihara M, Erster S, Zaika A, et al. p53 has a direct apoptogenic role at the mitochondria. Mol Cell 2003;11:577–590.

166. Mihara M, Moll UM. Detection of mitochondrial localization of p53. Methods Mol Biol 2003;234:203–209.

167. Alnemri ES, Livingston DJ, Nicholson DW, et al. Human ICE/CED-3 protease nomenclature. Cell 1996;87:171.

168. Green DR, Reed JC. Mitochondria and apoptosis. Science 1998;281:1309–1312.

169. Wolf BB, Green DR. Suicidal tendencies: apoptotic cell death by caspase family proteinases. J Biol Chem 1999;274:20,049–20,052.

170. Aouad SM, Cohen LY, Sharif-Askari E, Haddad EK, Alam A, Sekaly RP. Caspase-3 is a component of Fas death-inducing signaling complex in lipid rafts and its activity is required for complete caspase-8 activation during Fas-mediated cell death. J Immunol 2004;172:2316–2323.

171. Srinivasula SM, Ahmad M, Fernandes-Alnemri T, Alnemri ES. Autoactivation of procaspase-9 by Apaf-1-mediated oligomerization. Mol Cell 1998;1:949–957.

172. Keogh SA, Walczak H, Bouchier-Hayes L, Martin SJ. Failure of Bcl-2 to block cytochrome c redistribution during TRAIL-induced apoptosis. FEBS Lett 2000;471:93–98.

173. Kroemer G, Dallaporta B, Resche-Rigon M. The mitochondrial death/life regulator in apoptosis and necrosis. Ann Rev Physiol 1998;60:619–642.

174. Korsmeyer SJ. Programmed cell death and the regulation of homeostasis. Harvey Lect 1999;95:21–41.

175. Yin XM, Wang K, Gross A, et al. Bid-deficient mice are resistant to Fas-induced hepatocellular apoptosis. Nature 1999;400:886–891.

176. Cardone MH, Roy N, Stennicke HR, et al. Regulation of cell death protease caspase-9 by phosphorylation. Science 1998;282:1318–1321.

177. Roy N, Deveraux QL, Takahashi R, Salvesen GS, Reed JC. The c-IAP-1 and c-IAP-2 proteins are direct inhibitors of specific caspases. EMBO J 1997;16:6914–6925.

178. Deveraux QL, Roy N, Stennicke HR, et al. IAPs block apoptotic events induced by caspase-8 and cytochrome c by direct inhibition of distinct caspases. EMBO J 1998;17: 2215–2223.

179. Deveraux QL, Takahashi R, Salvesen GS, Reed JC. X-linked IAP is a direct inhibitor of cell-death proteases. Nature 1997;388:300-304.

180. Ekert PG, Silke J, Vaux DL. Caspase inhibitors. Cell Death Differ 1999;6:1081–1086.

181. Wang Z, Sampath J, Fukuda S, Pelus LM. Disruption of the inhibitor of apoptosis protein survivin sensitizes Bcr–abl-positive cells to STI571-induced apoptosis. Cancer Res 2005;65:8224–8232.

182. Yamaguchi Y, Shiraki K, Fuke H, et al. Targeting of X-linked inhibitor of apoptosis protein or survivin by short interfering RNAs sensitize hepatoma cells to TNF-related apoptosis-inducing ligand- and chemotherapeutic agent-induced cell death. Oncol Rep 2005;14: 1311–1316.

183. Bossy-Wetzel E, Newmeyer DD, Green DR. Mitochondrial cytochrome c release in apoptosis occurs upstream of DEVD-specific caspase activation and independently of mitochondrial transmembrane depolarization. EMBO J 1998;17:37–49.

184. Jurgensmeier JM, Xie Z, Deveraux Q, Ellerby L, Bredesen D, Reed JC. Bax directly induces release of cytochrome c from isolated mitochondria. Proc Natl Acad Sci USA 1998;95:4997–5002.

185. Duckett CS, Li F, Wang Y, Tomaselli KJ, Thompson CB, Armstrong RC. Human IAP-like protein regulates programmed cell death downstream of Bcl-xL and cytochrome c. Mol Cell Biol 1998;18:608–615.

186. Tamm I, Wang Y, Sausville E, et al. IAP-family protein survivin inhibits caspase activity and apoptosis induced by Fas (CD95), Bax, caspases, and anticancer drugs. Cancer Res 1998;58:5315–5320.

187. Hegde R, Srinivasula SM, Datta P, et al. The polypeptide chain-releasing factor GSPT1/eRF3 is proteolytically processed into an IAP-binding protein. J Biol Chem 2003;278:38,699–38,706.

188. Martins LM, Iaccarino I, Tenev T, et al. The serine protease Omi/HtrA2 regulates apoptosis by binding XIAP through a reaper-like motif. J Biol Chem 2002;277:439–444.

189. Srinivasula SM, Hegde R, Saleh A, et al. A conserved XIAP-interaction motif in caspase-9 and Smac/DIABLO regulates caspase activity and apoptosis. Nature 2001;410:112–116.

190. van Loo G, van Gurp M, Depuydt B, et al. The serine protease Omi/HtrA2 is released from mitochondria during apoptosis. Omi interacts with caspase-inhibitor XIAP and induces enhanced caspase activity. Cell Death Differ 2002;9:20–26.

191. Fulda S, Wick W, Weller M, Debatin KM. Smac agonists sensitize for Apo2L/TRAIL- or anticancer drug-induced apoptosis and induce regression of malignant glioma in vivo. Nat Med 2002;8:808–815.

192. Glass CK, Rosenfeld MG. The coregulator exchange in transcriptional functions of nuclear receptors. Genes Dev 2000;14:121–141.

193. Kouzarides T. Histone acetylases and deacetylases in cell proliferation. Curr Opin Genet Dev 1999;9:40–48.

194. Grignani F, De Matteis S, Nervi C, et al. Fusion proteins of the retinoic acid receptor-alpha recruit histone deacetylase in promyelocytic leukaemia. Nature 1998;391:815–818.

195. Lin RJ, Nagy L, Inoue S, Shao W, Miller WH, Jr, Evans RM. Role of the histone deacetylase complex in acute promyelocytic leukaemia. Nature 1998;391:811–814.

196. Strahl BD, Allis CD. The language of covalent histone modifications. Nature 2000;403: 41–45.

197. Van Lint C, Emiliani S, Verdin E. The expression of a small fraction of cellular genes is changed in response to histone hyperacetylation. Gene Expr 1996;5:245–253.

198. Lagger G, O'Carroll D, Rembold M, et al. Essential function of histone deacetylase 1 in proliferation control and CDK inhibitor repression. EMBO J 2002;21:2672–2681.

199. Marks PA, Miller T, Richon VM. Histone deacetylases. Curr Opin Pharmacol 2003;3:344–351.

200. Fandy TE, Shankar S, Ross DD, Sausville E, Srivastava RK. Interactive effects of HDAC inhibitors and TRAIL on apoptosis are associated with changes in mitochondrial functions and expressions of cell cycle regulatory genes in multiple myeloma. Neoplasia 2005;7:646–657.

201. Fang JY. Histone deacetylase inhibitors, anticancerous mechanism and therapy for gastrointestinal cancers. J Gastroenterol Hepatol 2005;20:988–994.

202. Rosato RR, Wang Z, Gopalkrishnan RV, Fisher PB, Grant S. Evidence of a functional role for the cyclin-dependent kinase-inhibitor p21WAF1/CIP1/MDA6 in promoting differentiation and preventing mitochondrial dysfunction and apoptosis induced by sodium butyrate in human myelomonocytic leukemia cells (U937). Int J Oncol 2001;19:181–191.

203. Gray SG, Ekstrom TJ. The human histone deacetylase family. Exp Cell Res 2001;262: 75–83.

204. McKinsey TA, Olson EN. Toward transcriptional therapies for the failing heart: chemical screens to modulate genes. J Clin Invest 2005;115:538–546.

205. Kwon SH, Ahn SH, Kim YK, et al. Apicidin, a histone deacetylase inhibitor, induces apoptosis and Fas/Fas ligand expression in human acute promyelocytic leukemia cells. J Biol Chem 2002;277:2073–2080.

206. Marks PA, Richon VM, Miller T, Kelly WK. Histone deacetylase inhibitors. Adv Cancer Res 2004;91:137–168.

207. Boyle GM, Martyn AC, Parsons PG. Histone deacetylase inhibitors and malignant melanoma. Pigment Cell Res 2005;18:160–166.

208. Butler LM, Agus DB, Scher HI, et al. Suberoylanilide hydroxamic acid, an inhibitor of histone deacetylase, suppresses the growth of prostate cancer cells in vitro and in vivo. Cancer Res 2000;60:5165–5170.

209. Tang XX, Robinson ME, Riceberg JS, et al. Favorable neuroblastoma genes and molecular therapeutics of neuroblastoma. Clin Cancer Res 2004;10:5837–5844.

210. Zhang Y, Adachi M, Zhao X, Kawamura R, Imai K. Histone deacetylase inhibitors FK228, N-(2-aminophenyl)-4-[N-(pyridin-3-yl-methoxycarbonyl)amino- methyl]benzamide and m-carboxycinnamic acid bis-hydroxamide augment radiation-induced cell death in gastrointestinal adenocarcinoma cells. Int J Cancer 2004;110:301–308.

211. Takimoto R, Kato J, Terui T, et al. Augmentation of Antitumor Effects of p53 Gene Therapy by Combination with HDAC Inhibitor. Cancer Biol Ther 2005;4:421–428.

212. Sakajiri S, Kumagai T, Kawamata N, Saitoh T, Said JW, Koeffler HP. Histone deacetylase inhibitors profoundly decrease proliferation of human lymphoid cancer cell lines. Exp Hematol 2005;33:53–61.

213. Bordin M, D'Atri F, Guillemot L, Citi S. Histone deacetylase inhibitors up-regulate the expression of tight junction proteins. Mol Cancer Res 2004;2:692–701.

214. Shao Y, Gao Z, Marks PA, Jiang X. Apoptotic and autophagic cell death induced by histone deacetylase inhibitors. Proc Natl Acad Sci USA 2004;101:18,030–18,035.

215. Park JH, Jung Y, Kim TY, et al. Class I histone deacetylase-selective novel synthetic inhibitors potently inhibit human tumor proliferation. Clin Cancer Res 2004;10: 5271–5281.

216. Marks PA, Richon VM, Breslow R, Rifkind RA. Histone deacetylase inhibitors as new cancer drugs. Curr Opin Oncol 2001;13:477–483.

217. McLaughlin F, La Thangue NB. Histone deacetylase inhibitors open new doors in cancer therapy. Biochem Pharmacol 2004;68:1139–1144.

218. Nebbioso A, Clarke N, Voltz E, et al. Tumor-selective action of HDAC inhibitors involves TRAIL induction in acute myeloid leukemia cells. Nat Med 2005;11:77–84.

219. Inoue S, MacFarlane M, Harper N, Wheat LM, Dyer MJ, Cohen GM. Histone deacetylase inhibitors potentiate TNF-related apoptosis-inducing ligand (TRAIL)-induced apoptosis in lymphoid malignancies. Cell Death Differ 2004;11 Suppl 2:S193–S206.

220. Facchetti F, Previdi S, Ballarini M, Minucci S, Perego P, La Porta CA. Modulation of pro- and anti-apoptotic factors in human melanoma cells exposed to histone deacetylase inhibitors. Apoptosis 2004;9:573–582.

221. Rosato RR, Almenara JA, Dai Y, Grant S. Simultaneous activation of the intrinsic and extrinsic pathways by histone deacetylase (HDAC) inhibitors and tumor necrosis factor-related apoptosis-inducing ligand (TRAIL) synergistically induces mitochondrial damage and apoptosis in human leukemia cells. Mol Cancer Ther 2003;2:1273–1284.

222. Shetty S, Graham BA, Brown JG, et al. Transcription factor NF-kappaB differentially regulates death receptor 5 expression involving histone deacetylase 1. Mol Cell Biol 2005;25:5404–5416.

223. Goldsmith KC, Hogarty MD. Targeting programmed cell death pathways with experimental therapeutics: opportunities in high-risk neuroblastoma. Cancer Lett 2005;228:133–141.

224. Vanoosten RL, Moore JM, Karacay B, Griffith TS. Histone Deacetylase Inhibitors Modulate Renal Cell Carcinoma Sensitivity to TRAIL/Apo-2L-induced Apoptosis by Enhancing TRAIL-R2 Expression. Cancer Biol Ther 2005;4:1104–1112.

225. Insinga A, Monestiroli S, Ronzoni S, et al. Inhibitors of histone deacetylases induce tumor-selective apoptosis through activation of the death receptor pathway. Nat Med 2005;11: 71–76.

226. Insinga A, Minucci S, Pelicci PG. Mechanisms of selective anticancer action of histone deacetylase inhibitors. Cell Cycle 2005;4:741–743.

227. Clarke N, Jimenez-Lara AM, Voltz E, Gronemeyer H. Tumor suppressor IRF-1 mediates retinoid and interferon anticancer signaling to death ligand TRAIL. EMBO J 2004;23: 3051–3060.

228. Ruiz de Almodovar C, Lopez-Rivas A, Ruiz-Ruiz C. Interferon-gamma and TRAIL in human breast tumor cells. Vitam Horm 2004;67:291–318.

229. Oehadian A, Koide N, Mu MM, et al. Interferon (IFN)-beta induces apoptotic cell death in DHL-4 diffuse large B cell lymphoma cells through tumor necrosis factor-related apoptosis-inducing ligand (TRAIL). Cancer Lett 2005;225:85–92.

230. Wang SH, Mezosi E, Wolf JM, et al. IFNgamma sensitization to TRAIL-induced apoptosis in human thyroid carcinoma cells by upregulating Bak expression. Oncogene 2004;23: 928–935.

231. Langaas V, Shahzidi S, Johnsen JI, Smedsrod B, Sveinbjornsson B. Interferon-gamma modulates TRAIL-mediated apoptosis in human colon carcinoma cells. Anticancer Res 2001;21:3733–3738.

232. Fulda S, Debatin KM. IFNgamma sensitizes for apoptosis by upregulating caspase-8 expression through the Stat1 pathway. Oncogene 2002;21:2295–2308.

233. Park SY, Seol JW, Lee YJ, et al. W. IFN-gamma enhances TRAIL-induced apoptosis through IRF-1. Eur J Biochem 2004;271:4222–4228.

234. Ortiz MA, Bayon Y, Lopez-Hernandez FJ, Piedrafita FJ. Retinoids in combination therapies for the treatment of cancer:mechanisms and perspectives. Drug Resist Updat 2002;5: 162–175.

235. Cuello M, Coats AO, Darko I, et al. N-(4-hydroxyphenyl) retinamide (4HPR) enhances TRAIL-mediated apoptosis through enhancement of a mitochondrial-dependent amplification loop in ovarian cancer cell lines. Cell Death Differ 2004;11:527–541.

236. Sun SY, Yue P, Hong WK, Lotan R. Augmentation of tumor necrosis factor-related apoptosis-inducing ligand (TRAIL)-induced apoptosis by the synthetic retinoid 6-[3-(1-adamantyl)-4-hydroxyphenyl]-2-naphthalene carboxylic acid (CD437) through up-regulation of TRAIL receptors in human lung cancer cells. Cancer Res 2000;60:7149–7155.

237. Sun SY, Yue P, Lotan R. Implication of multiple mechanisms in apoptosis induced by the synthetic retinoid CD437 in human prostate carcinoma cells. Oncogene 2000;19:4513–4522.

238. Altucci L, Rossin A, Raffelsberger W, Reitmair A, Chomienne C, Gronemeyer H. Retinoic acid-induced apoptosis in leukemia cells is mediated by paracrine action of tumor-selective death ligand TRAIL. Nat Med 2001;7:680–686.

239. Wajant H, Gerspach J, Pfizenmaier K. Tumor therapeutics by design: targeting and activation of death receptors. Cytokine Growth Factor Rev 2005;16:55–76.

240. Gliniak B, Le T. Tumor necrosis factor-related apoptosis-inducing ligand's antitumor activity in vivo is enhanced by the chemotherapeutic agent CPT-11. Cancer Res 1999;59:6153–6158.

241. Gibson SB, Oyer R, Spalding AC, Anderson SM, Johnson GL. Increased expression of death receptors 4 and 5 synergizes the apoptosis response to combined treatment with etoposide and TRAIL. Mol Cell Biol 2000;20:205–212.

242. Keane MM, Ettenberg SA, Nau MM, Russell EK, Lipkowitz S. Chemotherapy augments TRAIL-induced apoptosis in breast cell lines. Cancer Res 1999;59:734–741.

243. Hotta T, Suzuki H, Nagai S, et al. Chemotherapeutic agents sensitize sarcoma cell lines to tumor necrosis factor-related apoptosis-inducing ligand-induced caspase-8 activation, apoptosis and loss of mitochondrial membrane potential. J Orthop Res 2003;21:949–957.

244. Shankar S, Chen X, Srivastava RK. Effects of sequential treatments with chemotherapeutic drugs followed by TRAIL on prostate cancer in vitro and in vivo. Prostate 2005;62:165–186.

245. Shankar S, Srivastava RK. Enhancement of therapeutic potential of TRAIL by cancer chemotherapy and irradiation: mechanisms and clinical implications. Drug Resist Updat 2004;7:139–156.

246. Marini P, Schmid A, Jendrossek V, et al. Irradiation specifically sensitises solid tumour cell lines to TRAIL mediated apoptosis. BMC Cancer 2005;5:5.

247. Rezacova M, Vavrova J, Vokurkova D, Tichy A, Knizek J, Psutka J. The importance of abrogation of G(2)-phase arrest in combined effect of TRAIL and ionizing radiation. Acta Biochim Pol 2005;52:889–895.

248. Wendt J, von Haefen C, Hemmati P, Belka C, Dorken B, Daniel PT. TRAIL sensitizes for ionizing irradiation-induced apoptosis through an entirely Bax-dependent mitochondrial cell death pathway. Oncogene 2005;24:4052–4064.

249. Hamasu T, Inanami O, Asanuma T, Kuwabara M. Enhanced induction of apoptosis by combined treatment of human carcinoma cells with X rays and death receptor agonists. J Radiat Res (Tokyo) 2005;46:103–110.

250. Ciusani E, Croci D, Gelati M, et al. In vitro effects of topotecan and ionizing radiation on TRAIL/Apo2L-mediated apoptosis in malignant glioma. J Neurooncol 2005;71:19–25.

251. Morrison BH, Tang Z, Jacobs BS, Bauer JA, Lindner DJ. Apo2L/TRAIL induction and nuclear translocation of inositol hexakisphosphate kinase 2 during IFN-beta-induced apoptosis in ovarian carcinoma. Biochem J 2005;385:595–603.

252. Gong B, Chen Q, Endlich B, Mazumder S, Almasan A. Ionizing radiation-induced, Bax-mediated cell death is dependent on activation of cysteine and serine proteases. Cell Growth Differ 1999;10:491–502.

253. Yoshida T, Maeda A, Tani N, Sakai T. Promoter structure and transcription initiation sites of the human death receptor 5/TRAIL-R2 gene. FEBS Lett 2001;507:381–385.

254. Chen X, Kandasamy K, Srivastava RK. Differential roles of RelA (p65) and c-Rel subunits of nuclear factor kappa B in tumor necrosis factor-related apoptosis-inducing ligand signaling. Cancer Res 2003;63:1059–1066.

255. Park EJ, Pezzuto JM. Botanicals in cancer chemoprevention. Cancer Metastasis Rev 2002;21:231–255.

256. Cal C, Garban H, Jazirehi A, Yeh C, Mizutani Y, Bonavida B. Resveratrol and cancer: chemoprevention, apoptosis, and chemo-immunosensitizing activities. Curr Med Chem Anti-Canc Agents 2003;3:77–93.

257. Gusman J, Malonne H, Atassi G. A reappraisal of the potential chemopreventive and chemotherapeutic properties of resveratrol. Carcinogenesis 2001;22:1111–1117.

258. Fulda S, Debatin KM. Sensitization for tumor necrosis factor-related apoptosis-inducing ligand-induced apoptosis by the chemopreventive agent resveratrol. Cancer Res 2004;64:337–346.
259. Fulda S, Debatin KM. Resveratrol-mediated sensitisation to TRAIL-induced apoptosis depends on death receptor and mitochondrial signalling. Eur J Cancer 2005;41:786–798.
260. Delmas D, Jannin B, Latruffe N. Resveratrol: preventing properties against vascular alterations and ageing. Mol Nutr Food Res 2005;49:377–395.
261. Ulrich S, Wolter F, Stein JM. Molecular mechanisms of the chemopreventive effects of resveratrol and its analogs in carcinogenesis. Mol Nutr Food Res 2005;49:452–461.
262. Aggarwal BB, Bhardwaj A, Aggarwal RS, Seeram NP, Shishodia S, Takada Y. Role of resveratrol in prevention and therapy of cancer: preclinical and clinical studies. Anticancer Res 2004;24:2783–2840.
263. Mitchell SH, Zhu W, Young CY. Resveratrol inhibits the expression and function of the androgen receptor in LNCaP prostate cancer cells. Cancer Res 1999;59:5892–5895.
264. Hsieh TC, Wu JM. Grape-derived chemopreventive agent resveratrol decreases prostate-specific antigen (PSA) expression in LNCaP cells by an androgen receptor (AR)-independent mechanism. Anticancer Res 2000;20:225–228.
265. Bhat KP, Pezzuto JM. Resveratrol exhibits cytostatic and antiestrogenic properties with human endometrial adenocarcinoma (Ishikawa) cells. Cancer Res 2001;61:6137–6144.
266. Jang DS, Kang BS, Ryu SY, Chang IM, Min KR, Kim Y. Inhibitory effects of resveratrol analogs on unopsonized zymosan-induced oxygen radical production. Biochem Pharmacol 1999;57:705–712.
267. Narayanan BA, Narayanan NK, Re GG, Nixon DW. Differential expression of genes induced by resveratrol in LNCaP cells: P53-mediated molecular targets. Int J Cancer 2003;104:204–212.
268. Fulda S, Debatin KM. Sensitization for anticancer drug-induced apoptosis by the chemopreventive agent resveratrol. Oncogene 2004;23:6702–6711.
269. Deeb D, Jiang H, Gao X, et al. Curcumin sensitizes prostate cancer cells to tumor necrosis factor-related apoptosis-inducing ligand/Apo2L by inhibiting nuclear factor-kappaB through suppression of IkappaBalpha phosphorylation. Mol Cancer Ther 2004;3:803–812.
270. Deeb D, Xu YX, Jiang H, et al. Curcumin (diferuloyl-methane) enhances tumor necrosis factor-related apoptosis-inducing ligand-induced apoptosis in LNCaP prostate cancer cells. Mol Cancer Ther 2003;2:95–103.
271. Elsharkawy AM, Oakley F, Mann DA. The role and regulation of hepatic stellate cell apoptosis in reversal of liver fibrosis. Apoptosis 2005;10:927–939.
272. Jung EM, Lim JH, Lee TJ, Park JW, Choi KS, Kwon TK. Curcumin sensitizes tumor necrosis factor-related apoptosis-inducing ligand (TRAIL)-induced apoptosis through reactive oxygen species-mediated upregulation of death receptor 5 (DR5). Carcinogenesis 2005;26:1905–1913.
273. Ramachandran C, Rodriguez S, Ramachandran R, et al. Expression profiles of apoptotic genes induced by curcumin in human breast cancer and mammary epithelial cell lines. Anticancer Res 2005;25:3293–3302.
274. Sah NK, Munshi A, Kurland JF, McDonnell TJ, Su B, Meyn RE. Translation inhibitors sensitize prostate cancer cells to apoptosis induced by tumor necrosis factor-related apoptosis-inducing ligand (TRAIL) by activating c-Jun N-terminal kinase. J Biol Chem 2003;278:20,593–20,602.
275. Whanger PD. Selenium and its relationship to cancer: an update dagger. Br J Nutr 2004;91:11–28.
276. Sinha R, El-Bayoumy K. Apoptosis is a critical cellular event in cancer chemoprevention and chemotherapy by selenium compounds. Curr Cancer Drug Targets 2004;4:13–28.
277. Gopee NV, Johnson VJ, Sharma RP. Sodium selenite-induced apoptosis in murine B-lymphoma cells is associated with inhibition of protein kinase C-delta, nuclear factor kappaB, and inhibitor of apoptosis protein. Toxicol Sci 2004;78:204–214.

278. Klein EA. Selenium and vitamin E cancer prevention trial. Ann N Y Acad Sci 2004;1031: 234–241.

279. Klein EA. Selenium: epidemiology and basic science. J Urol 2004;171:S50–S53; discussion S53.

280. Klein EA, Thompson IM. Update on chemoprevention of prostate cancer. Curr Opin Urol 2004;14:143–149.

281. Dennis LK, Cohen MB. On the chemoprevention TRAIL. Cancer Biol Ther, 2002;1:291–292.

282. Allen JG, Steele P, Masters HG, D'Antuono MF. A study of nutritional myopathy in weaner sheep. Aust Vet J 1986;63:8–13.

283. Yamaguchi K, Uzzo RG, Pimkina J, et al. Methylseleninic acid sensitizes prostate cancer cells to TRAIL-mediated apoptosis. Oncogene 2005;24:5868–5877.

284. Shi RX, Ong CN, Shen HM. Protein kinase C inhibition and x-linked inhibitor of apoptosis protein degradation contribute to the sensitization effect of luteolin on tumor necrosis factor-related apoptosis-inducing ligand-induced apoptosis in cancer cells. Cancer Res 2005;65:7815–7823.

285. Leverkus M, Neumann M, Mengling T, et al. Regulation of tumor necrosis factor-related apoptosis-inducing ligand sensitivity in primary and transformed human keratinocytes. Cancer Res 2000;60:553–559.

286. Griffith TS, Chin WA, Jackson GC, Lynch DH, Kubin MZ. Intracellular regulation of TRAIL-induced apoptosis in human melanoma cells. J Immunol 1998;161:2833–2840.

287. Griffith TS, Rauch CT, Smolak PJ, et al. Functional analysis of TRAIL receptors using monoclonal antibodies. J Immunol 1999;162:2597–2605.

288. Kim CH, Gupta S. Expression of TRAIL (Apo2L), DR4 (TRAIL receptor 1), DR5 (TRAIL receptor 2) and TRID (TRAIL receptor 3) genes in multidrug resistant human acute myeloid leukemia cell lines that overexpress MDR 1 (HL60/Tax) or MRP (HL60/AR). Int J Oncol 2000;16:1137–1139.

289. Holen I, Cross SS, Neville-Webbe HL, et al. Osteoprotegerin (OPG) Expression by Breast Cancer Cells in vitro and Breast Tumours in vivo - A Role in Tumour Cell Survival? Breast Cancer Res Treat 2005;92:207–215.

290. Holen I, Croucher PI, Hamdy FC, Eaton CL. Osteoprotegerin (OPG) is a survival factor for human prostate cancer cells. Cancer Res 2002;62:1619–1623.

291. Shipman CM, Croucher PI. Osteoprotegerin is a soluble decoy receptor for tumor necrosis factor-related apoptosis-inducing ligand/Apo2 ligand and can function as a paracrine survival factor for human myeloma cells. Cancer Res 2003;63:912–916.

292. Kandasamy K, Srivastava RK. Role of the phosphatidylinositol 3′-kinase/PTEN/Akt kinase pathway in tumor necrosis factor-related apoptosis-inducing ligand-induced apoptosis in non-small cell lung cancer cells. Cancer Res 2002;62:4929–4937.

293. Wang J, Zheng L, Lobito A, et al. Inherited human Caspase-10 mutations underlie defective lymphocyte and dendritic cell apoptosis in autoimmune lymphoproliferative syndrome type II. Cell 1999;98:47–58.

294. Irmler M, Thome M, Hahne M, et al. Inhibition of death receptor signals by cellular FLIP. Nature 1997;388:190–195.

295. Zhang L, Zhu H, Teraishi F, et al. Accelerated degradation of caspase-8 protein correlates with TRAIL resistance in a DLD1 human colon cancer cell line. Neoplasia 2005;7: 594–602.

296. Rippo MR, Moretti S, Vescovi S, et al. FLIP overexpression inhibits death receptor-induced apoptosis in malignant mesothelial cells. Oncogene 2004;23:7753–7760.

297. Suh WS, Kim YS, Schimmer AD, et al. Synthetic triterpenoids activate a pathway for apoptosis in AML cells involving downregulation of FLIP and sensitization to TRAIL. Leukemia 2003;17:2122–2129.

298. Nam SY, Jung GA, Hur GC, et al. Upregulation of FLIP(S) by Akt, a possible inhibition mechanism of TRAIL-induced apoptosis in human gastric cancers. Cancer Sci 2003;94: 1066–1073.

299. Kim JH, Ajaz M, Lokshin A, Lee YJ. Role of antiapoptotic proteins in tumor necrosis factor-related apoptosis-inducing ligand and cisplatin-augmented apoptosis. Clin Cancer Res 2003;9:3134–3141.

300. Wajant H, Haas E, Schwenzer R, et al. Inhibition of death receptor-mediated gene induction by a cycloheximide-sensitive factor occurs at the level of or upstream of Fas-associated death domain protein (FADD). J Biol Chem 2000;275:24,357–24,366.

301. Ganten TM, Koschny R, Haas TL, et al. Proteasome inhibition sensitizes hepatocellular carcinoma cells, but not human hepatocytes, to TRAIL. Hepatology 2005;42:588–597.

302. Arch EM, Goodman BK, Van Wesep RA, et al. Deletion of PTEN in a patient with Bannayan-Riley-Ruvalcaba syndrome suggests allelism with Cowden disease. Am J Med Genet 1997;71:489–493.

303. Li J, Yen C, Liaw D, et al. PTEN, a putative protein tyrosine phosphatase gene mutated in human brain, breast, and prostate cancer. Science 1997;275:1943–1947.

304. Tashiro H, Blazes MS, Wu R, et al. Mutations in PTEN are frequent in endometrial carcinoma but rare in other common gynecological malignancies. Cancer Res 1997;57:3935–3940.

305. Ittmann MM. Chromosome 10 alterations in prostate adenocarcinoma (review). Oncol Rep 1998;5:1329–1335.

306. Whang YE, Wu X, Suzuki H, et al. Inactivation of the tumor suppressor PTEN/MMAC1 in advanced human prostate cancer through loss of expression. Proc Natl Acad Sci USA 1998;95:5246–5250.

307. Neri LM, Borgatti P, Tazzari PL, et al. The phosphoinositide 3-kinase/AKT1 pathway involvement in drug and all-trans-retinoic acid resistance of leukemia cells. Mol Cancer Res 2003;1:234–246.

308. Stambolic V, Suzuki A, de la Pompa JL, et al. Negative regulation of PKB/Akt-dependent cell survival by the tumor suppressor PTEN. Cell 1998;95:29–39.

309. Sekulic A, Hudson CC, Homme JL, et al. A direct linkage between the phosphoinositide 3-kinase-AKT signaling pathway and the mammalian target of rapamycin in mitogen-stimulated and transformed cells. Cancer Res 2000;60:3504–3513.

310. Bortul R, Tazzari PL, Cappellini A, et al. Constitutively active Akt1 protects HL60 leukemia cells from TRAIL-induced apoptosis through a mechanism involving NF-kappaB activation and cFLIP(L) up-regulation. Leukemia 2003;17:379–389.

311. Cappellini A, Tabellini G, Zweyer M, et al. The phosphoinositide 3-kinase/Akt pathway regulates cell cycle progression of HL60 human leukemia cells through cytoplasmic relocalization of the cyclin-dependent kinase inhibitor p27(Kip1) and control of cyclin D1 expression. Leukemia 2003;17:2157–2167.

312. Martelli AM, Tazzari PL, Tabellini G, et al. A new selective AKT pharmacological inhibitor reduces resistance to chemotherapeutic drugs, TRAIL, all-trans-retinoic acid, and ionizing radiation of human leukemia cells. Leukemia 2003;17:1794–1805.

313. Nesterov A, Lu X, Johnson M, Miller GJ, Ivashchenko Y, Kraft AS. Elevated AKT activity protects the prostate cancer cell line LNCaP from TRAIL-induced apoptosis. J Biol Chem 2001;276:10,767–10,774.

314. Puduvalli VK, Sampath D, Bruner JM, Nangia J, Xu R, Kyritsis AP. TRAIL-induced apoptosis in gliomas is enhanced by Akt-inhibition and is independent of JNK activation. Apoptosis 2005;10:233–243.

315. Yoshida S, Narita T, Koshida S, Ohta S, Takeuchi Y. TRAIL/Apo2L ligands induce apoptosis in malignant rhabdoid tumor cell lines. Pediatr Res 2003;54:709–717.

316. Kim KU, Wilson SM, Abayasiriwardana KS, et al. A Novel In Vitro Model of Human Mesothelioma for Studying Tumor Biology and Apoptotic Resistance. Am J Respir Cell Mol Biol 2005;33:541–548.

317. Kang JQ, Chong ZZ, Maiese K. Akt1 protects against inflammatory microglial activation through maintenance of membrane asymmetry and modulation of cysteine protease activity. J Neurosci Res 2003;74:37–51.

318. Alladina SJ, Song JH, Davidge ST, Hao C, Easton AS. TRAIL-induced apoptosis in human vascular endothelium is regulated by phosphatidylinositol 3-kinase/Akt through the short form of cellular FLIP and Bcl-2. J Vasc Res 2005;42:337–347.

319. Rychahou PG, Murillo CA, Evers BM. Targeted RNA interference of PI3K pathway components sensitizes colon cancer cells to TNF-related apoptosis-inducing ligand (TRAIL). Surgery 2005;138:391–397.

320. Yuan XJ, Whang YE. PTEN sensitizes prostate cancer cells to death receptor-mediated and drug-induced apoptosis through a FADD-dependent pathway. Oncogene 2002;21:319–327.

321. Verma IM, Stevenson JK, Schwarz EM, Van Antwerp D, Miyamoto S. Rel/NF-kappa B/I kappa B family: intimate tales of association and dissociation. Genes Dev 1995;9:2723–2735.

322. Baeuerle PA, Baltimore D. NF-kappa B: ten years after. Cell 1996;87:13–20.

323. Ghosh S, May MJ, Kopp EB. NF-kappa B and Rel proteins: evolutionarily conserved mediators of immune responses. Annu Rev Immunol 1998;16:225–260.

324. Ghosh S, Karin M. Missing pieces in the NF-kappaB puzzle. Cell 2002;109 Suppl:S81–S96.

325. DiDonato J, Mercurio F, Rosette C, et al. Mapping of the inducible IkappaB phosphorylation sites that signal its ubiquitination and degradation. Mol Cell Biol 1996;16:1295–1304.

326. Makarov SS. NF-kappa B in rheumatoid arthritis: a pivotal regulator of inflammation, hyperplasia, and tissue destruction. Arthritis Res 2001;3:200–206.

327. Vincenti MP, Coon CI, Brinckerhoff CE. Nuclear factor kappaB/p50 activates an element in the distal matrix metalloproteinase 1 promoter in interleukin-1beta-stimulated synovial fibroblasts. Arthritis Rheum 1998;41:1987–1994.

328. Vincenti MP, Brinckerhoff CE. Transcriptional regulation of collagenase (MMP-1, MMP-13) genes in arthritis: integration of complex signaling pathways for the recruitment of gene-specific transcription factors. Arthritis Res 2002;4:157–164.

329. Catley MC, Chivers JE, Cambridge LM, et al. IL-1beta-dependent activation of NF-kappaB mediates PGE2 release via the expression of cyclooxygenase-2 and microsomal prostaglandin E synthase. FEBS Lett 2003;547:75–79.

330. Crofford LJ, Tan B, McCarthy CJ, Hla T. Involvement of nuclear factor kappa B in the regulation of cyclooxygenase-2 expression by interleukin-1 in rheumatoid synoviocytes. Arthritis Rheum 1997;40:226–236.

331. Sherman MP, Aeberhard EE, Wong VZ, Griscavage JM, Ignarro LJ. Pyrrolidine dithiocarbamate inhibits induction of nitric oxide synthase activity in rat alveolar macrophages. Biochem Biophys Res Commun 1993;191:1301–1308.

332. Eberhardt W, Kunz D, Pfeilschifter J. Pyrrolidine dithiocarbamate differentially affects interleukin 1 beta- and cAMP-induced nitric oxide synthase expression in rat renal mesangial cells. Biochem Biophys Res Commun 1994;200:163–170.

333. Ravi R, Bedi GC, Engstrom LW, et al. Regulation of death receptor expression and TRAIL/Apo2L-induced apoptosis by NF-kappaB. Nat Cell Biol 2001;3:409–416.

334. Jeremias I, Debatin KM. TRAIL induces apoptosis and activation of NFkappaB. Eur Cytokine Netw 1998;9:687–688.

335. Oya M, Ohtsubo M, Takayanagi A, Tachibana M, Shimizu N, Murai M. Constitutive activation of nuclear factor-kappaB prevents TRAIL-induced apoptosis in renal cancer cells. Oncogene 2001;20:3888–3896.

336. Eid MA, Lewis RW, Abdel-Mageed AB, Kumar MV. Reduced response of prostate cancer cells to TRAIL is modulated by NFkappaB-mediated inhibition of caspases and Bid activation. Int J Oncol 2002;21:111–117.
337. Hu WH, Johnson H, Shu HB. Tumor necrosis factor-related apoptosis-inducing ligand receptors signal NF-kappaB and JNK activation and apoptosis through distinct pathways. J Biol Chem 1999;274:30,603–30,610.
338. Shetty S, Gladden JB, Henson ES, et al. Tumor necrosis factor-related apoptosis inducing ligand (TRAIL) up-regulates death receptor 5 (DR5) mediated by NFkappaB activation in epithelial derived cell lines. Apoptosis 2002;7: 413–420.
339. Ichikawa K, Liu W, Fleck M, et al. TRAIL-R2 (DR5) mediates apoptosis of synovial fibroblasts in rheumatoid arthritis. J Immunol 2003;171:1061–1069.
340. Kaliberov S, Stackhouse MA, Kaliberova L, Zhou T, Buchsbaum DJ. Enhanced apoptosis following treatment with TRA-8 anti-human DR5 monoclonal antibody and overexpression of exogenous Bax in human glioma cells. Gene Ther 2004;11:658–667.
341. Ohtsuka T, Buchsbaum D, Oliver P, Makhija S, Kimberly R, Zhou T. Synergistic induction of tumor cell apoptosis by death receptor antibody and chemotherapy agent through JNK/p38 and mitochondrial death pathway. Oncogene 2003;22:2034–2044.
342. Takeda K, Yamaguchi N, Akiba H, et al. Induction of tumor-specific T cell immunity by anti-DR5 antibody therapy. J Exp Med 2004;199:437–448.
343. Voelkel-Johnson C. An antibody against DR4 (TRAIL-R1) in combination with doxorubicin selectively kills malignant but not normal prostate cells. Cancer Biol Ther 2003;2: 283–290.
344. Georgakis GV, Li Y, Humphreys R, et al. Activity of selective fully human agonistic antibodies to the TRAIL death receptors TRAIL-R1 and TRAIL-R2 in primary and cultured lymphoma cells: induction of apoptosis and enhancement of doxorubicin- and bortezomib-induced cell death. Br J Haematol 2005;130:501–510.

12

DNA Damage-Dependent Apoptosis

Tomasz Skorski

Summary

DNA damage caused by endogenous and exogenous agents represents a serious survival challenge for cells. The cell-cycle checkpoints and DNA repair mechanisms are activated to allow more time for effective repair. However, if these mechanisms fail or the damage is irreparable, then apoptosis is induced to eliminate cells carrying DNA lesions and preserve genomic integrity. This chapter discusses how DNA damage-dependent apoptosis is induced and executed.

Key Words: DNA damage; DNA repair; apoptosis; checkpoint; p53; p73; mismatch.

1. Introduction

The integrity of genomic DNA is constantly challenged by the action of endogenous reactive oxygen species (ROS), by stochastic errors in replication or recombination, and by environmental and therapeutic genotoxic agents. In addition, massive DNA damage could be inflicted by genotoxic treatment with cytotoxic drugs and/or radiation (*see* Fig. 1). The precise type of DNA damage depends on the genotoxic agent used, and each type of damage represents a unique challenge for the cell (*see* Fig. 2). The DNA lesions sustained, and the mechanisms used to repair them, can depend on the cell's origin.

Mammalian cells developed several mechanisms to deal with the threat from DNA damaging agents. Damaged bases, mismatches and adducts are not usually lethal, unless their number reaches a certain threshold, but if unrepaired or mis-repaired they cause mutations in the genome *(1)*. By contrast, even small numbers of DNA double strand breaks (DSBs) can be lethal if unrepaired; and if not repaired correctly they can induce chromosomal aberrations *(2)*. Human cells may use at least seven different repair mechanisms to deal with DNA lesions that represent a "clear and present danger" to survival and genomic integrity (Table 1). These mechanisms—direct reversion, base-excision repair (BER), nucleotide-excision repair (NER), mismatch repair (MMR), homologous recombination repair (HRR), non-homologous end-joining (NHEJ), and single-strand annealing (SSA)—perform unique and/or partially overlapping functions.

Global DNA repair capability can be affected not only by the efficiency of repair mechanisms, but also by the time allowed for repair. DNA damage checkpoints (Table 2) control the length of the cell-cycle phase, the activation of DNA repair pathways and movement of DNA repair proteins to the sites of DNA damage, thus helping to integrate DNA repair with cell-cycle progression *(3,4)*. Disassociation of the checkpoints and DNA repair mechanisms can lead to apoptosis or genomic instability *(5–8)*.

From: *Apoptosis, Cell Signaling, and Human Diseases: Molecular Mechanisms, Volume 2*
Edited by R. Srivastava © Humana Press Inc., Totowa, NJ

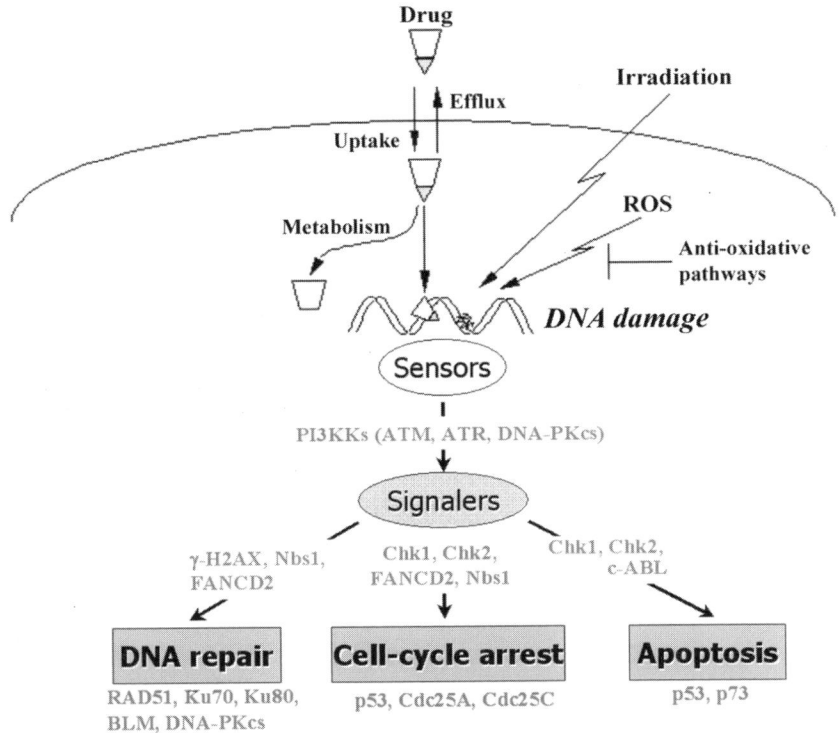

Fig. 1. The cellular events which decide whether "to be or not to be" after DNA damage. Drug metabolism and efflux may reduce the DNA damage. In addition, antioxidative pathways can minimize the damage caused by ROS generated by metabolic pathways or irradiation. Upon DNA damage the lesions are sensed by phosphatidylinositol 3-kinase–like kinases (PI3KKs) and the information is transferred via various signalers to the effectors regulating DNA repair, DNA damage-dependent cell-cycle checkpoints, and apoptosis. The final outcome (cell death/survival, genomic stability/instability) depends on the interplay between these mechanisms.

If the DNA damage is irreparable, then the apoptotic machinery is activated. It is clear that a coherent view of the apoptotic response to DNA damage must take into account events in both the nucleus and cytoplasm (*see* Fig. 3).

ATM and ATR serine/threonine kinases trigger a variety of DNA damage responses including proapoptotic pathways *(9,10)*. Targets of ATM/ATR and their downstream effectors Chk1 and Chk2 serine/threonine kinases include p53, in which phosphorylation on Ser15, Thr18, and Ser20 is essential for prevention of the Hdm2-p53 interaction and degradation of p53 *(11)*. In addition, DNA damage-activated c-ABL tyrosine kinase can phosphorylate Y394 of Hdm2 and prevent the ubiquitination and nuclear export of p53 *(12)*. In summary, accumulation of p53 in response to DNA damage is usually caused by abrogation of its Hdm2-dependent ubiquitination followed by proteasome degradation *(13)*. Altogether, accumulation of p53 is a key element in response to genotoxic stress *(14)*.

There is a broad consensus that one of the primary physiological roles of p53 in DNA damage response is to act as a transcriptional activator of genes encoding apoptotic effectors. In addition to DNA damage-induced ATM/ATR-dependent accumulation of

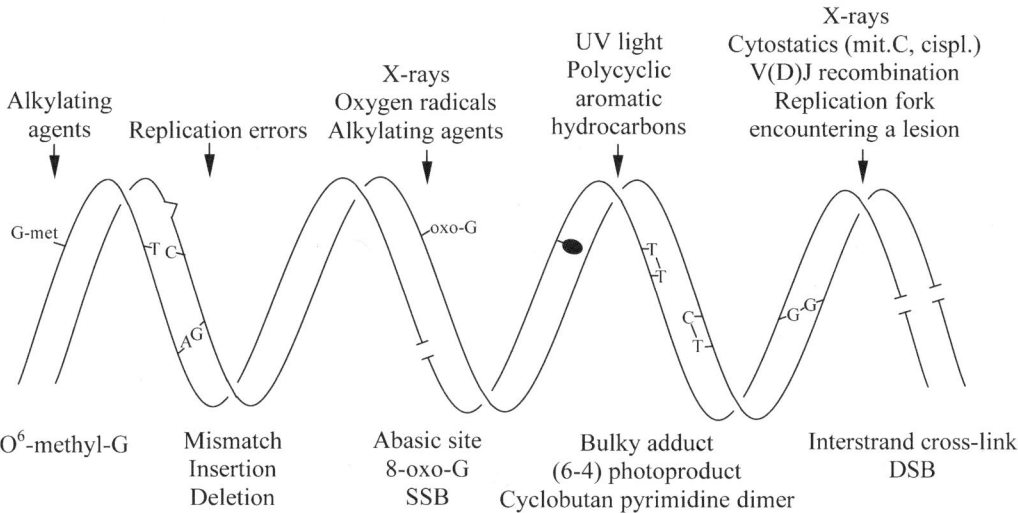

Fig. 2. Genomic DNA is bombarded by exogenous and endogenous damaging agents, which cause a variety of DNA lesions. These lesions trigger signaling pathways leading to cell-cycle checkpoint activation, DNA repair, and/or apoptosis.

p53 which enhances its total cellular transcriptional capability, c-ABL binds the C-terminus of the tetramerized p53 and enhances its transactivation ability *(15,16)*. p53 directly or indirectly regulates transcription of several genes encoding members of the Bcl-2 family (Table 3). This diverse family includes pro- and antiapoptotic proteins that together regulate mitochondrial permeability. The p53 targets considered most important in this respect are the proapoptotic Bax, Noxa and PUMA *(17)*. In addition, p53 may indirectly repress transactivation of the antiapoptotic Bcl-2 *(18)*. p53 can also down-regulate the antiapoptotic gene survivin, and up-regulate proapoptotic genes such as Fas, Apaf1, IGF-BP3, Killer/DR5, CD95, PERP, PIGs, and PIDD *(19)*.

Aside from transcriptional activation, several other roles for p53 have been identified that could contribute to its ability to induce DNA damage-dependent apoptosis. For example, p53 can directly regulate apotosis by translocation to the mitochondria *(20)*. Once at the mitochondrial outer membrane, p53 appears to antagonize the antiapoptotic Bcl-2 and Bcl-x_L proteins directly by binding to them, and induces oligomerization of the proapoptotic protein Bak, resulting in increased permeability of the outer mitochondrial membrane followed by release of cytochrome *c* and activation of caspase 3. Altogether, mitochondrial translocation of p53 triggers a rapid first wave of caspase 3 activation and cell death followed by a slower transcription-dependent p53 death wave *(21)*. Interestingly, p53 binds to Bcl-x_L via its DNA binding domain, which is often mutated in cancers. These tumor-derived transactivation-deficient p53 mutants lose or compromise their ability to interact with Bcl-x_L and to promote cytochrome *c* release. Thus, tumor-derived p53 mutations may represent "double hits," eliminating the transcriptional as well as the direct mitochondrial functions of p53.

p53 was also found to be required for the specific deamination of two asparagine residues in Bcl-x_L, neutralizing its ability to block the action of proapoptotic proteins such as Bax *(22)*. The direct mitochondrial function as well as the role in deamination

Table 1
Major DNA Damage Repair Pathways

DNA repair mechanisms	Description
Direct reversion	Methylguanine-DNA methyltransferase (MGMT) transfers the methyl group from the DNA to an internal cysteine residue on the MGMT protein *(40)*.
Base excision repair (BER)	BER is involved in the repair of "small" lesions such as base damage caused by hydroxylation, oxidation, methylation or deamination. DNA glycosylase recognizes abnormal DNA bases and catalyzes their cleavage, creating an abasic site. If the damage is limited to a single base, a "short patch" is created: an endonuclease, APE-1, then generates a single nucleotide gap at the abasic site. DNA polβ fills the gap, which is ligated by the XRCC1–DNA ligase 3 complex. If the damage is more extensive (2–10 bases) a "long patch" is created: the abasic site recruits the proliferating cell nuclear antigen (PCNA)–DNA polδ/ϵ complex for repair synthesis, followed by FEN1 endonuclease to remove the displaced DNA flap, and DNA ligase 1 for sealing *(1)*.
Nucleotide excision repair (NER)	Deals with a wide range of helix-distorting lesions. The damage recruits the XPA–RPA–XPC–TFIIH complex to form a preincision complex. After XPB and XPD have unwound about 20 bases around the lesion, the XPG and ERCC1–XPF endonucleases make 3′- and 5′-incisions, respectively. The excised fragment is removed and the gap is filled by PCNA–DNA polδ/ϵ and ligation *(1)*.
Mismatch repair (MMR)	Repairs mispaired/modified bases and insertion/deletion loops. MSH2–MSH6 binds to mismatches and single-base loops, whereas MSH2–MSH3 focuses on insertion/deletion loops. These protein–DNA complexes attract MLH1–PMS2 or MLH1–PMS1 heterodimers followed by exonucleases, which excise the new strand past the mismatch/loop. Resynthesis steps involve PCNA, RFC, RPA, and DNA polδ/ϵ *(1,41,42)*.
Homologous recombination repair (HRR)	Repairs DSBs preferentially in late S and G2 phases of the cell cycle. DSB ends are resected by the RAD50–MRE11–NBS1 complex to create single-stranded DNA tails. RAD51 paralogs, in collaboration with RPA, RAD52 and RAD54, promote the invasion of the single-stranded DNA to the intact sister chromatid to find a matching sequence, which is used as a template to heal the broken ends by DNA synthesis. Finally, the junctions between the homologous chromosomes (Holliday junctions) are untangled by resolvases *(1,2,43)*.
Non-homologous end joining (NHEJ)	Repairs double-strand breaks primarily in G1 phase. The ends of the breaks are not modified or are processed by the RAD50–MRE11–NBS1 complex and/or DNA polymerases, and are then simply linked together, without any template, using the end-binding KU70–KU80 complex and DNA-dependent protein kinase, followed by ligation by the XRCC4 ligase *(1,2)*.

(Continued)

Table 1 (*Continued*)

DNA repair mechanisms	Description
Single-strand annealing (SSA)	Repairs DSBs using homology between the ends of the joined sequence. In contrast to HRR, the homology is found not by invasion of the sister chromatid, but by the RAD50–MRE11–NBS1-mediated resection of the broken ends to create single-stranded DNA tails. When this resection reveals complementary sequences, the two DNA tails are annealed before being ligated by ligase 4. The over hanging tails are then trimmed by the XPF–ERCC1 endonuclease *(2,43)*.

Table 2
DNA Damage-Induced Cell-Cycle Checkpoints

Checkpoints	Description
G1/S	This checkpoint ensures that damaged DNA is not replicated. DNA damage activates protein kinases including ATM, ATR, and c-ABL. These lead to the phosphorylation of p53 and p73, resulting in upregulation of the expression of the cyclin-dependent kinase inhibitor WAF1 (also known as p21). WAF1 can bind several cyclin–CDK complexes and might mediate the G1/S checkpoint. Genotoxic stress also induces the rapid ubiquitinylation and degradation of Cdc25A, the protein phosphatase that dephosphorylates Cdk2. Dephosphorylated Cdk2 activates cyclin E-Cdk2 and promotes the G1/S progression, so this degradation of Cdc25A, which is dependent on the activation of another kinase, Chk1, provides another means of halting the cell cycle *(3,4,44)*.
S	This checkpoint prevents replication if DNA damage was not repaired in G1 phase or did not occur until in early S phase. The exact mechanism of this checkpoint is poorly understood. It seems that DNA damage-dependent activation of ATM is responsible for activation of Chk2, which phosphorylates Cdc25A on Ser123 and promotes its ubiquitinylation *(45)*.
G2	This checkpoint presumably allows for repair of DNA that was damaged in late S phase or G2 phase, and prevents damaged DNA from being segregated into daughter cells. It depends on the inhibition of cdc2 kinase activity, which is activated by phosphorylation at Thr161 by Cdk-activating kinase (CAK) and dephosphorylation on Tyr15 and Thr14 by Cdc25C phosphatase. DNA-damage-dependent stimulation of ATM/ATR activates Chk1 and/or Chk2, which phosphorylate cdc25C on Ser216. This creates a binding site for the 14-3-3 proteins, and results in nuclear export of cdc25c and its retention in the cytoplasm. Nuclear cdc2 remains phosphorylated in the absence of cdc25C and cells are arrested in the G2 phase. ATM/ATR also phosphorylates and stabilizes p53, causing up-regulation of 14-3-3, GADD45 and p21 proteins, thus increasing the cytoplasmic retention of cdc25C and inhibition of the cdc2–cyclin B complex. In addition, p53 represses the transcription of cdc2 and cyclin B *(3,4,7,46,47)*.

(*Continued*)

Table 2 (*Continued*)

Checkpoints	Description
M	This checkpoint controls mitotic spindle and metaphase-anaphase transition. Mitotic entry is associated with chromosome condensation, formation of the mitotic spindle, and activation of the anaphase-promoting complex (APC) responsible for initiation of the physical separation of sister chromatids during anaphase. If chromosomal errors occur mitotic spindle may not form and/or APC may not be activated. Cells arrested in M phase can decondense their chromosomes and return to G2 phase before re-entering M phase, perhaps allowing for better access of DNA repair machinery to the lesions *(48)*.

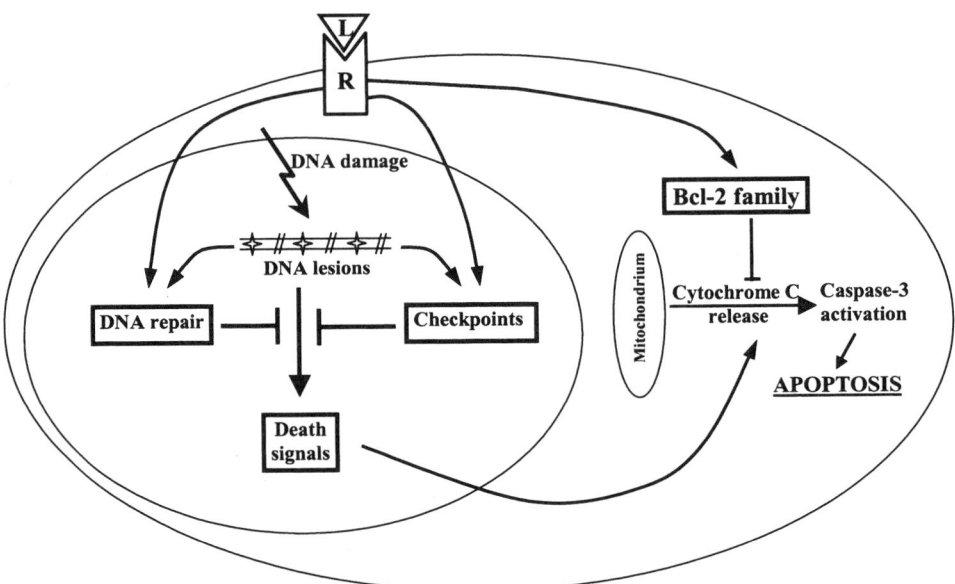

Fig. 3. A comprehensive model of DNA damage response. The expression of proteins involved in DNA damage responses (repair, checkpoint, apoptosis) is regulated by the physiological signaling mediated by a ligand (L) binding to a receptor (R). It can be further modified by the mechanisms induced by DNA damage (post-translational modification, [e.g., phosphorylation]). DNA repair and checkpoint activation work together to fix the problem (DNA damage) and prevent apoptosis. However, extensive DNA damage may not be efficiently repaired. As a result, unrepaired lesions trigger pronounced death-signaling pathways in the nucleus. Signals of this magnitude cannot be inhibited in the cytoplasm by the antiapoptotic Bcl-2 family proteins, resulting in the release of cytochrome *c* from mitochondria, activation of caspase-3 and apoptosis.

are consistent with early demonstrations that p53-mediated apoptosis does not necessarily involve new protein synthesis *(23)*.

Two p53 related genes—p63 and p73—have been discovered *(24,25)*. As might be predicted, both proteins share considerable homology and functional similarities with p53 *(26)*. p73, similarly to p53, is also regulated by DNA damage resulting in apoptosis,

Table 3
Bcl-2 Family Members

Antiapoptotic	Proapoptotic	Description
Bcl-2	Bax	An important site of activity of the Bcl-2 proteins
Bcl-x_L	Bak	is the mitochondrial membrane. Following a variety
Bcl-w	Bok	of death signals, members of proapoptotic Bcl-2-
Mcl-1	Bcl-x_S	family members undergo certain modifications.
A1	Bad	For example, Bax dimerizes, Bad is dephosphory-
	Bid	lated and Bid is cleaved. These modifications allow
	Bik	the proapoptotic Bcl-2-family members to trans
	Bim	locate to mitochondria and/or to heterodimerize
		with the antiapoptotic members of the family.
		Homodimerization (e.g., Bax–Bax) or hetero-
		dimerization (e.g., Bcl-xL–Bad) is an essential
		step for regulation of the release of cytochrome *c*
		from mitochondria. Cytosolic cytochrome *c*
		activates a proteolytic caspases cascade that leads
		to apoptotic death *(49,50)*.

albeit using a pathway distinct from that of p53. p73 is phosphorylated by c-Abl, which leads to the accumulation of p73 and apoptosis *(27–29)*. Interestingly, in response to DNA damage p53 may require p63 and p73 to activate promoters of apoptotic genes such as Noxa and Bax. Therefore, it is conceivable that p53, p63, and p73 work in concert to induce apoptosis after genotoxic treatment *(30)*.

In addition to the p53 family members, other proteins can regulate apoptotic response to DNA damage. Mismatch repair (MMR) proteins function to repair DNA lesions, which arise as a result of replication errors, and oxidative, alkylating or base crosslinking damage *(31)*. Mammalian cells have six MMR genes: MSH2, MSH6, MLH1, PMS1, PMS2, and MSH3 *(32)*. A role for the MMR system in eliciting apoptosis was suggested by the observation of reduced sensitivity to cytotoxic DNA damaging agents in MMR defective tumor cells *(33)*. Conversely, overexpression of MMR led to apoptosis *(34)*. The precise mechanism by which MMR mediates cell death remains unclear. One unexpected link between MMR proteins and apoptosis comes from recent studies of MBD4, a thymidine/uracil DNA glycosylase that interacts with MLH1 and Fas-associated death domain protein (FADD), an adaptor protein bridging death receptors with initiator caspases *(35)*. MBD4 controls nuclear export of FADD and affects Fas-mediated apoptosis. Interestingly, p53 undergoes rapid stabilization and phosphorylation on Ser15 in MMR-proficient cells treated with DNA damaging agents *(36,37)*. This phenomenon may depend on the MLH1-PMS1 heterodimer. In addition, stabilization of p73 and its nuclear localization in response to DNA damage was dependent on the interaction with PMS2 *(38)*.

E2F family of transcription factors can not only coordinate progress through the cell cycle, but also stimulate apoptosis *(39)*. The latter function is mediated by sequestering Mdm2 and stabilization of p53, and direct transcriptional activation of Apaf1 and p73. E2F1 activity is directly modulated in response to DNA damage, for example Chk2 can directly phosphorylate E2F1, which is important for DNA damage-dependent apoptosis.

2. Conclusions

In conclusion, although the role of p53 in DNA damage-induced apoptosis is well established, our knowledge about other signaling proteins in the DNA lesion—caspase axis is rather limited. Detailed information about this mechanism may be essential for better protection of normal cells against endogenous and exogenous genotoxic agents.

Acknowledgments

Tomasz Skorski is supported by the grants from National Institutes of Health, American Cancer Society, Department of Defense and a Scholarship from the Leukemia and Lymphoma Society.

References

1. Hoeijmakers JH. Genome maintenance mechanisms for preventing cancer. Nature 2001; 411:366–374.
2. Khanna KK, Jackson SP. DNA double-strand breaks: signaling, repair and the cancer connection. Nat Genet 2001;27:247–254.
3. Dasika GK, Lin SC, Zhao S, Sung P, Tomkinson A, Lee EY. DNA damage-induced cell cycle checkpoints and DNA strand break repair in development and tumorigenesis. Oncogene 1999;18:7883–7899.
4. Zhou BB, Elledge SJ. The DNA damage response: putting checkpoints in perspective. Nature 2000;408:433–439.
5. Shapiro GI, Harper JW. Anticancer drug targets: cell cycle and checkpoint control. J Clin Invest 1999;104:1645–1653.
6. Bracey TS, Williams AC, Paraskeva C. Inhibition of radiation-induced G2 delay potentiates cell death by apoptosis and/or the induction of giant cells in colorectal tumor cells with disrupted p53 function. Clin Cancer Res 1997;3:1371–1381.
7. Chan TA, Hermeking H, Lengauer C, Kinzler KW, Vogelstein B. 14-3-3 Sigma is required to prevent mitotic catastrophe after DNA damage. Nature 1999;401:616–620.
8. Elledge SJ. Cell cycle checkpoints: preventing an identity crisis. Science 1996; 274:1664–1672.
9. Shiloh Y. ATM and related protein kinases: safeguarding genome integrity. Nat Rev Cancer 2003;3:155–168.
10. Abraham RT. Cell cycle checkpoint signaling through the ATM and ATR kinases. Genes Dev 2001;15:2177–2196.
11. Michael D, Oren M. The p53-Mdm2 module and the ubiquitin system. Semin Cancer Biol 2003;13:49–58.
12. Goldberg Z, Vogt Sionov R, Berger M, et al. Tyrosine phosphorylation of Mdm2 by c-Abl: implications for p53 regulation. Embo J 2002;21:3715–3727.
13. Haupt Y, Maya R, Kazaz A, Oren M. Mdm2 promotes the rapid degradation of p53. Nature 1997;387:296–299.
14. Schuler M, Green DR. Transcription, apoptosis and p53: catch-22. Trends Genet 2005; 21:182–187.
15. Nie Y, Li HH, Bula CM, Liu X. Stimulation of p53 DNA binding by c-Abl requires the p53 C terminus and tetramerization. Mol Cell Biol 2000;20:741–748.
16. Goga A, Liu X, Hambuch TM, et al. p53 dependent growth suppression by the c-Abl nuclear tyrosine kinase. Oncogene 1995;11:791–799.
17. Norbury CJ, Zhivotovsky B. DNA damage-induced apoptosis. Oncogene 2004;23: 2797–2808.
18. Miyashita T, Harigai M, Hanada M, Reed JC. Identification of a p53-dependent negative response element in the bcl-2 gene. Cancer Res 1994;54:3131–3135.

19. Slee EA, O'Connor DJ, Lu X. To die or not to die: how does p53 decide? Oncogene 2004; 23:2809–2818.
20. Mihara M, Erster S, Zaika A, et al. p53 has a direct apoptogenic role at the mitochondria. Mol Cell 2003;11:577–590.
21. Erster S, Mihara M, Kim RH, Petrenko O, Moll UM. In vivo mitochondrial p53 translocation triggers a rapid first wave of cell death in response to DNA damage that can precede p53 target gene activation. Mol Cell Biol 2004;24:6728–6741.
22. Deverman BE, Cook BL, Manson SR, et al. Bcl-xL deamidation is a critical switch in the regulation of the response to DNA damage. Cell 2002;111:51–62.
23. Caelles C, Helmberg A, Karin M. p53-dependent apoptosis in the absence of transcriptional activation of p53-target genes. Nature 1994;370:220–223.
24. Kaghad M, Bonnet H, Yang A, et al. Monoallelically expressed gene related to p53 at 1p36, a region frequently deleted in neuroblastoma and other human cancers. Cell 1997;90:809–819.
25. Yang A, Kaghad M, Wang Y, et al. p63, a p53 homolog at 3q27-29, encodes multiple products with transactivating, death-inducing, and dominant-negative activities. Mol Cell 1998; 2:305–316.
26. Yang A, Kaghad M, Caput D, McKeon F. On the shoulders of giants: p63, p73 and the rise of p53. Trends Genet 2002;18:90–95.
27. Gong JG, Costanzo A, Yang HQ, et al. The tyrosine kinase c-Abl regulates p73 in apoptotic response to cisplatin-induced DNA damage. Nature 1999;399:806–809.
28. Agami R, Blandino G, Oren M, Shaul Y. Interaction of c-Abl and p73alpha and their collaboration to induce apoptosis. Nature 1999;399:809–813.
29. Yuan ZM, Shioya H, Ishiko T, et al. p73 is regulated by tyrosine kinase c-Abl in the apoptotic response to DNA damage. Nature 1999;399:814–817.
30. Flores ER, Tsai KY, Crowley D, et al. p63 and p73 are required for p53-dependent apoptosis in response to DNA damage. Nature 2002;416:560–564.
31. Meyers M, Hwang A, Wagner MW, Boothman DA. Role of DNA mismatch repair in apoptotic responses to therapeutic agents. Environ Mol Mutagen 2004;44:249–264.
32. Stojic L, Brun R, Jiricny J. Mismatch repair and DNA damage signalling. DNA Repair (Amst) 2004;3:1091–1101.
33. Fedier A, Fink D. Mutations in DNA mismatch repair genes: implications for DNA damage signaling and drug sensitivity (review). Int J Oncol 2004;24:1039–1047.
34. Zhang H, Richards B, Wilson T, et al. Apoptosis induced by overexpression of hMSH2 or hMLH1. Cancer Res 1999;59:3021–3027.
35. Bellacosa A, Cicchillitti L, Schepis F, et al. MED1, a novel human methyl-CpG-binding endonuclease, interacts with DNA mismatch repair protein MLH1. Proc Natl Acad Sci USA 1999;96:3969–3974.
36. Duckett DR, Bronstein SM, Taya Y, Modrich P. hMutSalpha- and hMutLalpha-dependent phosphorylation of p53 in response to DNA methylator damage. Proc Natl Acad Sci USA 1999;96:12,384–12,388.
37. Luo Y, Lin FT, Lin WC. ATM-mediated stabilization of hMutL DNA mismatch repair proteins augments p53 activation during DNA damage. Mol Cell Biol 2004;24:6430–6444.
38. Shimodaira H, Yoshioka-Yamashita A, Kolodner RD, Wang JY. Interaction of mismatch repair protein PMS2 and the p53-related transcription factor p73 in apoptosis response to cisplatin. Proc Natl Acad Sci USA 2003;100:2420–2425. Epub 2003 Feb 24.
39. Bell LA, Ryan KM. Life and death decisions by E2F-1. Cell Death Differ 2004;11:137–142.
40. Pegg AE. Repair of O(6)-alkylguanine by alkyltransferases. Mutat Res 2000;462:83–100.
41. Fishel R. Signaling mismatch repair in cancer. Nat Med 1999;5:1239–1241.
42. Kolodner RD, Marsischky GT. Eukaryotic DNA mismatch repair. Curr Opin Genet Dev 1999;9:89–96.

43. Norbury CJ, Hickson ID. Cellular responses to DNA damage. Annu Rev Pharmacol Toxicol 2001;41:367–401.
44. Mailand N, Falck J, Lukas C, et al. Rapid destruction of human Cdc25A in response to DNA damage. Science 2000;288:1425–1429.
45. Falck J, Mailand N, Syljuasen RG, Bartek J, Lukas J. The ATM-Chk2-Cdc25A checkpoint pathway guards against radioresistant DNA synthesis. Nature 2001;410:842–847.
46. Chan TA, Hwang PM, Hermeking H, Kinzler KW, Vogelstein B. Cooperative effects of genes controlling the G(2)/M checkpoint. Genes Dev 2000;14:1584–1588.
47. Taylor WR, Stark GR. Regulation of the G2/M transition by p53. Oncogene 2001;20: 1803–1815.
48. Shah JV, Cleveland DW. Waiting for anaphase: Mad2 and the spindle assembly checkpoint. Cell 2000;103:997–1000.
49. Antonsson B, Martinou JC. The Bcl-2 protein family. Exp Cell Res 2000;256:50–57.
50. Gross A, McDonnell JM, Korsmeyer SJ. BCL-2 family members and the mitochondria in apoptosis. Genes Dev 1999;13:1899–1911.

13

The Role of Proteasome in Apoptosis

Peter Low

Summary

The proteasome is an abundant multicatalytic enzyme complex that provides the main pathway for degradation of intracellular proteins in eukaryotic cells. It was previously considered a humble garbage collector, performing housekeeping duties to remove misfolded or spent proteins. Recent findings also substantiate a pivotal role of the proteasome in the regulation of apoptosis. Polyubiquitin conjugation of key pro- and antiapoptotic molecules targets these proteins for proteasomal degradation. Moreover, regulators of apoptosis themselves seem to have an active part in the proteolytic inactivation of death executors.

Proteasome inhibitors induce apoptosis in multiple cell types, whereas in other they are relatively harmless or even prevent apoptosis. The knowledge about the involvement of the ubiquitin–proteasome system in apoptosis is already clinically exploited, because proteasome inhibitors are being tested as experimental drugs in the treatment of cancer and other pathological conditions, where manipulation of apoptosis is desirable.

This chapter reviews the ubiquitin–proteasome system, highlighting the processes of apoptosis in which this pathway plays an important regulatory role and explains how the inhibitors of the proteasome, as well as substrate specific E3 ubiquitin ligases, might be used as new classes of anticancer drugs.

Key Words: Ubiquitin; proteasome; caspases; proteolysis.

1. Introduction

Apoptosis, a type of programmed cell death, is central to the development and homeostasis of metazoans. During development, apoptosis is used to sculpt or completely remove tissues *(1)*. Moreover, at later stages of development and during adult life, it is of central importance for the elimination of superfluous or harmful cells in a highly regulated manner. Not surprisingly, therefore, malfunction or dysregulation of this fundamental process can cause various diseases, ranging from autoimmune and immunodeficiency diseases to neurodegenerative disorders and cancer. In fact, every immunological or oncological disease can be traced to defects in apoptosis, thus generating a great interest and urgency in the understanding of mechanisms underlie apoptosis signaling. Since the concept of apoptosis was established in 1972 *(2)*, research efforts have identified hundreds of genes that control the initiation, execution, and regulation of apoptosis in several species *(3)*.

The ubiquitin–proteasome system is the main nonlysosomal route for intracellular protein degradation in eukaryotes. Recent findings also substantiate a pivotal role of this proteolytic pathway in the regulation of apoptosis. Regulatory molecules that are involved in programmed cell death have been identified as substrates of the proteasome *(4)*. This chapter provides a brief summary of the ubiquitin–proteasome system and

From: *Apoptosis, Cell Signaling, and Human Diseases: Molecular Mechanisms, Volume 2*
Edited by R. Srivastava © Humana Press Inc., Totowa, NJ

highlights processes of apoptosis in which this pathway plays an important role and finally discuss new therapeutic approaches for some human diseases that are linked with dysregulated apoptosis.

2. The Ubiquitin–Proteasome System for Protein Degradation

Nearly all proteins in mammalian cells are continually being degraded and replaced by *de novo* synthesis. The rates of degradation of individual cell constituents vary widely, with half-lives ranging from 10 min to several days or weeks. Degradation is also a means of eliminating misfolded, damaged, or mutant proteins with abnormal conformations whose accumulation might be harmful to the cell *(5)*.

Lysosomal and proteasomal degradation are the two major pathways for cellular protein turnover. Cell surface proteins that are taken up by endocytosis are degraded in the lysosome. Lysosomal degradation accounts for 10 to 20% of normal protein turnover. However, the bulk of cellular proteins (80%) is degraded by the proteasome in the cytoplasm and nucleus after being tagged with ubiquitin. Many apoptosis regulatory proteins have been identified as target substrates for ubiquitylation. In addition to being targets for ubiquitylation, some of them exhibit ubiquitin ligase activity *(6)*.

2.1. Ubiquitin

Proteins become tagged with ubiquitin polypeptides during ubiquitylation, which targets them for degradation by the proteasome. Ubiquitin is a small, 8.6-kDa globular protein present in all studied eukaryotic cells. It is coded either as polyubiquitin or as ribosomal fusion proteins. It is a highly conserved protein, amino acid sequence of the yeast and human ubiquitin differs only in 3 out of 76 amino acids *(7)*. Ubiquitin is covalently attached to substrates by an isopeptide bond between its C-terminal glycine and an ε-amino group of substrate (or another ubiquitin) lysine residue *(8)*. Classically, a chain of ubiquitins is formed as several ubiquitin molecules are processively added to the K48 of the preceding ubiquitin (*see* Fig. 1). Three enzymes mediate ubiquitin conjugation: In most organisms, a single E1 enzyme (ubiquitin-activating enzyme, UBA) becomes covalently linked to free ubiquitin through the C-terminal residue of ubiquitin, G76, in an energy-dependent manner. The activated ubiquitin is subsequently transferred to a cysteine on an E2 enzyme, (ubiquitin-conjugating enzyme, UBC, of which there are approx 100 in humans), which all contain a characteristic UBC domain. Finally E3 ubiquitin ligases (approx 500 in humans) recruit the ubiquitin-charged E2 to a substrate and the ubiquitin is transferred directly to the substrate *(9)* (*see* Fig. 1). The E3 ubiquitin ligase binds to specific protein substrates and promotes the transfer of ubiquitin from a thioester intermediate to amide linkages with proteins or polyubiquitin chains. E3 ubiquitin ligases can be classified into three major types based upon their domain structure and substrate recognition. The first class is the "N-end rule" ubiquitin ligases that target protein substrates bearing specific destabilizing N-terminal residues including Arg, Lys, His (type 1), Phe, Trp, Leu, Tyr, Ile (type 2), and Ala, Thr, Ser (type 3) *(10)*.

The second group of E3 is the HECT (homology to E6AP C-Terminus) domain family. The first member of this family is E6-AP (E6-associated protein) that in association with oncoprotein E6 promotes p53 ubiquitylation and degradation *(11)*. The HECT E3 ligases contain an approximately 350-amino-acid C-terminal region homologous to that

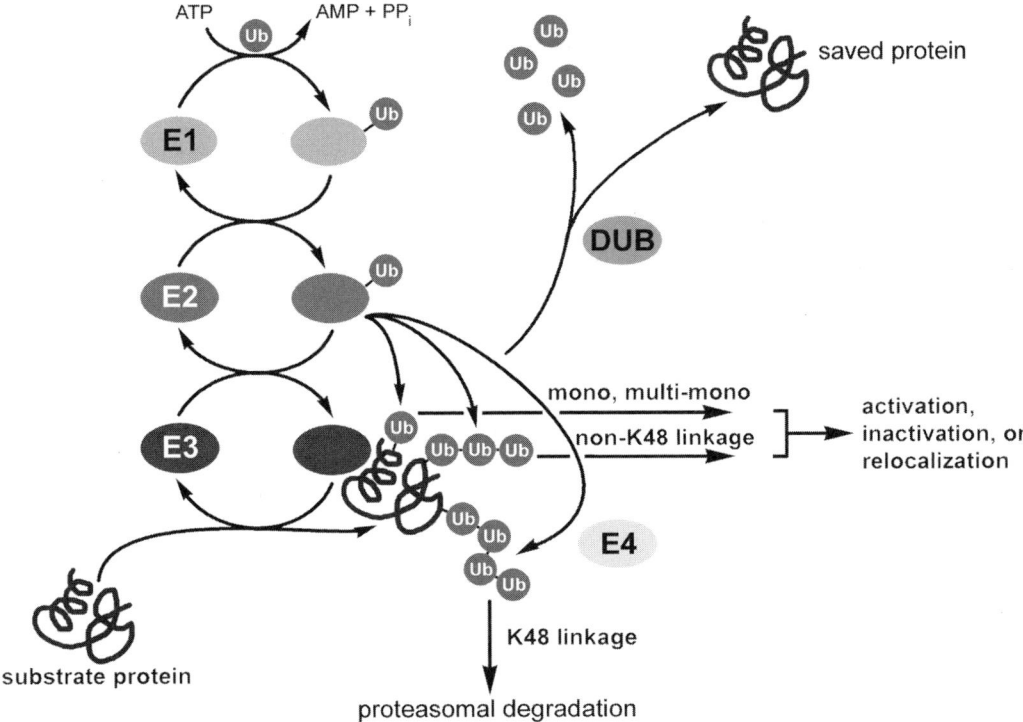

Fig. 1. The ubiquitylation cascade. Ubiquitin (Ub) is attached to substrates by an isopeptide bond between a substrate (or ubiquitin) lysine residue and the C-terminal glycine of ubiquitin. In most organisms, a single E1 enzyme becomes covalently linked to free ubiquitin through the free C-terminal residue of ubiquitin in an energy-dependent manner. The activated ubiquitin is subsequently transferred to a cysteine on an E2 ubiquitin-conjugating enzyme. E3 ligases recruit the ubiquitin-charged E2 to a substrate and the ubiquitin is transferred directly to the substrate. In some cases, an E4 might be required to promote polyubiquitylation. Substrates that are monoubiquitylated, multimonoubiquitylated or polyubiquitylated with chains linked through a ubiquitin lysine residue other than K48 are not degraded, and such modification might result in a change in the activation state of the protein or its localization. Substrates that are polyubiquitylated by a K48-linked chain of four or more ubiquitins are rapidly recruited to the proteasome and degraded. This process can be opposed by the action of deubiquitylating enzymes (DUB), which remove ubiquitin from substrate proteins, thus stabilizing them.

of E6-AP with a conserved active site cysteine residue, through which HECT domain E3s form thioester intermediates with ubiquitin. The N-terminal regions are highly variable and may be involved in substrate recognition *(12)*.

The third and the largest type of E3 ligase is the really interesting new gene (RING) family, containing a classic C3H2C3 or C3HC4 RING finger domain *(13)*. This domain has four pairs of metal binding residues with a characteristic linear sequence of Cys-X_2-Cys-X_{9-39}-Cys-X_{1-3}-His-X_{2-3}-Cys/His-X_2-Cys-X_{4-48}-Cys-X_2-Cys, where X can be any amino acid, although there are distinct preferences for particular types of amino acid at a particular position. The RING finger domain binds to two zinc atoms per molecule in a crossbraced system, where the first and third pairs of cysteines/histidine form the first binding site and the second and forth pairs of cysteines/histidine form the other. Almost all RING-containing proteins have the E3 ubiquitin ligase activity towards themselves as well as other protein substrates *(14,15)*.

The E3 ubiquitin ligases can exist and act as single peptides, examplified by Mdm2 and XIAP, or as multiple component complexes, such as SCF (Skp1-Cullin-F box protein) and anaphase promoting complex (APC). Through the covalent modification of a vast repertoire of cellular proteins with ubiquitin, E3 ubiquitin ligases regulate almost all aspects of eukaryotic cellular functions or biological processes, including cell-cycle progression, signal transduction, transcription regulation, DNA repair, endocytosis, transport, development, and, in pathological situations, oncogenesis. Importantly, E3 ligases determine the specificity of protein substrates through a specific E3-substrate binding and by this they define the specificity of the proteasomal degradation.

Substrates that are polyubiquitylated by a K48-linked chain of four or more ubiquitins are rapidly recruited to the proteasome and degraded. In some cases, an E4 (multi-ubiquitin chain-assembly factor) might be required to promote polyubiquitylation *(16)* (*see* Fig. 1). This process can be opposed by the action of deubiquitylating enzymes (DUB, formerly UCH, ubiquitin carboxy-terminal hydrolase), which remove ubiquitin from specific substrates, thus stabilizing them *(17)*.

Substrates that are monoubiquitylated, multimonoubiquitylated or polyubiquitylated with chains linked through a ubiquitin lysine residue other than K48 (K29 or K63) are not degraded, and such modification might result in a change in the activation state or localization of the protein *(18)* (*see* Fig. 1). Examples for such proteolysis-independent functions are ubiquitin-dependent endocytosis, transcriptional control, DNA repair, kinase activation, trafficking and other non-proteolytic activities *(19)*.

2.2. The 26S Proteasome

The 26S proteasome is a 2.4-MDa multifunctional ATP dependent proteolytic complex. It consists of a proteolytic core particle, the 20S (720 kDa) proteasome, sandwiched between two 19S (890 kDa) "cap" regulatory complexes *(20)*. These complexes associate together in an ATP-dependent manner (21). Core particle may also associate with the interferon-γ (IFN-γ)-induced heptameric ring 11S complex *(22)*, which is believed to stimulate production of antigenic peptides by proteasomes *(23)* (*see* Fig. 2).

2.2.1. The 20S Core Particle

The core structure of the proteasome is referred to as the 20S proteasome, according to its sedimentation rate *(24)*. The 20S proteasome is present in all three domains of life: archaea, bacteria and eukarya. The 20S proteasome is a hollow cylindrical particle consisting of four stacked rings. In eukarya, each outer ring is composed of seven different αsubunits (α ring), whereas each inner ring is composed of seven different β subunits (β ring) (*see* Fig. 2) *(25)*. Each β ring contains three different proteolytically active sites (on β1, β2, and β5 subunits). All these active sites face the inner chamber of the cylinder, and the only way for substrates to access this chamber is through the gated channels in the α rings *(26)*, which are too narrow to be traversed by tightly folded globular proteins *(27)*. Moreover, these channels are completely closed in the free 20S proteasomes. Thus, in contrast to the majority of other proteases that have easily accessible active sites, proteasomes have active sites that are confined to the inner cavity of the 20S core, thereby preventing uncontrolled destruction of cellular proteins.

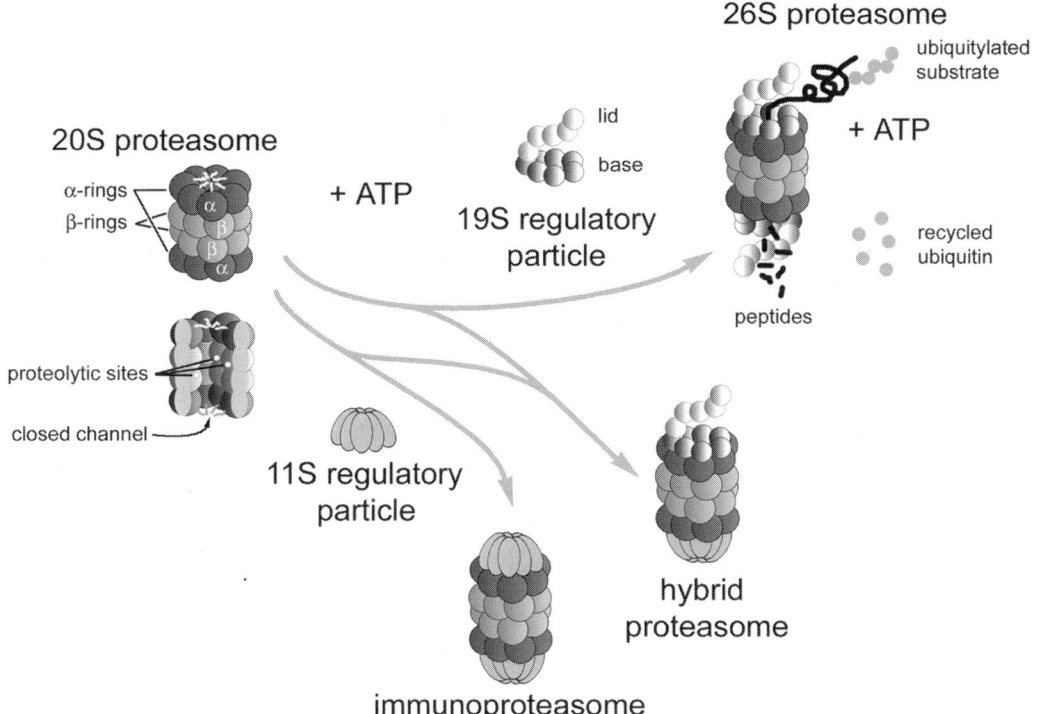

Fig. 2. The structure of the proteasome. The 26S proteasome is a multicatalytic ATP dependent proteolytic complex. It consists of a proteolytic core particle, the 20S proteasome, sandwiched between two 19S regulatory complexes. These complexes associate together in an ATP-dependent manner. 19S regulatory complexes consisting of a base and a lid, control the access of substrates into the proteolytic core. Proteasomes cut polypeptides at multiple sites, generating peptides with an average size of six residues. During this process the undegraded ubiquitin molecules are released by 19S particles for reuse in degradation of other substrates. Core particle may also associate with a heptameric ring 11S complex, which is believed to stimulate production of antigenic peptides by proteasomes.

Eukaryotic core particles contain six active sites, three on each of its two central β rings, and these proteolytic sites differ in their specificities (*see* Fig. 2). Two of them, termed "chymotrypsin-like," cut preferably after hydrophobic residues and have their catalytic residues located on the β5 subunits. Two sites, located on β2 subunits, are "trypsin-like" in cleaving after basic amino acids. The two remaining sites, located on β1 subunits, split peptide bonds preferentially after acidic residues *(28,29)*. These latter sites were traditionally termed "peptidyl glutamyl peptide hydrolase." However, it has been found that they cleave after aspartic acid residues faster than after glutamates, and therefore it was suggested that these sites be called "postacidic" or "caspase-like" *(30)*. These names indicate only general similarities to the substrate specificities of "classical" proteases, nevertheless they do not imply any similarity in catalytic mechanisms or physiological functions. The specificities of the proteasome's active sites are actually broader than reflected by their names.

2.2.2. Regulatory Complexes of Proteasome

2.2.2.1. THE 19S REGULATORY COMPLEX

19S regulatory complexes control the access of substrates into the proteolytic core. Each 19S particle consists of a base and a lid *(20,31)*. The lid, which contains at least nine polypeptides, is believed to bind to the polyubiquitin chain with high affinity and to cleave it away from the substrate. The base, which associates with the 20S particle, consists of eight polypeptides including six homologous ATPases of the AAA (ATPases Associated with a variety of cellular Activities) family *(32)*. These ATPases interact directly with α rings of the 20S core particle resulting in the ATP-dependent opening of the channel in the α rings *(33)*, which allows polypeptides access into the proteolytic chamber of the 20S particle.

ATPases of the 19S complexes are also likely to unfold the polypeptide and catalyze its translocation into the 20S proteasomes. Proteasomes cut polypeptides at multiple sites without releasing of polypeptide intermediates, thus generating peptides which range from 3 to 22 residues in length with an average size of 6 residues *(30)*. During this process the undegraded ubiquitin molecules are released by 19S particles for reuse in degradation of other substrates.

2.2.2.2. THE 11S REGULATOR

The 20S proteasome can also bind to 11S regulators, which only exist in higher eukaryotes. The 11S regulator does not enable the 20S proteasome to degrade intact proteins or ubiquitylated proteins. Rather, it enhances the degradation of peptides by the 20S proteasome, or enlarges the peptide repertoire in an ATP- and ubiquitin-independent fashion *(34)*. Moreover, the 11S regulator is drastically induced by IFN-γ. It is thus speculated that the 11S regulator is evolved to meet the need of antigen processing. Possibly, the 19S and 11S regulators attach to the 20S proteasome on each end, and such a structure might allow the proteasome to degrade proteins like the 26S, and generate short peptides in an 11S-regulated way. The physical existence of such a 19S/20S/11S entity, now termed "the hybrid proteasome," has recently been proven *(22)* (*see* Fig. 2).

3. Proteasome Function in Apoptosis

The ubiquitin-proteasome system plays a complex role in the regulation of apoptosis *(35,36)*. Apoptotic signaling pathways can be divided into two main routes, the death receptor (or extrinsic) pathway and the mitochondrial (or intrinsic) pathway, each of which involves the activation of specific initiator caspases such as caspase-8 and caspase-9, respectively (*see* Fig. 3). Both routes eventually converge at the level of caspase-3 activation. When apoptosis is invoked in response to an extrinsic or intrinsic signal, proapoptotic proteins (like Bcl-associated X protein (Bax) and caspases) are activated, synthesized, or translocated, and/or antiapoptotic proteins such as apoptosis inhibiting factor (AIF) and B-cell lymphoma 2 [Bcl-2]) are inactivated or destroyed. The ubiquitin-proteasome system is deeply involved in these processes, being able to degrade caspases, Bcl-2 family members, p53, IκB, and IAPs *(4,37)*.

Although, ubiquitylation is a reversible process and enzymes that remove ubiquitin from substrates (DUBs) might have key regulatory roles (*see* Fig. 1), proteasomal protein degradation is irreversible and is therefore ideally suited for controlling unidirectional

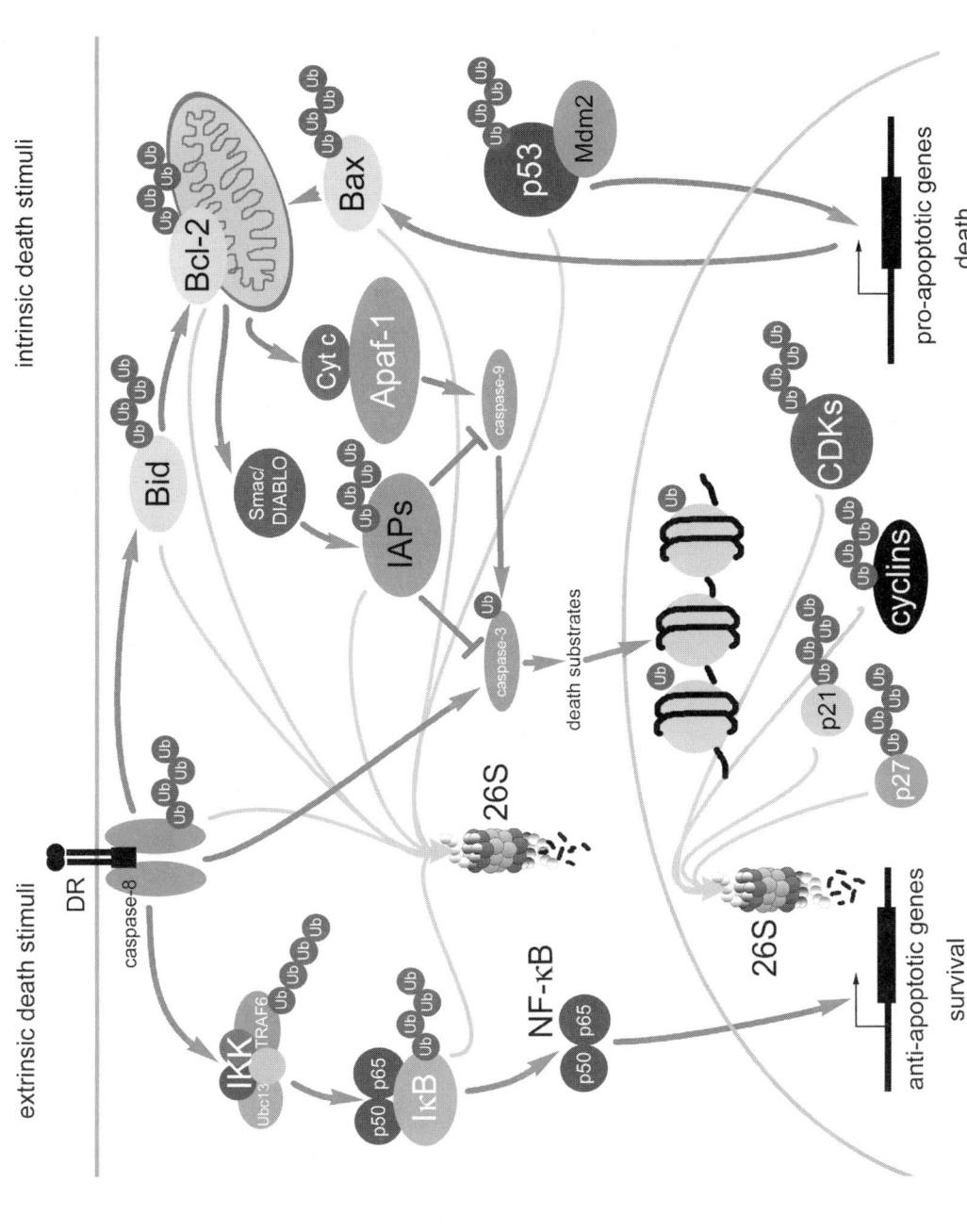

Fig. 3. Targets of ubiquitinylation in apoptotic pathways. Substrates of the ubiquitin-proteasome system are labelled by four ubiquitin signs (Ub) in a zigzag. IκB kinase (IKK) activation requires the assembly of K63-linked multiubiquitin chains, labelled by four ubiquitin signs in a straight. Monoubiquitinated targets are labelled by a single ubiquitin sign.

cellular pathways such as cell death. The proteasome is well placed to modulate the decision as to whether or not to undergo apoptosis, as both pro- and antiapoptotic proteins may be degraded by the proteasome. Proteasomal inhibition changes the cell susceptibility to apoptogenic stimuli, though the effect may be either positive or negative in a cell-type dependent manner *(38)*.

Mono- and multiubiquitylation have different effects on the fate of a protein as it is described above. In the following sections the functional consequences of the two kinds of modifications are discussed in connection with their protein targets involved in apoptosis.

3.1. Multiubiquitylation

3.1.1. IAPs

Inhibitor of apoptosis (IAP) proteins were first identified as viral products that were used by baculoviruses to inhibit defensive apoptosis of host cells and thereby provide the virus with more time to replicate. The IAP family members inhibit apoptosis by inactivating and degrading proapoptotic proteins and represent the only known subset of apoptosis-signaling molecules that contain E3 ligase activity *(39)*. Membership of the IAP family requires the presence of one or more characteristic baculoviral IAP repeat (BIR) motifs and the ability to suppress apoptosis. In contrast to Bcl-2 family proteins, which can inhibit only the mitochondrial branch of apoptosis, the IAPs can block both the intrinsic and extrinsic pathways. A principal function that requires the BIR motif(s) of IAPs is their ability to be regulated by binding upstream regulatory molecules (like Smac/DIABLO family of proteins), as well as to bind and reversibly inhibit specific downstream caspases. In addition, many IAPs have a carboxyterminal motif, such as a RING-finger-domain, that functions as ubiquitin ligase or a ubiquitin-conjugating (UBC)-domain, that is implicated in the process of ubiquitin conjugation *(40)*. The combination of BIR-mediated regulation and inactivation and RING-mediated destruction of proteins is central to the role of the IAP family in pro- and antiapoptotic decisions (*see* Fig. 3).

X-linked inhibitor of apoptosis (XIAP), cIAP1 and cIAP2 can also block caspases by binding to activated caspase-3, caspase-7 and caspase -9, thereby inhibiting their activities *(41)*. They can also interact with procaspase-9 and prevent its activation by apoptotic stimuli *(42)*.

IAPs C-terminal motif is functionally linked to ubiquitin/proteasome-mediated proteolysis. The BIR-repeat-containing ubiquitin-conjugating enzyme (BRUCE), a giant (528 kDa) IAP from mice harbors a UBC domain near its carboxyl-terminus and can form a covalent adduct (thioester) with ubiquitin in vitro, which indicates that this protein possesses ubiquitin-conjugating activity *(43)*. Other IAPs, including XIAP, cIAP1, and cIAP2, have the RING-finger domain of E3 ubiquitin ligases at their carboxyl-terminus *(44)*. In response to apoptotic stimuli, XIAP and cIAP1 undergo RING-domain dependent autoubiquitylation in vitro and in vivo, and are degraded in a proteasome-dependent manner *(45)*. XIAP can act as an ubiquitin-ligase for activated caspase-3 protease and promote its proteasomal degradation in living cells. Both the RING-finger domain of XIAP and the association of XIAP with caspase-3 are essential for caspase-3 ubiquitylation *(46)*.

Inhibition of caspase ubiquitylation activity of IAP during apoptosis depends on intermembrane space mitochondrial proteins. Binding of upstream regulatory molecules to the N-terminal BIR domains of XIAP, cIAP1 and cIAP2 to prevents IAPs to bind caspases, stimulates their autoubiquitylation and degradation *(45,47,48)*.

3.1.2. Bcl-2 Family Proteins

The Bcl-2 family of proteins comprises both antiapoptotic and proapoptotic members and has a pivotal role in controlling programmed cell death by regulating mitochondrial integrity and mitochondria-initiated caspase activation *(49)*. The Bcl-2 family of proteins is defined by sequence and structural homology to the Bcl-2 homology (BH) domain *(50)*. The C-terminal transmembrane region of the Bcl-2 protein functions to target these proteins to the mitochondrial outer membrane or endoplasmic reticulum (ER). The consequence of the membrane integration of proapoptotic members is the disruption of mitochondrial physiology: cytochrome *c* release, loss of mitochondrial membrane potential, and calcium release through the pores formed in the membrane *(51)*. On the contrary, relocalization of antiapoptotic members protects mitochondrial integrity by preventing these changes. Proteasomal degradation of proapoptotic Bcl-2 family members, such as BH3-interacting-domain death agonist (Bid) and Bax, can rescue the cell from suicide *(52–54)* (*see* Fig. 3).

3.1.3. NF-κB cascade

NF-κB was initially identified as a nuclear transcription factor binding to the B site of the intrinsic promoter of the κ light chain of immunoglobulins. NF-κB is an inducible transcription factor of the Rel family *(55)*. NF-κB exists in several dimeric forms, but the p50/p65 heterodimer is the predominant one. All heterodimers are sequestered in the cytoplasm by binding to inhibitors of the NF-κB (IκB) family of proteins that function to mask the nuclear localization signal of NF-κB. Upon activation of various upstream antiapoptotic or prosurvival signals, IκB undergoes phosphorylation by the IκB kinase (IKK) complex *(56)*. Phosphorylated IκB is recognized by the β-transducin repeat containing protein (β-TrCP) which promotes IκB ubiquitylation by the E3 ubiquitin ligase, SCF *(57)*. This results in proteasome mediated degradation of IκB (*see* Fig. 3). The free NF-κB then enters the nucleus and binds to its target sites (κB sites in the DNA), to initiate transcription of exceptionally large number of genes.

In addition to the activation of NF-κB, ubiquitin–proteasome system has two other targets in this pathway. Precursor protein of one of the NF-κB subunits is much larger (105 kDa) than the mature functional protein, p50. Post-translational processing of the precursor is brought about by an ubiquitin–proteasome-dependent reaction. Processing results in the complete degradation of the carboxy-terminal domains of the precursors, whereas the amino-terminal transcription-factor domains are left intact *(55,58,59)*. Ubiquitylation, on the other hand, is essential for the activation of IκB kinase (IKK). This activation requires the assembly of K63-linked multiubiquitin chains, which are catalyzed by the RING-finger protein tumor necrosis factor (TNF) receptor-associated factor 6 (TRAF6), together with the heterodimeric Ubc13/Uev1A ubiquitin-conjugating enzyme complex (also known as TRAF6-regulated IKK activator 1, TRIKA1) *(60)*.

In most cells, NF-κB activation protects from apoptosis, through the induction of survival genes. These include inhibitors of apoptosis molecules (cIAP1, cIAP2, and XIAP) and Bcl-2. Induction of these genes has been known to be involved in a number of diseases including cancer, AIDS, and inflammatory disorders. An approach to preventing

NF-κB activation is to inhibit proteasome activity, which stabilizes IκB and increases the susceptibility of malignant cells to chemotherapeutic agents or other cellular stresses, such as ionizing radiation *(61)*.

3.1.4. p53 and MDM2

The p53 tumor suppressor protein is one of the best-known proapoptotic proteins operating at the level of transcription *(62)*. It is maintained at low levels by constant ubiquitylation and proteasomal degradation. In response to intrinsic stressors—such as chemical- or radiation-induced DNA damage, hypoxia, and oncogene activation—ubiquitylation of p53 is inhibited through diverse pathways, depending on the nature of the stimulus and cell type. This leads to the accumulation of p53, which induces cell-cycle arrest and/or apoptosis to prevent cells from transformation. Wild-type p53 promotes the expression of Mdm2 (murine double minute 2), an E3 ubiquitin ligase for p53 which, in turn, ubiquitylates p53 and targets it for rapid degradation by the proteasome, so creating an efficient regulatory feedback loop *(63)* (*see* Fig. 3). Mdm2 was first identified as the gene responsible for the spontaneous transformation of an immortalized murine cell line BALB/c 3T3 *(64)*.

Many studies have indicated that defects of the p53 system are present in most, if not all, human tumor cells *(65)*. Using proteasome inhibitors to attack cancers that overexpress MDM2 (and therefore inactivate p53) might be a good approach to salvaging the p53 regulatory cascade and inducing apoptosis of these tumor cells. It is conceivable that new chemotherapeutic agents based on these studies will be generated in the not-so-distant future *(66)*.

3.1.5. Cell-Cycle Control

Proliferative and apoptotic pathways are stringently counterbalanced, and the cell-cycle checkpoint regulator p53 is probably the most prominent, but certainly not the only, cell-cycle regulator that operates in apoptotic pathways. The eukaryotic cell cycle is precisely coordinated by the interaction of short-lived regulatory proteins named cyclins with cyclin-dependent kinases (CDKs). Different cyclins act at different stages of the cell cycle. Cyclin D and cyclin E act during the G1 (gap) phase, cyclin E and cyclin A during the S (DNA synthesis) phase, and cyclin A and cyclin B during mitosis *(67)*. Cyclin levels vary throughout the cell cycle, and their rapid degradation by the proteasome is essential for cell-cycle progression *(68)* (*see* Fig. 3). We are only just beginning to understand the mechanisms by which these cell-cycle regulatory proteins affect apoptosis. Notably, modulation of proteasome mediated degradation of cyclins and CDKs frequently interferes with apoptosis *(69)*.

In terms of the ubiquitylation machinery, the targeting for turnover of cell-cycle regulatory proteins seems to be mediated through two distinct alternative strategies: activation of the target itself or activation of the E3 enzyme that transfers ubiquitin to a particular class of target. The former strategy allows for selectivity that is dependent on the regulatory context of individual target molecules, whereas the latter strategy allows for concerted and total destruction of populations of target molecules at particular points in the cell cycle. Most commonly, target activated destruction, in the context of the cell cycle, is carried out by a class of protein–ubiquitin ligase that is known collectively as SCF (for Skp1/Cullin/F-box protein) *(70)*. Alternative forms of another E3 ligase—that is known as the anaphase-promoting complex/cyclosome (APC/C)—are themselves activated through signaling pathways that are intrinsic to the cell cycle *(71)*. Rapidly proliferating cells, whether they are progenitor cells or cancer cells, generally show increased levels of expression of proteasome subunits *(72–74)*.

In addition to binding with an appropriate cyclin, activation of mammalian CDKs also requires dephosphorylation by a specific member of the CDC25 phosphatase family. The proteasome has also been implicated in regulating the stability of CDC25s during cell-cycle progression *(75–77)*. The interplay among CDKs, cyclins and CDC25s is highly regulated and includes feedback-loop mechanisms and the activity of CDK inhibitors such as p21 (also known as CIP1 and WAF1) and p27 (also known as KIP1) which are also degraded by the proteasome *(78)*. Inhibition of proteasome-mediated degradation arrests cells in G1 *(79–81)*, late S *(82)*, and G2/M phase of the cell cycle *(83)*. If this inhibition were specific to or preferentially targeted neoplastic cells, the differential effect could potentially be clinically relevant.

3.1.6. Caspases

Caspases (which are so-named as they are cysteine proteases that cleave after an aspartate residue in their substrates) are the central components of the apoptotic response *(6)*. Caspases form a conserved family of enzymes that irreversibly commit a cell to die. Although the first caspase, interleukin-1β-converting enzyme (ICE; also known as caspase-1), was identified in humans, the critical involvement of caspases in apoptosis was discovered in the nematode worm *Caenorhabditis elegans*, in which the indispensable gene *ced-3* (cell-death abnormality-3) was found to encode a cysteine protease that closely resembles the mammalian ICE *(84)*. Since then, at least 14 distinct mammalian caspases have been identified, of which 11 were identified in humans *(85)*. Over the past decade, many key events in caspase regulation have been documented at the molecular and cellular level *(86)*.

Precursors of caspases (procaspases) are constitutively expressed, but can be rapidly activated by a specific inducer. On the other hand, NF-κB can inactivate caspase-8 (e.g., in response to TNF stimulation) *(87)* *(see* Fig. 3). In malignant cells, caspase activation might be desirable to promote apoptosis. As inhibition of proteasome activity would prevent the activation of NF-κB, potentiate caspase activity and induce apoptosis in tumor cells.

3.2. Monoubiquitylation

3.2.1. Histones

The role of the ubiquitin-proteasome system is involved in transcription control *(88)*. The fundamental unit of eukaryotic chromatin—the nucleosome—consists of genomic DNA wrapped around an octamer of the conserved histone proteins H2A, H2B, H3, and H4. Post-translational modifications of C-terminal-tail domains of histones modulate chromatin structure and gene expression *(89)*. The first monoubiquitylated protein to be described was H2A *(90)*. Later studies revealed the monoubiquitylated form of H2B *(see* Fig. 3). This modification, which did not lead to the degradation of the histones, was demonstrated to be crucial in maintaining the chromatin in its transcriptionally active form. Recently, an interesting connection between H2A deubiquitylation and stress stimuli in human lymphocytes and breast carcinoma cells was observed *(91)*.

3.2.2. Caspases

Recent findings indicate that IAPs regulate not only their own levels, but also those of other substrates such as caspases *(92)*. In vitro ubiquitylation assays have shown that cIAP2 monoubiquitylates caspase-7 and caspase-3, whereas all other caspases tested did not undergo monoubiquitylation (*see* Fig. 3). The cIAP2 RING-finger motif itself seems to be sufficient to promote this substrate specificity. However, the functional consequences of this modification are currently undefined.

4. Targeting the Ubiquitin–Proteasome System for Therapy

After the release of the sequence of the human genome, it became quickly evident that the proteome is several orders of magnitudes larger than originally speculated, and that many diseases are caused by post-translational modification. Thus, today's efforts in biomedical research are primarily focused on the proteomic challenge. It is a delicate network of thousands of proteins and their interactions in the human cell are tightly regulated; imbalances in this network lead to diseases like cancer and various genetic disorders. The proteasome and its upstream system of ubiquitin-conjugating enzymes is responsible for 80% of the cell's protein degradation, and thereby has a major role in cellular homeostasis.

An ideal cancer target plays an essential role in carcinogenesis and/or in inhibition of apoptosis or inhibition of its expression or activity induces apoptosis in cancer cells. Last but not least it should be expressed at a very low level in normal cells so its inhibition would have a minimal effect on normal cell growth and function *(93)*. The fact that the proteasome is crucial for the execution of many normal cellular functions indicates that it would be difficult to use it as a target for chemotherapy and yet maintain a tolerable therapeutic index. However, empirical findings indicate that many types of actively proliferating malignant cells are more sensitive to proteasome blockade than noncancerous cells. One aspect of this differential susceptibility is that many types of malignant cells rapidly proliferate and have one or more aberrant cell-cycle checkpoints. Therefore, these cells might accumulate defective proteins at a much higher rate than normal cells, which increases their dependency on the proteasome as a disposal mechanism.

4.1. Proteasome Inhibitors

Proteasome forms a new class of proteolytic enzymes called threonine proteases. Unlike any other protease, all the proteolytic sites inside proteasome utilize N-terminal threonines of β subunits as the active site nucleophiles. Each of the proteolytic activities of the proteasome can be inhibited independently. Several recent reviews detailing the properties and mechanisms of action of proteasome inhibitors are available *(94,95)*. Inhibition of all active sites is not required to significantly reduce protein breakdown. In fact, inhibition of the chymotrypsin-like site (located on the $\beta5$ subunits) causes a large reduction in the rates of protein breakdown *(96)*, and therefore most new inhibitors were designed to block it. In contrast, inactivation of trypsin-like or caspase-like sites had little effect on overall proteolysis. Proteasome inhibitors might be natural or synthetic, and five important classes of proteasome inhibitors exist: peptide aldehydes, peptide vinyl sulphones, peptide boronates, peptide epoxiketones and β-lactones. However, at present, only two compounds are in clinical development because of their antineoplastic

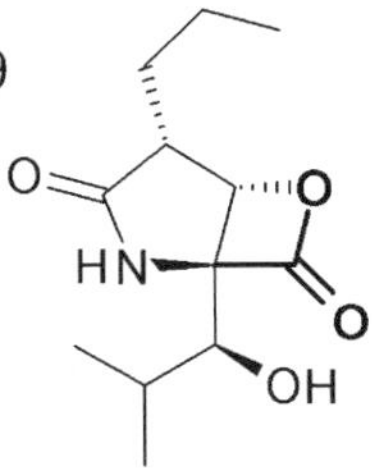

Fig. 4. Structure of proteasome inhibitors. Antineoplastic compound pyrazylcarbonyl-Phe-Leu-boronate (PS-341, bortezomib, Velcade™) and anti-inflammatory drug candidate, a lactacystin derivative, PS-519. Pharmacophores are bold.

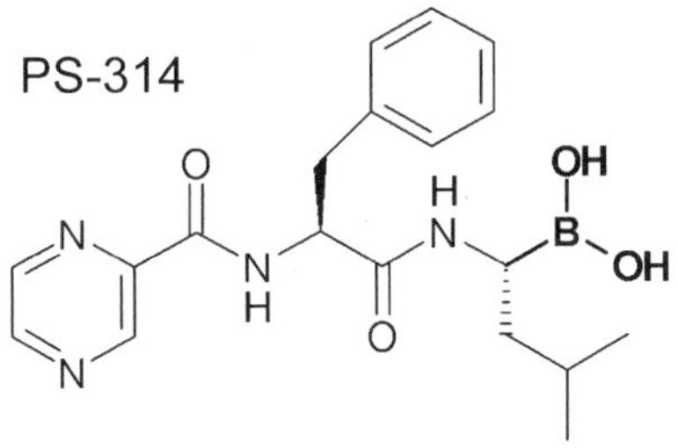

(pyrazylcarbonyl-Phe-Leu-boronate, PS-341, bortezomib, Velcadeí [Millennium Pharmaceuticals]) *(97)* and anti-inflammatory (a lactacystin derivative, PS-519, also known as MLN519) properties *(61)* (*see* Fig. 4). Human multiple-myeloma cells were found to be more susceptible than normal peripheral-blood mononuclear cells to growth inhibition and apoptosis that is induced by treatment with the proteasome inhibitor bortezomib *(98)*. A tolerable and efficacious dose and schedule of bortezomib in myeloma was defined in phase I and II studies *(99,100)*. Bortezomib has been safely administered as cyclical therapy (a usual cycle entails twice-weekly treatment for 2 wk every 3 wk) for up to 13 cycles in patients with myeloma *(101)*. Proteasome activity is maximally inhibited 1 h after dosing, so the rest periods between dosing allows for recovery of proteasome function *(102)*. The ultimate result of inhibiting proteasome activity is apoptosis, increased sensitivity and decreased resistance to standard chemotherapy and radiation therapy. Much still remains to be learned about the exact sequence of events and how these events are specific to a certain inhibitor and tumor type *(103)*.

4.2. New Class of Therapeutics: Substrate Specific E3 Ligases

E3 ubiquitin ligases play an essential role in regulation of many biological processes. Furthermore, E3s are enzymes that determine the specificity of protein substrates for targeted degradation via 26S proteasome. Upon those criteria they represent a class of drugable targets for pharmaceutical intervention. Although further validation is needed which requires a better understanding of biological functions of many E3 ligases, it has become clear that some E3 ubiquitin ligases are promising cancer targets. Targeting a specific E3 would have the potential to selectively stabilize specific cellular proteins regulated by this E3, thus avoiding any unwanted effects on other cellular proteins. This would, therefore, achieve a high level of specificity with less associated toxicity, in contrast with the general proteasome inhibitors. Three of the most promising E3s are the Mdm2/Hdm2, IAPs, and SCF. These E3 ligases or their components are over-expressed in many human cancers and their inhibition leads to growth suppression or apoptosis.

As described above for Mdm2, Hdm2 (human counterpart of Mdm2)—upon induction by p53—binds to p53 and acts as an E3 ubiquitin ligase and rapidly degrades p53 *(104)*. Whereas Mdm2 is normally expressed at a low level, Hdm2 was found overexpressed via gene amplification, increased transcription or enhanced translation in a variety of human cancers, including breast carcinomas, soft tissue sarcomas, esophageal carcinomas, lung carcinomas, glioblastomas, and malignant melanomas *(105)*. Thus, targeting Hdm2 is a promising approach to reactivate p53 to induce apoptosis in human cancer cells harboring a wild type p53 *(93)*.

IAP family proteins are characterized by containing one or several BIR domains that are required for suppression of apoptosis. In XIAP, BIR3 (the third BIR domain) potently inhibits the activity of the active caspase 9 whereas the linker region between the BIR1 and BIR2 selectively targets active caspase 3 and 7 *(106)*. In addition, some RING containing IAPs, such as cIAP-2 and XIAP, may also inhibit apoptosis through targeting caspase 3 for ubiquitin-dependent degradation *(46)*. Consistent with their anti-apoptotic function, IAPs, particularly *survivin*, were found overexpressed in most common human cancers but not in normal, terminally differentiated adult tissues *(107)*. In apoptotic cells, the caspase inhibition by IAPs is negatively regulated by the apoptosis inducer Smac/DIABLO. Smac physically interacts with multiple IAPs and relieves their inhibitory effect on caspases 3, 7, and 9. Interestingly, the IAPs were recently found to promote degradation of Smac themselves *(108,109)*. Targeting IAP's ubiquitin ligases could therefore be a feasible approach to inducing cancer cell apoptosis or rendering cancer cells susceptible to conventional cancer therapies by increasing cellular levels of caspases and Smac *(93)*.

The SCF complexes are the largest family of E3 ubiquitin ligases, consisting of Skp-1, Cullins, F-Box proteins, and the ROC/Rbx/SAG RING finger protein *(110)*. Overexpression of SCF components has been found in many human cancers (e.g., breast cancers *[111]*, oral squamous carcinoma, small cell lung carcinoma, and prostate cancer. Substrate specificity of SCF E3 ligases is determined by the F-Box proteins. In consequence, a general inhibitor may not have a desired specificity against any particular SCF complex.

Although targeting E3 ubiquitin ligases is still in its infancy, speedy approval of the general proteasome inhibitor, bortezomib, for the treatment of relapsed and refractory multiple myeloma suggests the promise of specific E3 inhibitors in anticancer therapy. It is expected that E3 ubiquitin ligases will represent an important new target platform for future mechanism-driven drug discovery *(93)*.

4.3. Influencing Cell Survival

In an other set of human diseases even excessive apoptosis causes the problem. These are neurodegenerative diseases such as Alzheimer's disease, Parkinson's disease, amyotrophic lateral sclerosis, and several hereditary diseases that are caused by extended polyglutamine repeats *(115–117)*. With regard to neurodegenerative diseases in particular, the precise identification of distinct errors that occur during the interplay of proteasomal function and neuronal apoptosis is certainly of great potential for devising specific treatments. Genetic data from various systems clearly indicate that there is a link between proteasomal dysfunction and neuronal disorders *(115,116)*. Moreover, an increasing body of evidence supports a crucial role for the apoptotic system in the manifestation of many of the neurodegenerative diseases *(117)*, although our knowledge about the mechanisms that are involved in these disorders is still incomplete.

5. Implications and Future Directions

It is difficult to outline a single and clearly defined role of the ubiquitin–proteasome system in apoptosis. Fortunately enough, rapidly proliferating cells with abnormal phenotypes (i.e., cancer cells) are the most sensitive to the proapoptogenic action of proteasome inhibitors, whereas normal cells are less sensitive or not sensitive at all, even when they form highly proliferating populations such as bone marrow cells or epithelia. This makes proteasome inhibitors very promising agents in cancer therapy, and they are already used in clinical trials.

At present it is difficult to establish the exact reasons why some cells are sensitive to proteasome inhibition and others are not. One mechanism is the rapid proliferation rate of many tumors, which might make these cells more dependent on proteasomes to remove increased number of misfolded or damaged proteins. Another more elegant mechanism is that inhibition of proteasome activity might reverse or bypass some of the effects of cell cycle or apoptosis checkpoint mutations that have caused the development or maintenance of the cancerous phenotype.

As they have a role in the production of antigenic peptides, proteasome inhibitors are also being studied as potential anti-inflammatory (for example, in arthritis, asthma, multiple sclerosis, and psoriasis) and antiviral agents. Another area of future research is the development of agents that target regulatory events occurring upstream from the activated proteasome. These potential targets might control phosphorylation or ubiquitylation of proteasome substrates and their regulators. The rewards of developing agents that can specifically interfere with the E3 ligase system and its interaction with its substrates will remarkably be high in the area of cancer treatment, given the need for a tight regulation of oncogenic and tumor suppressing proteins. Thus, the understanding of the complex ubiquitylation systems and its explicit interactions and mechanisms should lead to more innovative cancer therapies that will offer the hope for better and perhaps greater curative successes in malignant and other diseases.

Acknowledgment

I would like to thank Monika Lippai for critical comments on the manuscript. My apologies to those whose work could only be cited indirectly owing to space limitations.

References

1. Baehrecke EH. How death shapes life during development. Nat Rev Mol Cell Biol 2002;3:779–787.
2. Kerr JFR, Wyllie AH, Currie AR. Apoptosis: a basic biological phenomenon with wide-ranging implications in tissue kinetics. Br J Cancer 1972;26:239–257.
3. Danial NN, Korsmeyer SJ. Cell death: Critical control points. Cell 2004;116:205–219.
4. Jesenberger V, Jentsch S. Deadly encounter: Ubiquitin meets apoptosis. Nat Rev Mol Cell Biol 2002;3:112–121.

5. Mayer RJ. The meteoric rise of regulated intracellular proteolysis. Nat Rev Mol Cell Biol 2000;1:145–148.

6. Riedl SJ, Shi YG. Molecular mechanisms of caspase regulation during apoptosis. Nat Rev Mol Cell Biol 2004;5:897–907.

7. Jentsch S, Seufert W, Hauser HP. Genetic Analysis of the Ubiquitin System Biochim Biophys Acta 1991;1089:127–139.

8. Hershko A, Ciechanover A. The ubiquitin system. Annu Rev Biochem 1998;67:425–479.

9. Ciechanover A. The ubiquitin-proteasome pathway: on protein death and cell life. EMBO J 1998;17:7151–7160.

10. Varshavsky A. The N-end Rule. Cell 1992;69:725–735.

11. Scheffner M, Huibregtse JM, Vierstra RD, Howley PM. The HPV-16 E6 and E6-AP complex functions as a ubiquitin-protein ligase in the ubiquitination of p53. Cell 1993;75:495–505.

12. Schwarz SE, Rosa JL, Scheffner M. Characterization of human HECT domain family members and their interaction with UbcH5 and UbcH7. J Biol Chem 1998;273:12,148–12,154.

13. Freemont PS. Ubiquitination: RING for destruction? Curr Biol 2000;10:R84–R87.

14. Fang SY, Jensen JP, Ludwig RL, Vousden KH, Weissman AM. Mdm2 is a RING finger-dependent ubiquitin protein ligase for itself and p53. J Biol Chem 2000;275:8945–8951.

15. Fang S, Lorick KL, Jensen JP, Weissman AM. RING finger ubiquitin protein ligases: implications for tumorigenesis, metastasis and for molecular targets in cancer. Semin Cancer Biol 2003;13:5–14.

16. Koegl M, Hoppe T, Schlenker S, Ulrich HD, Mayer TU, Jentsch S. A novel ubiquitination factor, E4, is involved in multiubiquitin chain assembly. Cell 1999;96:635–644.

17. Fischer JA. Deubiquitinating enzymes: Their roles in development, differentiation, and disease. In: International Review Of Cytology — A Survey Of Cell Biology, Vol 229. 2003: 43–72.

18. Weissman AM. Themes and variations on ubiquitylation. Nat Rev Mol Cell Biol 2001; 2:169–178.

19. Pickart CM. Back to the future with ubiquitin. Cell 2004;116:181–190.

20. Voges D, Zwickl P, Baumeister W. The 26S proteasome: A molecular machine designed for controlled proteolysis. Ann Rev Biochem 1999;68:1015–1068.

21. Coux O, Tanaka K, Goldberg AL. Structure and functions of the 20S and 26S proteasomes. Annu Rev Biochem 1996;65:801–847.

22. Tanahashi N, Murakami Y, Minami Y, Shimbara N, Hendil KB, Tanaka K. Hybrid proteasomes — Induction by interferon-gamma and contribution to ATP-dependent proteolysis. J Biol Chem 2000;275:14,336–14,345.

23. Rechsteiner M, Realini C, Ustrell V. The proteasome activator 11 S REG (PA28) and Class I antigen presentation. Biochem J 2000;345:1–15.

24. Pickart CM, Cohen RE. Proteasomes and their kin: Proteases in the machine age. Nat Rev Mol Cell Biol 2004;5:177–187.

25. Baumeister W, Walz J, Zuhl F, Seemuller E. The proteasome: paradigm of a self-compartmentalizing protease. Cell 1998;92:367–380.

26. Groll M, Bajorek M, Kohler A, et al. A gated channel into the proteasome core particle. Nat Struct Biol 2000;7:1062–1067.

27. Wenzel T, Baumeister W. Conformational constraints in protein degradation by the 20S proteasome. Nat Struct Biol 1995;2:199–204.

28. Nussbaum AK, Dick TP, Keilholz W, et al. Cleavage motifs of the yeast 20S proteasome beta subunits deduced from digests of enolase 1. Proc Natl Acad Sci USA 1998;95:12,504–12,509.

29. Dick TP, Nussbaum AK, Deeg M, et al. Contribution of proteasomal beta-subunits to the cleavage of peptide substrates analyzed with yeast mutants. J Biol Chem 1998;273: 25,637–25,646.

30. Kisselev AF, Akopian TN, Castillo V, Goldberg AL. Proteasome active sites allosterically regulate each other, suggesting a cyclical bite-chew mechanism for protein breakdown. Mol Cell 1999;4:395–402.

31. Glickman MH, Rubin DM, Fried VA, Finley D. The regulatory particle of the *Saccharomyces cerevisiae* proteasome. Mol Cell Biol 1998;18:3149–3162.

32. Patel S, Latterich M. The AAA team: related ATPases with diverse functions. Trends Cell Biol 1998;8:65–71.

33. Köhler A, Cascio P, Leggett DS, Woo KM, Goldberg AL, Finley D. The axial channel of the proteasome core particle is gated by the Rpt2 ATPase and controls both substrate entry and product release. Mol Cell 2001;7:1143–1152.

34. Rechsteiner M, Hoffman L, Dubiel W. The Multicatalytic and 26S Proteases. J Biol Chem 1993;268:6065–6068.

35. Wojcik C. Regulation of apoptosis by the ubiquitin and proteasome pathway. J Cell Mol Med 2002;6:25–48.

36. Orlowski RZ. The role of the ubiquitin-proteasome pathway in apoptosis. Cell Death Differ 1999;6:303–313.

37. Lee JC, Peter ME. Regulation of apoptosis by ubiquitination. Immunol Rev 2003;193: 39–47.

38. Zhang HG, Wang JH, Yang XW, Hsu HC, Mountz JD. Regulation of apoptosis proteins in cancer cells by ubiquitin. Oncogene 2004;23:2009–2015.

39. Salvesen GS, Duckett CS. IAP proteins: Blocking the road to death's door. Nat Rev Mol Cell Biol 2002;3:401–410.

40. Vaux DL, Silke J. IAPs, RINGs, and ubiquitylation. Nat Rev Mol Cell Biol 2005;6: 287–297.

41. Deveraux QL, Takahashi R, Salvesen GS, Reed JC. X-linked IAP is a direct inhibitor of cell death proteases. Nature 1997;388:300–304.

42. Deveraux QL, Roy N, Stennicke HR, et al. IAPs block apoptotic events induced by caspase-8 and cytochrome c by direct inhibition of distinct caspases. EMBO J 1998;17: 2215–2223.

43. Hauser HP, Bardroff M, Pyrowolakis G, Jentsch S. A giant ubiquitin-conjugating enzyme related to IAP apoptosis inhibitors. J Cell Biol 1998;141:1415–1422.

44. Joazeiro CAP, Weissman AM. RING finger proteins: Mediators of ubiquitin ligase activity. Cell 2000;102:549–552.

45. Yang Y, Fang SY, Jensen JP, Weissman AM, Ashwell JD. Ubiquitin protein ligase activity of IAPs and their degradation in proteasomes in response to apoptotic stimuli. Science 2000;288:874–877.

46. Suzuki Y, Nakabayashi Y, Takahashi R. Ubiquitin-protein ligase activity of X-linked inhibitor of apoptosis protein promotes proteasomal degradation of caspase-3 and enhances its anti-apoptotic effect in Fas-induced cell death. Proc Natl Acad Sci USA 2001;98: 8662–8667.

47. Du CY, Fang M, Li YC, Li L, Wang XD. Smac, a mitochondrial protein that promotes cytochrome c-dependent caspase activation by eliminating IAP inhibition. Cell 2000;102: 33–42.

48. Srinivasula SM, Datta P, Fan XJ, Fernandes-Alnemri T, Huang ZW, Alnemri ES. Molecular determinants of the caspase-promoting activity of Smac/DIABLO and its role in the death receptor pathway. J Biol Chem 2000;275:36,152–36,157.

49. Adams JM, Cory S. Life-or-death decisions by the Bcl-2 protein family. Trends Biochem Sci 2001;26:61–66.

50. Marsden VS, Strasser A. Control of apoptosis in the immune system: Bcl-2, BH3-only proteins and more. Annu Rev Immunol 2003;21:71–105.

51. Ferri KF, Kroemer G. Organelle-specific initiation of cell death pathways. Nat Cell Biol 2001;3:E255–E263.

52. Breitschopf K, Zeiher AM, Dimmeler S. Ubiquitin-mediated degradation of the proapoptotic active form of Bid - A functional consequence on apoptosis induction. J Biol Chem 2000;275:21,648–21,652.

53. Li BY, Dou QP. Bax degradation by the ubiquitin/proteasome-dependent pathway: involvement in tumor survival and progression. Proc Natl Acad Sci USA 2000;97:3850–3855.

54. Thomas M, Banks L. Inhibition of Bak-induced apoptosis by HPV-18 E6. Oncogene 1998;17:2943–2954.

55. Karin M, Ben-Neriah Y. Phosphorylation meets ubiquitination: The control of NF-κB activity. Annu Rev Immunol 2000;18:621–623.

56. Regnier CH, Song HY, Gao X, Goeddel DV, Cao ZD, Rothe M. Identification and characterization of an IκB kinase. Cell 1997;90:373–383.

57. Yaron A, Hatzubai A, Davis M, et al. Identification of the receptor component of the IκB α-ubiquitin ligase. Nature 1998;396:590–594.

58. Palombella VJ, Rando OJ, Goldberg AL, Maniatis T. The ubiquitin-proteasome pathway is required for processing the NF-κB1 precursor protein and the activation of NF-κB. Cell 1994;78:773–785.

59. Coux O, Goldberg AL. Enzymes catalyzing ubiquitination and proteolytic processing of the p105 precursor of nuclear factor κB1. J Biol Chem 1998;273:8820–8828.

60. Deng L, Wang C, Spencer E, et al. Activation of the IκB kinase complex by TRAF6 requires a dimeric ubiquitin-conjugating enzyme complex and a unique polyubiquitin chain. Cell 2000;103:351–361.

61. Adams J. The proteasome: A suitable antineoplastic target. Nat Rev Cancer 2004;4:349–360.

62. Vousden KH, Lu X. Live or let die: The cell's response to p53. Nat Rev Cancer 2002;2:594–604.

63. Haupt Y, Maya R, Kazaz A, Oren M. Mdm2 promotes the rapid degradation of p53. Nature 1997;387:296–299.

64. Cahillysnyder L, Yangfeng T, Francke U, George DL. Molecular analysis and chromosomal mapping of amplified genes isolated from a transformed mouse 3T3-cell line. Somat. Cell Mol Genet 1987;13:235–244.

65. Liu MC, Gelmann EP. p53 gene mutations: Case study of a clinical marker for solid tumors. Semin Oncol 2002;29:246–257.

66. Yang YL, Li CCH, Weissman AM. Regulating the p53 system through ubiquitination. Oncogene 2004;23:2096–2106.

67. King RW, Deshaies RJ, Peters JM, Kirschner MW. How proteolysis drives the cell cycle. Science 1996;274:1652–1659.

68. Glotzer M, Murray AW, Kirschner MW. Cyclin is degraded by the ubiquitin pathway. Nature 1991;349:132–138.

69. Reed SI. Ratchets and clocks: The cell cycle, ubiquitylation and protein turnover. Nat Rev Mol Cell Biol 2003;4:855–864.

70. Willems AR, Goh T, Taylor L, Chernushevich I, Shevchenko A, Tyers M. SCF ubiquitin protein ligases and phosphorylation-dependent proteolysis. Philos Trans R Soc Lond Ser B-Biol Sci 1999;354:1533–1550.

71. Visintin R, Prinz S, Amon A. CDC20 and CDH1: A family of substrate-specific activators of APC-dependent proteolysis. Science 1997;278:460–463.

72. Pajonk F, Pajonk K, McBride WH. Apoptosis and radiosensitization of Hodgkin cells by proteasome inhibition. Int J Radiat Oncol Biol Phys 2000;47:1025–1032.

73. Kumatori A, Tanaka K, Inamura N, et al. Abnormally high expression of proteasomes in human leukemic cells. Proc Natl Acad Sci USA 1990;87:7071–7075.

74. Kanayama H, Tanaka K, Aki M, et al. Changes in expressions of proteasome and ubiquitin genes in human renal cancer cells. Cancer Res 1991;51:6677–6685.

75. Bernardi R, Liebermann DA, Hoffman B. Cdc25A stability is controlled by the ubiquitin-proteasome pathway during cell cycle progression and terminal differentiation. Oncogene 2000;19:2447–2454.

76. Baldin V, Cans C, Knibiehler M, Ducommun B. Phosphorylation of human CDC25B phosphatase by CDK1-cyclin A triggers its proteasome-dependent degradation. J Biol Chem 1997;272:32,731–32,734.

77. Chen F, Zhang Z, Bower J, et al. Arsenite-induced Cdc25C degradation is through the KEN-box and ubiquitin-proteasome pathway. Proc Natl Acad Sci USA 2002;99: 1990–1995.

78. Tam SW, Theodoras AM, Pagano M. Kip1 degradation via the ubiquitin–proteasome pathway. Leukemia 1997;11 (Suppl 3):363–366.

79. Kumeda SI, Deguchi A, Toi M, Omura S, Umezawa K. Induction of G1 arrest and selective growth inhibition by lactacystin in human umbilical vein endothelial cells. Anticancer Res 1999;19:3961–3968.

80. Hashemolhosseini S, Nagamine Y, Morley SJ, Desrivieres S, Mercep L, Ferrari S. Rapamycin inhibition of the G1 to S transition is mediated by effects on cyclin D1 mRNA and protein stability. J Biol Chem 1998;273:14,424–14,429.

81. Rao S, Porter DC, Chen XM, Herliczek T, Lowe M, Keyomarsi K. Lovastatin-mediated G1 arrest is through inhibition of the proteasome, independent of hydroxymethyl glutaryl-CoA reductase. Proc Natl Acad Sci USA 1999;96:7797–7802.

82. Machiels BM, Henfling MER, Gerards WLH, et al. Detailed analysis of cell cycle kinetics upon proteasome inhibition. Cytometry 1997;28:243–252.

83. Wojcik C, Schroeter D, Stoehr M, Wilk S, Paweletz N. An inhibitor of the chymotrypsin-like activity of the multicatalytic proteinase complex (20S proteasome) induces arrest in G2-phase and metaphase in HeLa cells. Eur J Cell Biol 1996;70:172–178.

84. Boyce M, Degterev A, Yuan J. Caspases: an ancient cellular sword of Damocles. Cell Death Diff 2004;11:29–37.

85. Degterev A, Boyce M, Yuan JY. A decade of caspases. Oncogene 2003;22:8543–8567.

86. Fesik SW. Insights into programmed cell death through structural biology. Cell 2000;103: 273–282.

87. Varfolomeev EE, Schuchmann M, Luria V, et al. Targeted disruption of the mouse Caspase 8 gene ablates cell death induction by the TNF receptors, Fas/Apo1, and DR3 and is lethal prenatally. Immunity 1998;9:267–276.

88. Muratani M, Tansey WR. How the ubiquitin-proteasome system controls transcription. Nat Rev Mol Cell Biol 2003;4:192–201.

89. Bach I, Ostendorff HP. Orchestrating nuclear functions: ubiquitin sets the rhythm. Trends Biochem Sci 2003;28:189–195.

90. Goldknopf IL, Busch H. Isopeptide linkage between nonhistone and histone 2A polypeptides of chromosomal conjugate protein A24. Proc Natl Acad Sci USA 1977;74:864–868.

91. Mimnaugh EG, Kayastha G, McGovern NB, et al. Caspase-dependent deubiquitination of monoubiquitinated nucleosomal histone H2A induced by diverse apoptogenic stimuli. Cell Death Differ 2001;8:1182–1196.

92. Huang P, Oliff A. Signaling pathways in apoptosis as potential targets for cancer therapy. Trends Cell Biol 2001;11:343–346.

93. Sun Y. Targeting E3 ubiquitin ligases for cancer therapy. Cancer Biol Ther 2003;2:623–629.

94. Kisselev AF, Goldberg AL. Proteasome inhibitors: from research tools to drug candidates. Chem Biol 2001;8:739–758.

95. Mykles DL. Proteinase families and their inhibitors. In: Methods In Cell Biology. 2001:247–287.

96. Rock KL, Gramm C, Rothstein L, et al. Inhibitors of the proteasome block the degradation of most cell-proteins and the generation of peptides presented on MHC class-I molecules. Cell 1994;78:761–771.

97. Albanell J, Adams J. Bortezomib, a proteasome inhibitor, in cancer therapy: from concept to clinic. Drug Future 2002;27:1079–1092.

98. Hideshima T, Richardson P, Chauhan D, et al. The proteasome inhibitor PS-341 inhibits growth, induces apoptosis, and overcomes drug resistance in human multiple myeloma cells. Cancer Res 2001;61:3071–3076.

99. Aghajanian C, Soignet S, Dizon DS, et al. A Phase I trial of the novel proteasome inhibitor PS-341 in advanced solid tumor malignancies. Clin Cancer Res 2002;8:2505–2511.

100. Richardson PG, Barlogie B, Berenson J, et al. A Phase II study of bortezomib in relapsed, refractory myeloma. N Engl J Med 2003;348:2609–2617.

101. Berenson JR, Jagannath S, Barlogie B, et al. Experience with long-term therapy using the proteasome inhibitor, bortezomib, in advanced multiple myeloma (MM). Proc Am Soc Clin Oncol 2003;22:581.

102. Orlowski RZ, Stinchcombe TE, Mitchell BS, et al. Phase I trial of the proteasome inhibitor PS-341 in patients with refractory hematologic malignancies. J Clin Oncol 2002;20: 4420–4427.

103. Adams J. The development of proteasome inhibitors as anticancer drugs. Cancer Cell 2004;5:417–421.

104. Freedman DA, Wu L, Levine AJ. Functions of the MDM2 oncoprotein. Cell Mol Life Sci 1999;55:96–107.

105. Zhang RW, Wang H. MDM2 oncogene as a novel target for human cancer therapy. Curr Pharm Design 2000;6:393–416.

106. Shi YG. Mechanisms of caspase activation and inhibition during apoptosis. Mol Cell 2002;9:459–470.

107. Altieri DC. Validating survivin as a cancer therapeutic target. Nat Rev Cancer 2003;3: 46–54.

108. Hu SM, Yang XL. Cellular inhibitor of apoptosis 1 and 2 are ubiquitin ligases for the apoptosis inducer Smac/DIABLO. J Biol Chem 2003;278:10,055–10,060.

109. MacFarlane M, Merrison W, Bratton SB, Cohen GM. Proteasome-mediated degradation of Smac during apoptosis: XIAP promotes Smac ubiquitination in vitro. J Biol Chem 2002;277:36,611–36,616.

110. Deshaies RJ. SCF and cullin/RING H2-based ubiquitin ligases. Annu Rev Cell Dev Biol 1999;15:435–467.

111. Chen LC, Manjeshwar S, Lu Y, et al. The human homologue for the *Caenorhabditis elegans* cul-4 gene is amplified and overexpressed in primary breast cancers. Cancer Res 1998;58: 3677–3683.

112. Gstaiger M, Jordan R, Lim M, et al. Skp2 is oncogenic and overexpressed in human cancers. Proc Natl Acad Sci USA 2001;98: 5043–5048.

113. Yokoi S, Yasui K, Saito-Ohara F, et al. A novel target gene, Skp2, within the 5p13 amplicon that is frequently detected in small cell lung cancers. Am J Pathol 2002;161:207–216.

114. Yang G, Ayala G, De Marzo A, et al. Elevated Skp2 protein expression in human prostate cancer: Association with loss of the cyclin-dependent kinase inhibitor p27 and PTEN and with reduced recurrence-free survival. Clin Cancer Res 2002;8:3419–3426.

115. Shimura H, Hattori N, Kubo S, et al. Familial Parkinson disease gene product, parkin, is a ubiquitin-protein ligase. Nat Genet 2000;25:302–305.

116. Cummings CJ, Reinstein E, Sun YL, et al. Mutation of the E6-AP ubiquitin ligase reduces nuclear inclusion frequency while accelerating polyglutamine-induced pathology in SCA1 mice. Neuron 1999;24:879–892.

117. Yuan JY, Yankner BA. Apoptosis in the nervous system. Nature 2000;407:802–809.

14

Apoptosis Induction in T Lymphocytes by HIV

Maria Saveria Gilardini Montani

Summary

Infection with human immunodeficiency virus (HIV) is characterized by a progressive decrease in CD4+ T-cell number with a consequent impairment of the immune response and the onset of opportunistic infections and neoplasms. The mechanism responsible of CD4 T-cell depletion is mainly the induction of apoptosis, which can be activated by HIV through various pathways. In this chapter direct and indirect mechanisms of HIV-induced apoptosis in infected and uninfected bystander cells are described.

Key Words: Apoptosis; HIV; T-lymphocytes; Env; Tat; Nef; Vpr.

1. Introduction

Human immunodeficiency virus (HIV) is a retrovirus that infects cells of the immune system. Such infection is characterized mainly by a progressive decline in both functionality and the number of CD4+ T-lymphocytes, thus resulting in the acquired immune deficiency syndrome (AIDS), which leads to the onset of opportunistic infections, neoplasm, and ultimately death *(1)*.

Viruses use different strategies to survive and to take advantage of the host. According to the strategy used to live and multiply, viruses may be labeled as "hit and run" or "hit and stay" viruses. Hit and run viruses are mainly cytolytic, destroying host cells with the aim to infect as many individuals as possible. Viruses, which propagate through this strategy, are influenza, rhinoviruses, and measles and are quickly cleared by the immune system. By contrast, once inside a host, retroviruses and herpesviruses persistently stay *(2)*. HIV is a hit and stay virus which, after an acute phase, establishes a long period of latency in which progeny viruses are produced but do not elicit an immune response *(3,4)*.

In order to stay, HIV has developed different strategies to evade the host immune system as rapid mutations, shedding of decoy antigens, modulation of host MHC class I molecules, to avoid recognition of infected target cells by cytotoxic T-lymphocytes (CTLs), and destruction of CTLs *(5)*. However, the principal mechanism of immune evasion used by HIV is the progressive destruction of both infected and uninfected bystander CD4+ T-helper (Th) cells. To affect the Th population means to disconnect the whole immune system, as Th cells play a central role in orchestrating the adaptive immune response. The cells involved in the adaptive response, mainly B- and T-lymphocytes, recognize foreign molecules via antigen-specific membrane receptors. Depending upon the type of infection, the adaptive immune system mounts either a cell-mediated response, which is necessary to eradicate intracellular microorganism infections, or a humoral immune response to target and eliminate extracellular pathogens. After they have encountered the antigen associated

From: *Apoptosis, Cell Signaling, and Human Diseases: Molecular Mechanisms, Volume 2*
Edited by R. Srivastava © Humana Press Inc., Totowa, NJ

with major histocompatibility complex (MHC) molecules on the surface of an antigen presenting cell (APC), T-helper cells coordinate this decision by secreting a typical pattern of cytokines which influences one type of response or the other.

Moreover loss of CD4 T-cells has been associated with pathogenesis development. This relationship was confirmed in animal models. In fact, HIV-infected chimpanzees support viral replication, but they do not loose CD4 T-cells and do not develop AIDS, whereas macaques infected with simian immunodeficiency virus (SIV) support both viral replication and CD4 T-lymphocyte depletion and develop AIDS *(6)*. Despite intensive studies, several crucial questions remain to be addressed about the mechanisms by which HIV infection induces the loss of CD4$^+$ T-cells and the cause is likely to be multifactorial. Among the hypotheses to explain this phenomenon there is the impaired production of T-cells by the thymus, homing of virus-specific T-cells to lymphoid tissues, altered T-cell turnover and apoptosis of both infected and bystander CD4$^+$ T-cells *(7–9)*. A number of evidence supports the concept that lymphocyte apoptosis is the main reason of cell loss and is pivotal for disease pathogenesis *(10–14)*.

Moreover, as apoptotic cells can be acquired from macrophages and dendritic cells and presented to T-cells inducing tolerance rather than immunity, presentation of apoptotic cells represents a further mechanism of immune evasion *(15)*.

2. Structural and Functional Organization of HIV Genome

HIV-1 encodes three structural genes, gag (group-specific antigen), pol (polymerase), and env (envelope) that are common to all replication-competent retroviruses. The product of *gag* is translated from unspliced mRNA as a precursor poly-protein. This precursor is cleaved by the viral protease (encoded by pol gene) into the structural proteins of the virion: matrix (Map21), capsid (Cap24), nucleocapsid (NCp7), and several additional small polypeptides, such as p1, p2, and p6, involved in the assembly and morphogenesis of mature capsids. The *pol* gene also encodes a polyprotein, and it is expressed as a fusion protein with *gag* upon ribosomal frameshifting. The Gag-Pol polyprotein is also processed by the viral protease. The cleavage of *pol* gives rise to three enzymes: protease, reverse transcriptase, and integrase. Reverse transcriptase contains three enzymatic activities: RNA-dependent DNA polymerase, RNAase H, and DNA-dependent DNA polymerase. Integrase is responsible for integration of the viral DNA in the cellular chromosome. The *env* gene is essential for viral binding and entry into the host cells. It encodes the precursor glycoprotein gp160, which is cleaved into a surface moiety gp120 and a transmembrane moiety gp41. The surface glycoprotein is required for binding to cellular receptors, whereas the transmembrane glycoprotein is responsible for the fusion with the cell membrane. In addition to the structural genes, HIV-1 also encodes six small open reading frames (ORFs). Two of these are regulatory genes, tat and rev, encoding transactivator proteins essential for viral replication. Tat is a transcriptional transactivator. Rev is a post-transcriptional transactivator that allows nuclear export of unspliced and singly spliced mRNAs encoding viral structural proteins. The remaining four small ORFs are also known as accessory genes and include *vif, vpr, vpu,* and *nef*. The definition of accessory genes denotes the fact that they are nonessential for virus replication in cell culture *(16,17)*.

3. HIV Infection

HIV-1 enters target cells by direct fusion of the viral and target cell membrane. The fusion reaction is mediated by the viral envelope glicoprotein (Env), the CD4 receptor, and the chemokine co-receptors CXCR4 and CCR5. The HIV Env consists of two non-covalently associated subunits generated by cleavage of the gp 160 precursor into gp 120 and gp 41. Binding of gp 120 to CD4 alone does not trigger membrane fusion, but causes conformational change in gp 120, which allows the binding to the chemokine coreceptor and exposition of gp41 to promote membrane fusion.

HIV infects mainly T-helper cells, monocytes/macrophages, dendritic cells (DCs), and microglial cells. This tropism is determined, at the level of viral entry, by the use of CD4 as a primary receptor and of two different chemokine receptors as coreceptors, thus establishing a strain-specificity. HIV R5 strain (M-tropic virus) uses the CC-chemokine receptor 5 (CCR5) as coreceptor and can therefore infect macrophages, DCs and memory T-cells, whereas HIV X4 strain (T-tropic virus) uses the CXCR4 as core-ceptor and can infect both naive and memory T-cells and thymocytes *(18–20)*. The target cells for viral infection change as the virus evolves to use different chemokine coreceptors for entry. It appears that the transmissible HIV-1 strain is mainly the nonsyncytia-inducing M-tropic strain of HIV-1 that uses CCR5 as a coreceptor. Hence, the initial targets in infection are cells expressing CD4 and CCR5. As the disease progresses, in infected individuals, the virus acquires the ability to use the CXCR4 receptor and hence to infect both naive and memory CD4 T-cells and to induce syncytia *(21)*. Therefore, HIV strains isolated from patients with AIDS are either capable of using both CCR5 and CXCR4 or using only CXCR4. Intriguingly, DCs express both CD4 and these chemokine receptors, even if immature DCs expresses high levels of CCR5 and low levels of CXCR4, whereas mature DCs express high levels of CXCR4. Because they encounter HIV at various tissue surfaces, migrate to lymphoid tissues, and interact with T-cells, dendritic cells may be major carriers that allow HIV to gain access to the lymphoid system *(22,23)*.

Once internalized, HIV is uncoated and its RNA genome is reverse transcribed to a double-stranded complementary DNA that, following the entrance into the nucleus through the nuclear pores, is integrated into one of the chromosomes of the target cell yielding the long terminal repeat (LTR)-flanked provirus. The LTR contains enhancer and promoter sequences with binding sites for several transcription factors and a polyadenylation signal. Once the provirus is newly transcribed into RNA, the RNA nuclear transport moves the RNA out of the host nucleus toward the inner surface of the cell membrane. Here protein synthesis occurs, followed by RNA packaging and virion assembly before that budding or cell lysis allows the spread out of virions. In resting T-cells HIV adopts a state of proviral latency.

4. HIV-Induced T-Cell Apoptosis

Recent studies have emphasized that a chronic, generalized activation of the immune system by HIV-1 is the prominent cause of CD4 T-cell depletion in HIV-infected patients *(24)*. This persistent activation of the immune system is caused by a high and continuous virion production and results in an increased turnover of T-lymphocytes as demonstrated by labeling with deuterate glucose the DNA of proliferating cells, in vivo, in HIV-1 infected individuals *(25)*. Further analyses have shown that HIV-1 infection tends to increase the rate of activated CD4 T-cell death and proliferation and to increase the rate of resting CD4 T-cell activation, but does not increase the fraction of activated CD4 T-cell, consistent with their preferential loss in HIV-1 infected individuals. In contrast, HIV infection does not lead to an increase in proliferation or death rate of

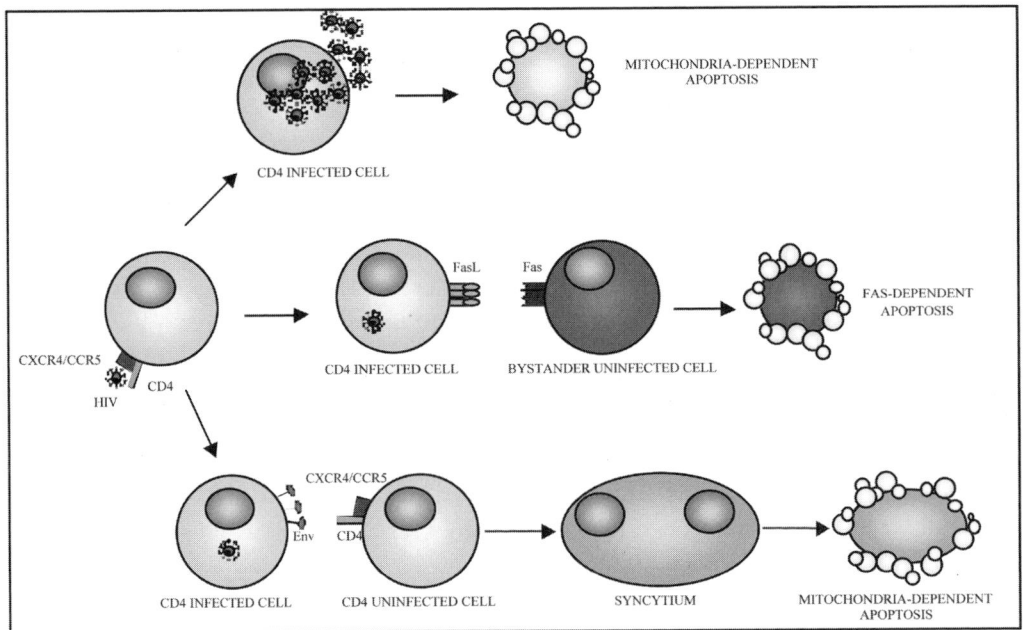

Fig. 1. HIV-induced apoptosis mechanisms. Infected cells may die by HIV replication or may kill bystander cells by Fas/FasL interaction or by cell fusion to form a ayncytium that dies by a mitochondria-dependent apoptosis.

activated CD8 T-cells but does increase the fraction of activated CD8 T-cells, consistent with these cells remaining in an activated state longer and undergoing more rounds of proliferation than CD4 T cells *(26)*.

It is unclear why HIV induces a nonspecific activation of lymphocytes in vivo, but among the factors that can contribute to this process there are the dysregulation of cytokine synthesis, resulting from a shift towards a Th 2 phenotype and the ability of HIV to activate transcription programs *(27)*. Moreover, in humans the HIV-induced immune activation is strongly predictive of disease progression and the lack of chronic immune activation in HIV-infected chimpanzees correlates with their resistance to apoptosis *(28)*. Generally, activation of the immune system results in an increased susceptibility of activated cells to apoptosis induction. This process is fundamental to maintain homeostasis and to induce the reduction of T-cell expansion following an antigen-driven proliferative response *(29)*.

CD4 T-lymphocytes from HIV-seropositive individuals exhibit enhanced spontaneous apoptosis and activation-induced apoptosis ex vivo compared with lymphocytes isolated from HIV-seronegative individuals *(30)*. However, this is not merely an ex vivo phenomenon, as demonstrated by the finding that, in lymph nodes from HIV-seropositive individuals, apoptotic lymphocytes are clearly apparent and, intriguingly, many of the apoptotic cells are not directly infected with HIV *(31)*. Moreover CD4 T-lymphocytes from HIV-seropositive individuals also exhibit enhanced susceptibility to Fas-mediated apoptosis in vitro as compared with those of HIV-seronegative controls *(32,33)*. The same enhanced susceptibility to Fas-mediated apoptosis can be induced in CD4 T lymphocytes from HIV-seronegative individuals by incubation with gp120 or CD4 crosslinking antibody *(34)*. Altogether these evidences suggest that HIV infection influences cell susceptibility to both the extrinsic and the intrinsic apoptotic pathways *(35,36)*. In fact, Fas-mediated apoptosis proceeds through the activation of the extrinsic pathway, whereas spontaneous apoptosis correlates with activation of the intrinsic pathway. The extrinsic pathway of apoptosis is activated by external ligands and is death receptor-dependent, because after binding between ligand and death receptor, receptor triggering induces the formation of the death inducing signaling complex (DISC) and activation of the initiator caspase-8. Downstream of caspase-8, two pathways have been reported; caspase-8 may either directly activate caspase-3 or cleave Bid, a proapoptotic member of the Bcl-2 family, which allows amplification of apoptotic signal through mitochondrial pathway activation *(37)*. The intrinsic pathway of apoptosis is activated by internal sensors; it proceeds through the loss of mitochondrial membrane potential, the release of proapoptotic factors, as cytochrome *c*, from mitochondria into the cytoplasm and activation of the initiator caspase-9 *(38)*. A key event in regulating the intrinsic pathway of apoptosis and the amplifying loop that may be required in extrinsic pathway is represented by the expression of different members of the Bcl-2 family, which regulate mitochondrial membrane permeabilization.

In peripheral blood mononuclear cells (PBMCs) from HIV+ patients the antiapoptotic protein Bcl-2 is down regulated, as it has been shown to happen in vivo in lymph nodes and blood. Cells that have down-regulated Bcl-2 have the phenotype of in vivo activated cells indicating that HIV primed cells in vivo for a spontaneous apoptosis. Moreover CD4 and CD8 cells, activated in vivo, increase expression of Fas, hence becoming more susceptible to Fas-mediated apoptosis. However HIV does not limit its action to sensitize cells to apoptosis, but does induce apoptosis through different mechanisms *(39,40)*.

First of all, HIV induces apoptosis through viral proteins. Numerous in vitro studies have demonstrated the ability of some viral proteins to activate both intrinsic and extrinsic apoptotic pathways. Moreover, because viral proteins, such as Env, Tat, Vpr, and Nef, can be released outside the infected cells, they can exert their effects in different ways and extent both in infected and uninfected bystander cells (*see* Table 1).

Beyond viral proteins, apoptosis can be specifically induced in infected cells by viral replication and by HIV-specific CTLs-mediated killing and in uninfected cells by infected cells-mediated killing and by fusion with infected cells. The effect of apoptotic viral proteins will be described first because viral proteins are the main inducers of apoptosis and since they act in infected and uninfected cells.

5. Effects of Viral-Proteins on Infected and Uninfected T-Cells

5.1. Env-Mediated Apoptosis Induction

As above mentioned, in order to enter a target cell HIV-1 must first bind, through its envelope protein, two cell surface molecules, the receptor CD4 and the coreceptor CXCR4 or CCR5. Although initial studies focused on and demonstrated the ability of gp120 to induce apoptosis by CD4-binding, additional analysis reveled that CXR4 and CCR5 actively participate to trigger apoptosis too.

Gp120 is present not only on viral particles, but also on the surface of infected cells and as soluble protein *(41,42)*. Crosslinking of bound gp120 on human CD4$^+$ T-cells, followed by signaling through the T-cell receptor (TcR) results in activation-dependent apoptosis *(43,44)*. The mechanism for gp120 induction of apoptosis has been reported to

Table 1
HIV Gene Products Involved in Apoptosis Induction and Proposed Mechanisms

HIV gene product	Mechanism	Reference
Env	Upregulation of Fas	46
	Upregulation of FasL	47
	Downregulation of FLIP	48
	Syncytia formation	54
	Priming for activation dependent apoptosis	43,44,51
	Downregulation of Bcl-2	49
	Upregulation of Bax	48
	Caspase-independent apoptosis through CXCR4	
	Caspase-8-dependent apoptosis through CCR5	
	Downregulation of CD4 molecules	99
Tat	Increased activity of NF-κB	57
	Increased activity of NF-AT	58
	Increased activity of AP-1	59
	Upregulation of FasL	60
	Upregulation of caspase-8 expression and activity	
	Upregulation of Bcl-2	63
	Downregulation of Bcl-2	62
	Downregulation of MHC class I	98
Vpr	Cell cycle arrest in G2	65,66,67
	Loss of mitochondrial potential	69,70,71
	Upregulation of Bcl-2	72
	Downregulation of Bax	72
Nef	Maintaining high virus load	75
	Downregulation of MHC class I	76,77,78
	Downregulation of CD4 molecules	77,78,79
	Upregulation of FasL	80,83
	Inhibition of ASK-1	84
	Inactivation of Bad	85
HIV protease	Inactivation of Bcl-2	86
	Activation of caspase-8	87
Vpu	Degradation of CD4	88
	Downregulation of MHC class I molecules	89
	Increased membrane permeability	90

be both Fas-dependent and Fas-independent *(45)*. In fact, much evidence has been reported that gp120 binding to CD4 resulted in an up-regulation of Fas *(46)* and FasL *(47)* and down-modulation of the regulatory protein FLIP *(48)*, events driving a Fas-mediated pathway. However, downregulation of bcl-2 and upregulation of the proapoptotic gene *bax*, with the consequent dissipation of the mitochondrial membrane potential, leading to mitochondria-dependent apoptosis has been reported as well *(48–50)*. It is quite likely that in vivo CD4 uninfected T-cells from HIV infected individuals were continuously primed by HIV-1 infected antigen presenting cells (APCs) through an antigenic stimulus and by a second signal provided by gp 120, in lymph nodes. When in periphery these CD4 primed uninfected T-cells encounter APCs, they receive stimulatory signals through the TcR, that leads to apoptosis *(51,52)*. Because it has been shown that the pattern of coreceptor involved determines a different mechanism of apoptosis triggering, the above mentioned data may find a rationale taking into consideration the coreceptor involved. In fact, it has been reported that binding of Env to CXCR4 receptor induces a caspase-independent death and to CCR5 receptor induces a caspase-8-dependent form of apoptosis *(53)*. Moreover, binding of Env, expressed on the surface of infected cells, to CD4/CXCR4, on the surface of uninfected cells, results in fusion and generation of a syncytium, that readily undergoes apoptosis induction *(54)*.

5.2. Tat-Mediated Apoptosis Induction

The viral protein Tat upregulates viral transcription at the level of elongation via interaction with the Tat activation region (TER) located at the 5′-end of all viral mRNAs. HIV-1 Tat can be secreted by virus-infected cells and can be taken up by bystander cells via endocytosis *(55,56)*. Tat increases the activity of a number of cellular transcription regulators including nuclear factor κ-B (NF-κB), nuclear factor of activated T-cells (NFAT) and AP-1 *(57–59)*. Because of the variety of the interactions with cellular genes the effect of Tat on the survival/apoptosis pathways are likely to be multiple. The activation of NF-κB, for example has been involved in both proapoptotic and in antiapoptotic phenomena. Moreover, Tat expression has been correlated with susceptibility to apoptosis via death receptors. In fact, Tat, secreted from infected cells, may upregulate FasL on uninfected cells *(60)* and sensitizes T-cells to Fas-mediated apoptosis *(34)*. Tat responsive sites in the Fas ligand promoter are identical to NF-κB binding sites, which suggests that up regulation of Fas-L by Tat is NF-κB dependent *(60)*. Moreover, Tat favors the activation of the extrinsic pathway of apoptosis upregulating caspase-8, which is the main initiator caspase of death receptor induced death. Barz and Emerman showed that upon expression of Tat, the levels of caspase-8 mRNA, protein and enzymatic activity increased *(61)*. The upregulation of caspase-8 renders cells susceptibile to apoptosis via Fas signaling. Although there is much evidence underlining the role of Tat as apoptosis-inducing molecule there is also evidence sustaining the contrary. In fact, Tat has been described both to decrease and to increase Bcl-2 expression *(62,63)*. Increasing Bcl-2, Tat could exert an anti-apoptotic effect. Altogether this evidence shows that Tat may have a dual role in the regulation of apoptosis in T-cells.

5.3. Vpr-Mediated Apoptosis Induction

The first reported role for Vpr was a moderate transactivation activity on the long terminal repeat (LTR) of the viral promoter *(64)*. In addition to its role in viral activation, Vpr has been reported to arrest cells in the G2 phase of the cell cycle and to cause apoptosis *(65–68)*. Because Vpr is cell permeable it induces the same effect in both infected and uninfected cells. Moreover Jacotot and coll have shown that exogenous Vpr can directly cause the loss of mitochondrial potential and trigger apoptotic changes in a cell-free system as well as in intact cells *(69)*. Specifically, Vpr has been shown to interact with the permeability transition pore complex (PTPC) causing increased ion permeability and swelling of mitochondria, leading to the rapid release of cytochrome *c* and induction of apoptosis *(69)*. These findings were further confirmed by a recent in vivo study in which a modified adenovirus carrying the HIV-1 Vpr gene was used *(70,71)*. However, it has been reported that Vpr may act as a negative regulator of apoptosis as well. A putative mechanism accounting for the protective effect of Vpr envisage the upregulation of Bcl-2 and downregulation of bax, since overexpression of Bcl-2 abrogates the apoptotic effect of Vpr on mitochondria *(72)*. Conti et al. suggested that in the course of acute infection Vpr plays an early cytoprotective role that allows viral replication, and increases apoptosis at later stages of infection *(73)*.

5.4. Nef-Mediated Apoptosis Induction

Nef is expressed early in the viral life cycle, even from nonintegrated HIV genomes *(74)*. Since the first studies, it has been clear that HIV-Nef is functional to HIV because it maintains high virus load in vivo during the course of persistent infection and it is required for full pathologic potential *(75)*. Later on it became evident that the role of Nef was relevant not only to increase viral replication rate but also to help infected cells not to be eliminated by the immune system through different mechanisms. To this aim Nef downregulates MHC class I and CD4 molecules and upregulates FasL on the surface of the infected cells. Down regulation of MHC allows for both survival of HIV-infected cells that are not recognized as target by CTLs and down regulation of CD4 molecules, preventing CD4 interaction with gp120 and avoiding superinfections and CD4-dependent apoptosis of infected cells *(76–79)*. On the other hand, Nef has the ability to induce FasL expression interacting with the TCR-ζ chain inside T-cells *(80)*. The mechanism by which Nef activates TCR-ζ is not clear. Following translocation to the plasma membrane lipid bilayer, Nef could undergo a conformational change that allows its interaction with signaling proteins, like Lck and LAT, which are present in the glycolipid-enriched microdomains (lipid rafts) where Nef is also found *(81)*. By binding to molecules of different compartments/rafts nef may function as an intracellular crosslinker and may therefore mimic a physiological TcR stimulation. Although this signaling is probably sufficient to activate a ζ-specific function (as FasL upregulation) it is not strong enough to initiate a proliferative signal *(82)*. FasL expressing T-cells may kill bystander cells through a Fas/FasL pathway and may further promote viral escape of immune response eliminating Fas positive CTLs too *(83)*. Infected cells are not destroyed themselves by FasL because they are resistant to apoptosis resulting from the presence of internal viral proteins. In this context it has been reported that Nef may inhibit apoptosis in directly infected cells by inhibiting the apoptosis signal-regulator kinase 1 (ASK-1), or by phosphorylating, hence inactivating, Bad a proapoptotic member of the Bcl-2 family protein *(84,85)*.

5.5. HIV Protease-Mediated Apoptosis Induction

The HIV protease can inactivate anti-apoptotic Bcl-2, through cleavage, and activate procaspase-8 rendering the cells more susceptible to the induction of both intrinsic and extrinsic apoptotic pathways *(86,87)*.

5.6. Vpu-Mediated Apoptosis Induction

The Vpu gene helps viral escape since it encodes a cytoplasmic viral protein that promotes degradation of CD4 in the endoplasmic reticulum of target cells *(88)* and down-regulation of MHC I molecules expression on the surface of infected cells *(89)*.

Furthermore Vpu contributes to induce apoptosis in infected cells since it can induce membrane permeability *(90)*.

6. Apoptosis of HIV-Infected T-Cells

The half-life of most HIV-infected T-cells in vivo is between 12 and 36 h *(91)*. Apoptosis of cells directly infected with HIV accounts for only a small percentage of the apoptotic cells observed, because the vast majority of cells undergoing apoptosis are uninfected *(31)*. Analysis of the mechanisms responsible of infected cell apoptosis has shown that, in these cells, cell death is dependent upon viral replication and is independent of the Fas/FasL pathway *(92,93)*.

There are several mechanisms by which HIV can directly induce cell death in infected cell. As above described, viral proteins play a major role in inducing apoptosis, and may cause the death of both infected and uninfected cells. However, the evidence that, soon after infection, the level of viral load correlates with the rate of CD4 T loss supports the idea that active HIV-1 replication could directly contribute to the depletion of CD4 T-cells. Recent findings indicate a crucial role of mitochondria in the regulation of HIV induced cell death. Petit and colleagues showed that HIV-1 induces mitochondrial membrane permeabilization leading to a caspase-independent cell death of primary CD4 T-cells and suggested that the targeting of Bax to the mitochondria may be a major contributory factor *(94)*. Moreover the levels of antiapoptotic Bcl-2 are significantly lower in HIV-infected individuals *(95)*.

Intracellular virus replication implies also terminal consequences because of cellular toxicity increases resulting from a build up of unintegrated linear viral DNA *(96)*. Furthermore, cell viability can be compromised because the plasma membrane becomes disrupted or more permeable due to the continuous budding of the virion *(97)* or because of specific HIV proteins, such as Vpu, that can induce membrane permeability *(90)*.

An additional mechanism involved in infected T-cell death is represented by the cytotoxic T cells (CTLs), the immune system effector cells. However, like many other viruses, HIV developed different strategies to prevent or, at least to delay, the death of infected cells. In fact, as CTLs recognize their targets through the binding of their TcR with peptide and MCH class I molecules on target cells, both down-regulation of MHC class I molecules and a continuous variability of peptide may contribute in avoiding recognition of infected cells. Down-regulation of MHC class I is caused by viral proteins such as Nef *(76)*, Tat *(98)*, and Vpu *(89)*. Moreover, Env, Vpu, and Nef have the ability to decrease CD4 on the surface of the infected cells *(88,99,100)*. This down regulation of CD4 has the role of preventing both a super infection and an Env-induced apoptosis.

7. Apoptosis of Uninfected T-Cells

During the asymptomatic phase of HIV infection apoptosis plays a major role in killing uninfected cells. There are two mechanisms by which uninfected cells maybe killed: either by HIV proteins released from infected cells acting on neighbouring uninfected cells, or by infected cells themselves (*see* Fig. 1).

7.1. Apoptosis of Uninfected T-Cells by Viral Proteins

HIV proteins such as gp 120, Tat, Nef, and Vpu have been shown to induce cell death in uninfected cells. Both CD4 and CD8 T-lymphocytes are more susceptible to Fas-induced apoptosis in HIV$^+$ individuals and this is related to the up-regulation of Fas and

FasL. Several viral proteins have been involved in increasing Fas and FasL expression, including Env *(46)*, Tat *(55,34)*, and Nef *(101)*.

Inappropriate signaling, through the binding of the HIV-1 envelope to the CD4, may induce abnormal programmed CD4 T-cell death. Crosslinking of the CD4 molecules, by anti-CD4 antibodies or by gp120 plus anti-gp120 antibodies, primes purified normal human CD4$^+$ T-cells for programmed cell death in response to subsequent T-cell receptor stimulation, or induces apoptosis in normal resting human CD4$^+$ T-cells in the absence of T-cell receptor stimulation. Engagement of CD4 molecules triggers the expression of Bax and the consequent dissipation of the mitochondrial membrane potential *(48,49)*.

7.2. Apoptosis of Uninfected T-Cells by Infected Cells: Bystander Effect

Because infected cells upregulate FasL, they may kill CD4 bystander uninfected cells in a Fas-dependent pathway. Moreover, in HIV-infected individuals not only CD4, but also CD8 T-cells are activated and this may correlate with an increased susceptibility to CD8$^+$ Fas-mediated apoptosis. This hypothesis is supported by the observation that, in vitro, activated T-cells expressing FasL can kill Fas expressing CD8 T-cells in a way that is independent of antigen recognition and dependent of Fas/FasL interaction *(47,102)*. Monocytes and macrophages may also kill bystander cells at times of interaction such as might occur during antigen presentation *(103,104)*.

7.3. Apoptosis of Uninfected T-Cells by Syncytia Formation

Syncytia are generated by the fusion of infected T-cells, expressing high levels of Env on their surface with uninfected neighboring cells expressing CD4 and CXCR4 *(105)*. Although syncytia formation is not necessary for the progression to AIDS, they have a short lifespan and dye by apoptosis contributing to the decrease of T-cells *(106)*. Moreover, in vitro studies have correlated this kind of death with the activation of the apoptosis intrinsic pathway, because it proceeds through mitochondrial membrane depolarization, release of cytochrome *c*, caspase activation, and nuclear chromatin condensation *(54)*. Further investigations outlined that this pathway is initiated by upregulation of the cyclin B-CDK1 (cyclin-dependent kinase 1) and by accumulation of mammalian target of rapamycin (MTOR) in the nucleus. This leads to MTOR-mediated serine 15 phosphorylation of p53, p53-dependent upregulation of expression of Bax, translocation of Bax to mitochondria, and activation of the mitochondrial pathway *(107)*. In the lymph nodes of HIV-positive individuals, syncytia express markers of early apoptosis, such as tissue transglutaminase *(108)* and show increased expression of cyclin B and MTOR *(107)*.

8. Apoptosis by Cytokines

The observation that resistance to apoptosis in both HIV and SIV infection is associated with a predominance of the Th1 phenotype and that, in progressive HIV infection, there is a shift towards a Th2 cytokine phenotype *(109)* let hypothesize that acting on cytokines production may restore the immune system. In vitro and in vivo interleukin (IL)-2 treatments have shown that IL-2 (that together with IFN-γ are the Th 1 cytokines) may prevent apoptosis of CD4 T-cells possibly through an antiapoptotic mechanism that has been correlated with increased Bcl-2 expression *(110)*. Interestingly, IL-15, whose receptor shares the β-chain and the γ-chain with the IL-2 receptor and use a common JAK3/STAT5 signaling pathway, can inhibit spontaneous apoptosis too *(111,112)*. As for IL-2, IL-15 inhibition of apoptosis was correlated with its ability to increase Bcl-2 expression. Moreover, similar results may be obtained blocking the action of Th2 cytokines as IL-4 and IL-10, with antibodies in cells from HIV-infected individuals *(113)*. Altogether these evidences indicate that spontaneous apoptosis in patients with HIV infection can be reduced by promoting a TH1 phenotype.

9. Inhibition of Apoptosis Through Therapy

Highly active antiretroviral therapy (HAART), commonly composed by nucleoside-analog reverse trascriptase inhibitors (NRTIs) in combination with protease inhibitors (PIs), produces a significant immune system reconstitution with sustained increase in circulating CD4 T-cells together with a rapid drop in plasma viral RNA levels and decrease in apoptosis *(114,115)*. The decrease in apoptosis is seen as early as 4 d after protease inhibitor therapy is initiated *(116)*. Importantly such changes were seen even before significant changes in plasma viremia were observed, suggesting that the anti-apoptotic effect was distinct from the antiviral effects *(117)*. The PI inhibition of apoptosis was associated with reduced rate of caspase-3 activation *(118)* and inhibition of mitochondrial pathway of apoptosis *(119)*. A number of evidence reported that apoptosis is exclusively inhibited in cells from HIV-infected patients *(117,120,121)*. HIV protease induces the proteolytic cleavage of a variety of cellular substrates that are involved in the regulation of apoptosis induction (procaspase-8 and bcl-2). Because HIV protease is found only in HIV-infected cells, it could be that inhibition of protease might inhibit apoptosis in infected but not in uninfected cells.

References

1. Lane HC, Fauci AS. Immunologic abnormalities in the acquired immunodeficiency syndrome Ann Rev Immunol 1985;3:477–500.
2. Oldstone MB. Viral persistence: mechanisms and consequences. Curr Opin Microbiol 1998 Aug;1(4):436–441.
3. Peterlin BM, Trono D. Hide, shield and strike back: how HIV-infected cells avoid immune eradication. Nat Rev Immunol 2003 Feb;3(2):97–107.
4. Hilleman MR. Strategies and mechanisms for host and pathogen survival in acute and persistent viral infections. Proc Natl Acad Sci USA 2004 Oct 5;101 Suppl 2:14,560–14,566.
5. Mosier DE. Virus and target cell evolution in human immunodeficiency virus type 1 infection. Immunol Res 2000;21(2-3):253–258.
6. Gougeon ML, Garcia S, Henney J, et al. Programmed cell death in AIDS-related HIV and SIV infections. AIDS Res Hum Retroviruses 1993;9:553–563.
7. Douek DC, McFarland RD, Keiser PH, et al. Changes in thymic function with age and during the treatment of HIV infection. Nature 1998;396:690–695.
8. Pakkar NG, Notermans DW, de Boer RJ, et al. Biphasic kinetics of peripheral blood T cells after triple combination therapy in HIV-1 infection: a composite of redistribution and proliferation. Nat Med 1998;4:208–214.
9. Hellerstein MK, McCune JM. T cell turnover in HIV-1 disease. Immunity 1997;7:583–589.
10. Gougeon ML, Laurent-Crawford AG, Hovanessian AG, Montagnier L. Direct and indirect mechanisms mediating apoptosis during HIV infection: contribution to in vivo CD4 T cell depletion Sem Immunol 1993 Jun;5(3):187–194.
11. Badley AD, Dockrell D, Pray CV. Apoptosis in AIDS Adv Pharmacol 1997;41:271–294.
12. Badley AD, Pilon AA, Landay A, Lynch DH. Mechanisms of HIV-associated lymphocyte apoptosis Blood 2000 Nov 1;96(9):2951–2964

13. Bell DJ, Dockrell DH. Apoptosis in HIV-1 infection. J Eur Acad Dermatol Venereol 2003 Mar;17(2):178–183.

14. Alimonti JB, Ball TB, Fowke KR. Mechanisms of CD4+ T lymphocyte cell death in human immunodeficiency virus infection and AIDS. J Gen Virol 2003 Jul;84(Pt 7):1649–1661.

15. Steinman RM, Turley S, Mellman I, Inaba K. The induction of tolerance by dendritic cells that have captured apoptotic cells. J Exp Med 2000 Feb 7;191(3):411–416.

16. Luciw PA. Human immunodeficiency viruses and their replication. In: Fields BN, Knippe DM, Howley PM, ed. Fields Virology. Philadelphia: Lippincott-Raven 1996;1881–1975.

17. Roshal M, Zhu Y, Planelles V. Apoptosis in AIDS. Apoptosis 2001 Feb-Apr;6(1-2):103–116.

18. Littman DR. Chemokine receptors: keys to AIDS pathogenesis? Cell 1998 May 29;93(5): 677–680.

19. Berger EA, Murphy PM, Farber JM. Chemokine receptors as HIV-1 coreceptors: roles in viral entry, tropism, and disease. Annu Rev Immunol 1999;17:657–700.

20. Blaak H, van't Wout AB, Brouwer M, Hooibrink B, Hovenkamp E, Schuitemaker H. In vivo HIV-1 infection of CD45RA(+)CD4(+) T cells is established primarily by syncytium-inducing variants and correlates with the rate of CD4(+) T cell decline Proc Natl Acad Sci USA 2000 Feb 1;97(3):1269–1274.

21. Schuitemaker H, Koot M, Kootstra NA, et al. Biological phenotype of human immunodeficiency virus type 1 clones at different stages of infection: progression of disease is associated with a shift from monocytotropic to T-cell-tropic virus population. J Virol 1992 Mar;66(3):1354–1360.

22. Granelli-Piperno A, Delgado E, Finkel V, Paxton W, Steinman RM. Immature dendritic cells selectively replicate macrophagetropic (M-tropic) human immunodeficiency virus type 1, while mature cells efficiently transmit both M- and T-tropic virus to T cells. J Virol 1998 Apr;72(4):2733–2737.

23. Rowland-Jones SL. HIV/ The deadly passenger in dendritic cells Curr Biol 1999;9: 248–250.

24. Hazenberg MD, Hamann D, Schuitemaker H, Miedema F. T cell depletion in HIV-1 infection: how CD4+ T cells go out of stock. Nat Immunol 2000 Oct;1(4):285–289.

25. Mohri H, Perelson AS, Tung K, et al. Increased turnover of T lymphocytes in HIV-1 infection and its reduction by antiretroviral therapy. J Exp Med 2001 Nov 5;194(9):1277–1287.

26. Ribeiro RM, Mohri H, Ho DD, Perelson AS. In vivo dynamics of T cell activation, proliferation, and death in HIV-1 infection: why are CD4+ but not CD8+ T cells depleted? Proc Natl Acad Sci USA 2002 Nov 26;99(24):15,572–15,577.

27. Gougeon ML. Apoptosis as an HIV strategy to escape immune attack. Nat Rev Immunol 2003 May;3(5):392–404.

28. Heeney J, Jonker R, Koornstra W, et al. The resistance of HIV-infected chimpanzees to progression to AIDS correlates with absence of HIV-related T-cell dysfunction. J Med Primatol 1993 Feb-May;22(2-3):194–200.

29. Lenardo M, Chan KM, Hornung F, et al. Mature T lymphocyte apoptosis—immune regulation in a dynamic and unpredictable antigenic environment. Annu Rev Immunol 1999;17: 221–253.

30. Meyaard L, Otto SA, Jonker RR, Mijnster MJ, Keep RP, Miedema F. Programmed cell death of T cells in HIV-1 infection. Science (Washington) 1992 Jul 10;257(5067):217–219.

31. Finkel TH, Tudor-Williams G, Banda NK. Apoptosis occurs predominantly in bystander cells and not in productively infected cells HIV- and SIV-infected lymph nodes. Nature Med 1995 Feb;1(2):129–134.

32. Katsikis PD, Wunderlich ES, Smith CA, Herzenberg LA. Fas antigen stimulation induces marked apoptosis of T Lymphocytes in human immunodeficiency virus-infected individuals. J Exp Med 1995 Jun 1;181(6):2029–2036.

33. Dockrell DH, Badley AD, Algeciras-Schimnich A. Activation-induced CD4 T cell death in HIV-positive individuals correlates with Fas susceptibility, CD4+ T cell count, and HIV plasma viral copy number AIDS Res Hum Retroviruses. 1999 Nov 20;15(17):1509–1518.

34. Westendorp MO, Frank R, Ochsenbauer C, et al. Sensitization of T cells to CD95-mediated apoptosis by HIV-1 Tat and gp120. Nature 1995 Jun 8;375(6531):497–500.

35. Benedict CA, Norris PS, Ware CF. To kill or be killed: viral evasion of apoptosis. Nat Immunol 2002;3(11):1013–1018.

36. Petit F, Arnoult D, Viollet L, Estaquier J. Intrinsic and extrinsic pathways signaling during HIV-1 mediated cell death. Biochimie 2003;85(8):795–811.

37. Scaffidi C, Fulda S, Srinivasan A, et al. CD95 (APO-1/Fas) signaling pathways. EMBO J 1998 Mar 16;17(6):1675–1687.

38. Green DR. Apoptotic pathways: paper wraps stone blunts scissors. Cell 2000;102:1–4.

39. Selliah N, Finkel TH. Biochemical mechanisms of HIV induced T cell apoptosis. Cell Death Differ 2001;8(2):127–136.

40. Arnoult D, Petit F, Lelievre JD, Estaquier J. Mitochondria in HIV-1-induced apoptosis. Biochem Biophys Res Commun 2003 May 9;304(3):561–574.

41. Chirmule N, Pahwa S. Envelope glycoproteins of human immunodeficiency virus type 1: profound influences on immune functions. Microbiol Re 1996;60(2):386–406.

42. Herbein G, Van Lint C, Lovett JL, Verdin E. Distinct mechanisms trigger apoptosis in human immunodeficiency virus type 1-infected and in uninfected bystander T lymphocytes J. Virol 1998;72(1):660–670.

43. Banda NK, Bernier J, Kurahara DK, et al. Crosslinking CD4 by human immunodeficiency virus gp120 primes T cells for activation-induced apoptosis. J Exp Med 1992;176(4): 1099–1106.

44. Tuosto L, Gilardini Montani MS, Lorenzetti S, et al. Differential susceptibility to monomeric HIV gp120-mediated apoptosis in antigen-activated CD4+ T cell populations. Eur J Immunol 1995;25(10):2907–2916.

45. Perfettini JL, Castedo M, Roumier T, et al. Mechanisms of apoptosis induction by the HIV-1 envelope. Cell Death Differ 2005;12(Suppl 1):916–923.

46. Oyaizu N, McCloskey TW, Than S, Hu R, Kalyanaraman VS, Pahwa S. Cross-linking of CD4 molecules upregulates Fas antigen expression in lymphocytes by inducing interferon-gamma and tumor necrosis factor-alpha secretion. Blood 1994;84(8):2622–2631.

47. Tateyama M, Oyaizu N, McCloskey TW, Than S, Pahwa S. CD4 T lymphocytes are primed to express Fas ligand by CD4 cross-linking and to contribute to CD8 T-cell apoptosis via Fas/FasL death signaling pathway. Blood 2000;96:195–202.

48. Somma F, Tuosto L, Gilardini Montani MS, Di Somma MM, Cundari E, Piccolella E. Engagement of CD4 before TCR triggering regulates both Bax- and Fas (CD95)-mediated apoptosis J Immunol 2000;164(10):5078–5087.

49. Hashimoto F, Oyaizu N, Kalyanaraman VS, Pahwa S. Modulation of Bcl-2 protein by CD4 cross-linking: a possible mechanism for lymphocyte apoptosis in human immuno-deficiency virus infection and for rescue of apoptosis by interleukin-2. Blood 1997;90(2): 745–53.

50. Arthos J, Cicala C, Selig SM, et al. The role of the CD4 receptor versus HIV coreceptors in envelope-mediated apoptosis in peripheral blood mononuclear cells. Virology 2002;292: 98–106.

51. Cottrez F, Manca F, Dalgleish AG, Arenzana-Seisdedos F, Capron A, Groux H. Priming of human CD4+ antigen-specific T cells to undergo apoptosis by HIV-infected monocytes. A two-step mechanism involving the gp120 molecule. J Clin Invest 1997;99(2):257–266.

52. Ahr B, Robert-Hebmann V, Devaux C, Biard-Piechaczyk M. Apoptosis of uninfected cells induced by HIV envelope glycoproteins. Retrovirology 2004;1(1):12.

53. Vlahakis SR, Algeciras-Schimnich A, Bou G, et al. Chemokine-receptor activation by env determines the mechanism of death in HIV-infected and uninfected T lymphocytes. J Clin Invest 2001;107(2):207–215.

54. Ferri KF, Jacotot E, Blanco J, et al. Apoptosis control in syncytia induced by the HIV type 1-envelope glycoprotein complex: role of mitochondria and caspases. J Exp Med 2000; 192(8):1081–1092.

55. Ensoli B, Barillari G, Salahuddin SZ, Gallo RC, Wong-Staal F. Tat protein of HIV-1 stimulates growth of cells derived from Kaposi's sarcoma lesions of AIDS patients. Nature 1990;345(6270):84–86.

56. Mann DA, Frankel AD. Endocytosis and targeting of exogenous HIV-1 Tat protein. EMBO J 1991;10(7):1733–1739.

57. Demarchi F, d'Adda di Fagagna F, Falaschi A, Giacca M. Activation of transcription factor NF-kappaB by the Tat protein of human immunodeficiency virus type 1. J Virol 1996; 70(7):4427–4437.

58. Macian F, Rao A. Reciprocal modulatory interaction between human immunodeficiency virus type 1 Tat and transcription factor NFAT1. Mol Cell Biol 1999;19(5):3645–3653.

59. Kumar A, Manna SK, Dhawan S, Aggarwal BB. HIV-Tat protein activates c-Jun N-terminal kinase and activator protein-1. J Immunol 1998;161(2):776–781.

60. Li-Weber M, Laur O, Dern K, Krammer PH. T cell activation-induced and HIV tat-enhanced CD95(APO-1/Fas) ligand transcription involves NF-κB Eur. J Immunol 2000; 30(2):661–670.

61. Bartz SR, Emerman M. Human immunodeficiency virus type 1 Tat induces apoptosis and increases sensitivity to apoptotic signals by up-regulating FLICE/caspase-8. J Virol 1999; 73(3):1956–1963.

62. Sastry KJ, Marin MC, Nehete PN, McConnell K, el-Naggar AK, McDonnell TJ. Expression of human immunodeficiency virus type I tat results in down-regulation of bcl-2 and induction of apoptosis in hematopoietic cells. Oncogene 1996;13(3):487–493.

63. Zauli G, Gibellini D, Caputo A, et al. The human immunodeficiency virus type-1 Tat protein upregulates Bcl-2 gene expression in Jurkat T-cell lines and primary peripheral blood mononuclear cells. Blood 1995;86(10):3823–3834.

64. Gummuluru S, Emerman M. Cell cycle- and Vpr-mediated regulation of human immunodeficiency virus type 1 expression in primary and transformed T-cell lines. J Virol 1999; 73(7):5422–5430.

65. He J, Choe S, Walker R, Di Marzio P, Morgan DO, Landau NR. Human immunodeficiency virus type 1 viral protein R (Vpr) arrests cells in the G2 phase of the cell cycle by inhibiting p34cdc2 activity. J Virol 1995;69(11):6705–6711.

66. Re F, Braaten D, Franke EK, Luban J. Human immunodeficiency virus type 1 Vpr arrests the cell cycle in G2 by inhibiting the activation of p34cdc2-cyclin B. J Virol 1995; 69(11):6859–6864.

67. Stewart SA, Poon B, Jowett JB, Chen IS. Human immunodeficiency virus type 1 Vpr induces apoptosis following cell cycle arrest. J Virol 1997;71(7):5579–5592.

68. Watanabe N, Yamaguchi T, Akimoto Y, Rattner JB, Hirano H, Nakauchi H. Induction of M-phase arrest and apoptosis after HIV-1 Vpr expression through uncoupling of nuclear and centrosomal cycle in HeLa cells. Exp Cell Res 2000;258(2):261–269.

69. Jacotot E, Ravagnan L, Loeffler M, et al. The HIV-1 viral protein R induces apoptosis via a direct effect on the mitochondrial permeability transition pore. J Exp Med 2000; 191(1):33–46.

70. Muthumani K, Zhang D, Hwang DS, et al. Adenovirus encoding HIV-1 Vpr activates caspase 9 and induces apoptotic cell death in both p53 positive and negative human tumor cell lines. Oncogene 2002;21(30):4613–4625.

71. Muthumani K, Choo AY, Hwang DS, et al. Mechanism of HIV-1 viral protein R-induced apoptosis. Biochem Biophys Res Commun 2003;304(3):583–592.

72. Conti L, Rainaldi G, Matarrese P, et al. The HIV-1 vpr protein acts as a negative regulator of apoptosis in a human lymphoblastoid T cell line: possible implications for the pathogenesis of AIDS J Exp M 1998;187(3):403–413.

73. Conti L, Matarrese P, Varano B, et al. Dual role of the HIV-1 vpr protein in the modulation of the apoptotic response of T cells. J Immunol 2000;165(6):3293–3300.

74. Wu Y, Marsh JW. Selective transcription and modulation of resting T cell activity by pre-integrated HIV DNA. Science 2001;293(5534):1503–1506.

75. Kestler HW 3rd, Ringler DJ, Mori K, Panicali DL, Sehgal PK, Daniel MD, Desrosiers RC. Importance of the nef gene for maintenance of high virus loads and for development of AIDS. Cell 1991;65(4):651–662.

76. Schwartz O, Marechal V, Le Gall S, Lemonnier F, Heard JM. Endocytosis of major histocompatibility complex class I molecules is induced by the HIV-1 Nef protein. Nat Med 1996;2(3):338–342.

77. Collins KL, Chen BK, Kalams SA, Walker BD, Baltimore D. HIV-1 Nef protein protects infected primary cells against killing by cytotoxic T lymphocytes. Nature 1998;391(6665): 397–401.

78. Collins KL. Resistance of HIV-infected cells to cytotoxic T lymphocytes. Microbes Infect 2004;6(5):494–500.

79. Garcia JV, Miller AD. Serine phosphorylation-independent downregulation of cell-surface CD4 by nef. Nature 1991;350(6318):508–511.

80. Xu XN, Laffert B, Screaton GR, et al. Induction of Fas ligand expression by HIV involves the interaction of Nef with the T cell receptor zeta chain. J Exp Med 1999;189(9):1489–1496.

81. Arold ST, Baur AS. Dynamic Nef and Nef dynamics: how structure could explain the complex activities of this small HIV protein. Trends Biochem Sci 2001;26(6):356–363.

82. Fackler OT, Baur AS. Live and let die: Nef functions beyond HIV replication. Immunity 2002;16(4):493–497.

83. Xu XN, Screaton GR, Gotch FM, et al. Evasion of cytotoxic T lymphocyte (CTL) responses by nef-dependent induction of Fas ligand (CD95L) expression on simian immunodeficiency virus-infected cells. J Exp Med 1997;186(1):7–16.

84. Geleziunas R, Xu W, Takeda K, Ichijo H, Greene WC. HIV-1 Nef inhibits ASK1-dependent death signalling providing a potential mechanism for protecting the infected host cell. Nature 2001;410(6830):834–838.

85. Wolf D, Witte V, Laffert B, Blume K, Stromer E, Trapp S, d'Aloja P, Schurmann A, Baur AS. HIV-1 Nef associated PAK and PI3-kinases stimulate Akt-independent Bad-phosphorylation to induce anti-apoptotic signals. Nat Med 2001;7(11):1217–1224.

86. Strack PR, Frey MW, Rizzo CJ, et al. Apoptosis mediated by HIV protease is preceded by cleavage of Bcl-2. Proc Natl Acad Sci USA 1996;93(18):9571–9576.

87. Nie Z, Phenix BN, Lum JJ, et al. HIV-1 protease processes procaspase 8 to cause mitochondrial release of cytochrome c, caspase cleavage and nuclear fragmentation. Cell Death Differ 2002;9(11):1172–1184.

88. Willey RL, Maldarelli F, Martin MA, Strebel K. Human immunodeficiency virus type 1 Vpu protein induces rapid degradation of CD4. J Virol 1992;66(12):7193–7200.

89. Kerkau T, Bacik I, Bennink JR, et al. The human immunodeficiency virus type 1 (HIV-1) Vpu protein interferes with an early step in the biosynthesis of major histocompatibility complex (MHC) class I molecules. J Exp Med 1997;185(7):1295–1305.

90. Gonzalez ME, Carrasco L. Human immunodeficiency virus type 1 VPU protein affects Sindbis virus glycoprotein processing and enhances membrane permeabilization. Virology 2001;279(1):201–209.

91. Perelson AS, Neumann AU, Markowitz M, Leonard JM, Ho DD. HIV-1 dynamics in vivo: virion clearance rate, infected cell life-span, and viral generation time. Science 1996; 271(5255):1582–1586.

92. Yagi T, Sugimoto A, Tanaka M, et al. Fas/FasL interaction is not involved in apoptosis of activated CD4+ T cells upon HIV-1 infection in vitro. J Acquir Immune Defic Syndr Hum Retrovirol 1998;18(4):307–315.

93. Gandhi RT, Chen BK, Straus SE, Dale JK, Lenard MJ, Baltimore D. HIV-1 directly kills CD4+ T cells by a Fas-independent mechanism. J Exp Med 1998;187:1113–1122.

94. Petit F, Arnoult D, Lelievre JD, et al. Productive HIV-1 infection of primary CD4+ T cells induces mitochondrial membrane permeabilization leading to a caspase-independent cell death. J Biol Chem 2002;277(2):1477–1487.

95. Re M, Gibellini D, Aschbacher R, et al. High levels of HIV-1 replication show a clear correlation with downmodulation of Bcl-2 protein in peripheral blood lymphocytes of HIV-1-seropositive subjects. J Med Virol 1998;56:66–73

96. Levy JA. Pathogenesis of human immunodeficiency virus infection. Microbiol Rev 1993; 57:183–289

97. Fauci AS. The human immunodeficiency virus: infectivity and mechanisms of pathogenesis. Science 1988;239(4840):617–622.

98. Howcroft TK, Strebel K, Martin MA, Singer DS. Repression of MHC class I gene promoter activity by two-exon Tat of HIV. Science 1993;260(5112):1320–1322.

99. Crise B, Buonocore L, Rose JK. CD4 is retained in the endoplasmic reticulum by the human immunodeficiency virus type 1 glycoprotein precursor. J Virol 1990;64(11):5585–5593.

100. Salghetti S, Mariani R, Skowronski J. Human immunodeficiency virus type 1 Nef and p56lck protein-tyrosine kinase interact with a common element in CD4 cytoplasmic tail. Proc Natl Acad Sci USA 1995;92(2):349–353.

101. Zauli G, Gibellini D, Secchiero P, et al. Human immunodeficiency virus type 1 Nef protein sensitizes CD4(+) T lymphoid cells to apoptosis via functional upregulation of the CD95/CD95 ligand pathway. Blood 1999;93(3):1000–1010.

102. Piazza C, Gilardini Montani MS, Moretti S, Cundari E, Piccolella E. Cutting edge: CD4+ T cells kill CD8+ T cells via Fas/Fas ligand-mediated apoptosis. Immunol 1997;158(4): 1503–1506.

103. Oyaizu N, Adachi Y, Hashimoto F, et al. Monocytes express Fas ligand upon CD4 cross-linking and induce CD4+ T cells apoptosis: a possible mechanism of bystander cell death in HIV infection. J Immunol 1997;158(5):2456–2463.

104. Badley AD, Dockrell D, Simpson M, et al. Macrophage-dependent apoptosis of CD4+ T lymphocytes from HIV-infected individuals is mediated by FasL and tumor necrosis factor. J Exp Med 1997;185(1):55–64.

105. Lifson JD, Reyes GR, McGrath MS, Stein BS, Engleman EG. AIDS retrovirus induced cytopathology: giant cell formation and involvement of CD4 antigen. Science 1986;232(4754):1123–1127.

106. Richman DD, Bozzette SA. The impact of the syncytium-inducing phenotype of human immunodeficiency virus on disease progression. J Infect Dis 1994;169(5):968–974.

107. Castedo M, Perfettini JL, Roumier T, Kroemer G. Cyclin-dependent kinase-1: linking apoptosis to cell cycle and mitotic catastrophe. Cell Death Differ 2002;9(12):1287–1293.

108. Amendola A, Gougeon ML, Poccia F, Bondurand A, Fesus L, Piacentini M. Induction of "tissue" transglutaminase in HIV pathogenesis: evidence for high rate of apoptosis of CD4+ T lymphocytes and accessory cells in lymphoid tissues. Proc Natl Acad Sci USA 1996;93(20):11,057–11,062.

109. Clerici M, Shearer GM. A TH1—>TH2 switch is a critical step in the etiology of HIV infection. Immunol Today 1993;14(3):107–111.

110. Adachi Y, Oyaizu N, Than S, McCloskey TW, Pahwa S. IL-2 rescues in vitro lymphocyte apoptosis in patients with HIV infection: correlation with its ability to block culture-induced down-modulation of Bcl-2. J Immunol 1996;157(9):4184–4193.

111. Waldmann T, Tagaya Y, Bamford R. Interleukin-2, interleukin-15, and their receptors. Int Rev Immunol 1998;16(3–4):205–226.

112. Naora H, Gougeon ML. Interleukin-15 is a potent survival factor in the prevention of spontaneous but not CD95-induced apoptosis in CD4 and CD8 T lymphocytes of HIV-infected individuals. Correlation with its ability to increase BCL-2 expression. Cell Death Differ 1999;6(10):1002–1011.

113. Clerici M, Sarin A, Berzofsky JA, et al. Antigen-stimulated apoptotic T-cell death in HIV infection is selective for CD4+ T cells, modulated by cytokines and effected by lymphotoxin. AIDS 1996;10(6):603–611.

114. Hammer SM. Advances in antiretroviral therapy and viral load monitoring. AIDS 1996;10 Suppl 3:S1–S11.

115. Johnson N, Parkin JM. Anti-retroviral therapy reverses HIV-associated abnormalities in lymphocyte apoptosis. Clin Exp Immunol 1998;113(2):229–234.

116. Badley AD, Parato K, Cameron DW, et al. Dynamic correlation of apoptosis and immune activation during treatment of HIV infection. Cell Death Differ 1999;6(5):420–432.

117. Phenix BN, Angel JB, Mandy F, et al. Decreased HIV-associated T cell apoptosis by HIV protease inhibitors. AIDS Res Hum Retroviruses 2000;16(6):559–567.

118. Sloand EM, Kumar PN, Kim S, Chaudhuri A, Weichold FF, Young NS. Human immunodeficiency virus type 1 protease inhibitor modulates activation of peripheral blood CD4(+) T cells and decreases their susceptibility to apoptosis in vitro and in vivo. Blood 1999; 94(3):1021–1027.

119. Phenix BN, Lum JJ, Nie Z, Sanchez-Dardon J, Badley AD. Antiapoptotic mechanism of HIV protease inhibitors: preventing mitochondrial transmembrane potential loss. Blood 2001;98(4):1078–1085.

120. Lu W, Andrieu JM. HIV protease inhibitors restore impaired T-cell proliferative response in vivo and in vitro: a viral-suppression-independent mechanism. Blood 2000;96(1):250–258.

121. Phenix BN, Cooper C, Owen C, Badley AD. Modulation of apoptosis by HIV protease inhibitors. Apoptosis 2002;7(4):295–312.

15

Inhibitor of Apoptosis Proteins and Caspases

Jai Y. Yu, John Silke, and Paul G. Ekert

Summary

Inhibitors of apoptosis proteins (IAPs) were first identified as insect viral proteins that block host cell apoptosis. Cellular homologs bearing the characteristic baculoviral IAP repeat (BIR) domain have now been found in all metazoans and have more diverse functions than their name suggests. Some cellular IAPs do, in fact, inhibit apoptosis, in part at least by directly inhibiting cysteine proteases, called caspases, that are ultimately responsible for killing a cell. The primary aim of this chapter is to explore the intricate regulation of caspases by IAPs, and the exquisite counter regulation of IAPs by antagonist proteins. Interest in IAPs extends beyond apoptosis pathways and recent evidence suggests that IAPs modulate cell-signaling pathways that affect growth and proliferation and some of these remarkable studies will be discussed.

Key Words: Inhibitor of apoptosis proteins; caspase; apoptosis; IAP antagonists.

1. Discovery of IAPs

IAPs were first identified in insect viruses as genes that could complement the loss of another antiapoptotic viral gene, p35 *(1)*. Insect cells infected with a baculovirus that lacked p35 underwent apoptosis and the virus failed to replicate, showing that cell death is an effective host cell response to limit viral replication. Thus, viruses that are able to prevent host cell apoptosis have a selective advantage over viruses that cannot and as a consequence virus genomes encode several antiapoptotic proteins. A screen for genes that functioned like p35 surprisingly turned up two completely unrelated genes, inhibitor of apoptosis proteins (IAP), one from *Cydia pomonella* granulosis virus (CpGV) *(2)* and another from *Orgyia pseudotsugata* nuclear polyhedrosis virus (OpMNPV) *(3)*, that contained a novel repeated zinc-finger like domain, christened a baculoviral IAP Repeat (BIR).

With the expansion of the family, the name "IAP" has become misleading because newer family members have no role in inhibiting apoptosis. Because the BIR is the signature domain of these proteins, but not all family members act to block cell death, the name BIR containing proteins (BIRps or BIRCs) has been adopted as the official terminology by the Human Genome Nomenclature Committee. In this chapter we will nevertheless use the term IAPs because the focus will be on those family members that have a role in inhibiting apoptosis. We will discuss the known mechanisms by which IAPs can inhibit caspases, the ways IAPs themselves are regulated, and then briefly discuss other possible functions.

Because baculoviral IAPs blocked apoptosis of insect cells, and other cell death genes have been shown to function in heterologous systems, the baculoviral IAPs were tested for functioned in mammalian cells. OpIAP inhibited apoptosis caused by overexpression of

From: *Apoptosis, Cell Signaling, and Human Diseases: Molecular Mechanisms, Volume 2*
Edited by: R. Srivastava © Humana Press Inc., Totowa, NJ

caspase-1 in HeLa cells, showing that an insect virus protein could inhibit mammalian cell death *(4)*. This confirmed that elements of mammalian and insect cell death pathways are conserved, and raised the possibility that mammalian cellular IAP homologs existed.

In *Drosophila*, IAPs (DIAP1 and DIAP2) were identified in a search for genes resembling baculoviral IAPs. Crossing of fly stocks mutant for DIAP1 (also known as *thread*) showed that they enhanced apoptosis induced by overexpression of *reaper (rpr)* and *head involution defective (hid)* in the *Drosophila* eye *(5)*.

Mammalian IAPs were identified using several experimental strategies, including genome database searches using the characteristic protein motifs of IAP proteins the BIR and RING domains *(6–8)* and by purification of proteins binding to the cytoplasmic domain of tumor necrosis factor receptor 2 (TNFR2) *(9)*.

Soon after the identification of IAPs in higher organisms, these proteins were demonstrated to directly inhibit caspases in several experimental systems. Recombinant XIAP, cIAP1, and cIAP2 were reported to prevent substrate cleavage by active caspases in vitro *(10–12)*. Overexpression of XIAP also inhibited apoptosis in mammalian cell lines in response to a variety of stimuli *(10)*.

Structural studies subsequently revealed the details of the molecular interactions between IAPs and caspases *(13–17)*.

2. Structure of IAPs

The signature structural element of an IAP is the BIR. Proteins containing the BIR motif have been found in insect and mammalian viruses, yeasts, nematodes, flies, and vertebrates *(18)*. However, the BIR containing proteins in yeasts and nematodes that resemble the mammalian protein Survivin are required for chromosome segregation, rather than inhibition of cell death. The structural elements of selected family members are shown in schematic form in Fig. 1.

The caspase inhibitory IAPs all bear at least one BIR domain at the N-terminus. The BIR is a zinc-coordinating domain with the motif $CX_2CH_{16}HX_{6-8}C$ and varies in length from approximately 70 to 100 residues. The structure of this domain comprises four or five α-helices and a β-sheet *(19–21)*. This domain and adjacent regions linking the IAPs play important roles in the caspase inhibitory function of these IAPs.

The C-terminus of IAPs varies. In baculoviral and many of the mammalian IAPs, the C-terminus bears a C_3HC_4 RING domain *(22,23)*. The RING domain is a zinc binding structure that functions as a ubiquitin E3 ligase *(24)*. The mammalian IAPs, cIAP1 and cIAP2, also contain another domain, a caspase recruitment domain (CARD), located between the BIRs and the RING. The role of the CARD in cIAP1 and cIAP2 is unknown and is neither necessary nor sufficient for the caspase inhibitory role of those proteins *(12)*. The CARD domain is a homotypic protein–protein interaction domain that interacts with other specific CARD domains. For example, the CARD of caspase-9 interacts with the CARD of the adaptor molecule Apaf-1. The CARD shares some structural similarities with several other protein interaction domains including death domains (DD), death effector domains (DED), and pyrin domains (PYD) *(25)*. CARD domains are not restricted to caspases or molecules that bind caspases and the presence of a CARD does not imply that the protein can recruit and activate a caspase.

The giant (approx 530 kD) IAP BRUCE/Apollon has a ubiquitin-conjugating enzyme (UBC) domain similar to that of E2 ubiquitin ligases *(26)*. The BIR containing

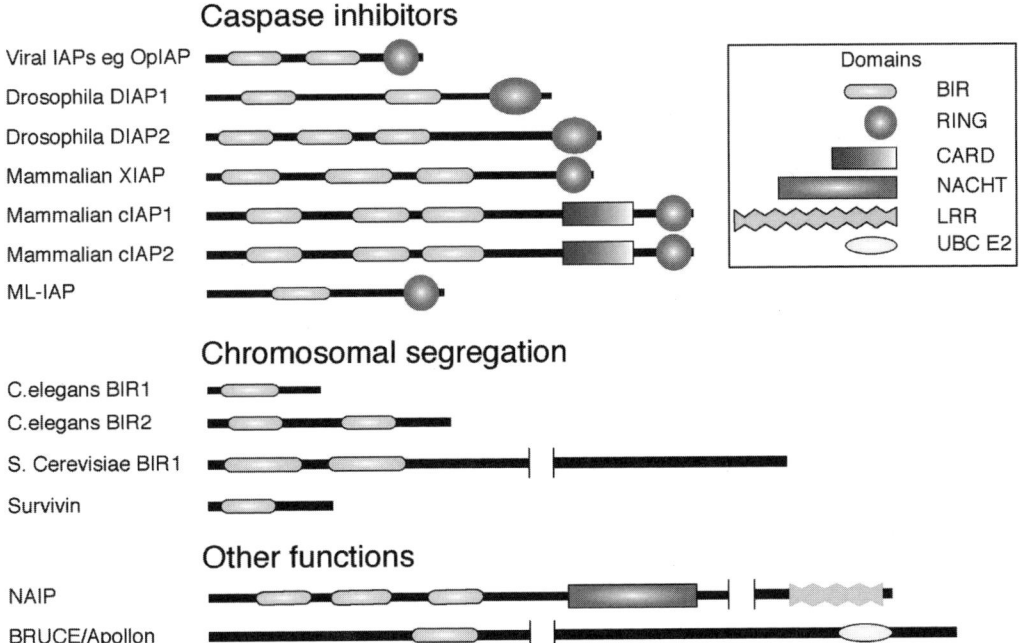

Fig. 1. Schematic representation of selected members of the BIRC family. Family members are functionally grouped into those that have a role in inhibiting apoptosis, those with a role in chromosomal segregation and cell division, and those with other, largely unknown, functions. The various protein domains are shown in the box.

proteins found in *Saccharomyces cerevisiae, S. pombe, Caenorhabditis elegans*, the *Drosophila* IAP deterin, and mammalian Survivin do not have really interesting new gene (RING), CARD, or UBC domains in addition to their BIRs.

2.1. Mechanisms of Caspase Activation

As a prelude to a discussion of how IAPs inhibit caspase activity, it is worth briefly reviewing the role of caspases and how these enzymes are activated. Caspases are a family of intracellular proteases with catalytic cysteines that cleave substrates after aspartate residues (with the anomalous exception of the *Drosophila* caspase Dronc that also cleaves after glutamate residues). Caspase specificity for substrates is determined primarily by the 3-amino acids immediately before the labile aspartate of the substrate. This aspartate and preceeding amino acids can be designated P_4-P_1. The pockets in the caspase that contact these four residues (including the labile aspartate) can be designated S_4 to S_1 (Fig. 2) *(28)*. Caspases have a stringent requirement for an aspartate residue to occupy the S_1 position *(29)*.

Caspases exist as zymogens within the cell. They consist of a large (p20) and a small (p10) subunit that make up the catalytic domain. Caspases also have prodomains, situated immediately before the p20 subunit of varying length, some of which contain protein–protein interaction domains, for example CARD or DEDs. Caspases can be subclassified either by substrate sequence preference, the nature of the prodomain, or phylogenetically and there is considerable overlap in these subclassifications *(28)*.

Caspases with long prodomains tend to be initiator caspases, which are activated by interaction with an adaptor molecule and that have as their main substrate other caspases. Caspase-9 interacts with Apaf-1. Oligomerization within the context of large multiprotein complexes with Apaf-1 (known as the apoptosome) is a key step in caspase-9 activation *(30,31)*. When activated, caspase-9 cleaves and activates other caspases, including caspase-3 and caspase-7. Caspases-3 and caspase-7 are examples of effector caspases. These are substrates of initiator caspases, and are activated by proteolysis between the p20 and p10 subunits by initiator caspases. Once activated, they function to cleave vital substrates within the cell. Effector caspases tend to have a short prodomains and a substrate sequence preference of D-X-X-D.

3. Caspase-IAP Binding

Biochemical studies have revealed the kinetics of inhibition of caspases-3, -7, and -9 by cIAP1, cIAP2, and XIAP. XIAP can directly inhibit caspase-3, caspase-7, and caspase-9 with a K_i of approx 0.6 nM for caspase-3 and 290 nM for caspase-9 using full length XIAP *(32)*. When the BIR2 alone of XIAP is used, it can directly inhibit caspase-3 or caspase-7 activity with a K_i of 2 to 5 nM *(33)*. cIAP1 is able to inhibit caspase-3 with a K_i of 0.1μM and caspase-7 with a K_i of 42 nM. cIAP2 is able to inhibit caspase-3 with a K_i of 35 nM caspase-7 with a K_i of 29 nM and caspase-9 with K_is of 100 nM *(10,12)*. Mutagenic and structural data have revealed the molecular mechanisms of inhibition of caspase-3, -7 and -9 by XIAP *(14,15,17,20,21,32,34–36)*. XIAP, cIAP1 and cIAP2 cannot inhibit caspase-8, caspase-1 or caspase-6 *(10,12)*. NAIP was unable to inhibit any of caspase-3, -7, -9, -1 or caspase-6 *(12)*. Because the mechanism of caspase inhibition is best characterized for the mammalian protein XIAP, much of the description of IAP-caspase interactions is based on this model.

There are three BIRs in XIAP and inhibition of initiator and effector caspases is achieved by different BIRs *(37)*. BIR3 is the minimal region of XIAP required to inhibit caspase-9, whereas BIR2 and the linker region between BIR2 and BIR1 is the minimal region required for caspase-3 and caspase-7 inhibition *(32,33)*. XIAP binds to and inhibits the processed form of caspase-9 through close interaction between its BIR3 domain and a short sequence of amino acids between the p20 and p10 subunits of caspase-9 known as the IAP binding motif (IBM) *(15,20,38)*. This tetrapeptide motif, ATPF, is exposed on the N-terminus of the small subunit (p12) after caspase-9 is cleaved *(38)*. Mutation or deletion of this linker region or just the first 2 residues significantly reduced the affinity of caspase-9 for the BIR3 domain of XIAP. Thus, this region of caspase-9 is essential for it and XIAP to bind.

The key residues on the BIR3 domain of XIAP responsible for interaction with caspase-9 were identified by mutagenesis and these contribute to a hydrophobic "groove" on the surface of BIR3 *(20)*. Their importance was confirmed by the structure of the XIAP BIR3 domain complexed with caspase-9 *(15)* and binding studies with full length XIAP *(34)*. This binding region on the BIR is sometimes referred to as the Smac binding pocket because it is also the region where the IAP antagonist DIABLO/Smac binds XIAP (*see* Subheading 4) *(16,39)*. This physical interaction between the BIR3 of XIAP and caspase-9 is distant from the active site of caspase-9, hence, BIR3 does not inhibit caspase-9 by occluding the active site. In fact BIR3 binding prevents caspase-9 activity by preventing it dimerizing. Dimerization is a key step in the activation of caspase-9 and dimerization is favored by the higher concentrations of caspase-9 achieved in the apoptosome as caspase-9 is recruited by Apaf-1 *(40)*. Prevention of caspase-9 dimerization by mutation of a single key residue required for caspase dimer formation (F390D) inhibits caspase-9 activity whereas promotion of dimerization using ammonium citrate (a kosmotropic ion that stabilises proteins and favors dimerization) activated recombinant caspase-9 zymogen *(31)*. XIAP binds to caspase-9 on the same surface that is required for caspase-9 dimerization and prevents this critical step from happening. As a result caspase-9 remains in a monomeric state and the active site is forced to adopt an inactive conformation *(15)* *(see* Fig. 3). It is interesting that in the case of caspase-3 where dimerization is not critical for activation, XIAP inhibits the active caspase site directly but in caspase-9 where dimerization is required for activation, this step is inhibited by XIAP. But it is still not clear why this different inhibition mechanism should have arisen.

An caspase-9 like IBM is found in other IAP binding molecules, in particular in IAP inhibitory molecules and in fact IBMs were first identified in *Drosophila* IAP antagonists and are the only conserved feature shared amongst all IAP antagonists *(see* Fig. 4) *(38)*. For the binding of caspase-9 and other molecules to the BIR3, the preferred first residue of an IBM is alanine, although serine is tolerated *(16)*. The hydrophobic groove (Smac pocket) on the BIR3 is also conserved on the BIR1 and BIR2 of XIAP and in other IAP molecules. As we shall see subtle differences do exist between Smac pockets of different BIRs and the type of IBM they accept.

The interaction between XIAP and caspases-3 and caspase-7 is distinct from the caspase-9-XIAP interaction because it involves a region flanking the BIR2 of XIAP in addition to the IBM-BIR interaction *(16)*. Mapping of the regions of XIAP required for caspase-3 inhibition was performed by in vitro caspase inhibition assays using various truncation mutants of XIAP and it was determined that the minimal region required was the BIR2 domain *(33)*, but this domain also contained the linker region preceding the BIR2 domain. Hints that the linker region may be involved in binding came from observations that mutations in residues from the linker region but not the BIR2 domain abolished caspase-3 binding *(21,32)*. Subsequent crystal structures suggested this linker region rather than the BIR2 domain makes intimate contact with the catalytic sites of active caspases-3 and caspase-7 *(14,17,35)*. The critical region of the linker has the amino acid sequence QVVD(148)ISD(151). The DISD sequence resembles a caspase-3 or caspase-7 substrate sequence (DXXD) and it was thought that this part of the linker region bound to the substrate binding pockets of caspase-7 or caspase-3 (S_4-S_1) and thereby blocked caspase activity *(21)*. If the linker functioned as a pseudo-substrate it was because caspases have an absolute requirement for an aspartate residue in the P_1 position of the substrate (that binds into the S_1 pocket of the catalytic site of the caspase *(see* Fig. 2) mutation of D151 in the linker should have abolished binding to the caspase. This turned out not to be the case with mutation of D151 having little effect on the ability of XIAP to inhibit caspase-3 whereas mutation of the D148 significantly diminished the ability of XIAP to inhibit caspase-3 *(21)*. Thus it was clear that the pseudosubstrate model of caspase inhibition by XIAP was not correct *(32)* but the real mechanism of inhibition, discovered by solving the structure of BIR2 bound to caspase-3 *(14)* was radically different to anything that had previously been proposed. Whereas the XIAP linker region does indeed obstruct the catalytic groove of caspase-3 and inhibits activity by blocking

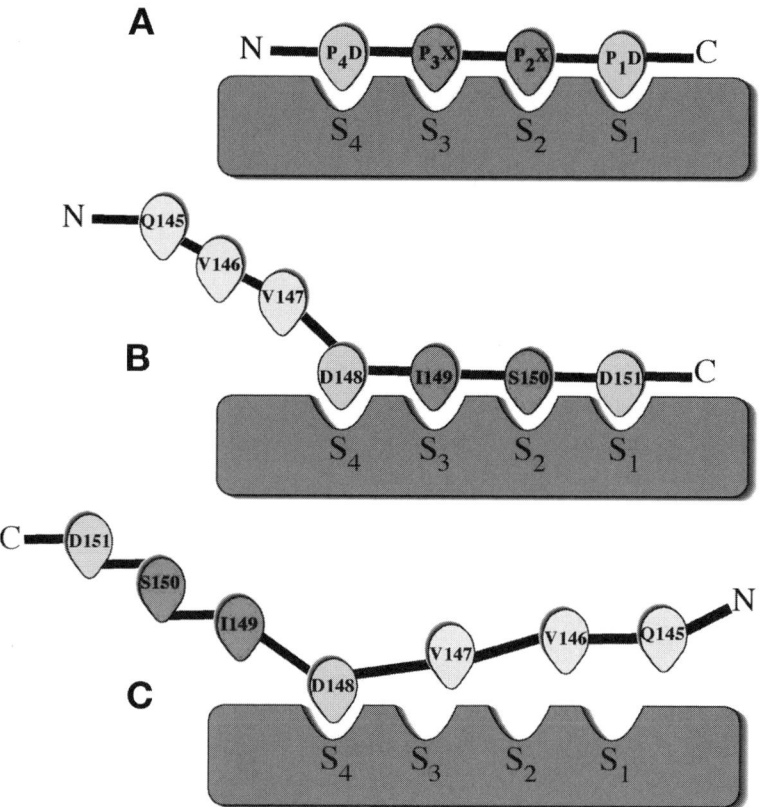

Fig. 2. Orientation of caspase substrates and XIAP linker region into the substrate binding region of caspase-3. **(A)** Orientation of a peptide substrate of caspase-3 or caspase-7. There is a stringent requirement for an aspartate in the P_1 position to bind into the S_1 substrate pocket of the caspase. **(B)** The DISD sequence of the linker region between BIR1 and BIR2 of XIAP conforms to a caspase-3 or caspase-7 substrate sequence and was thought to fit into the substrate-binding region of the caspase in a similar manner to an ideal substrate. However, mutation to Aspartate 148, not Aspartate 151 abolished inhibitory activity. **(C)** Correct binding of the linker region of XIAP into the substrate pocket showing the reverse orientation, Aspartate 148 binds in the S_4 substrate pocket and the rest of the linker region lies across the substrate binding region, obstructing access.

entry of substrates, the linker sequence does not bind into the substrate pockets of the caspase. Instead, the linker residues (L140, L141, and V146) flop over the substrate binding pocket in a reverse (C-N) orientation and are anchored by aspartate 148 that is wedged into the S_4 pocket. It is important to note that D148 is less important for caspase-7 inhibition than caspase-3 inhibition because caspase-7 also makes important IBM interactions with XIAP *(16,41)*.

The interaction between XIAP and caspase-3 and caspase-7 is not limited to the linker region because mutagenesis and structural experiments show that, like caspase-9, caspase-3, and XIAP interact via and IBM-Smac pocket interaction (Fig. 3). The IBM of caspase-3 is formed by the cleaved N-terminus of the small subunit *(16)*. Mutagenesis of both the putative IBM binding groove on the BIR2 and the linker residues on caspase-7

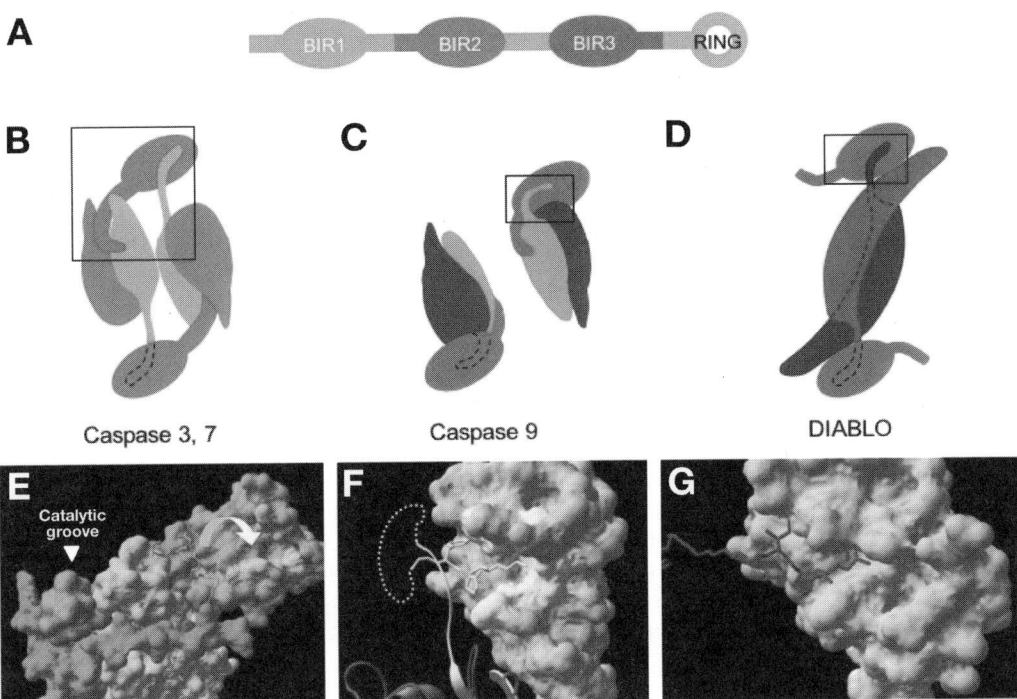

Fig. 3. Modes of binding between XIAP, caspases and DIABLO. (**A**) Schematic diagram of domains of XIAP showing the three BIR domains and the C-terminal RING domain. (**B**) Diagram of interaction between the BIR2 of XIAP and caspase-3 or caspase-7. The caspase large subunit is shown in dark orange whereas small subunit is shown in light orange. The BIR2 domain and the preceding linker region are shown in blue. The interaction involves two dimers of caspase-3 or -7. The linker region of preceding the BIR2 domain blocks the catalytic site of one dimer whereas the IBM of the small subunit from the opposing dimer binds to a groove on the BIR2. (**C**) Diagram of interaction between the BIR3 of XIAP and caspase-9. The caspase large subunit is shown in dark green whereas the small subunit is shown in light green. The BIR3 domain and the C-terminal linker region are shown in purple. The IBM at the N-terminus of the small subunit of caspase-9 binds to a groove on the BIR3 of XIAP while the linker C-terminal to the BIR3 binds to the dimerization surface of caspase-9. (**D**) Diagram of interaction between the BIR3 of XIAP and DIABLO. DIABLO is shown in red and maroon. The BIR3 domain and the C-terminal linker region are shown in purple. The IBM of DIABLO occupies the same groove on the BIR3 of XIAP as the caspase-9 IBM. (**E**) Structural model showing the boxed region in B with corresponding colour scheme. Caspase shown is caspase-3. Model based on PDB 1I3O. The N-terminus of the small subunit of the opposing caspase molecule is shown in stick mode. The proposed corresponding binding groove on XIAP is showing in orange. (**F**) Structural model showing the boxed region in C with corresponding colour scheme. Model based on PDB 1NW9. The N-terminus of the small subunit is shown in stick mode. The interacting residues on the BIR3 of XIAP are shown in green. (**G**) Structural model showing the boxed region in D with corresponding colour scheme. Model based on PDB 1G73. The IBM of DIA-BLO is shown in stick mode and the interacting residues on XIAP are shown in pink.

abolished binding between XIAP and caspase-7. The IBM sequence exposed on the N-terminus of caspase-7 has a serine residue in place of the alanine seen in caspase-9 and the IAP antagonists *(16)*. The BIR2 IBM binding groove appears to differ from that

IAP Antagonists					
Reaper	2		AVAFYIPDQAT	12	
HID	2		AVPFYLPEGGA	12	
Grim	2		AIAYFIPDQAQ	12	
DIABLO/Smac	53		AVPIAQKSEPH	63	
HtrA2/Omi	134		AVPAPPPTSPR	144	
Sickle	2		AIPFFEEEHAP	12	
JAFRAC2	18		AKPEDNESCYS	28	
Caspases					
DRONC	114		SRPPFISLNERR	125	*
Caspase-9(mouse)	350	SEPD	AVPYQEGP	361	
Caspase-9(human)	312	PEPD	ATPFQEG1	323	
Caspase-3(mouse)	181	IETD	SGVDDDMA	192	
Caspase-7(mouse)	195	IQAD	SGPINDID	206	

Fig. 4. Line up of IAP binding motifs (IBMs) in IAP antagonists and in caspases. The numerical values identify the positions of the shown sequence within each protein. The shaded amino acids show identical/similar residues. The caspase IBMs are shown in red and included the caspase cleavage site immediately before the IBM. Dronc differs from the other caspase IBMs in that a single amino acid (F118) binds into the IBM binding region of DIAP1.

in BIR3 in that it can tolerate either an alanine or a serine, whereas mutation of the alanine to a serine abolished binding to BIR3 *(38)*. Thus the BIR2 makes two types of contacts, one of which is similar to the interaction between the BIR3 domain of XIAP and the IBM of caspase-9. The proposed model of XIAP inhibition of the dimeric form of caspases-3 or caspase-7 includes the steric inhibition of the catalytic site of one monomer while anchoring to the IBM of the other *(see* Fig. 2). During apoptosis, the N-terminus of the small subunit of caspase-3 undergoes a second cleavage event (at D180) that removes the IBM and abolishes this binding and XIAP inhibition.

3.1. Other IAPs and Other Caspases

The IBM-Smac pocket interaction is the conserved mechanism by which IAPs bind caspases. The ability of the *Drosophila* IAP DIAP1 to inhibit *Drosophila* effector caspases (drICE and DCP1) is absolutely dependent on the exposure of an IBM in the *Drosophila* caspases *(42)*. Interestingly, DCP1 or drICE, when bound to DIAP1 retained catalytic activity because they were able to cleave a recombinant substrate, DIAP1 itself. Nevertheless, such DIAP1-bound caspases are not able to cleave a cellular substrate such as PARP or small peptide substrates *(16)*. Thus it would appear when DIAP1 binds caspases, it does not directly inhibit catalytic activity, but rather blocks the access of substrate. This is a subtle but clear distinction between mammalian XIAP and DIAP1, because caspases bound to XIAP do not have catalytic activity *(42)*. Yet their remains a consistency of action because both IAPs require the IBM-Smac pocket interaction and neither act as a pseudosubstrate to inhibit enzymatic activity. The inhibition of caspase-7 by cIAP1 is also dependent on an IBM-Smac pocket interaction and it in this instance caspase-7 retains catalytic activity but not the ability to cleave substrate when bound to cIAP1 *(42)*.

It is no surprise then, as has been previously pointed out *(16)*, that caspases that do not have an IBM motif are not inhibited by IAPs. Caspase-2, -6 and -8 do not have an IBM and are not inhibited by XIAP, cIAP1, or cIAP2. For instance, none of the mammalian IAPs block death directly induced by active caspase-2 in a yeast system *(43)*. Caspase-8 induced death can be inhibited by IAPs but this probably results from IAP inhibition of the downstream caspases-3, -7, and -9 because there is no physical interaction between caspase-8 and IAPs *(11,12)*. The physiological relevance of caspase inhibition by IAPs in the regulation of apoptosis signaled by death receptors such as Fas is open to question because death receptor induced apoptosis can proceed normally in *caspase-3$^{-/-}$* or *caspase-9$^{-/-}$* mice *(44–47)*. cIAP1 and cIAP2 both associate with the TNFR2 signaling complex *(9)*, and their role with in this complex may well have more to do with other functions of cIAP1 and cIAP2 that a potential to inhibit caspases (*see* Subheading 5.4).

Livin/ML-IAP/KIAP is a IAP containing only one BIR but also has a C-terminal RING *(48–50)* (*see* Fig. 1). It has been shown to inhibit caspase-3 with similar potency as cIAP1, but not caspase-7. There is also in vitro evidence for binding with caspase-9. The single BIR of Livin/ML-IAP/KIAP has conserved sequence suggesting it has a hydrophobic Smac-binding pocket that would bind an IBM. Mutation of residues in ML-IAP, such as D138A which is equivalent to the E314S mutation in the BIR3 of XIAP that abolished XIAP-caspase-9 binding *(34)*, abolish ML-IAP caspase inhibition and the ability of ML-IAP to block apoptosis in transfected cells *(48)*. Livin/ML-IAP/KIAP can potently bind the IAP inhibitor DIABLO/Smac *(51)*. The high affinity with which Livin/ML-IAP/KIAP binds DIABLO/Smac, higher even than the affinity with which it can bind caspase-9, has prompted the suggestion that the role of Livin/ML-IAP/KIAP in apoptosis regulation might be to bind IAP antagonists and mop them up before they reverse the inhibitory effect of other IAPs on caspases *(51)*. The evaluation of the physiological role of Livin/ML-IAP/KIAP will be further clarified when knockout mice are available.

Neuronal apoptosis inhibitory protein (NAIP) was first described as a candidate gene associated with spinal muscular atrophy *(52)* although it is now clear the principal genetic mutations giving rise to this disease are in the SMN genes *(53,54)*. NAIP contains three BIR domains, a nucleotide oligomerization domain and a leucine rich repeat domain at the C-terminal instead of a RING, and is a member of the CATERPILLER protein family *(25,55)*. There is conflicting evidence both supporting and refuting the ability of NAIP to inhibit caspases *(12,56)* although the effects observed on cell death inhibition were modest. There is strong evidence however to support a role for NAIP in the innate immune response, particularly to *Legionella* infection *(57,58)*.

BRUCE/Apollon is a 530 kD protein with one BIR domain and a ubiquitin-conjugating enzyme motif at the C-terminus *(59,60)*. It remains unclear whether the biological role of BRUCE/Apollon is specifically related to caspase inhibitory or ubiquitin-conjugating activity or both. Some data suggest BRUCE/Apollon is able to partially inhibit caspase-3 and caspase-7 activity in cell lysates and block apoptosis, an effect more dependent on the BIR than the UBC region of the molecule. Curiously, the BIR region of BRUCE/Apollon when expressed as a truncated protein was unable to block caspase activity *(61)*.

Survivin is a small (approx 16 kD) protein with a single BIR domain but no other recognizable domains *(62)*. Structurally its BIR domain may have a similar fold to the BIR2 of XIAP, although significant differences exist between the crystal structures of a truncated protein *(63)* and the full length protein *(64)*. Although some groups initially reported that recombinant Survivin inhibited caspase-3 and caspase-7 in vitro *(65–67)*, this has been challenged by other groups *(64,68)*. Survivin does not have the characteristic linker region before the BIR domain that is required for caspase-3 and caspase-7 inhibition but it cannot be excluded on the basis of sequence alone that Survivin might inhibit upstream caspases with an IBM. As will be discussed later, convincing evidence supports a role for Survivin in chromosomal segregation rather than regulation of apoptosis.

4. IAP Antagonists

The principal conserved mechanism by which IAP antagonists prevent IAPs from inhibiting caspases has already been introduced but IAP antagonists are quite diverse in sequence and merit some discussion in their own right.

In *Drosophila*, the proapoptotic regulators, REAPER, HID, and GRIM (collectively referred to as RHG), were discovered before *Drosophila* IAPs *(69–71)*. Two other IAP antagonists have since been identified, Sickle *(72,73)* and Jafrac2 *(74,75)*. RHG play a crucial role in normal *Drosophila* embryogenesis because deletion of these genes results in a significant reduction in developmentally programmed apoptosis and embryonic lethality *(69–71)*. Conversely, loss of DIAP1 results in massive apoptosis during embryogenesis that also results in embryonic lethality *(5)*. Strikingly, overexpression of any of the IAP antagonists is sufficient to trigger apoptosis, even in mammalian cells *(71,72,76,77)*. The *Drosophila* IAP antagonists interact directly with IAPs *(78–81)*. RHG, Sickle and Jafrac2 all contain an N-terminal IBM and their ability to induce death is to a large extent dependant upon the IBM. As expected from the mammalian studies, the *Drosophila* IBMs bind to the groove on IAP BIRs *(36)*. The BIR1 domain of DIAP1 is responsible for inhibiting the catalytic site of the effector caspase drICE and is also bound by the IBMs of Grim and Reaper but not HID *(36,74)*. The BIR2 domain of DIAP1 binds the initiator caspase Dronc and is also bound by Grim, HID and Reaper *(74,82)*. The interaction between Dronc and DIAP1 is atypical in several respects. A 12 amino acid sequence between the prodomain and the large subunit acts like an IBM, with a single residue, F118, making critical inhibitory contacts within the hydrophobic groove of BIR2. Although the "IBM" of Dronc does not conform to the typical IBM sequence *(see* Fig. 4), the F118 of Dronc occupies the same "pocket" as the F4 of a peptide corresponding to the IBM of HID. The "IBM" of Dronc binds in the reverse orientation to HID, reminiscent of the reverse orientation of the interaction between the linker region of XIAP and caspase-3 *(see* Fig. 3) *(82)*.

In mammals four IAP antagonists have been identified, these are DIABLO/Smac *(83,84)*, HtrA2/Omi *(85,86)*, GSPT1/eRF3 *(85)*, XAF-1 *(87)*, and VIAF *(88)*. Although these mammalian proteins are able to function in a similar manner to the *Drosophila* proteins in that they can bind and antagonize IAPs, they are dissimilar in their primary sequence and some have other important cellular functions (Fig. 4).

DIABLO/Smac was the first mammalian IAP regulator identified *(83,84)*, closely followed by HtrA2. Both DIABLO and HtrA2 are mitochondrial proteins and their N-terminal mitochondria importation signals are removed following importation into mitochondria to expose an N-terminal IBM *(38)*. In healthy cells, however, their IBM is unavailable for interaction with IAPs because they are closeted in the mitochondria. During apoptosis, however, DIABLO and HtrA2 are released into the cytoplasm (along with other proteins including cytochrome *c*) and are able to interact with and antagonize IAPs via an IBM–BIR interaction. The actual temporal sequence of events is still not clear and it is unknown whether in a dying cell DIABLO or HtrA2 prevent IAPs interacting with caspases or can actually liberate activated caspases held captive by IAPs.

It is quite probable both events occur. In a number of experimental paradigms, the mammalian IAP antagonists can displace caspases from IAPs. DIABLO/Smac has been shown to displace caspase-9 from XIAP by binding to the BIR3 domain of XIAP *(39,89)*. This mutually exclusive interaction releases bound caspase-9 *(38)*. HtrA2/Omi can bind to XIAP to alleviate its inhibition of caspase-3 *(86,90,91)*. XAF-1 is able to bind XIAP and prevent it from inhibiting caspase-3 activity *(87)*.

Neither DIABLO, HtrA2 or any of the other mammalian IAP antagonists are able to induce apoptosis just by overexpression alone, distinguishing them from their *Drosophila* counterparts. This suggests that *Drosophila* IAP antagonists must have triggered apoptosis in mammalian cells in part at least by some additional activity to IAP inhibition. It may be that the other regions of the *Drosophila* proteins have some activity that can promote apoptosis.

Whether the sole function of DIABLO is to antagonize IAPs remains unresolved, but what is clear is that DIABLO has a redundant role in the regulation of apoptosis in mammalian cells. Mice lacking DIABLO have no obvious phenotype, and no defect in apoptosis *(92)*. HtrA2, on the other hand, has a very important biological role, but one seemingly counter-intuitive to its presumed role in apoptosis regulation. A naturally occurring mouse mutant, *mnd2*, develops a muscle wasting neurodegenerative disease. This is a result of a missense mutation that destroys the serine protease activity of HtrA2 *(93)*, but importantly the mutant HtrA2 is still quite able to bind the BIR3 of XIAP. *Mnd2* mitochondria are more susceptible to changes in mitochondrial transition permeability in response to stress than wild type cells and *mnd2* MEFs more susceptible to stress-induced apoptosis, just the opposite effect to that which one might expect from cells lacking an IAP inhibitor. A targeted disruption of HtrA2 recapitulates the phenotype of the *mnd2* mouse *(94)* with cells from these mice also being more susceptible rather than less susceptible to apoptosis. When HtrA2 mice were crossed with mice lacking DIABLO/Smac, the double mutants had the same phenotype as *Htra2*$^{-/-}$ mice *(94)*. Thus, whereas the IAP antagonists thus far identified in mammals retain the ability to bind IAPs and reverse IAP-mediated caspase inhibition, their true biological roles appear to be, at best, redundant or even completely opposite of the anticipated phenotype. In this aspect at least, they stand in stark contrast with their insect counterparts.

5. Role of IAPs In Vivo

Although for some IAPs there exists substantial evidence to support the hypothesis that the biochemical activity of these molecules includes direct inhibition of caspases, this is not true of all IAP family members. Further, there are compelling reasons to carefully consider whether the ability to bind and inhibit caspases accounts for the varying biological roles of IAPs, even those IAPs that have an unequivocal caspase-inhibitory activity.

5.1. Viral IAPs

The baculoviral IAP OpIAP has an unquestioned ability to inhibit apoptosis in some insect cell lines in response to viral infection complementing for the loss of the p35 gene that blocks caspase activity, to block apoptosis induced by a range of stimuli in insect cells and to block apoptosis in mammalian cells in response to caspase transfection *(2–4,95–98)*. Yet, there exists some doubt as to whether the antiapoptotic activity of OpIAP in insect cells may be accounted for by direct caspase inhibition because some data suggest OpIAP cannot bind or inhibit recombinant *Drosophila* caspases in vitro (whereas still being able to bind a HID peptide) and truncated proteins consisting of the BIR domains alone may even enhance apoptosis *(97–99)*. In addition, mutations to the RING domain of OpIAP (which abolish E3 ligase activity) diminish the ability of OpIAP to block cell death *(95)*. Thus the ability of OpIAP to block apoptosis may, in part be attributable to other functions than caspase inhibition or to inhibition of other insect caspases than those examined to date.

5.2. Drosophila IAPs

The importance of the RHG proteins in *Drosophila* apoptosis during development has already been described as has the key role DIAP1 plays in regulating this cell death. When mutant flies lacking *grim*, *rpr*, and *hid* were crossed with mutant flies lacking *diap1*, the resulting phenotype was the same as those animals lacking the proapoptotic proteins alone *(81)*. This implies that *diap1* was epistatic to *grim*, *rpr*, and *hid*, meaning it functioned solely to suppress apoptosis induced by those proteins. DIAP1 interacts directly with *Drosophila* caspases. Ectopic expression of Dronc in the developing eye prevents full development, which is enhanced by DIAP1 loss of function mutation *(100,101)*. Similarly, DIAP1 can inhibit another caspase, drICE *(102)*. Together with the recent structural studies *(36)*, these findings support the hypothesis that the role of DIAP1 in vivo is to protect against cell death induced by RHG by inhibiting caspases. However, despite such convincing evidence, there is also good evidence that the ubiquitin ligase activity of the RING of DIAP1 plays an important role in IAP regulation of apoptosis, which may include targeting caspases for proteosomal degradation *(82)* (*see* Subheading 5.4).

5.3. Mammalian IAPs

Structural and biochemical evidence presented show convincingly that XIAP can physically interact with and inhibit caspases-3, -7, and -9. However, the lethal phenotype of DIAP1 deficient flies was not evident in mice lacking XIAP, nor were there any obvious enhancements of apoptosis *(103)*. A "compensatory" elevation of cIAP1 and cIAP2 in *XIAP*[−/−] embryonic fibroblasts was observed following TNF-α stimulation. The clear implication is that XIAP is neither required for normal development nor to regulate the apoptotic response in normal cells. It is also possible there is redundancy in IAP function between XIAP, cIAP1, and cIAP2 sufficient to compensate for the absence of XIAP, particularly if cIAP1 and cIAP2 are expressed at higher levels in the absence of XIAP. Mice lacking cIAP1 have been generated and, like XIAP knockout mice, develop normally and are normally fertile *(104)*. No increased sensitivity to apoptotic stimuli was reported in these mice. The loss of cIAP1 was, however, associated with an increased expression of cIAP2. It will be necessary to generate mice deficient in multiple IAPs to determine if they are physiologically necessary for development and apoptosis regulation. Thus, even the mammalian IAPs that have the most strongly established roles as caspase inhibitors are not individually required for normal development and their caspase inhibitory activity is dispensable in the regulation of apoptosis.

None of the other IAPs provide convincing evidence of important biological roles for IAP-dependent caspase inhibition under normal physiological circumstances. Mice with deletion of NAIP also develop normally. NAIP, as has been already alluded to, probably has a role in innate immunity and the response to *Legionella* infection. Apollon, the murine

BRUCE homolog, has a more striking knockout phenotype. These mice die before or immediately after birth displaying reduced angiogenesis *(105)*. Whether this phenotype is related to a caspase inhibitory activity of Apollon/BRUCE is unresolved. Apollon/BRUCE can inhibit apoptosis induced by overexpression of caspases and it can bind caspase-9 and DIABLO. However, Apollon/BRUCE has a UBC E2 domain at the C-terminus, rather than a RING, and can bind and ubiquitylate DIABLO and caspase-9. Transiently overexpressed DIABLO was able to kill *Apollon⁻/⁻* cells. It must be remembered that overexpressing DIABLO in the cytoplasmic compartment represents an unphysiological situation, therefore, the physiological relevance of these observations remains to be determined.

Under what circumstances might the caspase inhibitory activity of IAPs be biologically relevant? One answer may be that IAPs, and in particular XIAP, function to "mop up" low levels of inadvertent initiator and effector caspase activation and so prevent a cell committing to apoptosis in the absence of a legitimate death stimulus *(106)*. In addition, IAP regulation of active caspases may assume greater importance is in some cancer cells perhaps because IAPs may confer a survival advantage (contributing to oncogenic transformation) or increased resistance to chemotherapeutic drugs. Several lines of evidence support a role for IAPs in cancer cell biology and drug resistance. This includes observations demonstrating upregulated IAP expression in some cancers and cancer cell lines *(107–111)*. If it were so that some tumor cells have a greater reliance on IAP-mediated prosurvival activity than do untransformed cells, then it would follow that antagonizing IAPs would induce apoptosis in such cells. Further, agents that antagonized IAPs could have potential as anticancer agents. The results of experiments with recently developed IAP antagonist drugs provide some support for this hypothesis. These IAP antagonist drugs include peptides that bind into the IBM and prevent IAPs, in particular XIAP, blocking activated caspases *(112)*, small molecule DIABLO mimetics that also displace caspases from IAPs *(113)*, and a class of polyphenylurea molecules that inhibit IAP-dependent caspase inhibition but do not compete for Smac-IAP binding *(114)*. These drugs appear to function to allow effector caspase activation downstream of mitochondrial cytochrome c release and can function independently of antiapoptotic Bcl-2 family members. Remarkably, some of these agents can directly induce apoptosis in cancer cells suggesting that release of the IAP brake on activated caspases is indeed sufficient to induce apoptosis. These agents also potentiate the effect of other chemotherapeutic drugs. Some data suggest that untransformed cells (MEFs) are less susceptible to this kind of drug than transformed cells *(114)* supporting the hypothesis that cancer cells have a greater propensity for a "leakiness" in caspase activation, and XIAP-dependent caspase inhibition is important to the ability of such cells to clonally proliferate.

5.4. More Than Caspase Inhibitors

BIR containing proteins have significantly more diverse biological roles than apoptosis regulation, and mechanisms of action other than caspase inhibition. This is so even those BIRCs that can inhibit cell death. This will be briefly reviewed here and has been more comprehensively reviewed elsewhere *(115)*.

5.4.1. Ubiquitin Ligase Activity

RING domain proteins, such as IAPs, function as E3 ubiquitin ligases by recruiting an ubiquitin conjugating enzyme (UBC) to a substrate. RING containing proteins can promote ubiquitylation of other proteins or catalyse auto-ubiquitylation. Ubiquitylation of a protein with a chain of four or more ubiquitins linked in a Lys48 linkage targets proteins for proteosomal degradation. The C-terminal RING domains of IAPs can function as ubiquitin ligases. In *Drosophila*, heterozygotes for a deletion, or point mutations in the DIAP1 RING abolish DIAP1's ability to rescue the small eye phenotype caused by ectopic expression of the IAP antagonist *rpr* in the *Drosophila* eye *(80,116)*. This is despite the fact that Rpr still interacted with DIAP1. Thus, interaction is insufficient for DIAP1 to neutralize Rpr protein and the DIAP1 can auto-ubiquitylate in response to Grim and Rpr *(117–119)*. Dronc is also a target of DIAP1 ubiquitylation *(82,116)*. Therefore DIAP1 regulates apoptosis in part by ubiquitylating caspases such as Dronc although it is not clear whether this ubiquitylation targets Dronc for degradation. Furthermore, during the execution of apoptosis, upregulated RHG proteins bind to DIAP1 releasing Dronc and promoting auto-ubiquitylation and degradation of DIAP1. Loss of DIAP1 completely frees up Dronc and allows it to activate downstream caspases *(115,120)*.

Mammalian IAPs can ubiquitylate interacting proteins such as caspases, other IAPs, and interacting proteins within the TNF-receptor signaling complex *(104,121–123)*. As in *Drosophila* it remains unclear whether IAPs promote a K48 type of multi-ubiquitylation that targets proteins for proteosomal degradation, or some other type of ubiquitylation such as monoubiquitylation *(115)*. IAPs may crossregulate their own levels through a system of direct IAP-IAP interaction and ubiquitylation *(139)*. Such a system would explain the observation of increased cIAP2 protein expression in cIAP1 knockout mice *(104)*.

5.4.2. TNF Signaling

cIAP1 and cIAP2 were identified by virtue of their indirect association with the TNFR2 *(9,124)* via their direct interaction with TNF receptor associated factors (TRAFs). The ligands for TNF receptors are membrane proteins whose extracellular domains may undergo proteolysis to generate soluble ligand molecules, but they may also function as membrane bound ligands *(125)*. A subfamily of TNF receptors is called death receptors because they contain a cytoplasmic protein interaction domain (the Death Domain) and because these receptors can transduce an apoptotic signal. Adaptor proteins such as TRADD and FADD are recruited to the signaling complex, which then recruits and activates the effector caspase, caspase-8, which may result in apoptosis. However, antiapoptotic signaling through TNF receptor 1 (TNFR1) also occurs and in most situations this survival signal predominates. Binding of TNF-α to TNFR1 recruits the adaptor TRADD to the signaling complex, followed by TRAF1 and cIAP1 *(126)* and this complex results in the liberation of the transcription factor NF-κβ by promoting the degradation of IKβ. Many important details remain unknown, including the role of the RING domain of IAPs (and of course TRAFs, which also bear a RING domain) and the biological consequences of the ubiquitylation they promote.

5.4.3. Cell Division

Survivin contains a single N-terminal BIR domain without C-terminal domains *(62)*. Some evidence suggests overexpression of Survivin offered protection against apoptosis supporting its role as an antiapoptotic protein. Additionally, Survivin was highly expressed in some tumor cell lines, seemingly supporting the hypothesis that the antiapoptotic effect of Survivin overexpression could contribute to oncogenesis *(62,65,127)*. However, several compelling lines of evidence suggest Survivin does not inhibit apoptosis or caspases, and that the high expression of Survivin in tumors is because these cells are cycling rapidly. Significantly, expression of Survivin is restricted to the G2/M phase of the cell cycle *(127)*, providing an rationale for its high expression in rapidly dividing tumor cell lines. In G2/M phase, Survivin localizes to mitotic spindles where it acts in concert with INCENP and Aurora kinase to promote chromosome segregation *(128)*. Most importantly, Survivin null zygotes are not viable because of an absence of mitotic spindle formation causing failure of cell division *(129)*. Other Survivin-like IAPs have been identified in *S. cerevisiae* (ScBIR1P), *S. pombe* (SpBIR1P), *C. elegans* (CeBIR1 and CeBIR2), and *Drosophila* (Deterin) all of which function in mitosis *(130–132)*. The *C. elegans* CeBIR1 is required for cytokinesis and its absence can be complemented by human Survivin *(131)*. Similarly, deletion of yeast Survivin homologs, SpBIR1P, ScBIR1P, results in death from failure of cytokinesis *(130)*. The remarkably similar phenotype resulting from deficiency of Survivin and Survivin-like IAPs across such large phylogenetic differences confirms the primary role of Survivin is to regulate chromosomal segregation during cell division.

5.4.4. TGF Signalling

The association of IAPs with BMP receptors was first demonstrated in *Drosophila* where DIAP1 and DIAP2 were shown to interact with thick veined (Tkv), a TGF-β type 1 receptor for the BMP family signaling protein Decapentaplegic *(133)*. XIAP was subsequently shown to associate with the bone morphogenic protein type 1 receptor (BMP-R1A) and may serve as an adaptor molecule for signaling downstream to TAB1, the kinase TAK1 *(134)* and other members of the TGF-β receptor superfamily *(135)*. BMP belong to the TGF-β superfamily of cytokines that control the development and homeostasis of most tissues in metazoans *(136)*. XIAP, as well as NAIP and ML-IAP, have been shown to interact with TAK-1, a mitogen activated protein kinase (MAPK) *(137,138)* that functions in the BMP signaling pathway. This signaling cascade results in JNK activation and cell survival and is proposed to be an alternative antiapoptotic mechanism to caspase inhibition *(138)*. Although the interactions between IAPs and BMP receptors are conserved the absence of noticeable defects in the XIAP knockout mice argue that the role of XIAP role is redundant, at least in mammals.

6. Conclusion

The BIRC family is large and heterogeneous family defined by the presence of a BIR domain. IAPs, meaning those BIRCs that can inhibit apoptosis are distinct from other BIRC family members. Many apoptosis inhibiting IAPs can inhibit caspases, and therefore apoptosis, directly. Antiapoptotic IAPs that sequester IAP antagonists may prevent apoptosis indirectly by allowing other anticaspase IAPs to prevent apoptosis but may also inhibit apoptosis in an unanticipated fashion. Although we understand in impressive detail the molecular level at which IAPs interact with IAP antagonists and caspases, our understanding of the way these interactions play out in a physiological situation lags considerably. Whereas it is clear that IAPs are indispensable for *Drosophila* development, the fact that knock-out mice bearing deletions of individual IAPs are essentially normal raises questions about the significance of IAPs in mammals. One area where IAPs may be important in mammals is in cancer biology, given the early promise shown by IAP antagonist mimetics as specific anticancer drugs. There is sufficient evidence that the antiapoptotic IAPs have conserved roles at several cell surface receptor interfaces, such as TNF receptors, but what these roles might be is still largely unknown. An exciting area of IAP biology will be understanding how the RING E3 ligase function interfaces with the BIR part of the IAP, defining the substrates for the E3 ligase function and determining what types of ubiquitylation IAPs perform and how this affects substrate function. Obviously these small proteins still have many secrets to reveal.

References

1. Clem RJ, Fechheimer M, Miller LK. Prevention of apoptosis by a baculovirus gene during infection of insect cells. Science 1991;254:1388–1390
2. Crook NE, Clem RJ, Miller LK. An apoptosis inhibiting baculovirus gene with a zinc finger like motif. J Virol 1993;67:2168–2174.
3. Birnbaum MJ, Clem RJ, Miller LK. An apoptosis inhibiting gene from a nuclear polyhedrosis virus encoding a polypeptide with cys/his sequence motif. J Virol 1994;68:2521–2528.
4. Hawkins CJ, Uren AG, Hacker G, Medcalf RL, Vaux DL. Inhibition of interleukin 1-beta-converting enzyme-mediated apoptosis of mammalian cells by baculovirus IAP. Proc Natl Acad Sci USA 1996;93:13,786–13,790.
5. Hay BA, Wassarman DA, Rubin GM. Drosophila homologs of baculovirus inhibitor of apoptosis proteins function to block cell death. Cell 1995;83:1253–1262.
6. Liston P, Roy N, Tamai K, et al. Suppression of apoptosis in mammalian cells by NAIP and a related family of IAP genes. Nature 1996;379:349–353.
7. Duckett CS, Nava VE, Gedrich RW, et al. A conserved family of cellular genes related to the baculovirus IAP gene and encoding apoptosis inhibitors. EMBO J 1996;15:2685–2694.
8. Uren AG, Pakusch M, Hawkins CJ, Puls KL, Vaux DL. Cloning and expression of apoptosis inhibitory protein homologs that function to inhibit apoptosis and/or bind tumor necrosis factor receptor-associated factors. Proc Natl Acad Sci USA 1996;93:4974–4978.
9. Rothe M, Pan MG, Henzel WJ, Ayres TM, Goeddel DV. The TNFR2-TRAF signaling complex contains two novel proteins related to baculoviral-inhibitor of apoptosis proteins. Cell 1995;83:1243–1252.
10. Deveraux QL, Takahashi R, Salvesen GS, Reed JC. X-linked IAP is a direct inhibitor of cell-death proteases. Nature 1997;388:300–304.
11. Deveraux QL, Roy N, Stennicke HR, et al. IAPs block apoptotic events induced by caspase-8 and cytochrome c by direct inhibition of distinct caspases. EMBO J 1998;17:2215–2223.
12. Roy N, Deveraux QL, Takahashi R, Salvesen GS, Reed JC. The c-IAP-1 and c-IAP-2 proteins are direct inhibitors of specific caspases. EMBO J 1997;16:6914–6925.
13. Riedl SJ, Fuentes-Prior P, Renatus M, et al. Structural basis for the activation of human procaspase-7. Proc Natl Acad Sci USA 2001;98:14,790–14,795.
14. Riedl SJ, Renatus M, Schwarzenbacher R, et al. Structural basis for the inhibition of caspase-3 by XIAP. Cell 2001;104:791–800.
15. Shiozaki EN, Chai J, Rigotti DJ, et al. Mechanism of XIAP-mediated inhibition of caspase-9. Mol Cell 2003;11:519–527.
16. Scott FL, Denault JB, Riedl SJ, Shin H, Renatus M, Salvesen GS. XIAP inhibits caspase-3 and -7 using two binding sites: Evolutionarily conserved mechanism of IAPs. Embo J 2005;24:645–655.
17. Huang Y, Park YC, Rich RL, Segal D, Myszka DG, Wu H. Structural basis of caspase inhibition by XIAP: Differential roles of the linker versus the BIR domain. Cell 2001;104:781–790.

18. Uren AG, Coulson EJ, Vaux DL. Conservation of baculovirus inhibitor of apoptosis repeat proteins (BIRPs) in viruses, nematodes, vertebrates and yeasts. Trends Biochem Sci 1998;23:159–162.

19. Hinds MG, Norton RS, Vaux DL, Day CL. Solution structure of a baculoviral inhibitor of apoptosis (IAP) repeat. Nat Struct Biol 1999;6:648–651.

20. Sun C, Cai M, Meadows RP, et al. NMR structure and mutagenesis of the third BIR domain of the inhibitor of apoptosis protein XIAP. J Biol Chem 2000;275:33,777–33,781.

21. Sun C, Cai M, Gunasekera AH, et al. NMR structure and mutagenesis of the inhibitor-of-apoptosis protein XIAP. Nature 1999;401:818–822.

22. Barlow PN, Luisi B, Milner A, Elliott M, Everett R. Structure of the C3HC4 domain by 1h-nuclear magnetic resonance spectroscopy. A new structural class of zinc-finger. J Mol Biol 1994;237:201–211.

23. Borden KL, Boddy MN, Lally J, et al. The solution structure of the ring finger domain from the acute promyelocytic leukaemia proto-oncoprotein PML. EMBO J 1995;14:1532–1541.

24. Yang Y, Fang SY, Jensen JP, Weissman AM, Ashwell JD. Ubiquitin protein ligase activity of IAPs and their degradation in proteasomes in response to apoptotic stimuli. Science 2000;288:874–877.

25. Tschopp J, Martinon F, Burns K. Nalps: A novel protein family involved in inflammation. Nat Rev Mol Cell Biol 2003;4:95–104.

26. Hauser HP, Bardroff M, Pyrowolakis G, Jentsch S. A giant ubiquitin-conjugating enzyme related to IAP apoptosis inhibitors. J Cell Biol 1998;141:1415–1422.

27. Hawkins CJ, Yoo SJ, Peterson EP, Wang SL, Vernooy SY, Hay BA. The drosophila caspase Dronc cleaves following glutamate or aspartate and is regulated by DIAP1, HID, and GRIM. J Biol Chem 2000;275:27,084–27,093.

28. Nicholson DW. Caspase structure, proteolytic substrates, and function during apoptotic cell death. Cell Death Differ 1999;6:1028–1042.

29. Thornberry NA, Ranon TA, Pieterson EP, et al. A combinatorial approach defines specificities of members of the caspase family and granzyme B - functional, relationships established for key mediators of apoptosis. J Biol Chem 1997;272:17,907–17,911.

30. Stennicke HR, Deveraux QL, Humke EW, Reed JC, Dixit VM, Salvesen GS. Caspase-9 can be activated without proteolytic processing. J Biol Chem 1999;274:8359–8362.

31. Boatright KM, Renatus M, Scott FL, et al. A unified model for apical caspase activation. Mol Cell 2003;11:529–541.

32. Silke J, Ekert PG, Day CL, et al. Direct inhibition of caspase 3 is dispensable for the anti-apoptotic activity of XIAP. EMBO J 2001;20:3114–3123.

33. Takahashi R, Deveraux Q, Tamm I, et al. A single BIR domain of XIAP sufficient for inhibiting caspases. J Biol Chem 1998;273:7787–7790.

34. Silke J, Hawkins CJ, Ekert PG, et al. The anti-apoptotic activity of XIAP is retained upon mutation of both the caspase 3- and caspase 9-interacting sites. J Cell Biol 2002;157:115-124.

35. Chai J, Shiozaki E, Srinivasula SM, et al. Structural basis of caspase-7 inhibition by XIAP. Cell 2001;104:769–780.

36. Yan N, Wu JW, Chai J, Li W, Shi Y. Molecular mechanisms of Drice inhibition by DIAP1 and removal of inhibition by REAPER, HID and GRIM. Nat Struct Mol Biol 2004;11:420–428.

37. Deveraux QL, Leo E, Stennicke HR, Welsh K, Salvesen GS, Reed JC. Cleavage of human inhibitor of apoptosis protein XIAP results in fragments with distinct specificities for caspases. EMBO J 1999;18:5242–5251.

38. Srinivasula SM, Hegde R, Saleh A, et al. A conserved XIAP-interaction motif in caspase-9 and Smac/DIABLO regulates caspase activity and apoptosis. Nature 2001;410:112–116.

39. Wu G, Chai J, Suber TL, et al. Structural basis of IAP recognition by Smac/DIABLO. Nature 2000;408:1008–1012.

40. Renatus M, Stennicke HR, Scott FL, Liddington RC, Salvesen GS. Dimer formation drives the activation of the cell death protease caspase-9. Proc Natl Acad Sci USA 2001;98: 14,250–14,255.

41. Suzuki Y, Nakabayashi Y, Nakata K, Reed JC, Takahashi R. X-linked inhibitor of apoptosis protein (XIAP) inhibits caspase-3 and -7 in distinct modes. J Biol Chem 2001;276: 27,058–27,063.

42. Tenev T, Zachariou A, Wilson R, Ditzel M, Meier P. IAPs are functionally non-equivalent and regulate effector caspases through distinct mechanisms. Nat Cell Biol 2005;7:70–77.

43. Ho PK, Jabbour AM, Ekert PG, Hawkins CJ. Caspase-2 is resistant to inhibition by inhibitor of apoptosis proteins (IAPs) and can activate caspase-7. FEBS J 2005;272:1401–1414.

44. Woo M, Hakem R, Soengas MS, et al. Essential contribution of caspase 3 CPP32 to apoptosis and its associated nuclear changes. Genes Dev 1998;12:806–819.

45. Hakem R, Hakem A, Duncan GS, et al. Differential requirement for caspase 9 in apoptotic pathways in vivo. Cell 1998;94:339–352.

46. Kuida K, Haydar TF, Kuan CY, et al. Reduced apoptosis and cytochrome c-mediated caspase activation in mice lacking caspase-9. Cell 1998;94:325–337.

47. Yoshida H, Kong YY, Yoshida R, et al. Apaf1 is required for mitochondrial pathways of apoptosis and brain development. Cell 1998;94:739–750.

48. Vucic D, Stennicke HR, Pisabarro MT, Salvesen GS, Dixit VM. ML-IAP, a novel inhibitor of apoptosis that is preferentially expressed in human melanomas. Curr Biol 2000;10:1359–1366.

49. Kasof GM, Gomes BC. Livin, a novel inhibitor of apoptosis protein family member. J Biol Chem 2001;276:3238–3246.

50. Lin JH, Deng G, Huang Q, Morser J. Kiap, a novel member of the inhibitor of apoptosis protein family. Biochem Biophys Res Commun 2000;279:820–831.

51. Vucic D, Franklin MC, Wallweber HJ, et al. Engineering ML-IAP to produce an extraordinarily potent caspase-9 inhibitor: Implications for Smac-dependent anti-apoptotic activity of ML-IAP. Biochem J 2005;385:11–20.

52. Roy N, Mahadevan MS, Mclean M, et al. The gene for neuronal apoptosis inhibitory protein is partially deleted in individuals with spinal muscular atrophy. Cell 1995;80:167–178.

53. Chang JG, Jong YJ, Lin SP, et al. Molecular analysis of survival motor neuron (SMA) and neuronal apoptosis inhibitory protein (NAIP) genes of spinal muscular atrophy patients and their parents. Hum Genet 1997;100:577–581.

54. Campbell L, Potter A, Ignatius J, Dubowitz V, Davies K. Genomic variation and gene conversion in spinal muscular atrophy - implications for disease process and clinical phenotype. Am J Hum Genet 1997;61:40–50.

55. Harton JA, Linhoff MW, Zhang J, Ting JP. Cutting edge: Caterpiller: A large family of mammalian genes containing card, pyrin, nucleotide-binding, and leucine-rich repeat domains. J Immunol 2002;169:4088–4093.

56. Maier JK, Lahoua Z, Gendron NH, et al. The neuronal apoptosis inhibitory protein is a direct inhibitor of caspases-3 and -7. J Neurosci 2002;22:2035–2043.

57. Wright EK, Goodart SA, Growney JD, et al. NAIP5 affects host susceptibility to the intracellular pathogen legionella pneumophila. Curr Biol 2003;13:27–36.

58. Diez E, Lee SH, Gauthier S, et al. BIRc1e is the gene within the lgn1 locus associated with resistance to legionella pneumophila. Nat Genet 2003;33:55–60.

59. Hauser HP, Bardroff M, Pyrowolakis G, Jentsch S. A giant ubiquitin-conjugating enzyme related to IAP apoptosis inhibitors. J Cell Biol 1998;141:1415–1422.

60. Chen Z, Naito M, Hori S, Mashima T, Yamori T, Tsuruo T. A human IAP-family gene, apollon, expressed in human brain cancer cells. Biochem Biophys Res Commun 1999;264:847–854.

61. Bartke T, Pohl C, Pyrowolakis G, Jentsch S. Dual role of Bruce as an antiapoptotic IAP and a chimeric E2/E3 ubiquitin ligase. Mol Cell 2004;14:801–811.

62. Ambrosini G, Adida C, Altieri DC. A novel anti-apoptosis gene, Survivin, expressed in cancer and lymphoma. Nat Med 1997;3:917–921.

63. Muchmore SW, Chen J, Jakob C, et al. Crystal structure and mutagenic analysis of the inhibitor-of-apoptosis protein Survivin. Mol Cell 2000;6:173–182.

64. Verdecia MA, Huang H, Dutil E, Kaiser DA, Hunter T, Noel JP. Structure of the human anti-apoptotic protein Survivin reveals a dimeric arrangement. Nat Struct Biol 2000;7:602–608.

65. Tamm I, Wang Y, Sausville E, et al. Iap-family protein survivin inhibits caspase activity and apoptosis induced by Fas (CD95), Bax, caspases, and anticancer drugs. Cancer Res 1998;58:5315–5320.

66. Shin S, Sung BJ, Cho YS, et al. An anti-apoptotic protein human Survivin is a direct inhibitor of caspase-3 and -7. Biochemistry 2001;40:1117–1123.

67. Conway EM, Pollefeyt S, Cornelissen J, et al. Three differentially expressed Survivin cDNA variants encode proteins with distinct antiapoptotic functions. Blood 2000;95:1435–1442.

68. Banks DP, Plescia J, Altieri DC, et al. Survivin does not inhibit caspase-3 activity. Blood 2000;96:4002–4003.

69. White K, Grether ME, Abrams JM, Young L, Farrell K, Steller H. Genetic control of programmed cell death in *drosophila*. Science 1994;264:677–683.

70. Grether ME, Abrams JM, Agapite J, White K, Steller H. The head involution defective gene of drosophila melanogaster functions in programmed cell death. Genes Dev 1995;9:1694–1708.

71. Chen P, Nordstrom W, Gish B, Abrams JM. Grim, a novel cell death gene in drosophila. Genes Dev 1996;10:1773–1782.

72. Srinivasula SM, Datta P, Kobayashi M, et al. Sickle, a novel drosophila death gene in the reaper/hid/grim region, encodes an IAP-inhibitory protein. Curr Biol 2002;12:125–130.

73. Claveria C, Caminero E, Martinez-A C, Campuzano S, Torres M. GH3, a novel proapoptotic domain in drosophila GRIM, promotes a mitochondrial death pathway. EMBO J 2002;21:3327–3336.

74. Zachariou A, Tenev T, Goyal L, Agapite J, Steller H, Meier P. IAP-antagonists exhibit non-redundant modes of action through differential DIAP1 binding. EMBO J 2003;22:6642–6652.

75. Tenev T, Zachariou A, Wilson R, Paul A, Meier P. Jafrac2 is an IAP antagonist that promotes cell death by liberating Dronc from DIAP1. EMBO J 2002;21:5118–5129.

76. Haining WN, Carboy-Newcomb C, Wei CL, Steller H. The proapoptotic function of drosophila HID is conserved in mammalian cells. Proc Natl Acad Sci USA 1999;96:4936–4941.

77. Claveria C, Albar JP, Serrano A, et al. Drosophila GRIM induces apoptosis in mammalian cells. EMBO J 1998;17:7199–7208.

78. Vucic D, Kaiser WJ, Miller LK. Inhibitor of apoptosis proteins physically interact with and block apoptosis induced by drosophila proteins HID and GRIM. Mol Cell Biol 1998;18:3300–3309.

79. Mccarthy JV, Dixit VM. Apoptosis induced by drosophila Reaper and Grim in a human system - attenuation by inhibitor of apoptosis proteins (cIAPs). J Biol Chem 1998;273:24,009–24,015.

80. Goyal L, McCall K, Agapite J, Hartwieg E, Steller H. Induction of apoptosis by drosophila Reaper, HID and Grim through inhibition of IAP function. EMBO J 2000;19:589–597.

81. Wang SL, Hawkins CJ, Yoo SJ, Müller H-AJ, Hay BA. The drosophila caspase inhibitor DIAP1 is essential for cell survival and is negatively regulated by HID. Cell 1999;98:453–463.

82. Chai J, Yan N, Huh JR, et al. Molecular mechanism of Reaper-Grim-HID-mediated suppression of DIAP1-dependent Dronc ubiquitination. Nat Struct Biol 2003;10:892–898.

83. Du C, Fang M, Li Y, Li L, Wang X. Smac, a mitochondrial protein that promotes cytochrome c–dependent caspase activation by eliminating IAP inhibition. Cell 2000;102:33–42.

84. Verhagen A, Ekert PG, Silke J, et al. Identification of DIABLO, a mammalian protein that promotes apoptosis by binding to and antagonizing IAP proteins. Cell 2000;102:43–53.

85. Hegde R, Srinivasula SM, Datta P, et al. The polypeptide chain–releasing factor GSPT1/eRF3 is proteolytically processed into an IAP-binding protein. J Biol Chem 2003; 278:38,699–38,706.

86. Verhagen AM, Silke J, Ekert PG, et al. HTRA2 promotes cell death through its serine protease activity and its ability to antagonise inhibitor of apoptosis proteins. J Biol Chem 2001;277:445–454.

87. Liston P, Fong WG, Kelly NL, et al. Identification of XAF1 as an antagonist of XIAP anti-caspase activity. Nat Cell Biol 2001;3:128–133.

88. Wilkinson JC, Richter BW, Wilkinson AS, et al. VIAF, a conserved inhibitor of apoptosis (IAP)-interacting factor that modulates caspase activation. J Biol Chem 2004;279:51,091–51,099.

89. Ekert PG, Silke J, Hawkins CJ, Verhagen AM, Vaux DL. DIABLO promotes apoptosis by removing MIHA/XIAP from processed caspase 9. J Cell Biol 2001;152:483–490.

90. van Loo G, van Gurp M, Depuydt B, et al. The serine protease OMI/HTRA2 is released from mitochondria during apoptosis. OMI interacts with caspase-inhibitor XIAP and induces enhanced caspase activity. Cell Death Differ 2002;9:20–26.

91. Martins LM, Iaccarino I, Tenev T, et al. The serine protease OMI/HTRA2 regulates apoptosis by binding XIAP through a reaper-like motif. J Biol Chem 2001;277:439–444.

92. Okada H, Suh WK, Jin J, et al. Generation and characterization of Smac/DIABLO-deficient mice. Mol Cell Biol 2002;22:3509–3517.

93. Jones JM, Datta P, Srinivasula SM, et al. Loss of OMI mitochondrial protease activity causes the neuromuscular disorder of mnd2 mutant mice. Nature 2003;425:721–727.

94. Martins LM, Morrison A, Klupsch K, et al. Neuroprotective role of the reaper-related serine protease HTRA2/OMI revealed by targeted deletion in mice. Mol Cell Biol 2004;24:9848–9862.

95. Clem RJ, Miller LK. Control of programmed cell death by the baculovirus genes p35 and IAP. Molecular & Cellular Biology 1994;14:5212–5222.

96. Hawkins CJ, Ekert PG, Uren AG, Holmgreen SP, Vaux DL. Anti-apoptotic potential of insect cellular and viral IAPs in mammalian cells. Cell Death & Differentiation 1998;5:569–576.

97. Seshagiri S, Miller LK. Baculovirus inhibitors of apoptosis (IAPs) block activation of sf-caspase-1. Proc Natl Acad Sci USA 1997;94:13,606–13,611.

98. Harvey AJ, Soliman H, Kaiser WJ, Miller LK. Anti-and pro-apoptotic activities of baculovirus and drosophila IAPs in an insect cell line. Cell Death Differ 1997;4:733–744.

99. Wright CW, Means JC, Penabaz T, Clem RJ. The baculovirus anti-apoptotic protein Op-IAP does not inhibit drosophila caspases or apoptosis in drosophila S2 cells and instead sensitizes S2 cells to virus-induced apoptosis. Virology 2005;335:61–71.

100. Quinn LM, Dorstyn L, Mills K, et al. An essential role for the caspase Dronc in developmentally programmed cell death in drosophila. J Biol Chem 2000;275:40,416–40,424.

101. Meier P, Silke J, Leevers SJ, Evan GI. The drosophila caspase Dronc is regulated by DIAP1. EMBO J 2000;19:598–611.

102. Kaiser WJ, Vucic D, Miller LK. The drosophila inhibitor of apoptosis d-IAP1 suppresses cell death induced by the caspase Drice. FEBS Letters 1998;440:243–248.

103. Harlin H, Reffey SB, Duckett CS, Lindsten T, Thompson CB. Characterization of XIAP-deficient mice. Mol Cell Biol 2001;21:3604–3608.

104. Conze DB, Albert L, Ferrick DA, et al. Posttranscriptional downregulation of c-IAP2 by the ubiquitin protein ligase c-IAP1 in vivo. Mol Cell Biol 2005;25:3348–3356.

105. Hao Y, Sekine K, Kawabata A, et al. Apollon ubiquitinates Smac and caspase-9, and has an essential cytoprotection function. Nat Cell Biol 2004;6:849–860.

106. Vaux DL, Silke J. HTRA2/OMI, a sheep in wolf's clothing. Cell 2003;115:251–253.

107. Imoto I, Yang ZQ, Pimkhaokham A, et al. Identification of cIAP1 as a candidate target gene within an amplicon at 11q22 in esophageal squamous cell carcinomas. Cancer Res 2001;61:6629–6634.

108. Nakagawa Y, Hasegawa M, Kurata M, et al. Expression of IAP-family proteins in adult acute mixed lineage leukemia (AMLL). Am J Hematol 2005;78:173–180.

109. Yamamoto K, Abe S, Nakagawa Y, et al. Expression of IAP family proteins in myelodysplastic syndromes transforming to overt leukemia. Leuk Res 2004;28:1203–1211.

110. Nemoto T, Kitagawa M, Hasegawa M, et al. Expression of IAP family proteins in esophageal cancer. Exp Mol Pathol 2004;76:253–259.

111. Krajewska M, Krajewski S, Banares S, et al. Elevated expression of inhibitor of apoptosis proteins in prostate cancer. Clin Cancer Res 2003;9:4914–4925.

112. Oost TK, Sun C, Armstrong RC, et al. Discovery of potent antagonists of the antiapoptotic protein xiap for the treatment of cancer. J Med Chem 2004;47:4417–4426.

113. Li L, Thomas RM, Suzuki H, De Brabander JK, Wang X, Harran PG. A small molecule Smac mimic potentiates TRAIL- and TNFalpha-mediated cell death. Science 2004;305:1471–1474.

114. Schimmer AD, Welsh K, Pinilla C, et al. Small-molecule antagonists of apoptosis suppressor XIAP exhibit broad antitumor activity. Cancer Cell 2004;5:25–35.

115. Vaux DL, Silke J. IAPs, RINGs and ubiquitylation. Nat Rev Mol Cell Biol 2005;6:287–297.

116. Wilson R, Goyal L, Ditzel M, et al. The DIAP1 ring finger mediates ubiquitination of Dronc and is indispensable for regulating apoptosis. Nat Cell Biol 2002;4:445–450.

117. Holley CL, Olson MR, Colon-Ramos DA, Kornbluth S. Reaper eliminates IAP proteins through stimulated IAP degradation and generalized translational inhibition. Nat Cell Biol 2002;4:439–444.

118. Hays R, Wickline L, Cagan R. Morgue mediates apoptosis in the drosophila melanogaster retina by promoting degradation of DIAP1. Nat Cell Biol 2002;4:425–431.

119. Ryoo HD, Bergmann A, Gonen H, Ciechanover A, Steller H. Regulation of drosophila IAP1 degradation and apoptosis by reaper and UBCD1. Nat Cell Biol 2002;4:432–438.

120. Palaga T, Osborne B. The 3d's of apoptosis: Death, degradation and DIAPs. Nat Cell Biol 2002;4:E149–E151.

121. Huang Hk, Joazeiro CAP, Bonfoco E, Kamada S, Leverson JD, Hunter T. The inhibitor of apoptosis, cIAP2, functions as a ubiquitin-protein ligase and promotes in vitro monoubiquitination of caspases-3 and -7. J Biol Chem 2000;275:26,661–26,664.

122. Park SM, Yoon JB, Lee TH. Receptor interacting protein is ubiquitinated by cellular inhibitor of apoptosis proteins (c-IAP1 and c-IAP2) in vitro. FEBS Lett 2004;566:151–156.

123. Morizane Y, Honda R, Fukami K, Yasuda H. X-linked inhibitor of apoptosis functions as ubiquitin ligase toward mature caspase-9 and cytosolic Smac/DIABLO. J Biochem 2005;137:125–132.

124. Shu HB, Takeuchi M, Goeddel DV. The tumor necrosis factor receptor 2 signal transducers TRAF2 and c-IAP1 are components of the tumor necrosis factor receptor 1 signaling complex. Proc Natl Acad Sci USA 1996;93:13,973–13,978.

125. Bodmer JL, Schneider P, Tschopp J. The molecular architecture of the TNF superfamily. Trends Biochem Sci 2002;27:19–26.

126. Micheau O, Tschopp J. Induction of tnf receptor i-mediated apoptosis via two sequential signaling complexes. Cell 2003;114:181–190.

127. Li FZ, Ambrosini G, Chu EY, et al. Control of apoptosis and mitotic spindle checkpoint by Survivin. Nature 1998;396:580–584.

128. Skoufias DA, Mollinari C, Lacroix FB, Margolis RL. Human Survivin is a kinetochore-associated passenger protein. J Cell Biol 2000;151:1575–1582.

129. Uren AG, Wong L, Pakusch M, et al. Survivin and the inner centromere protein INCENP show similar cell-cycle localization and gene knockout phenotype. Curr Biol 2000;10:1319–1328.

130. Uren AG, Beilharz T, O'Connell MJ, et al. Role for yeast inhibitor of apoptosis (IAP)-like proteins in cell division. Proc Natl Acad Sci USA 1999;96:10,170–10,175.

131. Fraser AG, James C, Evan GI, Hengartner MO. Caenorhabditis elegans inhibitor of apoptosis protein (IAP) homologue BIR-1 plays a conserved role in cytokinesis. Curr Biol 1999;9:292–301.

132. Jones G, Jones D, Zhou L, Steller H, Chu Y. Deterin, a new inhibitor of apoptosis from drosophila melanogaster. J Biol Chem 2000;275:22,157–22,165.

133. Oeda E, Oka Y, Miyazono K, Kawabata M. Interaction of drosophila inhibitors of apoptosis with thick veins, a type I serine/threonine kinase receptor for decapentaplegic. J Biol Chem 1998;273:9353–9356.

134. Yamaguchi K, Nagai S, Ninomiya-Tsuji J, et al. XIAP, a cellular member of the inhibitor of apoptosis protein family, links the receptors to TAB1-TAK1 in the BMP signaling pathway. EMBO J 1999;18:179–187.

135. Birkey Reffey S, Wurthner JU, Parks WT, Roberts AB, Duckett CS. X-linked inhibitor of apoptosis protein functions as a cofactor in transforming growth factor-beta signaling. J Biol Chem 2001;276:26,542–26,549.

136. Massague J, Chen YG. Controlling TGF-beta signaling. Genes Dev 2000;14:627–644.

137. Sanna MG, Duckett CS, Richter B, Thompson CB, Ulevitch RJ. Selective activation of JNK1 is necessary for the anti-apoptotic activity of HILP. Proc Natl Acad Sci USA 1998;95:6015–6020.

138. Sanna MG, da Silva Correia J, Luo Y, et al. ILPIP, a novel anti-apoptotic protein that enhances XIAP-mediated activation of JNK1 and protection against apoptosis. J Biol Chem 2002;277:30,454–30,462.

139. Silke J, Kratina T, Chu D, et al. Determination of cell survival by RING-mediated regulation of IAP abundance. Proc Natl Acad Sci USA 2005;102:16,182–16,187.

16

Intracellular Pathways Involved in DNA Damage and Repair to Neuronal Apoptosis

Maurizio Memo

Summary

One of the most well known connections between abnormalities of the DNA damage response and neurodegeneration has been the human syndrome of ataxia telangiectasia. However, other syndromes associated with defective DNA damage response also include neurological symptoms as a primary feature of their phenotypes. This argues that defects in the repair of, or response to, DNA damage impact significantly on brain function. This chapter summarizes some recent data underlying the contribution to neuronal function and dysfunction of at least two transcription factors known to be involved in DNA damage sensing and repairing: the tumor suppressor p53 and the component of the DNA mismatch repair system MSH2. Both proteins participate in the cancer prevention machinery for the body as well as in the neurodegenerative process.

Key Words: Neurodegeneration; mismatch repair; ATM; p53; NF-κB.

1. Introduction

Preservation of genomic stability is an essential biological function. Cells very efficiently engage mechanisms involving DNA surveillance/repair proteins that work to maintaining inherited nucleotide sequence of genomic DNA over time. After DNA damage, that can arise either during duplication or after genotoxic stimuli, cells activate intracellular pathways which are able to recognize the damage, arrest cell cycle, recruit DNA repair factors, repair the damage, or induce apoptosis. This definitely relevant process is finalized to prevent the generation and the persistence of impaired cells which may ultimately be detrimental to the organism. Studies on the role of DNA damage sensors and repair factors in terminally differentiated, not proliferating cells, like neurons have recently attracted great interest.

It is well recognized that mutation of genes related to DNA damage repair are associated with specific cancer-prone syndromes. Interestingly, many human pathological conditions with genetic defects in DNA damage responses are also characterized by neurological deficits. These neurological deficits can manifest themselves during many stages of development, suggesting an important role for DNA repair during the development and maintenance of the brain.

This chapter summarizes some recent data underlying the contribution to neuronal function and dysfunction of at least two transcription factors known to be involved in DNA damage sensing and repairing: the tumor suppressor p53 and the component of the DNA mismatch repair system MSH2. Both proteins participate in the cancer prevention machinery for the body as well as in the neurodegenerative process.

From: *Apoptosis, Cell Signaling, and Human Diseases: Molecular Mechanisms, Volume 2*
Edited by R. Srivastava © Humana Press Inc., Totowa, NJ

DNA DAMAGE SOURCES

Endogenous

▸ Cellular metabolism

▸ Free radicals production during metabolism

▸ Replication or DNA repair errors

▸ High levels of reducing agents

Exogenous

▸ Mutagenic compounds

▸ UV ionizing radiations

▸ Free radicals

▸ Heavy metals

▸ Ultrasounds

DNA REPAIR SYSTEMS

Excision repair

▸ NER nucleotide excision repair

▸ BER base excision repair

▸ **MMR mismatch repair**

Recombinational repair

Damage reversion

Fig. 1. List of endogenous and exogenous sources of DNA damage *(left)* e DNA repair systems *(right)*.

2. The Tumor Suppressor p53

p53, is a transcription factor belonging to a large family, including p73 and p63, whose main function is to control cell-cycle progression and apoptotic process. p53 is also senses DNA damage and participates in DNA repair machinery *(1,2)*. Loss or inactivation of p53 gene occurs in almost half of all human solid tumors, and is considered a fundamental predisposing event in the pathogenesis of many types of cancer *(3)*. Patients with Li-Fraumeni syndrome, characterized by germ-line mutations in p53 gene, present with a high risk of developing a variety of tumors *(4)*, and mice deficient in p53 display precocious tumor development *(5)*.

p53 protein is upregulated in response to various cellular stresses. Different cellular injury—including DNA damage, hypoxia, oxidative stress, ribonucleotide depletion, and chemioterapeutic agents—were shown to stabilized p53 protein, which in turn can either cause growth arrest, permitting the induction of DNA repair process, or alternatively, can direct cells to undergo apoptosis *(6–7)*. The involvement of p53 in sensing damaged DNA and in controlling its repair, is supported by the observation that p53 directly transactivates the proliferating cell nuclear antigen (PCNA), that is involved in DNA replication and repair, and the GADD45 gene, whose product interacts with PCNA. Furthermore, p53 was shown to bind several transcription factor IIH-associated proteins including the DNA helicase, XPB (ERCC3), which are involved in DNA damage repair machinery.

Up to now, it has been well recognized that, in proliferating cells, versatility in p53 activity as a response to exogenous or endogenous signal serves as a control mechanism to preserve the genome integrity. Remain an open question whether p53 may exert the same role in neuron cells by regulating DNA repair process or inducing apoptosis.

2.1. Contribution of p53 to Neuronal Degeneration

In the last decade an emerging body of evidence underscores the particular relevance of the tumor suppressor gene p53 in the central nervous system, because diverse form of neuronal damage have been associated with its induction. For example, damage resulting from neuronal stimulation by excitatory amino acid has been strongly associated with p53 accumulation. In particular, is has been found that treatment of cerebellar granule cells with a p53-specific antisense oligonucleotide prevented both glutamate-induced p53 expression and apoptosis *(8)*. In vivo studies, using kainic acid systemic administration or quinolinic acid intracerebral injection, further support the involvement of p53 in neurodegeneration by showing that an increase of p53 mRNA in vulnerable brain regions, which displayed signs of apoptosis, including DNA fragmentation *(9–10)*. Similarly, elevated p53 mRNA has been also detected in adrenalectomy-induced degeneration of hippocampal dentate granule cells *(11)*. Furthermore, with respect to ischemic cell death, enhanced p53 immunoreactivity has been observed in damaged cortical and striatal neurons 12 h following 2 h focal cerebral ischemia *(12)*, whereas p53 null mice have been reputed to be more resistant to focal cerebral ischemia *(13)*.

The role of p53 in neuronal death was further confirmed by using animals models of human neurodegenerative disease. Transgenic mice for APP gene, which overexpressed β-amyloid 1-42 peptide ($A\beta_{1-42}$) and represent an experimental model for studying Alzheimer's disease (AD) neuropathology, displayed an intensive p53 immunoreactivity in neurons undergoing apoptosis *(14)*. A correlation between p53 expression and Aβ injury was also found in vitro, using primary cultures of cortical neurons *(15)*. In particular, $A\beta_{25-35}$ neurotoxic peptide caused in these neurons an increase of p53 expression, which was observed from 8 h up to, at least, 20 h following the insult.

An additional set of experiments was done in SH-SY5Y neuroblastoma cells differentiated by retinoic acid treatment to adopt a neuronal-like phenotype. In these cells, different stimuli, such as Aβ and H_2O_2, lead to an increase of p53 expression and apoposis *(16)*. It is interesting to note that p53 induction was very fast following H_2O_2 pulse (5 min), whereas longer time was required for detecting an increased p53 expression after glutamate or Aβ exposure. These finding suggest that divergent cellular insults may converge to a common pathway that initiate with elevation of p53 protein levels. In this regards, it should be noted that both glutamate and Aβ may lead, although throughout different intracellular pathways, to the generation of free radicals as well as H_2O_2 is a reactive oxygen species producer. Indeed, it is well established that free radicals induce damage of cellular components such as lipids, proteins, and DNA and that accumulation of DNA damage is a well known stimulus for elevating p53 protein levels and for activating p53-mediated signaling pathways.

As recently pointed out by Morrison and Kinoshita *(17)*, alteration in p53 expression has been associated with neuronal damage in a variety of in vivo model systems. A remarkable example is the finding that p53 immunoreactivity has been detected in brain tissue from patients that have been diagnosed with Alzheimer's disease *(18)*.

2.2. p53 and Cell Senescence

As the number of different roles that p53 plays continues to proliferate, the question of the human relevance comes to the fore. For example, p53-deficient mice show behavioural

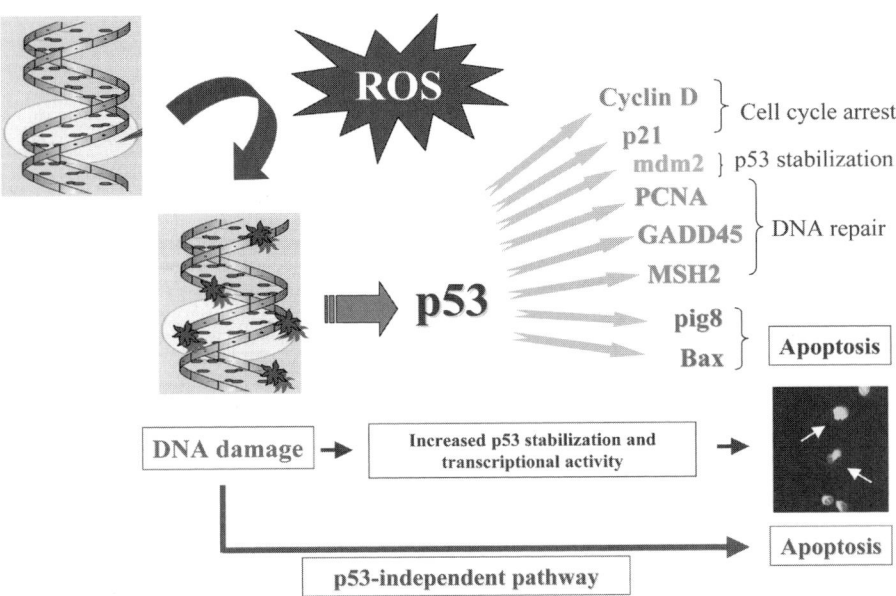

Fig. 2. Simplified scheme of oxidative stress-induced p53 activation.

alterations *(19)* whereas p53 mutant mice were shown to express a phenoptype characterized by early aging *(20)*. The correlation between tumor-suppressive mechanisms and senescence has been recent matter of investigation *(21)*. The signal transduction pathways that activate cellular senescence are now better understood. Division of telomerase-negative cells causes the erosion of the telomeric single-stranded 3′-extension, preventing extension of the double-stranded region and, perhaps more important, causing functional uncapping of the telomeres. The chromosomal DNA ends are then exposed and recognized by the cell as a double-stranded break. As expected, when a double-stranded break forms, DNA repair and damage checkpoint factors are recruited to the site of damaged DNA and a p53-dependent checkpoint is initiated. So replicative senescence is a p53-dependent checkpoint response to DNA damage *(22–23)*. p53 is activated by a variety of stressful cellular conditions, including DNA damage, oxidative stress, and oncogenic signals. It is not surprising then that senescence program can be activated by the same stresses, even in telomerase-positive cells. If p53 contribute to aging, it would be tricky to improve tumor-suppressive mechanisms without accelerating aging, and to retard aging without accelerating tumor formation.

2.3. p53 and NF-κB

The mechanisms by which p53 specifies the neuronal response to injury is not completely understood. One of the members of the Bcl2 family named Bax appears to be essential for p53-mediated neuronal death. Bax-deficient are indeed protected from cell death induced by DNA damaging agents adenovirus-mediated p53 overexpression *(24–25)*. Controversial reports exist on the role of the transcription factors NF-κB in p53-mediated apoptosis *(26–28)*. Pizzi et al. *(29)* demonstrated that different subunit composition of the NF-κB dimers may address opposing signals to cell function and recent works by Culmsee et al. *(26)* underlined the contribution of the transcriptional cofactor p300 in balancing opposite roles of NF-κB in modulating cell function. This additional nuclear modulator, coordinately with NF-κB, might extend the number and/or amplify the level of transcription of NF-κB target genes to finally alter the cellular response. The checkpoint, which discriminates between NF-κB-mediated physiological and pathological responses, might be cell-specific, influenced by the nature and the intensity of the extracellular stimulus, and continuously modifiable by the constitutive metabolic and functional context of the transcriptional machinery.

Interestingly, aberrant NF-κB regulation, specifically a high-constitutive activation, has been found in fibroblast from ataxia teleangectasia patients and demonstrated to be implicated in their high sensitivity to ionizing radiation *(30–31)*. Ataxia teleangectasia is a human disease characterized by neurological, immunological, and radiobiological problems, caused by a mutation in the ATM protein kinase gene. Recently, electron microscopic evidence of neuronal degeneration in the cerebellar cortex of ATM knockout mice was reported *(32)*. In light of these observations, a possible correlation among tumor development, neurological deficiencies and NF-κB activation would certainly deserve deeper investigation.

3. Contribution of the Mismatch Repair System to Neuronal Degeneration

The mismatch repair (MMR) system is an important member of the DNA checkpoint, that includes a number of proteins aimed to control genomic stability through cell-cycle arrest, DNA repair, and apoptosis *(33–34)*. Repair of damaged genes is the prominent role of this system, that is shared with the other mayor repair systems, such as the nucleotide excision repair (NER) and the base excision repair (BER).

Several studies have led to a biochemical model for postreplication mismatch repair in *Escherichia Coli*. Initiation of a MMR event occurs when MutS recognizes and binds mispaired nucleotides that result from polymerase misincorporation errors *(35)*. In eukaryotic there are at least seven MutS homologues (MSH1 to MSH7) and five MutL homologues designated MLH1, MLH2, MLH3, PMS1, and PMS2. With the exception of the homodimer MSH1-MSH1, involved in mitochondrial genomic stability in yeast and plant, the eukaryotic MutS and MutL homologs typically form heterodimers. MSH heterodimers are specialized for different but overlapping classes of mismatches: the most important in mammals are MSH2-MSH6 (MutSα) heterodimers, that generally recognize base mispairs (e.g., G/T, A/C), and MSH2-MSH3 (MutSβ), that bind to short insertion-deletion loops. MutS proteins interacts also with other DNA lesions; small alterations such as O^6-methylguanine, 8-oxoguanine and thymine glycol, major ultraviolet light photoproducts such as Cyclobutane pyrimidine dimers, and 1,2 intrastrand G-G intrastrands crosslinks produced by cysplatin are all targets of the MMR.

There are a number of other proteins involved in MMR and these include DNA polymerase δ, replication protein A, replication facto C, exonuclease1, RAD27, and DNA polymerase d and e associated exonucleases *(33–35)*.

PCNA has recently been added to the list of members of the MMR system; this protein was originally thought to interact with DNA polymerases and increase their activity. Recently, it was found that yeast MSH3 and MSH6 contain N-terminal sequence motifs characteristics of proteins that bind to PCNA *(36)* and that human PCNA coimmunoprecipitates in a complex containing MLH1 and PMS2 *(37)*. These evidences suggest that PCNA has also been implicated in MMR before the DNA synthesis step,

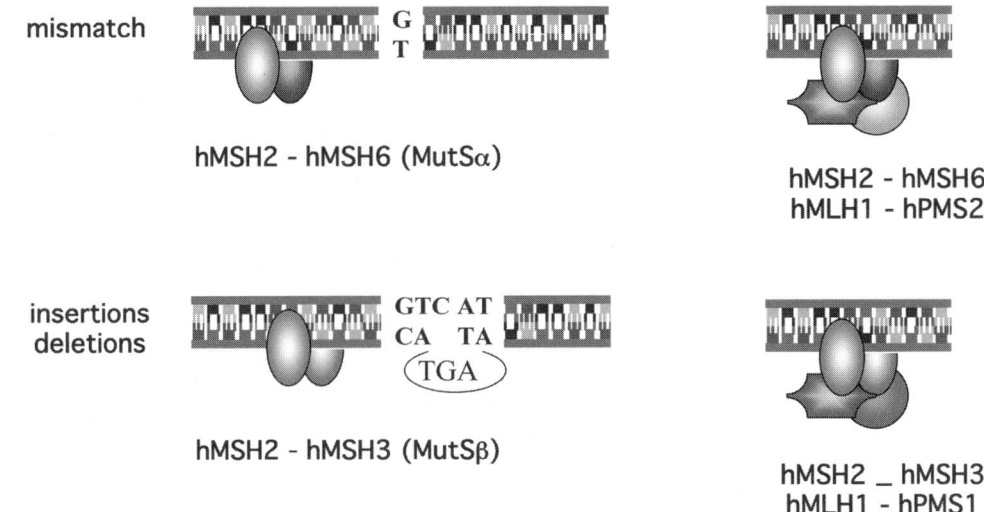

Fig. 3. Simplified scheme of MMR proteins interaction and function.

directly in mispair recognition, although the biochemical basis for this additional role of PCNA remains unclear *(38)*.

Once the DNA mismatch is recognized and the MutS and MutL hetorodimers complexes have combined with it, repair of the mismatch DNA proceeds by activating exonuclease-mediated degradation of DNA from a nick that is a distance of up to 1 to 2 kilobases from the mismatch. Degradation continues until the mismatched base is removed. The resulting long excision tract is filled in by polymerase d which inserts the correct nucleotide in to the sequence.

DNA repair is not the only function of MMR system; it is likely part of the DNA checkpoint involved in choosing cell fate. It has recently been shown that key proteins, which are involved in DNA repair, could play additional roles, in excess of DNA damage, and become proapoptotic proteins. Proteins of MMR system (i.e., MSH2 and MSH6, appear to play this dual role) *(39–41)*. In this line, hMutSα and hMutLα increase the phosphorylation of p53 in response to DNA methylator damage *(42)*. Human epithelial and mouse embryo fibroblast cell lines lacking MLH1 protein are more resistant to hydrogen peroxide, likely via the dysregulation of apoptosis. In fact, MLH1 proficient cells treated with hydrogen peroxide underwent apoptotic death that involves mitochondrial permeability and caspase 3 activation *(43,44)*.

3.1. Historical Background of Hereditary Nonpolyposis Colorectal Cancer

Long before molecular genetics had given us insight into the etiology of colorectal cancer, Aldred Warthin *(45)* had described several families who appeared to have predisposition to cancer. Throughout the 1970s and early 1980s there remained a great scepticism that cancer could have a strong hereditary component and the previous work was seen as anecdotal. However, by the 1980s many reports of a "cancer family syndrome" were appearing in the medical literature *(46)*. Cancer family syndrome then become subdivided into Lynch syndrome I (families characterized by mainly colorectal cancers at early age) and Lynch syndrome II (families with colorectal and extracolonic cancer). All of this different terminology was eventually clarified with the introduction of the term hereditary nonpolyposis colorectal cancer (HNPCC) to emphasise the lack of multiple colonic polyps and to separate it from the polyposis syndromes. Following the studies of large kindreds using linkage analysis, the HNPCC susceptibility loci were mapped to chromosome 2p16 and chromosome 3p21. Expanded microsatellites were found in HNPCC rather than regions of loss and this was termed microsatellite instability. Microsatellite instability had already been studied extensively in bacteria and yeast and this led to positional cloning strategies identifying the human homologous of the *mutS* gene, namely *hMSH2*, on chromosome 2p, followed closely by the identification of the human homologous of the *mutL* gene, namely *hMLH1*, on chromosome 3p.

Mutations in *hMSH2* and *hMLH1* account for the majority of reported HNPCC cases. Based on positional cloning and linkage analysis, it is now generally accepted that inactivation of one of the *MMR* genes is responsible for the microsatellite instability or replication errors seen in more than 90% of HNPCC *(47)*.

3.2. MMR System in the SNC

Very little is know about the possible role of the MMR system in the brain. Marietta et al. *(48)* demonstrated that proteins of MutSα complex were expressed in developing and adult rat brain tissues, suggesting that not only dividing cells but adult brain cell as as well, have the capacity to carry out DNA mismatch repair. Nuclear extracts from adult rat brain neurons were also found to be able to repair DNA mismatches *(49)*.

In this contest, we studied the expression of MMR proteins in neurons by in vivo and in vitro studies. The distribution of MSH2, one of the key protein involved in recognition of damaged DNA, was evaluated by immunohistochemistry in the adult rat brain. The results from these experiments showed that MSH2 was expressed in several areas of the brain including hippocampus, cerebellum, cortex, striatum, substantia nigra, and in spinal cord. MSH2 was detected only in neuronal cells with nuclear localization, whereas glial cells were not positive *(50)*. The involvement of MSH2 in neuron degeneration was studied in rats treated with kainic acid. This is a well known experimental paradigm of excitotoxicity, characterized by specific cell loss in CA3/CA4 hippocampal subfields. Kainate injection resulted in a dose-dependent increase of MSH2 expression specifically in the pyramidal cells of CA3/CA4 subfield *(50)*. MSH2 increased expression could be interpreted as a part of a proapoptotic signaling. This is in line with previous data showing overexpression of MSH2 and MLH1 proteins in cells undergoing apoptosis *(41)*.

Involvement of MMR in the repair of base oxidation induced by H_2O_2 was explored in human neuronal cell line *(51,52)*. To this aim, human SH-SY5Y neuroblastoma cells were differentiated to a neuron-like phenotype by treatment with retinoic acid and then treated with a pulse of H_2O_2. The oxidative lesion was carried out in a way to induce a submaximal lesion resulting in no more that 20 to 30 cell loss, at 24 h after the lesion. This lesion also caused a marked oxidative DNA damage, as detected by 8-OH-deoxyguanosine immunoreactivity, within 15 min and lasted until 2 h after the H_2O_2 pulse. The calculated number of DNA lesioned cells reached 50 % of the total at 2 h after the treatment and then declined. This observation suggested that a given number of neurons were able to recognize and repair the oxidized DNA whereas the others underwent apoptosis. In a time-frame between 30 min and 2 h, H_2O_2 also induced MSH2 nuclear translocation of several members of the MMR system, including MSH2, MSH6, and MLH1. The results indicate that, like nonproliferating cells, neurons possess the machinery to recognize and repair DNA damage, Additional evidence for MMR activity in neurons in response to DNA damage was obtained in functional studies from both undifferentiated and differentiated Ntera2 extracts (53).

3.3. MSH2 and Experimental Models of Neurodegenerative Diseases

Huntington disease (HD) is an autosomal dominant, progressive neurodegenerative disorder that is caused by an expanded CAG repeat sequence leading to an increase in the number of glutamine residues in the encoded protein (54). The normal CAG repeat range is 5 to 36, whereas 38 or more repeats are found in the disease state; the severity of the disease is roughly proportional to the number of CAG repeats (1–5). With increasing repeat number, the protein changes conformation and becomes increasingly prone to aggregation, suggesting important functional correlations between length and pathology, including severity and onset age. Because of their role CAG microsatellite instability, it was attractive to study the contribution of MMR proteins in HD. Wheeler et al. (55) recently reported that MSH2 hastens by many months the timing, but not the neuron specificity, of early disease in a precise genetic HD mouse model, implicating MMR-dependent pathways in modifying pathogenesis in man.

4. Conclusions

Historically, one of the most well known connections between abnormalities of the DNA damage response and neurodegeneration has been the human syndrome of ataxia telangiectasia. However, other syndromes associated with defective DNA damage response also include neurological symptoms as a primary feature of their phenotypes (56). This argues that defects in the repair of, or response to, DNA damage impact significantly on brain function. We speculate that the list of neuropathologies associated with an impairment of the DNA damage repair system could be even larger including chronic and progressive neurodegenerative diseases like AD. In this regard, quantitative and histopathological markers of oxidative DNA damage have been found the brain of Alzheimer's patients (57) suggesting that the neuronal machinery devoted to restore damaged DNA might be impaired in AD brain.

Future studies focussing on these and other aspects of β-amyloid neurotoxicity will be important for the molecular description of specific DNA processing and repair mechanisms in brain as well as to define whether cancer and neurodegenerative disease share common genetic risk factors for the development and progression of the disease. In this regard, DNA repair genes are good candidate for such a dual role. In fact, the activity of the DNA repair systems in the brain decreases as a function of age so that the capability to recognize and repair various types of DNA damage is reduced in the aging brain. One of the most intriguing aspects of this topic is whether or not neuronal cells lacking DNA repair factors are able to survive after DNA damage. One possibility is that they do survive and acquire a novel phenotype, which is a neuron with altered DNA. This could be relevant or not. If DNA damage is located in a region not important for transcription or in a coding region of a protein which is not important for cell function, they may survive without showing signs of impairment. On the other hand, DNA damage may be located in crucial sites along the DNA filament, thus altering transcription of proteins deeply involved in cell function. In this case, unrepaired DNA damage will generate "dysfunctional neurons" (perhaps for a certain period of time) and eventually (later) activate a death program.

There are two additional concepts to be taken into consideration: (1) accumulation of DNA damage over time, and (2) gene-specific vulnerability for mutations.

The decrease in efficiency of the DNA repair systems occurring in aging brain may increase the risk to generate and accumulate gene mutation within individual neurons and, as consequence, to generate cells with functional alterations (dysfunctional neurons). Moreover, there is a hierarchy in the sensitivity of the genes for mutation. Genes displaying elevated risk of mutation are those with high rate of transcription, being exposed to the nuclear environment for prolonged periods of time, free of histone protection. Moreover, based on lessons from oncology, almost all genes involved in DNA damage recognition and repair are particularly sensitive to mutations. Thus, it might be envisaged a self-strengthening process based on time-dependent accumulation of DNA damage which involves more and more (cell-function relevant) genes.

This view takes into consideration at least two features of many neurodegenerative diseases: anatomy (cell specificity) and timing (years for the manifestation of the disease).

References

1. Almong N, Rotter V. Involvement of p53 in cell differentiation and development, Biochem Biophys Acta 1997;1333, F1–F27.
2. Jacobs WB, Walsh GS, Miller FD. Neuronal survival and p73/p63/p53: a family affair. Neuroscientist 2004;10(5):443–455.
3. Purdie CA, Harrison DJ, Peter A, et al. Tumour incidence, spectrum and ploidy in mice with a large deletion in the p53 gene. Oncogene 1994;9:603–609.
4. Malkin D, Li FP, Strong LC, et al. Germ line p53 mutations in a familial syndrome of breast cancer, sarcomas, and other neoplasms. Science 1990;250:1233–1238.
5. Donehower LA, Harvey M, Slagle BT, et al. Mice deficient for p53 are developmentally normal but susceptible to spontaneous tumours. Nature 1992;356:215–221.
6. Ko LJ, Prives C. p53: puzzle and paradigm. Genes Dev 1996;10:1054–1072.
7. Giaccia AJ, Kastan MB. The complexity of p53 modulation: emerging patterns from divergent signals. Genes Dev 1998;12:2973–2983.
8. Uberti D, Belloni M, Grilli M, Spano P, Memo M. Induction of tumour-suppressor phosphoprotein p53 in the apoptosis of cultured rat cerebellar neurones triggered by excitatory amino acids. Eur J Neurosci 1998;10:246–254.
9. Sakhi S, Bruce A, Sun N, Tocco G, Baudry M, Schreiber SS. p53 induction is associated with neuronal damage in the central nervous system. Proc Natl Acad Sci USA 1994;89: 12,028–12,032.
10. Hughes PE, Alexi,T, Yoshida T, Schreiber SS, Knusel B. Excitotoxic lesion of rat brain with quinolinic acid induces expression of p53 messenger RNA and protein and p53-inducible genes Bax and Gadd-45 in brain areas showing DNA fragmentation. Neuroscience 1996;74: 1143–1160.
11. Schreiber SS, Sakhi S, Dugich-Djordjevic MM, Nichols NR. Tumor suppressor p53 induction and DNA damage in hippocampal granule cells after adrenalectomy. Exp Neurol 1994;130:368–377.
12. Chopp M, Li Y, Zhang ZG, Freytag SO. p53 expression in brain after middle cerebral artery occlusion in the rat. Biochem Biophys Res Commun 1992;182:1201–1207.

13. Crumrine RC, Thomas AL, Morgan PF. Attenuation of p53 expression protects against focal ischemic damage in transgenic mice. J Cereb Blood Flow Metab 1994;14:887–891.

14. LaFerla FM, Hall CK, Ngo L, Jay G. Extracellular deposition of beta-amyloid upon p53-dependent neuronal cell death in transgenic mice. J Clin Invest 1996;98:1626–1632.

15. Copani A, Uberti D, Sortino MA, Bruno V, Nicoletti F, Memo M. Activation of cell-cycle-associated proteins in neuronal death: a mandatory or dispensable path? Trends Neurosci 2001;24(1):25–31.

16. Uberti D, Grilli M, Memo M. Contribution of NF-kappaB and p53 in the glutamate-induced apoptosis. Int J Dev Neurosci 2000;18(4–5):447–454.

17. Morrison RS, Kinoshita Y. The role of p53 in neuronal cell death. Cell Death Differ 2000;7(10):868–879.

18. De la Monte SM, Sohn YK, Wands JR. Correlates of p53- and Fas (CD95)-mediated apoptosis in Alzheimer's disease. J Neurol Sci 1997;152: 73–83.

19. Amson R, Lassalle J-M, Halley H, et al. Behavioral alterations associated with apoptosis and down-regulation of presenilin 1 in the brains of p53-deficient mice. Proc Natl Acad Sci USA 2000;97:5346–5350.

20. Tyner SD, Venkatachalam S, Choi J, et al. p53 mutant mice that display early ageing-associated phonotypes. Nature 2002;415:45–53.

21. Pelicci PG. Do tumor-suppressive mechanisms contribute to organism aging by inducing stem cell senescence? J Clin Invest 2004;113:4–7.

22. Niida H, et al. Severe growth defect in mouse cells lacking the telomerase RNA component. Nat Gen 1998;19:203–206.

23. Ben-Porath I, Weinberg RA. When cells get stressed: an integrative view of cellular senescence. J Clin Invest 2004;113:8–13.

24. Xiang H, Hochman DW, Saya H, Fujiwara T, Schwartzkroin PA, Morrison RS. Evidence for p53-mediated modulation of neuronal viability. J Neurosci 1996 Nov 1;16(21):6753–6765.

25. Cregan SP, Arbour NA, Maclaurin JG, et al. p53 activation domain 1 is essential for PUMA upregulation and p53-mediated neuronal cell death. J Neurosci 2004;24(44): 10,003–10,012.

26. Culmsee C, Siewe J, Junker V, et al. Reciprocal inhibition of p53 and nuclear factor-kappaB transcriptional activities determines cell survival or death in neurons. J Neurosci 2003;23:8586–8595.

27. Grilli M, Memo M. Possible role of NF-κB and p53 in glutamate-induced pro-apoptotic neuronal pathway. Cell Death Diff 1999;6:22–27.

28. Grilli M, Memo M. Nuclear factor-kappaB/Rel proteins: a point of convergence of signalling pathways relevant in neuronal function and dysfunction. Biochem Pharmacol 1999; 57:1–7.

29. Pizzi M, Goffi F, Boroni F, et al. Opposing roles for NF-kappa B/Rel factors p65 and c-Rel in the modulation of neuron survival elicited by glutamate and interleukin-1beta. J Biol Chem 2002;277(23):20,717–20,723.

30. Savitsky K, Bar-Shira A, Gilad S, Rotman G, Ziv Y. A single ataxia teleangectasia gene with a product similar to PI-3 kinase. Science 1995;268:1749–1753.

31. Jung M, Zhang Y, Lee S, Dritschilo A. Correction of radiation sensitivity in ataxia teleangectasia cells by truncated IκBα. Science 1995;268:1619–1621.

32. Ro K, Xu Y, Aguila MC, Baltimore D. Degeneration of neurons, synapses, and neuropil and glial activation in a murine Atm knockout model of ataxia-teleangectasia. Proc Natl Acad Sci USA 1997;94:12,688–12,693.

33. Syngal S, Fox EA, Li C, et al. Interpretation of genetic tests results for hereditary nonpolyposis colorectal cancer. Implications for clinical predisposition testing. JAMA 1999;282: 247–253.

34. Sancar A. Excision repair invades the territory of mismatch repair. Nat Gen 1999;21: 247–249.
35. Su S-S, Modrich P. *Escherichia Coli mutS*-encoded protein binds to mismatched DNA base pairs. Proc Natl Acad Sci 1986;83:5057–5061.
36. Clark AB, Valle F, Drotschmann K, Gary RK, Kunkel TF. Functional interaction of proliferating cell nuclear antigen with MSH2-MSH6 and MSH2 and MSH3 complexes. J Biol Chem 2000;275(47):36,498–36,501.
37. Gu L, Hong Y, McCulloch S, Watanabe H, Li GM. GATP-dependent interaction of human mismatch repair proteins and dual role of PCNA in mismatch repair. Nucleic Acid Res 1998; 26(5):1173–1178.
38. Umar A, Buermeyer AB, Simon JA, et al. Requirement for PCNA in DNA Mismatch Repair at a Step Preceding DNA Resynthesis. Cell 1996;87:65–73.
39. Bernstein C, Bernstein H, Payne CM, Garewal H. DNA repair /pro-apoptotic dual role proteins in five major DNA repair pathways: fail-safe protection against carcinogenesis. Mutat Res 2002;511(2):145–178.
40. Hickman MJ, Samson LD. Role of DNA mismatch repair and p53 in signaling induction of apoptosis by alkylating agents. Proc Natl Acad Sci USA 1999;96:10,764–10,769.
41. Zhang H, Richards B, Wilson T, et al. Apoptosis induced by overexpression of hMSH2 or hMLH1. Cancer Res 1999;59:3021–3027.
42. Duckett DR, Bronstein SM, Taya Y, Modrich P. hMutSα and hMutLα-dependent phosphorylation of p53 in response to DNA methylator damage. Proc Natl Acad Sci USA 1999;96:12,384–12,388.
43. Hardman RA, Afshari CA, Barrett JC. Involvement of mammalian MLH1 in the apoptotic response to peroxide-induced oxidative stress. Cancer Res 2001;61:1392–1397.
44. De Weese TL, Shipman JM, Larrier NA, et al. Mouse embrionic stem cells carryng one or two defective MSH2 alleles respond abnormally to oxidative stress inflicted by low-level radiation. Proc Natl Acad Sci USA 1998;95:11,915–11,920.
45. Warthin AS. Heredity with reference to carcinoma. Arch Intern Med 1913;12:546–555.
46. Lynch HT, Show MW, Magnuson CW, Larsen AL, Krush AJ. Hereditary factors in cancer. Study of two large Midwestern kindreds. Arch Intern Med 1966;117:206–212.
47. Wheeler JMD, Bodmer WF, Mc Mortensen NJ. DNA mismatch repair genes and colorectal cancer. Gut 2000;47:148–153.
48. Marietta C, Palombo F, Gallinari P, Jiricny J, Brooks PJ. Expression of long-patch and short-patch DNA mismatch repair proteins in the embryonic and adult mammalian brain. Mol Brain Res 1998;53:317–320.
49. Brooks PJ, Marietta C, Goldman D. DNA mismatch repair and DNA methylation in adult brain neurons. J Neurosci 1996;16(3):939–945.
50. Belloni M, Uberti D, Rizzini C, et al. Distribution and kainate-mediated induction of the DNA mismatch repair protein MSH2 in rat brain. Neuroscience 1999;94:1323–1331.
51. Belloni M, Uberti D, Rizzini C, Jiricny J, Memo M. Induction of Two DNA Mismatch Repair Proteins, MSH2 and MSH6, in Differentiated Human Neuroblastoma SH-SY5Y Cells Exposed to Doxorubicin. J Neurochem 1999;72:974–979.
52. Uberti D, Ferrari Toninelli G, Memo M. Involvement of DNA damage and repair systems in neurodegenerative process. Toxicol Lett 2003;139:99–105.
53. David P. DNA replication and postreplication mismatch repair in cell-free exstracts from cultured human neuroblastoma and fibroblast cells, J Neurosci 1997;17: 8711–8720.
54. Huntington's Disease Collaborative Research Study. A novel gene containing a trinucleotide repeat that is expanded and unstable on Huntington's disease chromosomes. Cell 1993;72: 971–983.

55. Wheeler VC, Lebel L-A, Vrbanac V, Teed A, te Riele H, MacDonald ME. Mismatch repair gene MSH2 modifies the timing of early disease in HdhQ111 striatum. Hum Mol Gen 2003; 12:273–281.

56. Rolig RL, McKinnon PJ. Linking DNA damage and neurodegeneration. Trends Neurosci 2000;23:417–424.

57. Pratico D, Delanty N. Oxidative injury in diseases of the central nervous system: focus on Alzheimer's disease, Am J Med 2000;109:577–585.

17

The Biology of Caspases

Tasman James Daish and Sharad Kumar

Summary

Programmed cell death (PCD) is a vital part of the normal development of metazoans. PCD requires a conserved class of cysteine proteases, termed caspases, to affect the controlled demise of cells. When the regulatory pathways controlling cell death are perturbed through mutation, disease states, including cancers and neurodegenerative disorders, can occur. Therefore, caspases and the molecules regulating their function are potential candidates for therapeutic targets. This chapter describes caspase structure, regulation, and function in the context of biological function and the specific roles they play in the protection of living systems. A section is also dedicated to the emerging field concerning the nonapoptotic functions of "apoptotic" caspases in development.

Key Words: Apoptosis; caspases; oligomerisation; transcription; development.

1. Introduction

Life requires death, and the programmed form of cell death (PCD), or apoptosis, requires caspase activation to effect the tightly regulated cell or tissue destruction so critical to normal development and tissue integrity maintenance. PCD also functions in the defence against DNA damage, trauma or toxic insult, and immune/pathogenic challenge. Caspases are cysteine proteases that cleave their substrate targets following an aspartate residue. When an apoptotic program becomes deregulated through mutation, a host of disease states can occur including various cancers and immune or neurodegenerative disorders (1–5). Consequently, caspases can be studied as potential candidates for therapeutic targets to modify or prevent aberrant apoptosis. Discovery of an evolutionarily conserved PCD pathway involving caspases came with the identification of the *ced-3* gene in the nematode *Caenorhabditis elegans* that is required for the death of a discrete population of cells during development (6,7). Many caspases have now been identified in diverse species with 11 present in humans, 10 in mice, 4 in the chicken and zebrafish, and seven in *Drosophila* (8). This chapter describes the function, activation and regulation of caspases.

2. Caspase Structure

Caspases are divided into two general initiator and effector groups. Initiator caspases are characterized by long N-terminal prodomains containing protein–protein interaction motifs. Long prodomain-containing caspases have death effector domains (DEDs) or caspase recruitment domains (CARDs). Effector caspases that require activation by initiator caspases lack long prodomains or similar binding motifs (9,10). DED-containing caspases include caspase-8 and -10 in mammals and DREDD in *Drosophila* whereas the CARD-

From: *Apoptosis, Cell Signaling, and Human Diseases: Molecular Mechanisms, Volume 2*
Edited by R. Srivastava © Humana Press Inc., Totowa, NJ

containing caspases are caspase-1, -2, -4, -5, -9, -11, and -12 in mammals, DRONC and CED-3 in *Drosophila* and *C. elegans*, respectively *(11)*. Effector caspase members include caspase-3, -6, and -7 in mammals and DRICE, DCP-1, DECAY, and DAMM in *Drosophila (11,12)*. The *Drosophila* caspase STRICA, while having a long serine/threonine rich prodomain, lacks either DED or CARD domains and its function remains unclear *(13)*. The catalytic part of caspases consists of small and large subunits (often termed the p10 and p20), with the catalytic residue positioned in the large subunit. Activation of the inactive caspase zymogen often involves processing into a heterotetramer composed of two large and small subunit heterodimers *(14)*. The key difference between the caspase classes is their respective modes of activation. Whereas effector caspases require cleavage by initiator caspases for activation, initiator caspases have the inherent property of being able to autoactivate through adaptor-mediated dimerisation/oligomerisation. This proximity-induced dimerization is facilitated by the formation of large multiprotein complexes that bring initiator caspase molecules into close proximity *(15–17)*.

3. Caspase Activation Pathways

3.1. Extrinsic Activation Pathway

Death signaling arising from extracellular death ligands binding to their cognate cell surface receptors is termed the extrinsic death pathway. The extrinsic pathway involves formation of a death-inducing signaling complex (DISC) that leads to caspase-8 and then effector caspase activation *(18)*. DISC formation results from Fas ligand binding to the Fas receptor, then recruitment of the adaptor molecule FADD, which in turn recruits pro-caspase-8, triggering its proximity-induced autoactivation *(19)*. Activated caspase-8 also links to the intrinsic pathway by cleaving and activating the BH3-only protein BID *(20)*. Cleaved BID translocates to the mitochondria to effect cytochrome *c* release *(20)*.

3.2. Intrinsic Activation Pathway

Death signaling arising from stimuli within cells through activation of developmentally regulated death programs, cell stress caused by deprivation of nutrients, or cytotoxic damage is termed the intrinsic or mitochondrial death pathway. This pathway induces activation of caspase-9. In mammals, caspase-9 activation requires mitochondrial cytochrome *c* release which results in assemblage of an approximately 1 MDa multiprotein oligomeric complex termed the Apaf-1 apoptosome *(21,22)*. A number of other molecules are concurrently liberated from the mitochondria along with cytochrome *c* such as Smac/DIABLO, apoptosis inducing factor (AIF), and Omi/HtrA2 *(23–25)*. Cytochrome *c*, once released from the mitochondria, binds Apaf-1 to effect a conformational change in Apaf-1, which allows the binding of dATP to occur *(22)*. This in turn induces apoptosome formation and recruitment of caspase-9 to this complex which is then able to activate effector caspase-3 and -7 *(21,22)*.

In flies, the apical caspase DRONC is activated through interaction with the *ced-4/apaf-1* homolog adaptor molecule DARK/Hac-1, similar to mammalian caspase-9/Apaf-1 and *C. elegans* CED-3/CED-4 interactions *(26–28)*. The role of cytochrome *c* in the *Drosophila* intrinsic death pathway and in particular in DRONC activation is less clear than its observed role in apoptosome formation in mammals *(29)*. The death activator proteins RPR, HID, and GRIM sequester the DIAP1 inhibitory protein from its ubiquitinating action on DRONC allowing DRONC/DARK binding resulting in DRONC activation *(30–34)*. Similar to the activation of caspase-3 by caspase-9, active DRONC cleaves and activates the effector caspases DRICE and DCP-1 *(29,35)*.

3.3. Regulation of Caspase Activation by Bcl-2 Family Members

A critical group of molecules regulating the intrinsic apoptotic pathway are the Bcl-2 family proteins. This family is characterized by the presence of Bcl-2 homology domains (BH1-4) and include pro (Bax, Bok, and Bak) and antiapoptotic (Bcl-2, Bcl-x_L, Mcl-1, Bcl-w, and Bfl-1) members. Bcl-2 is the functional homolog of the antiapoptotic CED-9 *(36)* and the BH domains are required for protein–protein interactions between pro- and antiapoptotic Bcl-2 family members to modulate their respective functions. A third group, the *C. elegans* EGL-1 homologs, defined by the presence of the BH3 domain only (termed BH3-only members), are exclusively proapoptotic (Puma, Bim, Bid, Bad, and Noxa). BH3-only proteins are responsive to diverse stimuli and are tightly regulated by ubiquitination, phosphorylation, proteolytic processing as well as transcriptional controls *(37)* and function through proapoptotic Bax and Bak *(38,39)*. The absolute requirement of Bax and Bak in apoptosis is highlighted by the fact that Bax and Bak deficiency prevents the death-inducing potency of the proapoptotic BH3-only molecules *(38,40–43)*. The BH domains serve as the adaptor binding motifs that mediate Bcl-2 and BH-3-only protein interactions, and when bound, also serve to modify conformation and consequently cellular localisation and function. For example, the activation by oligomerisation of Bax can be induced by truncated (via caspase-8 cleavage) proapoptotic Bid (tBid) through BH3-mediated interactions *(20,38)*. Bax activation/oligomerisation occurs through BH3 binding domain accessibility to regulatory proteins, determined by its conformational status. Oligomerization of Bax leads to changes in mitochondrial membrane permeabilization causing Smac/DIABLO and cytochrome *c* release *(38)*. Thus, regulation of Bax can occur by a BH3-mediated binding competition for Bid association and sequestration between the antiapoptotic Bcl-2 members like Bcl-2 and Bcl-x_L and the proapoptotic BH3-only proteins, like Bid, Bad, or Noxa. Bak activation, in contrast, is prevented by constitutive BH3-mediated binding to antiapoptotic Mcl-1 and Bcl-x_L, with the former interaction occurring in the mitochondrial membrane under nonapoptotic conditions *(44,45)*. In response to specific death stimuli, the Mcl-1/Bak and Bcl-x_L/Bak complexes dissociate and Mcl-1 is degraded by ubiquitination *(44,46)*. Mcl-1/Bak dissociation is induced by the BH3-mediated binding to Mcl-1 by the proapoptotic BH3-only protein, Noxa *(45)*.

Regulation of apoptosis by Bcl-2 family proteins is complex as binding preferences/affinities of the respective BH domains, in particular the BH3-only protein domains, to the multiple pathway members can be quite specific (Noxa) or relatively nondiscriminate (Bim), resulting in profoundly different modulations *(39,45,47–49)*.

3.4. Extrinsic Pathway Regulation by FLIP

The caspase-8 homolog FLICE-like inhibitory protein (FLIP) differs from caspase-8 in that it lacks the critical catalytic cysteine residue. In normal cells, the c-FLIP$_L$ variant acts to enhance DISC-mediated caspase-8 activation following receptor-mediated death signaling *(50,51)*. The heterodimerisation of c-FLIP$_L$ with caspase-8 following c-FLIP$_L$ recruitment to the DISC induces caspase-8 activation, an interaction which imparts greater inherent catalytic potential than homodimerisation of caspase-8 *(51)*.

3.5. Regulation by Inhibitor of Apoptosis Proteins

Inhibitor of apoptosis proteins (IAPs) inhibit caspase function through the presence of up to three BIRs which are zinc finger domains of approx 80 amino acid residues and a C-terminal RING finger domain which imparts E3 ubiquitin ligase activity *(52,53)*. In *Drosophila*, IAP/caspase interaction is mediated via two amino-terminal BIRs *(54–56)*. The death activator proteins, RPR, HID, GRIM, and SICKLE bind the *Drosophila* IAPs (DIAP1 and DIAP2) and antagonize the IAP-caspase interaction through a short N-terminal RHG motif *(57,58)*. This interaction results in caspase activation and cell death by directing the sequestered IAPs to the ubiquitination pathway. The *diap1* mutant *thread* has deregulated developmental PCD because of unchecked processing and activation of DCP-1, DRICE and DRONC *(59)*. DRONC is rapidly processed in response to *diap1* RNAi followed by DRONC-dependent DRICE processing and activation *(60,61)*. The DRONC binding motif was identified to be a 12-residue sequence (residues 114–125) sited between the prodomain and caspase domain and required for BIR2-mediated binding with DIAP1 *(56)*. The interaction of DRONC with the IAP-BIR2 was shown to occupy the same binding pocket as other caspases and the N-terminal region of RPR, HID, GRIM, and SICKLE suggesting a competition between the RHG proteins and IAPs exists to regulate DRONC activation *(56)*. Mutation of the DIAP1 RING finger, while failing to prevent interactions with RPR, HID, or DRONC, completely blocks DRONC ubiquitination *(30)*. Although the precise role of ubiquitination in the turnover rate of DRONC in vivo remains unclear, the *Drosophila* IAPs function to limit the accumulation of DARK-dependent active forms of DRONC, a process countered by the RHG death activator proteins.

In mammals, apart from inhibiting caspases, IAPs can function in diverse cellular roles such as signal-transduction pathway cofactors and participants in the process of cell division *(62,63)*. In a process similar to the regulation of DRONC, the mammalian *dronc* homolog caspase-9 is inhibited by XIAP through BIR3-mediated binding *(64)*. Critical to this interaction are three XIAP residues, Trp310, Glu314, and His343 required for binding to the caspase-9 p10 subunit *(65)*. XIAP binding prevents assemblage of the active tetrapeptide form of caspase-9, an interaction which is not possible by c-IAP1 and c-IAP2 as the result of nonconservation of these critical residues *(64)*. The mechanism of IAP inhibition of caspase-9 differs to that of DRONC by DIAP1 in that caspase-9 is rendered catalytically impotent by XIAP binding whereas DRONC remains unprocessed when bound by DIAP1 because of its inability to interact with DARK. Thus, although there are eight mammalian IAP proteins, caspases-9 is regulated primarily by XIAP whereas caspase-3 and -7 are subject to inhibition by multiple IAP types as well as XIAP. The site of regulation of caspase-3 and -7 by XIAP is sited between the BIR1 and BIR2 binding domains, an interaction which prevents their recognition/binding to substrates *(66–69)*. As in *Drosophila*, the caspase inhibitory function of IAPs is antagonized through BIR3 binding by the mitochondrial protein Smac/DIABLO, which is released from the mitochondrial intermembrane space following cleavage of its anchoring domain *(24,70)*. This interaction occurs through a similar tetrapeptide motif (Ala-Thr-Pro-Phe) as is present in caspase-9 and effectively competes for binding with the XIAP BIR3 to release caspase-9.

3.6. Transcriptional Regulation of Caspases in Drosophila

Caspases were thought to exist in cells at relatively static levels as inactive zymogens, effectively priming the cell for death in readiness for receipt of a death signal. However, it is becoming increasingly evident that key death pathway components, such as Bcl-2 homologs and caspases, are transcriptionally regulated in a stage- and tissue-specific manner throughout development *(71)*. Although transcriptional regulation of caspases has been demonstrated in mammals, the system best exploited for analyzing the transcriptional control of the apoptotic machinery has been the *Drosophila* larval salivary glands and midgut, which both undergo rapid and developmentally coordinated destruction in response to a caspase-dependent hormone induced genetic program *(72–77)*.

Drosophila larvae enter metamorphosis in response to a late larval increase in the titre of the steroid hormone 20-hydroxyecdysone (ecdysone), which induces proliferation of progenitor cell clusters or imaginal discs destined to form adult structures, and the rapid removal via apoptosis of the redundant larval midgut *(78–80)*. About 10 h after this late larval ecdysone peak, a second late prepupal rise in ecdysone levels induces eversion of the forming adult head and entry into the pupal stage followed by the rapid destruction of the larval salivary glands. Analysis of the genetic cascades occurring in these tissues reveals rapid induction of a hierarchy of transcription factors and cell death machinery components including caspases, most markedly the apical CARD-containing caspase DRONC *(73,76,77,81)*.

The primary transducer of the ecdysone signal in *Drosophila* is the ecdysone receptor heterodimer complex (EcR), composed of the ecdysone receptor and the RXR homolog Ultraspiracle (Usp). The EcR has three isoforms, each with different expression profiles and distinct biological functions *(82)*. All three isoforms have Ultraspiracle as its binding partner, an interaction that increases the binding affinity of the hormone to the receptor complex, which in turn stabilizes the complex to increase DNA binding affinity *(83)*. Once bound by ecdysone, the EcR induces the transcription of a set of early response transcription factors. EcR-induced early response genes include the zinc finger-containing *Broad Complex (BR-C)*, the ETS-like *E74* transcription factor along with the orphan nuclear receptor *E75*, and a novel nuclear protein *E93 (81,84–87)*.

BR-C, E74, E75, βFTZ-F1, and E93 are transcription factors playing essential roles in the removal of the larval midgut and salivary glands *(76,77,81,87–91)*. The relationship of these transcription factors to core cell death genes like *rpr*, *hid*, *grim*, *dark*, *dronc*, and *crq* is complex and differs significantly between the salivary glands and midguts. *EcR*, *BR-C*, and *E74A* are all induced in the salivary glands at puparium formation but their expression does not result in *rpr*, *hid*, or *dronc* transcription or histolysis of this tissue, as occurs in the midgut at this time *(77)*. In the midgut, *E93* mutants impact *dronc* transcription whereas *BR-C* mutants have altered transcription of *rpr*, *hid*, and *crq* but have normal *dark* and *dronc* expression. *rpr* is directly upregulated by the EcR at the late prepupal ecdysone pulse *(87,92)*. *rpr* and *hid* expression is dependent on the BR-C Z1 isoform as their expression is reduced in salivary glands from animals mutant for this isoform whereas expression of βFTZ-F1 and *E93* is maintained. Expression of *hid*, but not *rpr*, is dependent on E74A *(87)*. The competence factor βFTZ-F1 is required for *diap2* expression, which is then repressed by E75 resulting in a temporal restriction of *diap2* expression in the late prepupal salivary glands.

Thus a complex temporally and spatially regulated transcriptional hierarchy coordinates the destruction of redundant larval tissues via nuclear receptor-mediated transcriptional upregulation of death activators and caspases, a cascade coordinating caspase activation with repression of apoptosis inhibiting proteins.

4. Caspases in Development

4.1. Nonvertebrate Animals

Mutation of the *C. elegans ced-3*, as well as the adaptor *ced-4*, blocks the developmental cell deaths of a discreet population of 131 cells during morphogenesis *(6,93)*. Even though the pattern of PCD is completely disrupted in these animals, they remain viable and otherwise normal *(6)*.

In *Drosophila*, *dronc* null mutants exhibit differential requirements for DRONC in the execution of developmental and stress induced cell deaths *(73,94,95)*. As *dronc* is the only CARD-containing caspase in *Drosophila*, it was predicted to be essential for all effector caspase activation and therefore absolutely required for all developmental cell deaths. Isolated *dronc* null mutants showed that *dronc* is necessary for development as homozygous mutants arrest as early pupa with a diverse range of predicted and intriguing phenotypes. Firstly, there is a differential requirement for DRONC for the destruction of the larval midgut and salivary glands, with the latter failing to undergo histolysis whereas the midguts of *dronc* mutant homozygotes show typically apoptotic hallmarks such as DNA fragmentation and effector caspase substrate activity *(73)*. Stress-induced cell death is blocked in imaginal tissues and CNS however, generation of cell-type specific mutant clones revealed only minor defects in the formation of the adult wing and eye *(94)*. Interestingly, *dronc* mutants have greatly increased circulating blood cells, which when cultured demonstrate resistance to multiple death-inducing agents *(94)*. Most embryonic cell deaths are *dronc*-dependent with mutants showing head involution defects, additional neuronal cell complements, and a failure to hatch at the end of embryogenesis *(28,94)*.

The only other *Drosophila* caspase mutants isolated are for *dcp-1* and *dredd (96,97)*. DCP-1 is required for stress-induced germ cell death at the mid oogenesis stage and for normal DRICE activation levels and localisation in egg chambers *(96)*. In a genetic screen using an immunocompromized *Drosophila* model to identify immune response pathway components, DREDD was shown to be required for immunity in response to bacterial infection *(97)*. The role of DREDD in a nonapoptotic process was demonstrated by the observation that antibacterial peptide gene expression was blocked in *dredd* mutants and DREDD specifically was involved in the response to gram-negative bacterial infection *(98)*. Overexpression of *strica* in the fly eye induces a rough eye phenotype which is reduced by coexpression of *diap1*, *p35* and to a lesser extent, *diap2 (99)*. Overexpression of *strica* in mammalian cells induces death, which is marginally repressed by *diap1* coexpression, however, STRICA uniquely has been shown to interact with DIAP2 suggestive of a role distinct to that of other initiator caspases *(13)*. As no *strica* mutant has been isolated, its precise requirement in *Drosophila* development or PCD in vivo remains predictive. DRICE processing is observed in response to RPR- or chemically-induced cell death in *Drosophila* cells and immunodepletion of DRICE in insect cell lysates results in a failure to respond to death stimuli indicating that DRICE is essential for apoptosis in this context *(100)*. *drice* is transcriptionally upregulated in the larval salivary glands prior to their removal suggestive of a role in developmental cell death as well as abundant DRICE being present in the larval salivary glands and midguts prior to their destruction *(73,101)*. *damm* overexpression induces death in cultured cells, however the precise physiological function of DAMM in vivo remains unclear as it does for *decay*, the effector caspase with high sequence homology with the mammalian caspase-3 family of proteins *(102,103)*.

4.2. Mammals

Mutation of caspases in mammalian models or in humans presents a complex relationship existing between function and compensatory redundancies, an example being caspase-2. Whereas caspase-2 is widely expressed in multiple tissue types, *caspase-2* knockout (KO) mice are viable, fertile, and show no overt phenotype *(104,105)*. The redundancy occurring *in caspase-2$^{-/-}$* animals may be partly explained by the compensatory activity of caspase-9 in *caspase-2* null neuronal apoptosis following growth factor withdrawal *(106)*. The observation that female *caspase-2* null mice have an increased complement of oocytes reveals a nonredundant role specifically in the regulation of germ cell attrition however this does not negatively impact fertility *(104,107)*.

Nonredundant caspase function is more obvious in mice deficient for *caspase-3* or -9, which are embryonic or perinatally lethal with significant disruptions to nervous system cell deaths resulting in hyperplasia of multiple brain structures *(108–110)*. The brain hyperplasic foci are more severe in *caspase-9$^{-/-}$* animals and this can be attributed to the caspase-9-dependent activation of the effector caspase-7 in specific neural structures *(109,110)*. Interestingly, the *caspase-3$^{-/-}$* phenotype was murine strain dependent *(111)*. *Caspase-3* null mice have an additional enlargement of the retinal neuroepithelia indicating a nonredundant role in this tissue *(108)*. These phenotypes evidence the essential and nonredundant roles for caspase-3 and -9 in CNS development, and furthermore, indicate their respective biological functions and activation occur through similar pathways. This is confirmed by the loss of caspase-3 processing in brain tissue mutant for caspase-9 *(109)*.

Caspase-1 and *-11* deficient mice have no significant developmental or cell death defects but do exhibit defective IL-1β, IL-18, IL-1α, IL-6, TNF-α and interferon-γ processing/production *(112–116)*. *Caspase-11$^{-/-}$* mice have reduced caspase-1 activation demonstrating its position upstream of caspase-1 in the proinflammatory pathway *(117)*.

Caspase-8$^{-/-}$ mice also demonstrate nonredundant functions in diverse developmental processes *(118,119)*. Mice lacking *caspase-8* are embryonic lethal and smaller in size at death than WT animals of equivalent age, have defective myocardial development and erythrocyte hyperaemia *(118)*. Furthermore, apoptosis in response to receptor-mediated death signaling is blocked in *caspase-8$^{-/-}$* fibroblasts and hepatocytes, the latter demonstrated convincingly in vivo by hepatocyte-specific conditional *caspase-8* KO in conjunction with anti-Fas antibody injection *(118,119)*. Conditional knockout of *caspase-8* specifically in bone marrow cells leads to a block in hematopoietic progenitor functional potential *(119)*. Consequently, aspects of the physiological role of caspase-8 appear independent of TNF family-mediated signalling as no comparable phenotype is observed in animals mutant for these receptors, however, animals mutant for the caspase-8 adaptor molecule FADD do have similar phenotypes *(119)*.

5. Caspase Substrates

Known caspase substrate targets in cells currently number nearly 400 proteins involved in most cell control systems/pathways including cytoskeletal or structural components, signaling molecules, cell repair and cell-cycle molecules as well as proteins regulating transcription and cell death *(8,120)*. Cleavage of such substrates, generally carried out by the effector class of caspases, function to effect cell death by targeting specific molecules which lead to the typical apoptotic phenotypes such as DNA fragmentation, membrane blebbing, and nuclear condensation. Complicating collation of a uniform set of rules for caspase targets is the fact that some substrates are cleaved in cell- or tissue-specific contexts, dependent on relative caspase expression levels, or nonconserved substrate recognition sequences exist between species. However, given the typical features of apoptotic cell death appear conserved from *C. elegans* and *Drosophila* to mammals, caspase substrates and specificities in this process are likely to be highly conserved.

Cleavage of Gelsolin and ROCK-1 are involved in the process of membrane blebbing, the former activated by caspase-3 to effect F-actin modifications whereas caspase-mediated activation of the latter induces myosin light chain phosphorylation *(121,122)*. Multiple caspase-mediated modifications to cell-cell interaction foci (E-cadheren) *(123)* and cell adhesion complexes like adherens junctions (β- and γ-catenin) *(124)* and desmosomes (Plakophilin-1) *(125)* lead to cell rounding or detachment, prevention of cell junction communications, or the blocking of signaling networks.

Apoptotic DNA fragmentation occurs through the DNase action of caspase activated DNase (CAD), a cytoplasmic molecule kept inactive through binding of the inhibitor ICAD *(126,127)*. Upon cleavage of ICAD by caspase-3, CAD moves into the nucleus and cleaves DNA, a process responsible for the typical "laddering" of genomic DNA from dying cells when electrophoresed *(127,128)*. Similarly, many caspase targets, once cleaved, impact regulation of chromatin structural modifications *(129,130)* and disable elements of the DNA repair machinery *(131)*.

As stated, caspase cleavage of certain targets can fundamentally change their biological function, as is the case for the antiapoptotic molecules Bcl-2 and Bcl-x_L which, following cleavage of their N-terminal BH4 domains, renders them with a proapoptotic function *(132)*. Bid cleavage by caspase-8 produces a proapoptotic C-terminal fragment which induces cytochrome *c* release from the mitochondria *(133)*. Such changes interfere with the balance between pro and antiapoptotic machinery components and tip the scale toward commitment to death as well as triggering an amplification of death signaling.

6. Nonapoptotic Roles of "Apoptotic" Caspases

The concept that caspases are constitutively present in cells effectively priming them for activation by death signalling induction, has been developed to one of complex transcriptional regulation as outlined above. More recently, our understanding of the role of caspases in biology has again developed from purely a destructive, all or nothing removal of cells or tissues, to having unexpected and diverse roles in developmental processes other than cell death. Such nonapoptotic functions include regulation of blood cell maturation/proliferation, roles in oocyte maturation and *Drosophila* spermatogenesis, and most recently an involvement in cellular migration and compensatory proliferation.

In mammals, some caspases are involved in maintenance of lymphocyte homeostasis but they are also required for the proliferation of primary human T-cells, with caspase-8 specifically shown to be involved in this process *(134–136)*. Cell cycle regulators like Wee1 have also been identified as caspase cleavage targets *(134,137)*. Differentiation of blood cells during erythropoiesis and keratinocyte formation are also caspase-dependent processes, the former involving caspase-3 and -9 activation and the latter specifically involving caspase-14 *(138,139)*. Caspase-3 and Apaf-1 are required for the physiological death of granulosa cells surrounding the murine oocyte during follicle regression, however, the germ cell environment in *Drosophila* appears unique in that DIAP1 removal does not necessarily result in effector caspase activation *(140,141)*. Caspase-8 has been shown to be required for normal endothelial cell function which, in *caspase-8* mutants, leads to circulatory failures involving cardiac related abnormalities *(119)*. As well as defects in hematopoietic progenitor cell function, macrophage precursor differentiation had significant caspase-8 dependence *(119)*. *Caspase-3* KO mice show defects in skeletal muscle fibre differentiation possibly as a result of disruption of actin fiber reorganization *(142)*. Therefore a key question to be answered is what cellular mechanisms determine how a cell can utilize aspects of caspase function while preventing cell death, and how this occurs in the context of normal development and tissue formation.

The apoptotic machinery is also an integral process in the regulation of hematopoiesis. When activated by the erythropoietin (Epo) receptor, the Stat5 transcription factor enters the nucleus and transcriptionally upregulates antiapoptotic Bcl-x$_L$ to ensure survival and maturation of erythroid progenitor cells *(143,144)*. Blocking Stat5 function by dominant negative expression or KO results in anaemia and increases in apoptosis of the erythroid precursor population *(145)*. Furthermore, Epo withdrawal induces caspase-3 activation which has been shown to have Bcl-x$_L$ as a substrate, thus explaining the apoptotic response to Epo deprivation *(146–148)*. As well as caspase-3 being involved in the terminal differentiation of lens epithelial cells, caspase activation occurs during the maturation process of monocytic, megakaryocytic, and erythroid cell lineages, in particular caspase-3 and -9 *(149–151)*. *Caspase-8* deficient mice have embryonic erythroid hyperplasia, however, in humans this is not observed because of a hypothesized redundancy with caspase-10, not present in mice *(136,152)*. Blockage of caspase activity on the GATA-1 transcription factor or inhibition of caspase activity under Epo withdrawal conditions prevents erythroid maturational arrest demonstrating a role for death molecules in the regulation of blood cell maturation through the targeting of GATA-1 *(153)*. A critical role for transient activation of multiple caspases in erythroid differentiation has also been demonstrated *(154)*. In addition to the functions stated above, the proinflammatory caspases (*see* Subheading 4.2.) function primarily in activating cytokines by controlling their cleavage-mediated activation.

In *Drosophila*, the process of spermatogenesis requires extrusion of the spermatid cytoplasm and organelles, which forms a "waste bag" through a cyst-forming process termed individualisation *(155,156)*. As occurs in hematopoiesis, death signaling resulting in caspase activation in the absence of cell death is also a requirement for normal spermatid maturation *(157,158)*. Prevention of caspase activation by overexpression of IAPs or the baculoviral pan-caspase inhibitor p35, or ablation of the effector caspase DCP-1 by RNAi in spermatids disrupts the individualisation process *(157,158)*. Furthermore, reductions in DRONC and/or DARK, which normally are present at specific foci around the forming cyst, also result in aberrant individualization *(157)*. Interestingly, DRICE activation was refractory to inhibition of the upstream death pathway components in this tissue. Surprisingly, there was a complete blockage of caspase-3/DRICE-like immunoreactivity in the spermatids of cytochrome c type d (cyt-c-d) mutants (*bln¹*) resulting in individualisation complex (IC) defects, indicating that cyt-c-d is required for caspase activation specifically in the *Drosophila* testis *(158)*.

The GTPase Rac was recently shown to be required for the migration of anterior epithelial cells through the nurse cell population of the *Drosophila* egg chamber follicle to the oocyte/nurse cell interface during oogenesis *(141)*. A screen to identify suppressors of a dominant negative (RacN17) Rac-induced border cell migration defect showed that *diap1* overexpression is able to suppress this phenotype, as did mutation of *dark*. This demonstrates that inhibition of DRONC activation, either by increasing DIAP1 or by removing functional DARK protein, restores normal border cell migration. Furthermore, loss of DIAP1 function, which in other tissues causes apoptosis by deregulation of caspase activation, results in border cell migration defects but not apoptosis.

The observation that up to 60% of cells in the developing *Drosophila* wing disc can be destroyed but a normal adult wing still formed demonstrates that a compensatory mechanism to adjust cell complements exists to prevent lethality in the face of disrupted developmental programs *(159,160)*. This process was recently shown to involve the growth mitogen Wingless, whose nonautonomous proliferative action was dependent on the apical caspase DRONC *(161)*. It was also demonstrated that the DRONC-dependent proliferation signal was not acting through the effector caspases DCP-1 or DRICE, as the experiments were carried out in "death signalling activated" cells, but under "death inhibiting" conditions (i.e., HID-induced cell death signaling was initiated in cells in which the baculoviral effector caspase inhibitor p35 was also expressed), effectively blocking DCP-1 and DRICE activation. Thus, DRONC is required for sending the proliferation signal from a cell undergoing apoptosis to a neighbouring healthy cell to maintain net cell complements.

References

1. Vaux DL, Flavell RA. Apoptosis genes and autoimmunity. Curr Opin Immunol 2000; 12(6):719–724.
2. Fearnhead HO. Getting back on track, or what to do when apoptosis is de-railed: recoupling oncogenes to the apoptotic machinery. Cancer Biol Ther 2004;3(1):21–28.
3. Bickler PE, Donohoe PH. Adaptive responses of vertebrate neurons to hypoxia. J Exp Biol 2002;205(Pt 23):3579–3586.
4. Waldmeier PC, Tatton WG. Interrupting apoptosis in neurodegenerative disease: potential for effective therapy? Drug Discov Today 2004;9(5):210–218.
5. Yuan J, Yankner BA. Apoptosis in the nervous system. Nature 2000;407(6805):802–809.
6. Ellis HM, Horvitz HR. Genetic control of programmed cell death in the nematode C. elegans. Cell 1986;44(6):817–829.
7. Yuan J, Shaham S, Ledoux S, et al. The C. elegans cell death gene ced-3 encodes a protein similar to mammalian interleukin-1 beta-converting enzyme. Cell 1993;75(4):641–652.
8. Lamkanfi M, Declercq W, Kalai M, et al. Alice in caspase land. A phylogenetic analysis of caspases from worm to man. Cell Death Differ 2002;9(4):358–361.
9. Kumar S. Mechanisms mediating caspase activation in cell death. Cell Death Differ 1999; 6(11):1060–1066.
10. Shi Y. Mechanisms of caspase activation and inhibition during apoptosis. Mol Cell 2002; 9(3):459–470.
11. Nicholson DW. Caspase structure, proteolytic substrates, and function during apoptotic cell death. Cell Death Differ 1999;6(11):1028–1042.
12. Kumar S, Doumanis J. The fly caspases. Cell Death Differ 2000;7(11):1039–1044.
13. Doumanis J, Quinn L, Richardson H, et al. STRICA, a novel Drosophila melanogaster caspase with an unusual serine/threonine-rich prodomain, interacts with DIAP1 and DIAP2. Cell Death Differ 2001;8(4):387–394.
14. Shi Y. A structural view of mitochondria-mediated apoptosis. Nat Struct Biol 2001;8(5): 394–401.

15. Salvesen GS, Dixit VM. Caspase activation: the induced-proximity model. Proc Natl Acad Sci USA 1999;96(20):10,964–10,967.

16. Boatright KM, Salvesen GS. Mechanisms of caspase activation. Curr Opin Cell Biol 2003; 15(6):725–731.

17. Boatright KM, Renatus M, Scott FL, et al. A unified model for apical caspase activation. Mol Cell 2003;11(2):529–541.

18. Peter ME, Krammer PH. The CD95(APO-1/Fas) DISC and beyond. Cell Death Differ 2003;10(1):26–35.

19. Donepudi M, Mac Sweeney A, Briand C, et al. Insights into the regulatory mechanism for caspase-8 activation. Mol Cell 2003;11(2):543–549.

20. Wei MC, Lindsten T, Mootha VK, et al. tBID, a membrane-targeted death ligand, oligomerizes BAK to release cytochrome c. Genes Dev 2000;14(16):2060–2071.

21. Rodriguez J, Lazebnik Y. Caspase-9 and APAF-1 form an active holoenzyme. Genes Dev 1999;13(24):3179–3184.

22. Li P, Nijhawan D, Budihardjo I, et al. Cytochrome c and dATP-dependent formation of Apaf-1/caspase-9 complex initiates an apoptotic protease cascade. Cell 1997;91(4): 479–489.

23. Hegde R, Srinivasula SM, Zhang Z, et al. Identification of Omi/HtrA2 as a mitochondrial apoptotic serine protease that disrupts inhibitor of apoptosis protein-caspase interaction. J Biol Chem 2002;277(1):432–438.

24. Du C, Fang M, Li Y, et al. Smac, a mitochondrial protein that promotes cytochrome c-dependent caspase activation by eliminating IAP inhibition. Cell 2000;102(1):33–42.

25. van Gurp M, Festjens N, van Loo G, et al. Mitochondrial intermembrane proteins in cell death. Biochem Biophys Res Commun 2003;304(3):487–497.

26. Chinnaiyan AM, O'Rourke K, Lane BR, et al. Interaction of CED-4 with CED-3 and CED-9: a molecular framework for cell death. Science 1997;275(5303):1122–1126.

27. Chinnaiyan AM, Chaudhary D, O'Rourke K, et al. Role of CED-4 in the activation of CED-3. Nature 1997;388(6644):728–729.

28. Quinn LM, Dorstyn L, Mills K, et al. An essential role for the caspase dronc in developmentally programmed cell death in Drosophila. J Biol Chem 2000;275(51):40,416–40,424.

29. Dorstyn L, Read S, Cakouros D, et al. The role of cytochrome c in caspase activation in Drosophila melanogaster cells. J Cell Biol 2002;156(6):1089–1098.

30. Wilson R, Goyal L, Ditzel M, et al. The DIAP1 RING finger mediates ubiquitination of Dronc and is indispensable for regulating apoptosis. Nat Cell Biol 2002;4(6):445–450.

31. Hay BA. Understanding IAP function and regulation: a view from Drosophila. Cell Death Differ 2000;7(11):1045–1056.

32. Wang SL, Hawkins CJ, Yoo SJ, et al. The Drosophila caspase inhibitor DIAP1 is essential for cell survival and is negatively regulated by HID. Cell 1999;98(4):453–463.

33. Goyal L, McCall K, Agapite J, et al. Induction of apoptosis by Drosophila reaper, hid and grim through inhibition of IAP function. EMBO J 2000;19(4):589–597.

34. Chai J, Du C, Wu JW, et al. Structural and biochemical basis of apoptotic activation by Smac/DIABLO. Nature 2000;406(6798):855–862.

35. Hawkins CJ, Yoo SJ, Peterson EP, et al. The Drosophila caspase DRONC cleaves following glutamate or aspartate and is regulated by DIAP1, HID, and GRIM. J Biol Chem 2000;275(35):27,084–27,093.

36. Hengartner MO, Horvitz HR. C. elegans cell survival gene ced-9 encodes a functional homolog of the mammalian proto-oncogene bcl-2. Cell 1994;76(4):665–676.

37. Puthalakath H, Strasser A. Keeping killers on a tight leash: transcriptional and post-translational control of the pro-apoptotic activity of BH3-only proteins. Cell Death Differ 2002;9(5):505–512.

38. Wei MC, Zong WX, Cheng EH, et al. Proapoptotic BAX and BAK: a requisite gateway to mitochondrial dysfunction and death. Science 2001;292(5517):727–730.

39. Cheng EH, Wei MC, Weiler S, et al. BCL-2, BCL-X(L) sequester BH3 domain-only molecules preventing BAX- and BAK-mediated mitochondrial apoptosis. Mol Cell 2001; 8(3):705–711.

40. Lindsten T, Ross AJ, King A, et al. The combined functions of proapoptotic Bcl-2 family members bak and bax are essential for normal development of multiple tissues. Mol Cell 2000;6(6):1389–1399.

41. Zong WX, Lindsten T, Ross AJ, et al. BH3-only proteins that bind pro-survival Bcl-2 family members fail to induce apoptosis in the absence of Bax and Bak. Genes Dev 2001;15(12):1481–1486.

42. Degenhardt K, Sundararajan R, Lindsten T, et al. Bax and Bak independently promote cytochrome c release from mitochondria. J Biol Chem 2002;277(16):14,127–14,134.

43. Degenhardt K, Chen G, Lindsten T, et al. BAX and BAK mediate p53-independent suppression of tumorigenesis. Cancer Cell 2002;2(3):193–203.

44. Cuconati A, Mukherjee C, Perez D, et al. DNA damage response and MCL-1 destruction initiate apoptosis in adenovirus-infected cells. Genes Dev 2003;17(23):2922–2932.

45. Willis SN, Chen L, Dewson G, et al. Proapoptotic Bak is sequestered by Mcl-1 and Bcl-xL, but not Bcl-2, until displaced by BH3-only proteins. Genes Dev 2005;19(11):1294–1305.

46. Nijhawan D, Fang M, Traer E, et al. Elimination of Mcl-1 is required for the initiation of apoptosis following ultraviolet irradiation. Genes Dev 2003;17(12):1475–1486.

47. Chen L, Willis SN, Wei A, et al. Differential targeting of prosurvival Bcl-2 proteins by their BH3-only ligands allows complementary apoptotic function. Mol Cell 2005;17(3): 393–403.

48. Kuwana T, Bouchier-Hayes L, Chipuk JE, et al. BH3 domains of BH3-only proteins differentially regulate Bax-mediated mitochondrial membrane permeabilization both directly and indirectly. Mol Cell 2005;17(4):525–535.

49. Letai A, Bassik MC, Walensky LD, et al. Distinct BH3 domains either sensitize or activate mitochondrial apoptosis, serving as prototype cancer therapeutics. Cancer Cell 2002;2(3): 183–192.

50. Micheau O, Thome M, Schneider P, et al. The long form of FLIP is an activator of caspase-8 at the Fas death-inducing signaling complex. J Biol Chem 2002;277(47): 45,162–45,171.

51. Chang DW, Xing Z, Pan Y, et al. c-FLIP(L) is a dual function regulator for caspase-8 activation and CD95-mediated apoptosis. EMBO J 2002;21(14):3704–3714.

52. Riedl SJ, Shi Y. Molecular mechanisms of caspase regulation during apoptosis. Nat Rev Mol Cell Biol 2004;5(11):897–907.

53. Shiozaki EN, Shi Y. Caspases, IAPs and Smac/DIABLO: mechanisms from structural biology. Trends Biochem Sci 2004;29(9):486–494.

54. Martin SJ. Destabilizing influences in apoptosis: sowing the seeds of IAP destruction. Cell 2002;109(7):793–796.

55. Yan N, Wu JW, Chai J, et al. Molecular mechanisms of DrICE inhibition by DIAP1 and removal of inhibition by Reaper, Hid and Grim. Nat Struct Mol Biol 2004;11(5): 420–428.

56. Chai J, Yan N, Huh JR, et al. Molecular mechanism of Reaper-Grim-Hid-mediated suppression of DIAP1-dependent Dronc ubiquitination. Nat Struct Biol 2003;10(11):892–898.

57. Wing J, Zhou L, Schwartz L, et al. Distinct cell killing properties of the Drosophila reaper, head involution defective, and grim genes. Cell Death Differ 1999;6(2):212–213.

58. Christich A, Kauppila S, Chen P, et al. The damage-responsive Drosophila gene sickle encodes a novel IAP binding protein similar to but distinct from reaper, grim, and hid. Curr Biol 2002;12(2):137–140.

59. Hay BA, Wassarman DA, Rubin GM. Drosophila homologs of baculovirus inhibitor of apoptosis proteins function to block cell death. Cell 1995;83(7):1253–1262.

60. Muro I, Hay BA, Clem RJ. The Drosophila DIAP1 protein is required to prevent accumulation of a continuously generated, processed form of the apical caspase DRONC. J Biol Chem 2002;277(51):49,644–49,650.

61. Yin VP, Thummel CS. A balance between the diap1 death inhibitor and reaper and hid death inducers controls steroid-triggered cell death in Drosophila. Proc Natl Acad Sci USA 2004;101(21):8022–8027.

62. Bolton DL, Hahn BI, Park EA, et al. Death of CD4(+) T-cell lines caused by human immunodeficiency virus type 1 does not depend on caspases or apoptosis. J Virol 2002; 76(10):5094–5107.

63. Harlin H, Reffey SB, Duckett CS, et al. Characterization of XIAP-deficient mice. Mol Cell Biol 2001;21(10): 3604–3608.

64. Shiozaki EN, Chai J, Rigotti DJ, et al. Mechanism of XIAP-mediated inhibition of caspase-9. Mol Cell 2003;11(2):519–527.

65. Sun C, Cai M, Meadows RP, et al. NMR structure and mutagenesis of the third Bir domain of the inhibitor of apoptosis protein XIAP. J Biol Chem 2000;275(43):33,777–33,781.

66. Deveraux QL, Roy N, Stennicke HR, et al. IAPs block apoptotic events induced by caspase-8 and cytochrome c by direct inhibition of distinct caspases. EMBO J 1998;17(8): 2215–2223.

67. Chai J, Shiozaki E, Srinivasula SM, et al. Structural basis of caspase-7 inhibition by XIAP. Cell 2001;104(5):769–780.

68. Huang Y, Park YC, Rich RL, et al Structural basis of caspase inhibition by XIAP: differential roles of the linker versus the BIR domain. Cell 2001;104(5):781–790.

69. Riedl SJ, Renatus M, Schwarzenbacher R, et al. Structural basis for the inhibition of caspase-3 by XIAP. Cell 2001;104(5):791–800.

70. Verhagen AM, Ekert PG, Pakusch M, et al. Identification of DIABLO, a mammalian protein that promotes apoptosis by binding to and antagonizing IAP proteins. Cell 2000; 102(1):43–53.

71. Kumar S, Cakouros D. Transcriptional control of the core cell-death machinery. Trends Biochem Sci 2004;29(4):193–199.

72. Daish TJ, Cakouros D, Kumar S. Distinct promoter regions regulate spatial and temporal expression of the Drosophila caspase dronc. Cell Death Differ 2003;10(12):1348–1356.

73. Daish TJ, Mills K, Kumar S. Drosophila Caspase DRONC is Required for Specific Developmental Cell Death Pathways and Stress-Induced Apoptosis. Developmental Cell 2004;7.

74. Thummel CS. Files on steroids—Drosophila metamorphosis and the mechanisms of steroid hormone action. Trends Genet 1996;12(8):306–310.

75. Baehrecke EH. Steroid regulation of programmed cell death during Drosophila development. Cell Death Differ 2000;7(11):1057–1062.

76. Lee CY, Simon CR, Woodard CT, et al. Genetic mechanism for the stage- and tissue-specific regulation of steroid triggered programmed cell death in Drosophila. Dev Biol 2002;252(1):138–148.

77. Lee CY, Cooksey BA, Baehrecke EH. Steroid regulation of midgut cell death during Drosophila development. Dev Biol 2002;250(1):101–111.

78. Bodenstein D. The postembryonic development of Drosophila. In: Biology of *Drosophila*. Demerec M, ed. Hafner Publishing Company: New York, 1965, p. 275–267.

79. Robertson CW. The metamorphosis of Drosophila melanogaster, including an accurately timed account of the principle morphological changes. J Morphol 1936;59:351–399.

80. Riddiford LM. Hormone receptors and the regulation of insect metamorphosis. Receptor 1993;3(3):203–209.

81. Lee CY, Wendel DP, Reid P, et al. E93 directs steroid-triggered programmed cell death in Drosophila. Mol Cell 2000;6(2):433–443.

82. Talbot WS, Swyryd EA, Hogness DS. Drosophila tissues with different metamorphic responses to ecdysone express different ecdysone receptor isoforms. Cell 1993;73(7):1323–1337.

83. Hall BL, Thummel CS. The RXR homolog ultraspiracle is an essential component of the Drosophila ecdysone receptor. Development 1998;125(23):4709–4717.

84. DiBello PR, Withers DA, Bayer CA, et al. The Drosophila Broad-Complex encodes a family of related proteins containing zinc fingers. Genetics 1991;129(2):385–397.

85. Burtis KC, Thummel CS, Jones CW, et al. The Drosophila 74EF early puff contains E74, a complex ecdysone-inducible gene that encodes two ets-related proteins. Cell 1990; 61(1):85–99.

86. Segraves WA, Hogness DS. The E75 ecdysone-inducible gene responsible for the 75B early puff in Drosophila encodes two new members of the steroid receptor superfamily. Genes Dev 1990;4(2):204–219.

87. Jiang C, Lamblin AF, Steller H, et al. A steroid-triggered transcriptional hierarchy controls salivary gland cell death during Drosophila metamorphosis. Mol Cell 2000;5(3):445–455.

88. Broadus J, McCabe JR, Endrizzi B, et al. The Drosophila beta FTZ-F1 orphan nuclear receptor provides competence for stage-specific responses to the steroid hormone ecdysone. Mol Cell 1999;3(2):143–149.

89. Lee CY, Clough EA, Yellon P, et al. Genome-wide analyses of steroid- and radiation-triggered programmed cell death in Drosophila. Curr Biol 2003;13(4):350–357.

90. Lee CY, Baehrecke EH. Genetic regulation of programmed cell death in Drosophila. Cell Res 2000;10(3):193–204.

91. Restifo LL, White K. Mutations in a steroid hormone-regulated gene disrupt the metamorphosis of the central nervous system in Drosophila. Dev Biol 1991;148(1):174–194.

92. Jiang C, Baehrecke EH, Thummel CS. Steroid regulated programmed cell death during Drosophila metamorphosis. Development 1997;124(22):4673–4683.

93. Stergiou L, Hengartner MO. Death and more: DNA damage response pathways in the nematode *C. elegans*. Cell Death Differ 2004;11(1):21–28.

94. Chew SK, Akdemir F, Chen P, et al. The Apical Caspase dronc Governs Programmed and Unprogrammed Cell Death in Drosophila. Developmental Cell 2004;7(December): 1–20.

95. Xu D, Li Y, Arcaro M, et al. The CARD-carrying caspase Dronc is essential for most, but not all, developmental cell death in Drosophila. Development 2005;132(9):2125–2134.

96. Laundrie B, Peterson JS, Baum JS, et al. Germline cell death is inhibited by P-element insertions disrupting the dcp-1/pita nested gene pair in *Drosophila*. Genetics 2003; 165(4):1881–1888.

97. Elrod-Erickson M, Mishra S, Schneider D. Interactions between the cellular and humoral immune responses in Drosophila. Curr Biol 2000;10(13):781–784.

98. Leulier F, Rodriguez A, Khush RS, et al. The Drosophila caspase Dredd is required to resist gram-negative bacterial infection. EMBO Rep 2000;1(4):353–358.

99. Doumanis J. Characterisation of a novel caspase STRICA and the Bcl-2 homologues BUFFY and DEBCL in Drosophila melanogaster., in PhD Thesis. 2004.

100. Fraser AG, McCarthy NJ, Evan GI. drICE is an essential caspase required for apoptotic activity in Drosophila cells. EMBO J 1997;16(20):6192–6199.

101. Martin DN, Baehrecke EH. Caspases function in autophagic programmed cell death in Drosophila. Development 2004;131(2):275–284.

102. Dorstyn L, Read SH, Quinn LM, et al. DECAY, a novel Drosophila caspase related to mammalian caspase-3 and caspase-7. J Biol Chem 1999;274(43):30,778–30,783.

103. Harvey NL, Daish T, Mills K, et al. Characterization of the Drosophila caspase, DAMM. J Biol Chem 2001;276(27):25,342–25,350.

104. Bergeron L, Perez GI, Macdonald G, et al. Defects in regulation of apoptosis in caspase-2-deficient mice. Genes Dev 1998;12(9):1304–1314.

105. O'Reilly LA, Ekert P, Harvey N, et al. Caspase-2 is not required for thymocyte or neuronal apoptosis even though cleavage of caspase-2 is dependent on both Apaf-1 and caspase-9. Cell Death Differ 2002;9(8):832–841.

106. Troy CM, Rabacchi SA, Hohl JB, et al. Death in the balance: alternative participation of the caspase-2 and -9 pathways in neuronal death induced by nerve growth factor deprivation. J Neurosci 2001;21(14):5007–5016.

107. Morita Y, Maravei DV, Bergeron L, et al. Caspase-2 deficiency prevents programmed germ cell death resulting from cytokine insufficiency but not meiotic defects caused by loss of ataxia telangiectasia-mutated (Atm) gene function. Cell Death Differ 2001;8(6):614–620.

108. Kuida K, Zheng TS, Na S, et al. Decreased apoptosis in the brain and premature lethality in CPP32-deficient mice. Nature 1996;384(6607):368–372.

109. Kuida K, Haydar TF, Kuan CY, et al. Reduced apoptosis and cytochrome c-mediated caspase activation in mice lacking caspase 9. Cell 1998;94(3):325–337.

110. Woo M, Hakem R, Soengas MS, et al. Essential contribution of caspase 3/CPP32 to apoptosis and its associated nuclear changes. Genes Dev 1998;12(6):806–819.

111. Leonard JR, Klocke BJ, D'Sa C, et al. Strain-dependent neurodevelopmental abnormalities in caspase-3-deficient mice. J Neuropathol Exp Neurol 2002;61(8):673–677.

112. Kuida K, Lippke JA, Ku G, et al. Altered cytokine export and apoptosis in mice deficient in interleukin-1 beta converting enzyme. Science 1995;267(5206):2000–2003.

113. Gu Y, Kuida K, Tsutsui H, et al. Activation of interferon-gamma inducing factor mediated by interleukin-1beta converting enzyme. Science 1997;275(5297):206–209.

114. Fantuzzi G, Puren AJ, Harding MW, et al. Interleukin-18 regulation of interferon gamma production and cell proliferation as shown in interleukin-1beta-converting enzyme (caspase-1)-deficient mice. Blood 1998;91(6):2118–2125.

115. Ghayur T, Banerjee S, Hugunin M, et al. Caspase-1 processes IFN-gamma-inducing factor and regulates LPS-induced IFN-gamma production. Nature 1997;386(6625):619–623.

116. Li P, Allen H, Banerjee S, et al. Mice deficient in IL-1 beta-converting enzyme are defective in production of mature IL-1 beta and resistant to endotoxic shock. Cell 1995; 80(3):401–411.

117. Wang S, Miura M, Jung YK, et al. Murine caspase-11, an ICE-interacting protease, is essential for the activation of ICE. Cell 1998;92(4):501–509.

118. Varfolomeev EE, Schuchmann M, Luria V, et al. Targeted disruption of the mouse Caspase 8 gene ablates cell death induction by the TNF receptors, Fas/Apo1, and DR3 and is lethal prenatally. Immunity 1998;9(2):267–276.

119. Kang TB, Ben-Moshe T, Varfolomeev EE, et al. Caspase-8 serves both apoptotic and nonapoptotic roles. J Immunol 2004;173(5):2976–2984.

120. Fischer U, Janicke RU, Schulze-Osthoff K. Many cuts to ruin: a comprehensive update of caspase substrates. Cell Death Differ 2003;10(1):76–100.

121. Kothakota S, Azuma T, Reinhard C, et al. Caspase-3-generated fragment of gelsolin: effector of morphological change in apoptosis. Science 1997;278(5336):294–298.

122. Coleman ML, Sahai EA, Yeo M, et al. Membrane blebbing during apoptosis results from caspase-mediated activation of ROCK I. Nat Cell Biol 2001;3(4):339–345.

123. Steinhusen U, Weiske J, Badock V, et al. Cleavage and shedding of E-cadherin after induction of apoptosis. J Biol Chem 2001;276(7):4972–4980.

124. Brancolini C, Sgorbissa A, Schneider C. Proteolytic processing of the adherens junctions components beta-catenin and gamma-catenin/plakoglobin during apoptosis. Cell Death Differ 1998;5(12):1042–1050.

125. Weiske J, Schoneberg T, Schroder W, et al. The fate of desmosomal proteins in apoptotic cells. J Biol Chem 2001;276(44):41,175–41,181.

126. Enari M, Sakahira H, Yokoyama H, et al. A caspase-activated DNase that degrades DNA during apoptosis, and its inhibitor ICAD. Nature 1998;391(6662):43–50.

127. Sakahira H, Enari M, Nagata S. Cleavage of CAD inhibitor in CAD activation and DNA degradation during apoptosis. Nature 1998;391(6662):96–99.

128. Liu X, Zou H, Slaughter C, et al. DFF, a heterodimeric protein that functions downstream of caspase-3 to trigger DNA fragmentation during apoptosis. Cell 1997;89(2):175–184.

129. Sahara S, Aoto M, Eguchi Y, et al. Acinus is a caspase-3-activated protein required for apoptotic chromatin condensation. Nature 1999;401(6749):168–173.

130. Kovacsovics M, Martinon F, Micheau O, et al. Overexpression of Helicard, a CARD-containing helicase cleaved during apoptosis, accelerates DNA degradation. Curr Biol 2002;12(10):838–843.

131. Lazebnik YA, Kaufmann SH, Desnoyers S, et al. Cleavage of poly(ADP-ribose) polymerase by a proteinase with properties like ICE. Nature 1994;371(6495):346–347.

132. Clem RJ, Cheng EH, Karp CL, et al. Modulation of cell death by Bcl-XL through caspase interaction. Proc Natl Acad Sci USA 1998;95(2):554–559.

133. Li H, Zhu H, Xu CJ, et al. Cleavage of BID by caspase 8 mediates the mitochondrial damage in the Fas pathway of apoptosis. Cell 1998;94(4):491–501.

134. Alam A, Cohen LY, Aouad S, et al. Early activation of caspases during T lymphocyte stimulation results in selective substrate cleavage in nonapoptotic cells. J Exp Med 1999;190(12):1879–1890.

135. Kennedy NJ, Kataoka T, Tschopp J, et al. Caspase activation is required for T cell proliferation. J Exp Med 1999;190(12):1891–1896.

136. Chun HJ, Zheng L, Ahmad M, et al. Pleiotropic defects in lymphocyte activation caused by caspase-8 mutations lead to human immunodeficiency. Nature 2002;419(6905):395–399.

137. Zhou BB, Li H, Yuan J, et al. Caspase-dependent activation of cyclin-dependent kinases during Fas-induced apoptosis in Jurkat cells. Proc Natl Acad Sci USA 1998;95(12): 6785–6790.

138. Schwerk C, Schulze-Osthoff K. Non-apoptotic functions of caspases in cellular proliferation and differentiation. Biochem Pharmacol 2003;66(8):1453–1458.

139. Eckhart L, Declercq W, Ban J, et al. Terminal differentiation of human keratinocytes and stratum corneum formation is associated with caspase-14 activation. J Invest Dermatol 2000;115(6):1148–1151.

140. Robles R, Tao XJ, Trbovich AM, et al. Localization, regulation and possible consequences of apoptotic protease-activating factor-1 (Apaf-1) expression in granulosa cells of the mouse ovary. Endocrinology 1999;140(6):2641–2644.

141. Geisbrecht ER, Montell DJ. A role for Drosophila IAP1-mediated caspase inhibition in Rac-dependent cell migration. Cell 2004;118(1):111–125.

142. Fernando P, Kelly JF, Balazsi K, et al. Caspase 3 activity is required for skeletal muscle differentiation. Proc Natl Acad Sci USA 2002;99(17):11,025–11,030.

143. Socolovsky M, Fallon AE, Wang S, et al. Fetal anemia and apoptosis of red cell progenitors in Stat5a-/-5b-/- mice: a direct role for Stat5 in Bcl-X(L) induction. Cell 1999; 98(2):181–191.

144. Nosaka T, Kawashima T, Misawa K, et al. STAT5 as a molecular regulator of proliferation, differentiation and apoptosis in hematopoietic cells. EMBO J 1999;18(17):4754–4765.

145. Chida D, Miura O, Yoshimura A, et al. Role of cytokine signaling molecules in erythroid differentiation of mouse fetal liver hematopoietic cells: functional analysis of signaling molecules by retrovirus-mediated expression. Blood 1999;93(5):1567–1578.

146. Gregoli PA, Bondurant MC. The roles of Bcl-X(L) and apopain in the control of erythropoiesis by erythropoietin. Blood 1997;90(2):630–640.

147. Gregoli PA, Bondurant MC. Function of caspases in regulating apoptosis caused by erythropoietin deprivation in erythroid progenitors. J Cell Physiol 1999;178(2):133–143.

148. Negoro S, Oh H, Tone E, et al. Glycoprotein 130 regulates cardiac myocyte survival in doxorubicin-induced apoptosis through phosphatidylinositol 3-kinase/Akt phosphorylation and Bcl-xL/caspase-3 interaction. Circulation 2001;103(4):555–561.

149. Sordet O, Rebe C, Plenchette S, et al. Specific involvement of caspases in the differentiation of monocytes into macrophages. Blood 2002;100(13):4446–4453.

150. De Botton S, Sabri S, Daugas E, et al. Platelet formation is the consequence of caspase activation within megakaryocytes. Blood 2002;100(4):1310–1317.

151. Ishizaki Y, Jacobson MD, Raff MC. A role for caspases in lens fiber differentiation. J Cell Biol 1998;140(1):153–158.

152. Testa U. Apoptotic mechanisms in the control of erythropoiesis. Leukemia 2004;18(7): 1176–1199.

153. De Maria R, Zeuner A, Eramo A, et al. Negative regulation of erythropoiesis by caspase-mediated cleavage of GATA-1. Nature 1999;401(6752):489–493.

154. Zermati Y, Garrido C, Amsellem S, et al. Caspase activation is required for terminal erythroid differentiation. J Exp Med 2001;193(2):247–254.

155. Fabrizio JJ, Hime G, Lemmon SK, et al. Genetic dissection of sperm individualization in Drosophila melanogaster. Development 1998;125(10):1833–1843.

156. Fuller MT. Spermatogenesis in Drosophila. In: The Development of *Drosophila melanogaster*. Bate M, Arias AM, eds., New York: Cold Spring Harbor Laboratory Press, 1993, p. 71–147.

157. Huh JR, Vernooy SY, Yu H, et al. Multiple apoptotic caspase cascades are required in nonapoptotic roles for Drosophila spermatid individualization. PLoS Biol 2004;2(1):E15.

158. Arama E, Agapite J, Steller H. Caspase activity and a specific cytochrome C are required for sperm differentiation in Drosophila. Dev Cell 2003;4(5):687–697.

159. Haynie JL, Bryant PJ. The effects of X-rays on the proliferation dynamics of cells in the imaginal wing disc of Drosophila melanogaster. Wilhelm Roux's Archives of Developmental Biology, 1977;183:85–100.

160. Milan M, Campuzano S, Garcia-Bellido A. Developmental parameters of cell death in the wing disc of Drosophila. Proc Natl Acad Sci USA 1997;94(11):5691–5696.

161. Huh JR, Guo M, Hay BA. Compensatory proliferation induced by cell death in the Drosophila wing disc requires activity of the apical cell death caspase Dronc in a nonapoptotic role. Curr Biol 2004;14(14):1262–1266.

18

Apoptosis and Human Diseases: Molecular Mechanisms
Oxidative Stress in the Pathogenesis of Diabetic Neuropathy

Mahdieh Sadidi, Ann Marie Sastry, Christian M. Lastoskie, Andrea M. Vincent, Kelli A. Sullivan, and Eva L. Feldman

Summary

Diabetes is a world wide epidemic with a growing rate of incidence in the United States. Diabetes is characterized by chronic hyperglycemia, leading to oxidative stress and diabetic complications such as diabetic neuropathy. The role of oxidative stress in diabetes is pervasive and complex. Oxidative stress stimulated by diabetic conditions is involved in many pathways that end in cell damage and programmed cell death. Considering the importance of oxidative stress in diabetic complications, it is important to explore antioxidant therapies as ways in which to protect cells. In this chapter, the conditions that lead to oxidative stress, the effects of such stress, and the importance of antioxidants as a major potential treatment of diabetic neuropathy will be explored in detail.

Key Words: Reactive oxygen species (ROS); reactive nitrogen species (RNS); mitochondria; apoptosis; advanced glycosylation end product (AGE); polyol pathway; antioxidant.

1. Introduction

Diabetes is an epidemic in the developed world. Approximately twenty million Americans are diabetic and the prevalence is increasing at a rate of 5%/yr. Diabetes is characterized by chronic hyperglycemia that produces dysregulation of cellular metabolism. Hyperglycemia-induced oxidative stress is strongly linked to peripheral neuropathy (DN), the most common diabetic complication, affecting approx 60% of all diabetic patients *(1,2)*.

Hyperglycemia results in the increased production of reactive oxygen and nitrogen species (ROS/RNS). An imbalance in the oxidant/antioxidant system, either resulting from excessive ROS/RNS production and/or antioxidant system impairment, leads to oxidative stress. Oxidative stress plays a major role in the pathogenesis of diabetic complications *(3–8)*. Over time and unchecked, ROS/RNS-induced damage impairs nerve function, resulting in peripheral and autonomic nervous system injury *(9–13)*.

Oxidative stress damages lipids and proteins, resulting in protein modification, lipid peroxidation, DNA fragmentation, and cell death. ROS/RNS-induced modification of macromolecules creates a footprint of toxicity. These footprint "biomarkers" are used to assess disease state and investigate effects of antioxidants on disease progression and therapeutic intervention. Biomarkers are important not only to identify the source of the oxidative/nitrosative stress but also to determine the extent of the oxidative/nitrosative insult.

From: *Apoptosis, Cell Signaling, and Human Diseases: Molecular Mechanisms, Volume 2*
Edited by R. Srivastava © Humana Press Inc., Totowa, NJ

The central role that oxidative stress plays in the development and progression of diabetic microvascular complications, including neuropathy, suggests the importance of antioxidant therapies for their prevention and control. Antioxidants affect reactive species levels and, as a result, control their influence on cellular function. Consequently, research to understand the mechanisms of oxidative stress and antioxidant therapies in the context of diabetic neuropathy is tremendously important.

2. The Chemistry of Oxidative Stress

There are two major categories of reactive species: the reactive oxygen species (ROS) including oxygen radicals such as superoxide (O_2^-) and the hydroxyl radical (·OH), and nonradical derivatives of molecular oxygen such as hydrogen peroxide (H_2O_2) and the reactive nitrogen species (RNS), including nitric oxide (NO·) and nitrogen dioxide (NO_2·) radicals, and peroxynitrite ($ONOO^-$), dinitrogen trioxide (N_2O_3), dinitrogen tetroxide (N_2O_4), and nitrous acid (HNO_2). Both ROS and RNS are products of cellular metabolism and are involved in signaling pathways under normal physiological conditions *(14)*. Activity of these species ranges from highly reactive OH that quickly reacts with broad spectrum of molecules, to H_2O_2, NO·, and O_2^- that react selectively with cellular macromolecules; to moderately reactive species such as NO_2·, and $ONOO^-$.

The concentration of cellular ROS/RNS is tightly regulated by the natural detoxification processes of the cell. When this control mechanism is disturbed, an imbalance between oxidant and antioxidant systems develops that leads to the modification and damage of cellular proteins, DNA, and lipids. This damage impairs cell function and eventually results in cell death. Below we will explore in more detail each of the major oxidative elements and how they function in the cell.

2.1. Superoxide (O_2^-)

Molecular oxygen, having two unpaired electrons with similar spin in the outer shell (also called triplet state) is not very reactive, but if one of the electrons is excited (singlet state), it becomes a highly reactive species. One-electron reduction of oxygen results in O_2^- production; a relatively stable intermediate. O_2^- is produced enzymatically and nonenzymatically in vivo. Examples of enzymatic sources of O_2^- production include NADPH oxidase[15] and P450-dependant oxygenases *(16)*. The nonenzymatic source of O_2^- is the direct transfer of a single electron to oxygen by reduced enzymes or prosthetic groups such as flavoproteins, iron-sulfur clusters, and ubi-semiquinone. In the mitochondria, the transfer of an electron to oxygen by the respiratory chain redox centers is the source of O_2^- *(17)*. This will be discussed in more detail in section IVe.

O_2^- is involved in the regulation of vascular function, cell division *(18,19)*, apoptosis *(20)*, inflammation *(21)*, and the antibacterial activity of neutrophils *(22)*. O_2^- is associated with many important physiological functions, such as smooth muscle relaxation, signal transduction, control of ventilation, enhancement of immunological functions, control of erythropoietin production, and other hypoxia-inducible functions *(23)*.

The intracellular concentration of O_2^- is tightly regulated in vivo. Dismutation of excessive O_2^- by superoxide dismutase (SOD) produces H_2O_2, which is reduced by catalase or glutathione peroxidase (Gpx) to H_2O. In this process, catalase reduces two molecules of H_2O_2 to H_2O and O_2. However, Gpx reduction of H_2O_2 is coupled to the oxidation of two molecules of glutathione (GSH) to (GSSG).

2.2. Hydrogen Peroxide (H_2O_2)

H_2O_2 is formed by both the dismutation of $O_2^{.-}$ in a reaction catalyzed by SOD and the activity of oxidase enzymes such as monoamine oxidases *(24)*. In the first case, SOD is present in organelles where $O_2^{.-}$ is formed (e.g., in mitochondria), the resulting H_2O_2 diffuses across intracellular membranes and through the cytosol, whereas the $O_2^{.-}$ remains at the original site of production. It is likely that most human cells and tissues are exposed to H_2O_2, but mitochondria have a higher than average amount of H_2O_2 in vitro and possibly in vivo *(25)*. H_2O_2 on its own is poorly reactive and cannot oxidize lipids, DNA, or proteins easily, however, it reacts well with proteins that contain highly reactive thiols or methionine residues *(26,27)*. H_2O_2 also reacts with ascorbate, heme proteins, and protein thiols *(28)*.

The redox status of cells regulates signal transduction, cell proliferation, apoptosis, and necrosis. H_2O_2, being a mild oxidant is diffusible across cellular compartments, and plays a major role in the redox-regulated processes of cells. There is growing evidence to support the role of H_2O_2 as an inter- and intracellular signaling molecule *(29–33)*. H_2O_2 is a second messenger in the activation of NF-κB in some cell types *(32)*. The inflammatory process is also modulated by H_2O_2, generated by activated phagocytes *(34–36)*.

H_2O_2 concentration in cells is controlled by a series of antioxidants such as glutathione peroxidase (Gpx), catalase and thioredoxin (Trx), which convert H_2O_2 to H_2O. H_2O_2 is removed by Gpx, in a process that involves an oxidation-reduction cycle of the active center, using GSH as the reductant.

$$H_2O_2 + Gpx_{red} + H^+ \longrightarrow Gpx_{ox} + H_2O$$

$$Gpx_{ox} + GSH \longrightarrow GS\text{-}Gpx + H_2O$$

$$GS\text{-}Gpx + \longrightarrow GSSG + Gpx_{red} + H^+$$

Removal of H_2O_2 by catalase occurs in two steps:

$$\text{Catalase-Fe(II)} + H_2O_2 \longrightarrow \text{Compound (I)} + H_2O$$

$$\text{Compound (I)} + H_2O_2 \longrightarrow \text{Catalase Fe (III)} + H_2O + O_2$$

In this process two molecules of H_2O_2 are converted to H_2O.

H_2O_2, in the presence of metals (ferrous iron, Fe^{+2}), is converted to the hydroxyl radical ($\cdot OH$) via the Fenton reaction.

$$Fe^{+2} + H_2O_2 \longrightarrow Fe^{+3} + \cdot OH + {}^-OH$$

Hydroxyl radicals are highly reactive species and cause DNA damage, lipid peroxidation, and protein modification *(37)*.

2.3. Nitric Oxide (NO)

NO is generated from arginine by nitric oxide synthase (NOS) in the presence of molecular oxygen, tetrahydrobiopterin (BH_4), and other cofactors. NO is an important messenger molecule in inflammation, blood flow, and neurotransmission *(38)*. Despite the natural physiological roles played by NO, the overproduction of NO is implicated in the pathogenesis of neurological disorders.

NO is a free radical with a short half-life as a result of its high reactivity with other reactants such as $O_2\cdot^-$. NO and $O_2\cdot^-$ are also thought to be mediators of tissue injury. High levels of NO are generated in tissues upon invasion by activated neutrophils and macrophages *(39)*. NO is highly reactive toward $O_2\cdot^-$, and can even out-compete SOD for superoxide *(40,41)*. The chemical reaction between NO and $O_2\cdot^-$ results in the formation of $ONOO^-$, a potent cytotoxin. There is increasing evidence that NO and $ONOO^-$ inhibit complex I of the mitochondrial respiratory chain, leading to energy depletion and cell death *(42)*. The effect of $ONOO^-$ on macromolecules is discussed in depth in the next section. Understanding of the physiological chemistry of $NO/O_2\cdot^-$-dependent reactions is important, and could help in the design of new therapeutic strategies for treating or preventing diseases caused by oxidative stress.

2.3.1. ONOO⁻

Peroxynitrite ($ONOO^-$), is formed in the chemical reaction between nitric oxide (NO) and superoxide ($O_2\cdot^-$), and is best known for its ability to nitrate free tyrosine and tyrosine residues in proteins. As a major mediator of oxidative stress, $ONOO^-$ is gaining attention in the neuroscience community. In fact, measures of nitrotyrosine immuno-reactivity are used increasingly as evidence of $ONOO^-$ production and participation in various neurotoxic conditions and neurodegenerative diseases *(43,44)*.

Exposure of purified DNA to $ONOO^-$ causes base modification and strand breaks *(45)*. $ONOO^-$ is also capable of oxidizing amino acids such as cysteine, methionine, and tryptophan as well as causing tyrosine nitration *(46)*. The presence of $ONOO^-$ leads to the formation of dityrosine in proteins and causes structural distortion and protein aggregation *(47)*.

Several groups of investigators proposed that the oxidative and cytotoxic potential of $ONOO^-$ is initiated by free radical intermediates from the decomposition of $ONOO^-$ *(48)*. The decomposition of $ONOO^-$ results in the formation of $\cdot NO_2$, $\cdot OH$, and activated trans-$ONOO^-$ ONOOH*, all of which oxidize target molecules. Possible routes of $ONOO^-$ formation and decomposition are shown below.

$$\cdot NO + O_2\cdot^- \longrightarrow ONOO^-$$

$$ONOO^- + H^+ \longleftrightarrow \text{cis-} ONOOH$$

$$\text{cis-}ONOOH \longleftrightarrow \text{trans-}ONOOH*$$

$$\text{trans-}ONOOH* \longleftrightarrow [NO_2][{}^-OH]$$

The reaction of both $ONOO^-$ and NO_2 with tyrosine results in 3-nitrotyrosine and dityrosine formation *(28,49,50)*. The nitration and dimerization of tyrosine residues by $ONOO^-$ suggests a radical mechanism in which a tyrosyl radical combines with NO_2 to form 3-nitrotyrosine *(51,52)*.

2.3.2. NO₂

In addition to nitration by $ONOO^-$, several other pathways result in protein tyrosine nitration. Eiserich et al. reported that nitration of tyrosine may occur via the peroxidase-catalyzed H_2O_2-dependent oxidation of NO_2^-, independent of the intermediate $ONOO^-$ *(53)*. One pathway that leads to the nitration of tyrosine involves heme-containing peroxidases. These peroxidase enzymes *(54)* catalyze the H_2O_2-mediated oxidation of NO_2^-. It is thought that NO_2^- at its physiological concentrations is oxidized by the Fenton reaction to NO_2. In this process, hydroxyl radicals ($OH\cdot$), formed from Fe^{+2}-dependent decomposition of H_2O_2, oxidize NO_2^- to NO_2.

The oxidation of NO_2^- to NO_2, in the presence of H_2O_2 by horseradish peroxidase (HRP) occurs according to the following reactions:

$$HRP + H_2O_2 \longrightarrow Compound\ I + H_2O$$

$$Compound\ I + NO_2^- \longrightarrow Compound\ II + NO_2$$

$$Compound\ II + NO_2^- \longrightarrow HRP + NO_2$$

Hydrogen peroxide converts the resting-state ferric heme of peroxidases to the hypervalent compound I. Compound I oxidizes NO_2^- to give NO_2 and compound II. Finally, compound II is reduced by another NO_2^- molecule to regenerate the resting state (HRP) and the second NO_2 molecule.

Peroxidases also cause the oxidation of tyrosine residues and result in tyrosyl radical formation. Intermediate tyrosyl radicals undergo rapid reactions resulting in dityrosine formation.

$$Tyr\cdot + Tyr \longrightarrow Tyr\text{-}Tyr$$

The nitrating species nitrate tyrosyl residues in proteins to form nitrotyrosine.

$$Tyr\cdot + NO_2 \longrightarrow Tyr\text{-}NO_2$$

Tyrosyl radicals in proteins are more long-lived than free tyrosyl radicals and may contribute to the nitration of proteins *(49)*. Reaction of NO_2 with tyrosine residues, similar to $ONOO^-$, results in 3-nitrotyrosine and dityrosine formation *(28,49,50,55)*.

Nitration of tyrosine residues with NO_2 by HRP through the Fenton reaction suggests a major role for the involvement of metals under pathological conditions. Increased iron stores are linked to the development of type 2 diabetes *(56–59)*, whereas iron depletion is reported to be protective *(59,60)*. The generation of NO_2 by the Fenton reaction is important in the pathogenesis of neurodegenerative disorders such as Alzheimer's disease *(61)*. Similarly, increases in the iron content of Parkinson's patients are considered as an alternative source of free radical generation and neurodegeneration.

Tyrosine nitration may contribute to cell and tissue injury during overproduction of NO and under oxidative/nitrosative stress. The possibility of the involvement of any species capable of Fenton-type chemistry, such as metal ions, peroxidases, and free heme, provides broader alternatives to $ONOO^-$ pathways for macromolecular modifications under pathological conditions.

2.4. The Products of RNS/ROS and GSH Interactions: Posttranslational Modification of Proteins

Intracellular antioxidants, including low molecular weight thiols such as glutathione and cysteine are scavengers of oxidizing species. These biological antioxidants are natural molecules, that prevent the uncontrolled production of reactive species, and protect biological structures from the deleterious effects of free radicals *(62)*. Low-molecular-weight thiol pools are important in preventing oxidant-mediated cell injury and in regulation of redox signaling cascades *(63)*. The participation of low-molecular-weight thiols in such events results from their ability to react with oxidized thiols of protein through *S*-thiolation *(64)*. Decreases in GSH pools lead to cellular damage by depriving cells of their antioxidant properties.

In biological systems, reactions involving NO occur through competing reactions with O_2^{-}, molecular oxygen, transition metals, and thiols *(65,66)*. The discovery of NO as a messenger for a variety of cellular functions led to considerable attention to the biological chemistry of *S*-nitrosothiols. It is thought that *S*-nitrosothiols are reservoirs of NO in both intracellular and extracellular spaces. Reactions of different reactive nitrogen species (RNS) such as $ONOO^-$, NO_2, and dinitrogen trioxide (N_2O_3) with proteins yield nitrosyl-metal complexes, nitrosothiols, and nitrothiols, that affect protein function. The mechanisms involved in the posttranslational modifications of proteins are discussed in more detail in the following sections.

NO and other RNS react with a small molecular thiol antioxidant such as GSH and cysteine, to yield *S*-nitrosothiols such as *S*-nitrosocysteine (CysNO) and *S*-nitrosoglutathione (GSNO) as well as nitrothiols such as $CysNO_2$ and $GSNO_2$ *(67)*. Interaction of these *S*-nitrosothiols with proteins results in protein *S*-nitrosation.

$$\text{GSNO + Protein-SH} \rightarrow \text{Protein-SNO + GSH}$$

S-nitrosothiols are found in a number of tissues and have been identified in the regulation of cellular functions through their ability to cause the oxidation of critical cysteine residues in target proteins *(41,68)*. *S*-nitrosylation of glyceraldehyde-3-phosphate dehydrogenase (GAPDH) by NO regulates apoptosis *(69)*. *S*-nitrosylation of proteins in mitochondria is reported by Stamler et al. *(70)*. In a similar study, *S*-nitrosylation of proteins such as catalase, malate dehydrogenase, sarcosine dehydrogenase was identified by mass spectrometry.

It is also known that reactions of *S*-nitrosylated species with glutathione results in *S*-glutathiolation. This is emerging as a novel mechanism by which GSNO modifies functionally important protein thiols. Several mechanisms are proposed for protein S-thiolation including the following:

1. Nucleophilic attack of protein thiolate (S^-) on *S*-nitrosylated compounds (e.g., CysNO, GSNO), which lead to mixed disulfide formation.

$$\text{Pro-S}^- + \text{GSNO} \longrightarrow \text{Pro-SS-G + HNO}$$

2. Oxidation of protein sulfhydryls or thiol-containing compounds by oxyradicals (e.g., H2O2) to form sulfenic acid (Pro-SOH, CysOH, GSOH) and/or thiyl radicals (e.g., ProS, Cys, GS·) and their subsequent interactions with sulfhydryl containing species results in disulfide formation (49).

$$\text{Pro-SOH +} \longrightarrow \text{Pro-SS-G + H}_2\text{O}$$

$$\text{GSOH + Pro-SH} \longrightarrow \text{Pro-SS-G + H}_2\text{O}$$

$$\text{Pro-S· + GSH} \longrightarrow \text{Pro-SS-G}^{·-}$$

$$\text{Pro-SS-G}^{·-} \quad \underset{O_2^{·-}}{\overset{O_2}{\rightsquigarrow}} \quad \text{Pro-SS-G}$$

Thiol-disulfide exchange between oxidized low molecular weight thiols (e.g., GSSG) and protein thiols*(71)* cause protein *S*-glutathiolation.

$$\text{Pro-SH} + \text{GSSG} \longrightarrow \text{Pro-SS-G} + \text{GSH.}$$

Nitrosation of protein thiols by S-nitrosylated compound through transnitrosylation, followed by interaction with GSH to produce mixed disulfides *(72)*.

$$N_2O_3 + \text{Pro-SH} \longrightarrow \text{Pro-SNO}$$

$$\text{Pro-SNO} + \text{GSH} \longrightarrow \text{Pro-SS-G} + \text{HNO}$$

S-glutathiolation by GSNO was suggested for the first time by Park et al. on the NO/GSH modification of yeast alcohol dehydrogenase *(73)*. This was further supported by the work on NO/GSH dependent S-gluthiolation and inhibition of aldose reductase by GSNO through mixed-disulfide formation *(74)*. S-glutathiolation of carbonic anhydrase *(75)*, GAPDH *(76)*, and cysteine protease caspase-3 *(77)* are also reported. Klat et al. demonstrated that the transcription factors c-Jun and p50 bind to GSNO through mixed-disulfide formation *(78)*. S-glutathiolation of c-jun and p50 involves a cysteine residue that is located in the DNA-binding domains of these transcriptional factors.

Considering the generation of ROS and RNS in the cell, the production of S-nitrosothiol and the subsequent S-glutathiolation of proteins may be a mechanism by which proteins are modified. The following are a few examples of modified proteins that may be used as footprints of oxidative stress. Inactivation and tyrosine niotration of succinyl-CoA: 3-oxoacid CoA-transferase (SCOT) in hearts from streptozotocin-treated rats is reported by Turko et al. *(79)* Inactivation of caspases by S-nitrosylation on their essential SH-group prevents apoptosis, and the removal of S-nitroso group activates the caspase, leading to apoptosis. S-nitrosation of p50 subunit also blocks the NF-κB pathway, and is reversed when the reductive system has recovered. Apoptosis is also regulated by the S-nitrosylation of GAPDH by NO *(69)*. The cellular redox status is regulated in part by thioredoxin system (thioredoxin, thioredoxin reductase, NADPH), glutaredoxin and GSH. Modifications of the enzymes involved in cellular redox maintenance disrupt redox systems and result in further protein modifications. These reversible modifications could protect proteins from irreversible inhibition, highlighting the potential therapeutic benefits of free radical scavengers and antioxidants against cumulative oxidative damage of proteins. An overview of ROS/RNS generation and posttranslational modification of proteins is illustrated in Fig. 1.

3. Footprints of ROS/RNS-Induced Toxicity as Potential Biomarkers

The term biomarker refers to a molecule that indicates changes in a subject's physiological state resulting from drug treatment, disease, toxins, or other environmental stresses. Biomarkers play an important role in medicine, particularly in drug discovery and development. Establishment of biomarkers is beneficial in the early detection of diseases and is useful in drug safety and efficacy measurements.

The extensive body of evidence pointing to the role of ROS/RNS in a variety of diseases, including diabetes, motivates research to identify new biomarkers in the area of diabetic neuropathy. Biomarkers related to oxidative stress could be used as an early indication or measure of the progression of a disease, in the identification of the source of oxidative/nitrosative stress, the determination of the extent of the oxidative/nitrosative

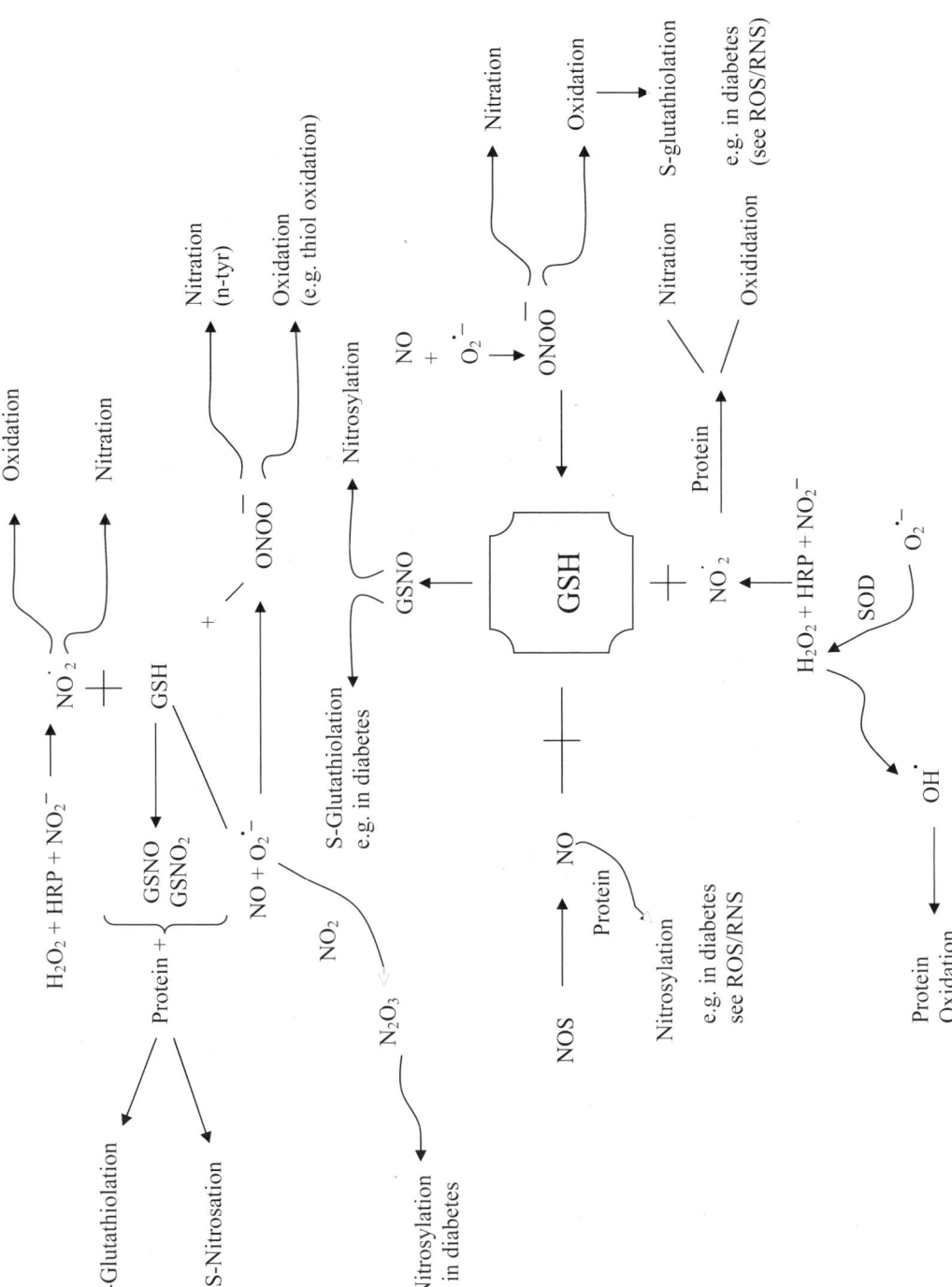

Fig. 1. RNS/ROS reaction pathways that lead to posttranslational modifications of proteins in cells.

insult, and the efficacy of antioxidant therapy for disease intervention. It is difficult to measure the presence of ROS/RNS directly, because they have very short half-lives as a result of their high reactivity. However, modification of macromolecules in cells/tissues or body fluids are used as the footprints of ROS/RNS generation. There is a significant body of evidence that the overproduction of ROS/RNS leads to irreversible modifications of proteins and cause tissue and cell damage. Identification of modified proteins would provide a valuable resource to understand human disease. ROS/RNS-induced posttranslational modifications of proteins such as the protein carbonyls, nitration of tyrosine residues and the S-nitrosation and S-glutathiolation of thiols (discussed in more detail in section II) are used as footprints of oxidative/nitrosative stress, which makes their study important in the context of diabetic neuropathy.

The process for establishing such biomarkers involves three major steps: The first step is identification. Combinations of available analytical tools enable researchers to identify and monitor a broad range of proteins in body fluids such as serum, urine, cerebrospinal fluid, and in cells and tissues. Alterations in protein concentration and posttranslational modifications of proteins, yield useful clues to the identification of new biomarkers. A biomarker should meet the following criteria: (1) a genuine indicator of oxidative stress; (2) stable in the same subject under similar conditions; (3) predictable (under similar conditions) with minimal variability among subjects; (4) not affected by external conditions such as diet and exercise; (5) not lost during preparation, storage or detection nor should it be formed as an artifact during these processes; (6) responsive to antioxidants under reasonable physiological conditions; and (7) suitable techniques for detection as well as further monitoring after treatment, particularly in vivo, should be available. Proteomic technologies are the best tools known to date. Considering the usefulness of biomarkers and the fact that those currently available are far from ideal *(80)* makes the search for biomarkers that satisfy the above criteria vitally important.

The identification step must be followed by validation. A valid biomarker should be a stable marker that is not subject to loss or artifactual induction, is not affected by dietary intake, is relevant to disease development, is detectable by a sensitive and reproducible assay, and is detectable using available and reliable analytical tools *(81)*. One should consider all factors that may affect markers. This is done by matching control subject groups with same age, gender, and other variables that are important in each case.

The final step, developing therapies and monitoring treatment and disease progression, is where biomarkers have tremendous application. Monitoring biomarkers (e.g., prostate serum antigen [PSA]) is a useful tool in screening for prostate cancer. Biomarkers may also be beneficial for detecting the progress of a disease and to predict drug responsiveness and efficacy; important tasks in drug design and development *(82)*.

4. Production of ROS in Diabetes

ROS/RNS generation and imbalance of cellular redox state are linked to many diseases, including diabetes, cancer, and neurodegenerative disease. There is a close link between oxidative stress and the development of diabetic complications. This hypothesis is reinforced by recent work, both in vivo and in vitro *(83)*. In type 1 diabetes, oxidative stress is apparent within a few years of diagnosis *(84)*. Type 2 diabetic patients have decreased plasma GSH levels and GSH-metabolizing enzymes, and increased lipid peroxidation compared with age-matched control subjects, both of which relate directly to the rate of development of complications *(85–87)*. This section describes diabetes-mediated oxidative stress mechanisms, as well as related cellular responses to oxidative stress, including biomarkers of diabetic complications as well as therapeutic targets. Figure 2 outlines the potential mechanisms of activation of signaling pathways by high glucose in cells.

4.1. Advanced Glycosylation End Product (AGE)-Mediated ROS Formation

Glycation is the result of the nonenzymatic addition of glucose or other saccharides to proteins, lipids and nucleotides *(88,89)*. Glucose initially binds to protein amino groups first forming a Schiff base then progressing to Amadori products *(90–92)*. Dehydration of the Amadori products results in the formation of AGEs and the subsequent crosslinking of proteins. AGE formation occurs normally over time; however, in diabetes, excess glucose results in accelerated AGE production *(89,91)*.

AGE modified substrates exert their effects by binding to cell surface receptors *(90–92)* for advanced glycation endproducts (RAGE) *(90,91)*. RAGE has definite downstream signaling targets and is the main receptor through which AGE signaling is mediated *(90,93–100)*.

In mesangial *(101)*, endothelial cells and sensory neurons activation of RAGE by AGEs or S100 results in a burst of ROS production. The exact mechanism for this is unknown but is thought to involve NADPH oxidase *(97)*. This event alone could contribute to cellular oxidative stress and dysfunction. In addition, RAGE signals via phosphatidylinositol-3 kinase (PI-3 kinase), Ki-Ras and ERK *(101)* which initiate and sustain the translocation of NF-κB from the cytoplasm to the nucleus in a number of cell types including circulating monocytes and endothelial cells *(102)*. The RAGE receptor gene contains two NF-κB binding sites within its promoter region, therefore, activation of RAGE leading to translocation of NF-κB results in the amplification of RAGE and promotes a cycle of damage and continued oxidative stress *(103)*.

4.2. The Polyol Pathway

The first enzyme in the polyol pathway is aldose reductase, a cytosolic oxidoreductase that uses NADPH as a cofactor to catalyze the reduction of carbonyl compounds such as glucose. Under normal nondiabetic conditions, aldose reductase has low affinity for glucose, resulting in limited metabolism of glucose through this pathway. However, under hyperglycemic conditions, glucose is converted to sorbitol by aldose reductase activity. Enzymatic activity of aldose reductase requires the conversion of NADPH to $NADP^+$, which results in the depletion of glutathione (GSH) and an increase in ROS production. Enzymatic activity of aldose reductase is shown in Fig. 3.

The putative mechanisms for polyol pathway-induced damage are as follows: reduction in (Na/K^+) ATPase activity, osmotic stress induced by high sorbitol concentration, and decreased in cytosolic NADPH that depletes GSH. Using transgenic mice overexpressing aldose reductase (AR), Alan et al. demonstrated oxidative stress in the lens tissues of the eye following glucose flux via the polyol pathway *(104,105)*. Decreased GSH level with increased lipid peroxidation in the transgenic mice suggest that enzymatic reduction of glucose to sorbitol contributes greatly to oxidative stress by decreasing cytosolic NADPH level and subsequent depletion of GSH. This conclusion is

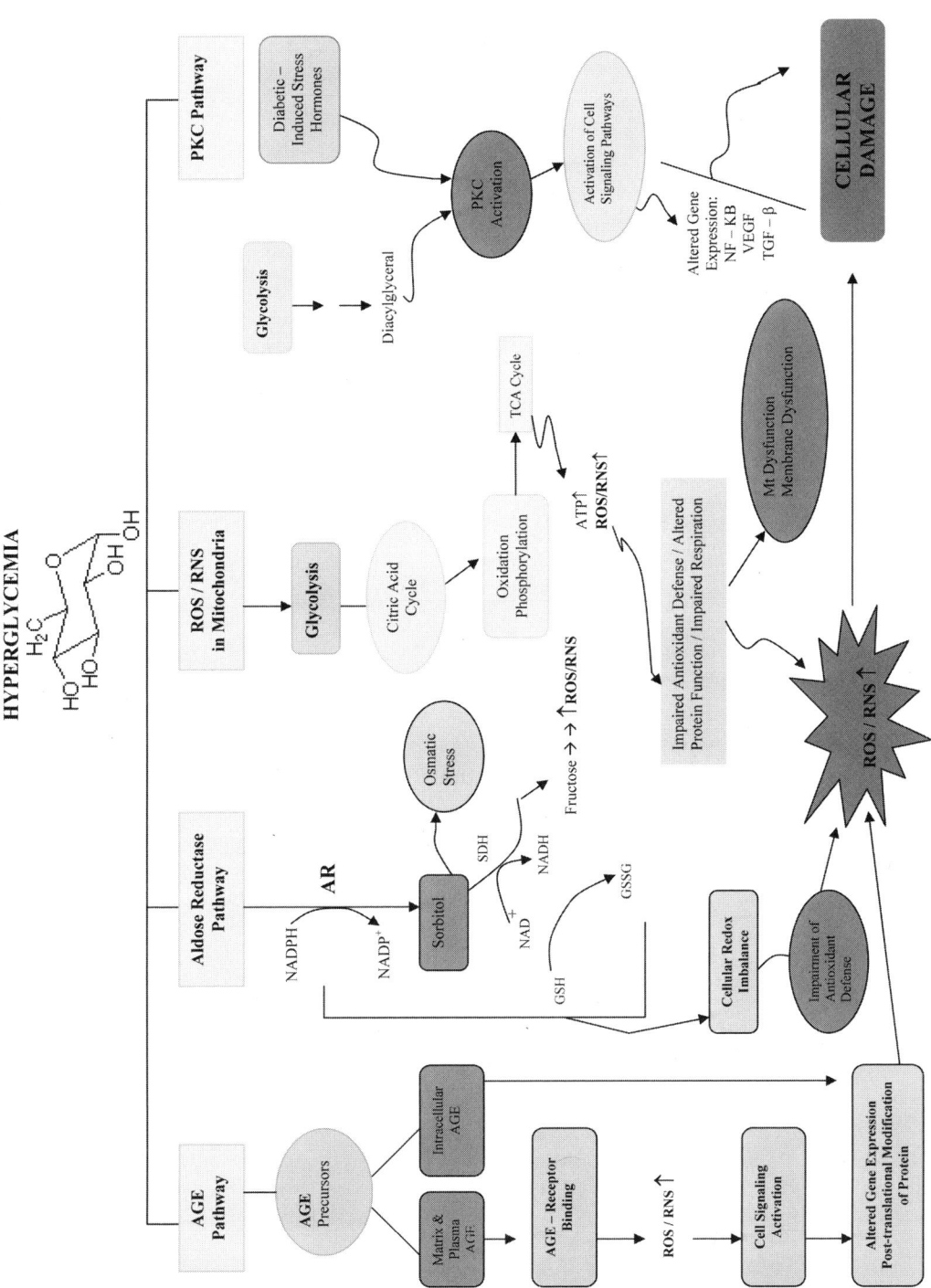

Fig. 2. Schematic diagram of pathways contributing to oxidative stress in response to hyperglycemia.

375

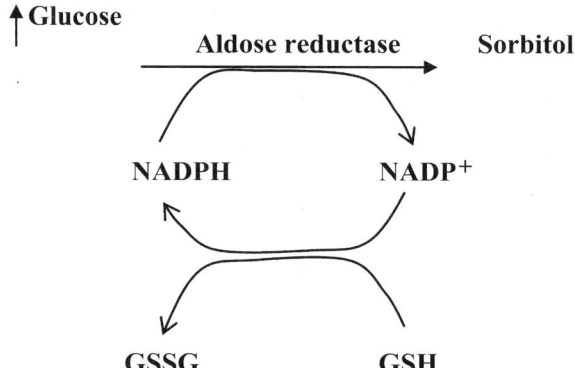

Fig. 3. Enzymatic activity of aldose reductase.

supported by use of the aldose reductase inhibitors (ARIs). Inhibition of AR restored the levels of GSH and decreased lipid peroxidase products. These findings confirm the involvement of polyol pathway in diabetic complications.

4.3. Protein Kinase C (PKC) Activation

PKC belongs to a family of phospholipid-dependent Ser-Thr kinases, involved in the regulation of cell growth, death, and stress-responsive signaling pathways *(106)*. PKCs are structurally unique, which make them sensitive to cellular redox changes *(107)*. PKCs contain an N-terminal regulatory domain with cysteine rich motifs that are easily oxidized by peroxide leading to PKC activation. In addition, the C-terminal catalytic domain contains reactive cysteines that are modified by antioxidants to cause PKC inactivation *(108,109)*.

PKC activation by hyperglycemia plays an important role in diabetic complications. Increased PKC activity is found in the retina, kidney, and microvasculature of diabetic rats *(110)*. The majority of PKC isomers are activated by diacylglycerol (DAG), which is synthesized as a messenger in a chain of reactions initiated by hyperglycemia *(106,110)*. PKC in turn activates mitogen activated protein kinase (MAPKs), which leads to increases in stress gene expression, such as heat shock proteins and c-Jun kinases, causing apoptosis or vascular atherosclerosis *(111)*.

PKC induces nerve injury by altering nerve blood flow. In diabetic rats, inhibition of PKC-β normalizes blood flow and nerve conduction deficits and also reduces oxidative stress in diabetic rats *(112,113)*. Inhibitors of PKC prevent high glucose-induced NF-κB activation, thus limiting diabetic complications *(114)*. These studies indicate a role for PKC in hyperglycemic induced nerve damage.

4.4. MAPK Activity

The generation of ROS/RNS triggers specific signaling pathways that alter cellular function. MAPK and other stress- responsive protein kinases are affected by alteration of the cellular redox state. The pro- or antiapoptotic actions of these kinases depend on the phosphorylation of the key elements involved in cell fate, leading to cell survival or death.

All subgroups of the MAPK family, such as MAPK, ERK1/2, JNK, and p38, are activated in the dorsal root ganglia (DRG) of a diabetic rat *(111)*. Although, the subgroups are structurally similar, they are functionally different. MAPK regulate cell fate by phosphorylating downstream molecules such as transcription factors, cell cycle regulators, and proteins involved in apoptosis (e.g., p53, Bcl2 family, and caspases). Moreover, most of these pathways interact with each other. ERK1/2 activation by oxidative stress is mediated by growth factor receptors such as epidermal growth factor (EGF) receptor and platelet derived growth factor (PDGF) receptor. These receptors are phosphorylated in response to oxidative stress (e.g., H_2O_2 and UV irradiation) *(115,116)*. Other stress-activated MAPK, JNK, and p38 MAPK are activated by pro-oxidants (e.g., JNK activation by H_2O_2) *(117–119)*. Also, NO activates P38 MAPK and results in apoptosis *(120)*. Activated PKC activates MAPK, during the development of diabetic neuropathy and nephropathy. However, the molecular mechanism by which MAPK contributes to diabetic complications is unknown.

4.5. ROS/RNS Formation in the Mitochondria

In the process of the oxidative metabolism of glucose, energy stored in the carbon bonds of glucose is transferred to the third phosphate bond of ATP. The electrons within the carbon bonds are transferred to the dinucleotide electron carrier (NADH) and flavin adenine dinucleotide ($FADH_2$). These electrons enter into the electron transport chain and flow from complex I to IV in the inner membrane of a mitochondria. Ultimately, molecular oxygen is reduced to H_2O by these electrons. In this process, the energy carried by electrons is used by complexes I, III, and IV, which are reduction and oxidation pumps, to pump protons out of the mitochondrial matrix. The protons pumped out of the matrix create an electrochemical gradient across the mitochondrial inner membrane. ATP synthase uses the electrochemical gradient to drive the synthesis of ADP to ATP. In mitochondria, increased ATP synthesis is regulated by uncoupling proteins. When uncoupling proteins (UCP) are activated, protons leak across the inner membrane and "uncouple" oxidative metabolism from ATP synthase, leading to loss of ATP production. When UCP are overexpressed, basal and hyperglycemia-induced ROS formation decreases in DRG neurons *(121)*. $O_2^{\cdot-}$ activates UCP in skeletal muscle and is considered a protective mechanism against the generation of ROS in hyperglycemia. Activation of UCP by $O_2^{\cdot-}$, increases mitochondrial membrane permeability resulting in decreased electrochemical potential and further decreases $O_2^{\cdot-}$ generation. The protective effect of UCP may result from a mild mitochondrial depolarization that limits Ca^{2+} accumulation and reduces reactive species generation (e.g., by limiting NOS activity) *(122)*. Otherwise, accumulation of Ca^{2+} may trigger many pathophysiological events that lead to cell death.

The presence of mitochondrial NOS (Mt NOS) and its activity were reported by Ghafouri and Richter in 1997 *(123)*. Mt NOS produces NO enzymatically and is associated with the matrix face of the mitochondrial inner membrane. The activity of Mt NOS depends on Ca^{2+}, and is increased with higher intramitochondrial Ca^{2+} concentration $[Ca^{2+}]m$ *(123)*. Increasing the concentration of Ca^{2+} leads to reduced mitochondrial transmembrane potential ($\Delta\psi$) and decreased O_2 consumption. Ca^{2+} homeostasis is maintained via a feedback mechanism that involves Mt NOS. This process regulates the Ca^{2+} load in the mitochondria and protects against the consequences of Ca^{2+} overload. $[Ca^{2+}]m$ is kept low in the mitochondria by two mechanisms: first, mitochondria precipitate the $[Ca^{2+}]m$ to nonionized calcium, granules, consisting mainly of tricalcium

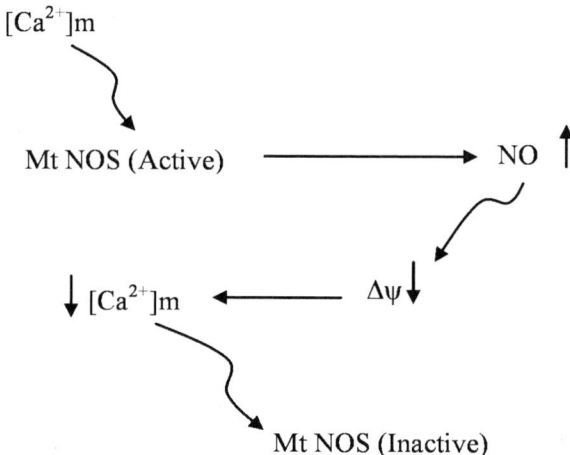

Fig. 4. Regulation of mitochondrial NOS activity.

phosphate and hydroxyapatite *(124,125)*. Second, Ca^{2+} leaves mitochondria in exchange for H^+ and Na^+ when $\Delta\psi$ is decreased. In summary, NO formation by Mt NOS is regulated, and the elevation of $[Ca^{2+}]m$ increases NO production and leads to reduction in $\Delta\psi$. Decrease in $\Delta\psi$ releases Ca^{2+} from the mitochondria and results in Mt NOS inactivation. Figure 4 shows the proposed mechanism by which mitochondrial NOS activity is regulated.

4.5.1. ONOO⁻ Production in Mitochondria

In the respiratory chain, 2 to 5% of electrons leak from electron flow and produce O_2^{-} *(126)*. Peroxynitrite ($ONOO^-$) is generated from the O_2^{-} produced from the electron leakage through the inner membrane respiratory chain, and NO, which is produced by the inner membrane Mt NOS *(see Fig. 5)*. The rate of peroxynitrite formation is 9.5×10^{-8} M/s⁻, exceeds the interaction of NO with cytochrome c oxidase (0.8×10^{-8} M/s) *(127)*.

Ghafourifar et al. reported that peroxynitrite-induced stress promotes cytochrome *c* release from the mitochondria and results in apoptosis *(128)*. Mt NOS is also involved in mitochondrial dysfunction and contractile failure in human skeletal muscle *(129)*. Nitration of the tyrosine residues of proteins occurs in the mitochondrial compartment *(79,130,131)*. S-nitrosation of protein thiols by NO is a very important reaction in the mitochondria *(65)*, which is more favorable in the inner membrane and inter membrane space.

Nitrosation of caspase-3 [65] and complex I (inner membrane protein) *(132)* highlight the involvement of NO in the regulation of mitochondrial function. Inactivation of caspase-3 by S-nitrosation protects cells from apoptosis and indicates an anti-apoptotic role for NO. However, S-nitrosation of proteins such as complex I and complex IV result in an impaired respiratory chain and O_2^{-} accumulation.

Generation of O_2^{-} in the presence of NO results in $ONOO^-$ formation *(see Fig. 5)*, leading to protein modifications. The nitrated proteins in vivo under diabetic conditions are involved in major mitochondrial functions, such as energy production (succinyl-CoA: 3-oxoacid CoA transferase and creatine kinase), antioxidants (peroxiredoxin 3 and glutathione S-transferase), and apoptosis (voltage-dependent anion channel-1). These

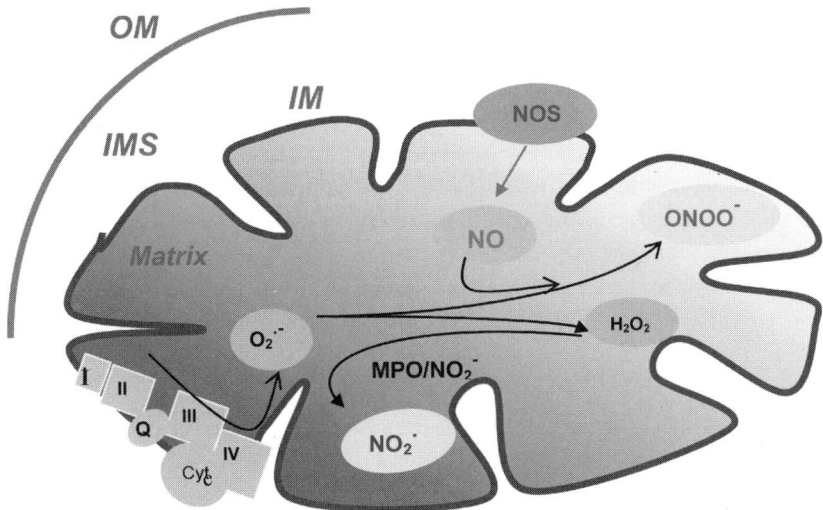

Fig. 5. Overview of ROS/ RNS formation in the mitochondria.

findings support the idea that ROS/RNS-induced posttranslational modifications of proteins are involved in mitochondrial dysfunction *(133)*.

5. Neuronal Response to Oxidative Stress

As we saw in previous sections, oxidative stress induced by hyperglycemia is directly involved in neuronal degeneration. In this section we will closely examine the response of neurons to oxidative stress in order to better understand the preceding discussion at the cellular level.

Neurons obtain glucose through facilitated concentration-dependent transport, which makes them very susceptible to high glucose induced oxidative stress. The antioxidant defense system of a cell is normally tightly regulated, however, in the presence of oxidative stress, it adapts to new environmental pressures. In the short term, the up-regulation of SOD, GSH peroxidases, and catalase, through increased gene expression, is detected in endothelial cell cultures (after 3–10 d) *(134)*. However, under sustained oxidative stress, antioxidant enzymes lose function. Indeed, diabetes is associated with a decline in antioxidant defense components such as SOD, Gprx, and GSH *(135)*, an increase in lipid peroxidation and a reduction in total radical antioxidant parameters (TRAP).

Hess et al. reported that NO donors such as SIN-1 and S-nitrosocysteine, rapidly and reversibly inhibit growth of neurites of DRG neurons in vitro. Additionally, they show that thioester-linked long-chain fatty acylation of neuronal proteins are inhibited by NO, possibly by cysteine modification *(136)*. Zochodne et al., demonstrated that inhibition of NOS improves peripheral nerve regeneration in mice *(137)*. Nitrotyrosine, a footprint of NO toxicity, is found in experimental diabetic neuropathy *(138)*. Diabetic peripheral nerves are susceptible to oxidative and nitrosative stress. NO overproduction and RNS generation (e.g., peroxynitrite) results in inhibition of neurite growth in diabetic regenerating neurons by yet unknown mechanism *(139)*.

Several mechanisms may contribute to the impairment of nerve regeneration in diabetes. Experiments demonstrate that in early diabetes, the vascular supply to peripheral trunks as well as neuronal responses to injury were considerably weakened *(140)*.

However, in nondiabetic nerves, the inflammatory and repair response to injury is associated with persistent sustained increases in blood flow. Increased blood flow may help in regeneration of the injury site by providing necessary components in the process such as blood-borne macrophages and other regulation associated molecules. In diabetic rats (STZ), an overall impairment of the inflammatory response is reported *(137)*. However the significance of these findings and their contribution in neuronal regeneration requires further clarification. The polyol pathway contributes to diabetic neuropathy and is also associated with deficits in axonal regeneration following damage. Therefore improvement of neuronal regeneration is considered as a major therapeutic approach in diabetic neuropathy.

6. Mitochondrial/Nerve Structure Changes

Aberrant mitochondrial morphology in the peripheral nerves of diabetic rodents was detected first by transmission electron microscopy (TEM) *(141–143)*. Mitochondria appear as swollen vesicles with only the remnants of intact cristae. Disrupted mitochondria are detected in sensory and sympathetic neurons and peripheral nerve fibers of experimental animals *(141,143,144)* and in human nerve biopsy material *(145)*.

Work in the Feldman laboratory established that glucose is uniquely toxic to transformed neurons (SH-SY5Y cells) *(146)*. Analysis of primary sensory neurons treated with excess glucose reveals the same mitochondrial changes reported in earlier work. In addition, deranged mitochondria are present in cells undergoing apoptosis as defined by both anatomical criteria and DNA end-labeling (TUNEL). The detection of fragmented DNA and disrupted mitochondria lead to the hypothesis that excess glucose directly or indirectly causes mitochondrial damage and programmed cell death. Russell et al. demonstrated that the mitochondria of sensory neurons exposed to 40 mM glucose lose membrane potential and swell *(147)*. This effect could be a result of the opening/formation of the voltage-dependent, anion-selective channel porin (VDAC) found in mitochondrial outer membranes *(148)*. The VDAC may coalesce into a megapore, the permeability transition pore (PTP) or, close entirely, resulting in heightened internal mitochondrial pressure. Either event would cause mitochondrial rupture.

This observation was examined further by th Sastry laboratory who detected the direct effects of glucose on mitochondrial structure *(148)*. Mitochondria from sensory neurons, control or exposed to 40mM glucose, were examined with atomic force microscopy (AFM). AFM confirms the formation of large megapores in the mitochondrial membrane (*see* Fig. 6). These pores contribute to mitochondrial depolarization, allow cytochrome *c* to flow into the cytosol thus initiating the death cascade.

As outlined above, high glucose impairs the mitochondrial electron transfer chain. Continued hyperglycemia results in ATP depletion, ROS formation, megapore formation, cytochrome *c* release into the cytosol and activation of the caspase cascade. It is now clearly established that glucose-mediated oxidative stress is an inciting event in the development of DN *(6,9–11,13,83,149–154)*.

7. Therapy for Diabetic Complications Neuropathy

Glycemic control is the most effective way of controlling diabetic complications in type 1 diabetic patients *(10,155)*. Treatment of type 2 diabetic patients with glibenclamide or glicaxide helps glycemic control and reverses early hyperglycemia-induced

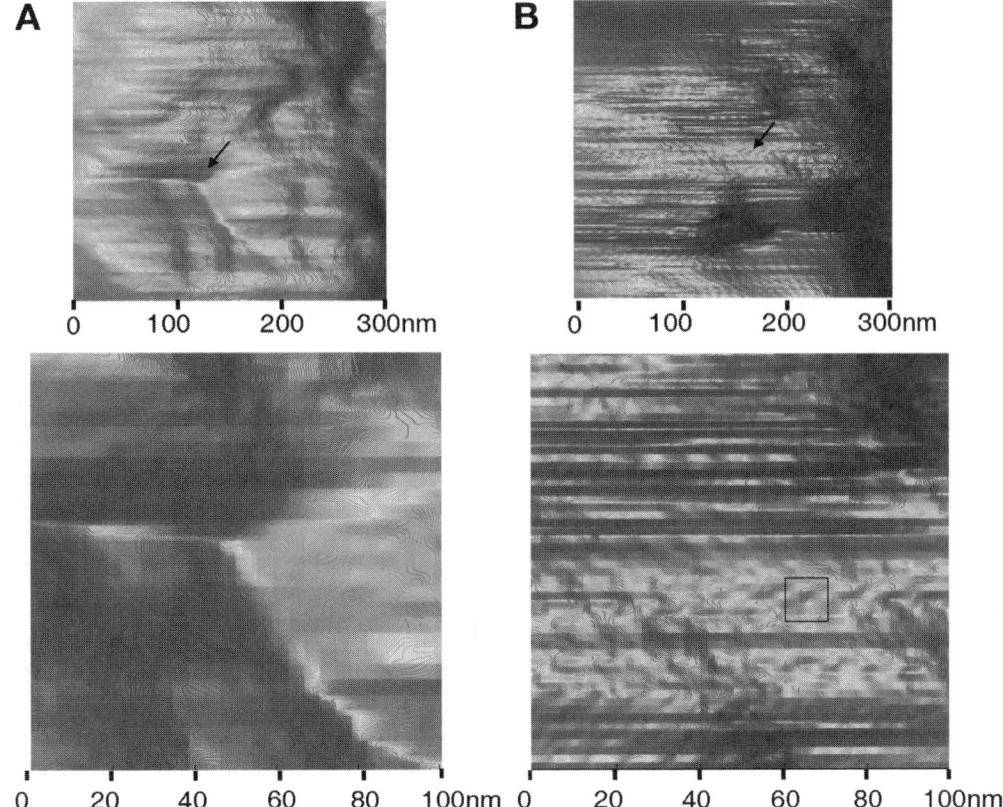

Fig. 6. Fluid-tapping atomic force microscopy (AFM) images of fixed, control, and glucose-challenged Sprague-Dawley embryonic DRG mitochondria. **(A)** The 256 × 256 pixel (approx 4 nm/pixel) image of an unchallenged outer membrane, revealing features of 10 to 100 nm. The indentation in the middle of the upper image may reveal topography of the inner membrane; a cropped portion of the upper image is shown in the lower image. **(B)** The 512 × 512 (approx 2 nm/pixel) image of a glucose-challenged outer membrane, revealing invaginated features that may be VDAC pores; a cropped portion of the upper image is shown in the lower image. DNB, 1,3-dinitrobenzene; VDAC, voltage-dependent, anion-selective channel porin.

complications *(156)*. Long-term control of diabetic complications is important to maintain quality of life for patients. Considering the important role played by oxidative stress in mediating vascular and microvascular complications, antioxidant therapy may be an important in intervention or prevention of diabetic complications. In addition to antioxidants, ARI and growth factor therapies may also protect against oxidative stress-induced complications.

7.1. Aldose Reductase Inhibitors

As discussed earlier, aldose reductase is an initial rate-limiting enzyme that converts glucose to sorbitol, and activation of the aldose reductase polyol pathway results in oxidative stress in animal models of diabetes. In the past few decades, ARI were developed and tested on animals with some efficacy *(157)*. However, the clinical potential of ARIs in preventing diabetic complications is controversial. The lack of efficacy in human clinical trials, in addition to unacceptable side effects associated with ARI led to abandonment of these drugs *(158)*. Sorbinil, restores normal levels of glutathione in the lenses of diabetic rats *(159)*, reverses the early deficits in peripheral nerve function and restores the antioxidant defense system *(160)*. The ARI WAY-121509 attenuates diabetic complications and corrects sciatic nerve conduction velocity (NCV) and endoneurial blood flow *(161)*. However, WAY-135706, an inhibitor of sorbitol dehydrogenase, had no beneficial effect on nerve function *(161)*. In human diabetes, Zenarestat was effective in type 1 *(162)* and type 2 *(163)* diabetic animals. Zenarestat with greater than 80% sorbitol suppression, improved small-nerve fiber density and nerve conduction velocity slowing in a 52-wk phase II clinical trial *(164)*. However, the study was suspended in phase III trial because of renal function disorders *(164)*. To date, Fidarestat is the only ARI that may still prove effective in slowing the progression of DN.

7.2. Nerve Growth Factor

Growth factors are involved in neuronal survival, health, and maintenance. Nerve growth factor (NGF) was the first neurotrophin discovered and characterized *(165)*. NGF is involved in the differentiation, growth and survival of neurons *(165)*. Evidence showing changes in the expression of growth factors in diabetic neuropathy as well as their role in neuronal degeneration in in vitro and in vivo models of diabetic neuropathy, led to their use in a clinical trial to combat neurological diseases *(165,166)*. NGF prevents and reverses biochemical deficits in several models of peripheral neuropathy. It was recently demonstrated that NGF prevented neuronal oxidative stress by increasing the GSH level *(167)*. Despite promising early clinical trails for diabetic neuropathy, later phase III trails failed *(168)*.

8. Clinical Trials of Antioxidant Therapy in Diabetes Complications

The trials of antioxidants in both animal models and human patients are the best indicators of the key role of oxidative stress in diabetic neuropathy. One of the limitations of animal models that should be considered is their short life span compared with long-term disease progression in human diabetes *(169)*. The complexity of the interactions of antioxidant reagents beyond their antioxidant activity, requires cautious data analysis in clinical trials *(170–172)*. This section discusses α-lipoic acid and vitamins C and E as major antioxidants tested in clinical trials.

8.1. α-Lipoic Acid

α-lipoic acid (1,2-dithiolane-3-pentanoic acid, LA) is a powerful antioxidant found in plant, and animal tissues *(173)*. LA is taken up in the diet, crosses the blood–brain barrier, and is reduced to dihydrolipoic acid (DHLA) in cells. Molecular structure of LA is shown in Fig. 7. LA scavenges peroxynitrite, hypochlorous acid, and hydryl radicals, however, its reduced form dihydrolipoic (*DHLP) acid, scavenges superoxide and peroxyl radicals. LA and DHLA are widely used antioxidants, with broad ROS/RNS scavenging properties *(174)*. LA and DHLA protect against metal induced damage by metal chelating which prevents free radical generation (e.g., OH) *(175)*. DHLA interacts with other antioxidants such as glutathione (GSH), ascorbate, coenzyme Q, and vitamin E, to regenerate their reduced forms *(174,176)*. Thus, DHLA plays a major role in the antioxidant network by regenerating antioxidant defense.

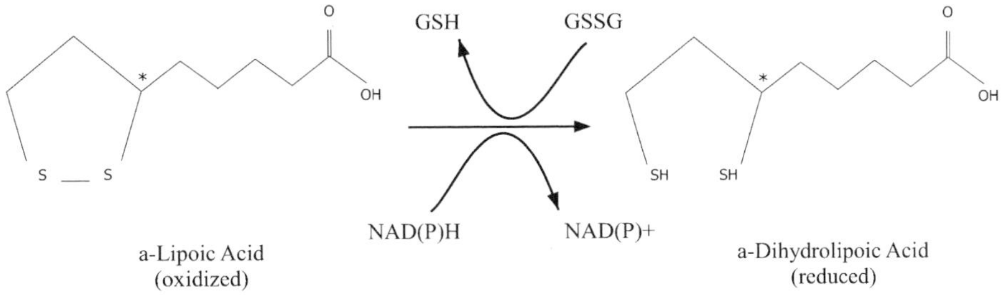

Fig. 7. Molecular structure of α-lipoic acid and its reduced form α-(DHLA).

LA prevents the development of nerve conduction and blood flow deficits *(177–179)*, maintains retina and lens antioxidants and energy status *(180)*, and prevents lipid peroxidation in the retina of rat models of diabetes. There is sufficient evidence to support the beneficial effects of LA in reducing oxidative stress in multiple tissues *(181,182)* in healthy adults and in diabetic patients *(183)*. Daily supplementation of 600 mg of LA for 3 modecreases lipid peroxidation. LA also prevents hypertension, hyperglycemia, and hyperinsulinemia, even in patients with poor glycemic control *(184)*. LA is licensed for use in diabetic patients in Germany, because this compound improves the microcirculation and significantly improves the sensory symptoms in the patients with diabetic neuropathy *(185)*. LA is in a phase III clinical trial (in the Sydney trial) in the treatment of diabetic complications in the United States *(186)*.

8.2. Vitamins E and C

The clinical trials of antioxidants strongly support the importance of oxidative stress in the development of diabetic neuropathy. Therapeutic strategies in the treatment of diabetic complications may block glucose induced oxidative stress by lowering hyperglycemia, by scavenging ROS/RNS, or finally by preventing the cellular damage resulting from oxidative stress (e.g., blocking signaling pathways that are activated by ROS/RNS directly or indirectly). However, it may not be a distinct pathway leading to diabetic neuropathy, therefore, the design of therapeutic strategy is complicated.

Vitamin E, also known as α-tocopheral, improves nerve conduction velocity in diabetic patients with moderate neuropathy *(187)*. Dietary vitamin E supplementation decreases lipid peroxidation and fatty acid levels in some tissues of diabetic rats *(188)*. α-tocopheral also prevents LDL (low density lipoprotein) peroxidation in vitro and in vivo *(189)*, and improves blood flow in the diabetic rat heart *(190)*. However, it is important to note that there are eight forms of naturally occurring Vitamin E, and α-tocopheral is only one form. The antioxidant properties of the other forms of Vitamin E are starting to be recognized. γ-tocopheral may be more effective in scavenging RNS than ROS *(191)*. The various forms of Vitamin E alter cellular signaling, but the molecular mechanism needs more clarification.

Vitamin C is an exogenous lipophilic antioxidant that decreases glycated hemoglobin and improves insulin action. Combined oral vitamin C and E therapy improves endothelial function in type 1 but not type 2 diabetes *(192)* and reduces oxidative stress in the eye *(193)*. Vitamin C, an endogenous antioxidant, protects the lipoproteins from peroxidative damage induced by peroxyl radicals under oxidative stress *(194)*. Vitamin C (ascorbate) is a critical antioxidant that prevents lipid peroxidation in hyperglycemia *(194)*. In type 2 diabetic patients, Vitamin C supplementation of 1 g/d for a 4-mo period increases cellular GSH levels and decreases reactive species (e.g., peroxyl radicals and other oxidants) in plasma *(195)*. In general, Vitamin C is used in combination therapies with other antioxidants.

8.3. The Future of Antioxidant Therapy in Clinical Trials

To improve future clinical trials, a better understanding of the processes by which ROS/RNS modulate cellular function and/or cause cellular damage is required. Among other things, the effects of antioxidants on oxidant-regulated gene expression and cell signaling should be thoroughly examined. Antioxidants may interfere with redox-regulated cellular functions and disturb the cellular prooxidant-antioxidant balance *(98,196)*. Cellular redox homeostasis may be influenced by a variety of environmental factors as well as dietary and endogenous antioxidant thiols. For example, GSH is one of the major tissue antioxidants, which plays an important role in maintaining cellular redox state. Increasing cellular GSH level by targeting the GSH-synthesizing enzymes *(197)*, as well as increasing its precursor may be used as a therapeutic approach. In general, the overall cellular energy level, antioxidant pools, and oxidizing agents are determinants of cellular fate. Future research should address and clarify questions of redox-regulated gene expression patterns using microarrays and modified proteins using proteomics techniques. These modifications could be monitored following the application of antioxidants or antioxidant mixtures. These techniques will provide valuable information that will improve our understanding of the mechanisms of oxidative-induced changes.

9. Conclusions

Hyperglycemia related oxidative stress plays a significant role in the development of vascular and neuronal damage, both of which contribute to DN. In fact, both hypo- and hyperglycemia lead to oxidative stress. Because maintaining euglycemia is not yet possible for all patients, antioxidant therapy may provide a useful way to protect cells from fluctuating glucose levels. Treatments that target oxidative stress associated with diabetes would help slow or prevent the progression of neuropathy as well as other diabetic complications. This promise invites research and development of new antioxidant therapies to minimize the effects of hyperglycemia.

Acknowledgments

The authors thank Ms. Judith Boldt for expert secretarial assistance. This work was supported by grants from the National Institutes of Health (NS38849), the Juvenile Diabetes Research Foundation Center for the Study of Complications in Diabetes and the Program for Understanding Neurological Diseases (PFUND).

References

1. Feldman EL SM, Russell JW, Greene DA. Diabetic neuropathy. In: Principles and Practice of Endocrinology and Metabolism. Becker KL (ed.). Lippincott, Williams & Wilkins, Philadelphia, PA, 2001:1391–1399.
2. Feldman EL SM, Russell JW, Greene DA. Somatosensory neuropathy. In: Ellenberg and Rifkin's Diabetes Mellitus. Porre Jr D, Sherwin RS, and Baron A (eds.). McGraw Hill. New York, NY, 2002;771–788.

3. van Golen CM, Soules ME, Grauman AR, Feldman EL. N-Myc overexpression leads to decreased beta1 integrin expression and increased apoptosis in human neuroblastoma cells. Oncogene 2003;22(17):2664–2673.

4. Russell JW, Golovoy D, Vincent AM, et al. High glucose-induced oxidative stress and mitochondrial dysfunction in neurons. FASEB J 2002;16(13):1738–1748.

5. Vincent A, Feldman E. New Insights into the Mechanisms of Diabetic Neuropathy. Rev Endocrine Metabol Dis 2004;5(3):227.

6. Vincent AM, McLean LL, Backus C, Feldman EL. Short-term hyperglycemia produces oxidative damage and apoptosis in neurons. FASEB J 2005:04-2513fje.

7. Vincent AM, Russell JW, Low P, Feldman EL. Oxidative Stress in the Pathogenesis of Diabetic Neuropathy. Endocr Rev 2004;25(4):612–628.

8. Maritim AC, Sanders RA, Watkins 3rd JB. Diabetes, oxidative stress, and antioxidants: A review. J Biochem Mol Toxicol 2003;17(1):24–38.

9. Stevens MJ, Obrosova I, Pop-Busui R, Greene DA, and Feldman EL. Pathogenesis of diabetic neuropathy. In *Ellenberg and Rifkin's Diabetes Mellitus*. Porre Jr D, Sherwin RS, Baron A (eds.). McGraw Hill. New York, NY, 2002:747–770.

10. Windebank AJ, Feldman EL. Diabetes and the nervous system. In *Neurobiology and general medicine*. Aminoff MJ (ed.). Churchill Livingstone. Philadelphia, PA, 2001;341–364.

11. Brownlee M. Biochemistry and molecular cell biology of diabetic complications. Nature 2001;414(6865):813–820.

12. Cameron NE, Eaton SE, Cotter MA, Tesfaye S. Vascular factors and metabolic interactions in the pathogenesis of diabetic neuropathy. Diabetologia 2001;44(11):1973–1988.

13. Feldman EL. Oxidative stress and diabetic neuropathy: a new understanding of an old problem. J Clin Invest 2003;111(4):431–433.

14. Hensley K, Robinson KA, Gabbita SP, Salsman S, Floyd RA. Reactive oxygen species, cell signaling, and cell injury. Free Radic Biol Med 2000;28(10):1456–1462.

15. Babior BM. The NADPH oxidase of endothelial cells. IUBMB Life 2000;50(4-5):267–269.

16. Coon MJ, Ding XX, Pernecky SJ, Vaz AD. Cytochrome P450: progress and predictions. FASEB J 1992;6(2):669–673.

17. Turrens JF. Mitochondrial formation of reactive oxygen species. J Physiol (Lond) 2003;552(2):335–344.

18. Uemura S, Matsushita H, Li W, et al. Diabetes Mellitus Enhances Vascular Matrix Metalloproteinase Activity: Role of Oxidative Stress. Circ Res 2001;88(12):1291–1298.

19. Burdon RH. Superoxide and hydrogen peroxide in relation to mammalian cell proliferation. Free Radic Biol Med 1995;18(4):775–794.

20. Lee M, Hyun D-H, Jenner P, Halliwell B. Effect of proteasome inhibition on cellular oxidative damage, antioxidant defences and nitric oxide production. J Neurochem 2001;78(1):32–41.

21. Khodr B, Khalil Z. Modulation of inflammation by reactive oxygen species: implications for aging and tissue repair. Free Radic Biol Med 2001;30(1):1–8.

22. Vazifeh D, Abdelghaffar H, Labro MT. Effect of Telithromycin (HMR 3647) on Polymorphonuclear Neutrophil Killing of Staphylococcus aureus in Comparison with Roxithromycin. Antimicrob Agents Chemother 2002;46(5):1364–1374.

23. Droge W. Free Radicals in the Physiological Control of Cell Function. Physiol Rev 2002;82(1):47–95.

24. Chance B, Sies H, Boveris A. Hydroperoxide metabolism in mammalian organs. Physiological Reviews 1979;59(3):527–605.

25. Kwong LK, Sohal RS. Substrate and Site Specificity of Hydrogen Peroxide Generation in Mouse Mitochondria. Arch Biochem Biophys 1998;350(1):118–126.

26. Hampton MB, Orrenius S. Dual regulation of caspase activity by hydrogen peroxide: implications for apoptosis. FEBS Letters 1997;414(3):552–556.

27. Levine RL, Berlett BS, Moskovitz J, Mosoni L, Stadtman ER. Methionine residues may protect proteins from critical oxidative damage. Mech Age Dev 1999;107(3):323–332.
28. van der Vliet A, Hu ML, O'Neill CA, Cross CE, Halliwell B. Interactions of human blood plasma with hydrogen peroxide and hypochlorous acid. J Lab Clin Med 1994;124 (5):701–707.
29. Abe J, Berk BC. Fyn and JAK2 mediate Ras activation by reactive oxygen species. The J Biol Chem 1999;274(30):21,003–21,010.
30. Wang X, Martindale JL, Liu Y, Holbrook NJ. The cellular response to oxidative stress: influences of mitogen-activated protein kinase signalling pathways on cell survival. Biochem J 1998;333 (Pt 2):291–300.
31. Dalton TP, Shertzer HG, Puga A. Regulation of gene expression by reactive oxygen. Ann Rev Pharmacol Toxicol 1999;39:67–101.
32. Schreck R, Rieber P, Baeuerle PA. Reactive oxygen intermediates as apparently widely used messengers in the activation of the NF-kappa B transcription factor and HIV-1. EMBO J 1991;10(8):2247–2258.
33. Griendling KK, Harrison DG. Dual role of reactive oxygen species in vascular growth. Circ Res 1999;85(6):562–563.
34. Blouin E, Halbwachs-Mecarelli L, Rieu P. Redox regulation of [beta]2-integrin CD11b/CD18 activation. Eur J Immunol 1999;29(11):3419–3431.
35. Roebuck KA. Oxidant stress regulation of IL-8 and ICAM-1 gene expression: differential activation and binding of the transcription factors AP-1 and NF-kappaB (Review). Int J Mol Med 1999;4(3):223–230.
36. Naseem KM, Bruckdorfer KR. Hydrogen peroxide at low concentrations strongly enhances the inhibitory effect of nitric oxide on platelets. Biochem J 1995;310(1):149–153.
37. Hawkins CL, Davies MJ. Generation and propagation of radical reactions on proteins. Biochim Biophys Acta 2001;1504(2-3):196–219.
38. Moncada S, Palmer RM, Higgs EA. Nitric oxide: physiology, pathophysiology, and pharmacology. Pharmacol Rev 1991;43(2):109–142.
39. Moncada S, Higgs A. The L-arginine-nitric oxide pathway. N Engl J Med 1993;329(27): 2002–2012.
40. Saran M, Bors W. Signalling by O2-. and NO.: how far can either radical, or any specific reaction product, transmit a message under in vivo conditions? Chem Biol Interact 1994;90 (1):35–45.
41. Huie RE, Padmaja S. The reaction of no with superoxide. Free Radic Res Commun 1993;18(4):195–199.
42. Jackson MJ, Papa S, Bolanos J, et al. Antioxidants, reactive oxygen and nitrogen species, gene induction and mitochondrial function. Mol Aspects Med 2002;23(1-3):209–285.
43. Ara J, Przedborski S, Naini AB, et al. Inactivation of tyrosine hydroxylase by nitration following exposure to peroxynitrite and 1-methyl-4-phenyl-1,2,3,6-tetrahydropyridine (MPTP). Proc Natl Acad Sci USA 1998;95(13):7659–7663.
44. Murray J, Taylor SW, Zhang B, Ghosh SS, Capaldi RA. Oxidative damage to mitochondrial complex I due to peroxynitrite: identification of reactive tyrosines by mass spectrometry. J Biol Chem 2003;278(39):37,223–37,230.
45. King PA, Jamison E, Strahs D, Anderson VE, Brenowitz M. 'Footprinting' proteins on DNA with peroxonitrous acid. Nucleic Acids Res 1993;21(10):2473–2478.
46. Pryor WA, Squadrito GL. The chemistry of peroxynitrite: a product from the reaction of nitric oxide with superoxide. Am J Physiol 1995;268(5 Pt 1):L699–L722.
47. Radi R, Beckman JS, Bush KM, Freeman BA. Peroxynitrite-induced membrane lipid peroxidation: the cytotoxic potential of superoxide and nitric oxide. Arch Biochem Biophys 1991;288(2):481–487.

48. Goldstein S, Czapski G, Lind J, Merenyi G. Tyrosine nitration by simultaneous generation of NO and O-(2) under physiological conditions. How the radicals do the job. J Biol Chem 2000;275(5):3031–3036.

49. Prutz WA, Monig H, Butler J, Land EJ. Reactions of nitrogen dioxide in aqueous model systems: oxidation of tyrosine units in peptides and proteins. Arch Biochem Biophys 1985;243(1):125–134.

50. Ando K, Beppu M, Kikugawa K, Nagai R, Horiuchi S. Membrane proteins of human erythrocytes are modified by advanced glycation end products during aging in the circulation. Biochem Biophys Res Commun 1999;258(1):123–127.

51. Pfeiffer S, Lass A, Schmidt K, Mayer B. Protein tyrosine nitration in mouse peritoneal macrophages activated in vitro and in vivo: evidence against an essential role of peroxynitrite. Faseb J 2001;15(13):2355–2364.

52. Pfeiffer S, Schmidt K, Mayer B. Dityrosine formation outcompetes tyrosine nitration at low steady-state concentrations of peroxynitrite. Implications for tyrosine modification by nitric oxide/superoxide in vivo. J Biol Chem 2000;275(9):6346–6352.

53. van der Vliet A, Jenner A, Eiserich JP, Cross CE, Halliwell B. Analysis of aromatic nitration, chlorination, and hydroxylation by gas chromatography-mass spectrometry. Methods Enzymol 1999;301:471–483.

54. Sampson JB, Ye Y, Rosen H, Beckman JS. Myeloperoxidase and horseradish peroxidase catalyze tyrosine nitration in proteins from nitrite and hydrogen peroxide. Arch Biochem Biophys 1998;356(2):207–213.

55. van der Vliet A, Eiserich JP, Halliwell B, Cross CE. Formation of reactive nitrogen species during peroxidase-catalyzed oxidation of nitrite. A potential additional mechanism of nitric oxide-dependent toxicity. J Biol Chem 1997;272(12):7617–7625.

56. Salonen JT, Tuomainen T-P, Nyyssonen K, Lakka H-M, Punnonen K. Relation between iron stores and non-insulin dependent diabetes in men: case-control study. BMJ 1998;317 (7160):727–730.

57. Ford ES, Cogswell ME. Diabetes and serum ferritin concentration among U.S. adults. Diabetes Care 1999;22(12):1978–1983.

58. Tuomainen TP, Nyyssonen K, Salonen R, et al. Body iron stores are associated with serum insulin and blood glucose concentrations. Population study in 1,013 eastern Finnish men. Diabetes Care 1997;20(3):426–428.

59. Fernandez-Real JM, Lopez-Bermejo A, Ricart W. Cross-Talk Between Iron Metabolism and Diabetes. Diabetes 2002;51(8):2348–2354.

60. Pieper GM, Siebeneich W. Diabetes-induced endothelial dysfunction is prevented by long-term treatment with the modified iron chelator, hydroxyethyl starch conjugated-deferoxamine. J Cardiovasc Pharmacol 1997;30(6):734–738.

61. Campbell A, Smith MA, Sayre LM, Bondy SC, Perry G. Mechanisms by which metals promote events connected to neurodegenerative diseases. Brain Res Bull 2001;55(2):125–132.

62. Chaudiere J, Ferrari-Iliou R. Intracellular antioxidants: from chemical to biochemical mechanisms. Food Chem Toxicol 1999;37(9-10):949–962.

63. Lander HM, Ogiste JS, Levi R, Novogrodsky A. Nitric Oxide-stimulated Guanine Nucleotide Exchange on p21. J Biol Chem 1995;270(13):7017–7020.

64. Thomas JA, Poland B, Honzatko R. Protein sulfhydryls and their role in the antioxidant function of protein S-thiolation. Arch Biochem Biophys 1995;319(1):1–9.

65. Mannick JB, Schonhoff C, Papeta N, et al. S-Nitrosylation of mitochondrial caspases. J Cell Biol 2001;154(6):1111–1116.

66. Wink DA, Mitchell JB. Chemical biology of nitric oxide: Insights into regulatory, cytotoxic, and cytoprotective mechanisms of nitric oxide. Free Radic Biol Med 1998;25 (4–5):434–456.

67. Akaike T, Inoue K, Okamoto T, et al. Nanomolar quantification and identification of various nitrosothiols by high performance liquid chromatography coupled with flow reactors of metals and Griess reagent. J Biochem 1997;122(2):459–466.

68. Benhar M, Stamler JS. A central role for S-nitrosylation in apoptosis. Nat Cell Biol 2005;7(7):645–646.

69. Hara MR, Agrawal N, Kim SF, et al. S-nitrosylated GAPDH initiates apoptotic cell death by nuclear translocation following Siah1 binding. Nat Cell Biol 2005;7(7):665–674.

70. Foster MW, Stamler JS. New insights into protein S-nitrosylation. Mitochondria as a model system. J Biol Chem 2004;279(24):25,891–25,897.

71. Gilbert HF. Molecular and cellular aspects of thiol-disulfide exchange. Adv Enzymol Relat Areas Mol Biol 1990;63:169–172.

72. Padgett CM, Whorton AR. Cellular responses to nitric oxide: role of protein S-thiolation/dethiolation. Arch Biochem Biophys 1998;358(2):232–242.

73. Park BG, Yoo CI, Kim HT, Kwon CH, Kim YK. Role of mitogen-activated protein kinases in hydrogen peroxide-induced cell death in osteoblastic cells. Toxicology 2005;215(1–2):115–125.

74. Chandra A, Srivastava S, Petrash JM, Bhatnagar A, Srivastava SK. Modification of aldose reductase by S-nitrosoglutathione. Biochemistry 1997;36(50):15,801–15,809.

75. Shi B, Triebe D, Kajiji S, Iwata KK, Bruskin A, Mahajna J. Identification and Characterization of Bax[epsilon], a Novel Bax Variant Missing the BH2 and the Transmembrane Domains. Biochemical and Biophysical Research Communications 1999;254(3):779–785.

76. Mohr S, Hallak H, de Boitte A, Lapetina EG, Brune B. Nitric Oxide-induced S-Glutathionylation and Inactivation of Glyceraldehyde-3-phosphate Dehydrogenase. J Biol Chem 1999;274(14):9427–9430.

77. Zech B, Wilm M, van Eldik R, Brune B. Mass spectrometric analysis of nitric oxide-modified caspase-3. J Biol Chem 1999;274(30):20,931–20,936.

78. Klatt P, Lamas S. c-Jun regulation by S-glutathionylation. Methods Enzymol 2002;348:157–174.

79. Turko IV, Marcondes S, Murad F. Diabetes-associated nitration of tyrosine and inactivation of succinyl-CoA:3-oxoacid CoA-transferase. Am J Physiol Heart Circ Physiol 2001;281(6):H2289–H2294.

80. Halliwell B, Whiteman M. Measuring reactive species and oxidative damage in vivo and in cell culture: how should you do it and what do the results mean? Br J Pharmacol 2004;142(2):231–255.

81. Griffiths HR, Moller L, Bartosz G, et al. Biomarkers. Mol Aspects Med 2002;23(1–3):101–208.

82. Zolg JW, Langen H. How industry is approaching the search for new diagnostic markers and biomarkers. Mol Cell Proteomics 2004;3(4):345–354.

83. Russell JW, Sullivan KA, Windebank AJ, Herrmann DN, Feldman EL. Neurons undergo apoptosis in animal and cell culture models of diabetes. Neurobiol Dis 1999;6(5):347–363.

84. Tsai EC, Hirsch IB, Brunzell JD, Chait A. Reduced plasma peroxyl radical trapping capacity and increased susceptibility of LDL to oxidation in poorly controlled IDDM. Diabetes 1994;43(8):1010–1014.

85. Altomare E, Vendemiale G, Chicco D, Procacci V, Cirelli F. Increased lipid peroxidation in type 2 poorly controlled diabetic patients. Diabete Metab 1992;18(4):264–271.

86. Zaltzberg H, Kanter Y, Aviram M, Levy Y. Increased plasma oxidizability and decreased erythrocyte and plasma antioxidative capacity in patients with NIDDM. Isr Med Assoc J 1999;1(4):228–231.

87. Sundaram RK, Bhaskar A, Vijayalingam S, Viswanathan M, Mohan R, Shanmugasundaram KR. Antioxidant status and lipid peroxidation in type II diabetes mellitus with and without complications. Clin Sci (Lond) 1996;90(4):255–260.

88. Lee AT, Cerami A. Nonenzymatic glycosylation of DNA by reducing sugars. Prog Clin Biol Res 1989;304:291–299.

89. Brownlee M. Negative consequences of glycation. Metabolism 2000;49:9–13.

90. Singh R, Barden A, Mori T, Beilin L. Advanced glycation end-products: a review. Diabetologia 2001;44:129–146.

91. Thornalley PJ. Glycation in diabetic neuropathy: characteristics, consequences, causes, and therapeutic options. Int Rev Neurobiol 2002;50:37–57.

92. Singh R, Barden A, Mori T, Beilin L. Advanced glycation end-products: a review. Diabetologia 2001;44(2):129–146.

93. Zill H, Bek S, Hofmann T, Huber J, Frank O, Lindenmeier M, et al. RAGE-mediated MAPK activation by food-derived AGE and non-AGE products. 2003 Jan 10;300:311–315. mediated MAPK activation by food-derived AGE and non-AGE products. Biochem Biophys Res Commun 2003;300:311–315.

94. Ishihara K, Tsutsumi K, Kawane S, Nakajima M, Kasaoka T. The receptor for advanced glycation end-products (RAGE) directly binds to ERK by a D-domain-like docking site. FEBS Lett 2003;550:107–113.

95. Zill H, Gunther R, Erbersdobler HF, Folsch UR, Faist V. RAGE expression and AGE-induced MAP kinase activation in Caco-2 cells. Biochem Biophys Res Commun 2001;288: 1108–1111.

96. Huang JS, Guh JY, Chen HC, Hung WC, Lai YH, Chuang LY. Role of receptor for advanced glycation end-product (RAGE) and the JAK/STAT-signaling pathway in AGE-induced collagen production in NRK-49F cells. J Cell Biochem 2001;81:102–113.

97. Wautier MP, Chappey O, Corda S, Stern DM, Schmidt AM, Wautier JL. Activation of NADPH oxidase by AGE links oxidant stress to altered gene expression via RAGE. Am J Physiol Endocrinol Metab 2001;280:E685–E694.

98. Li JH, Huang XR, Zhu HJ, Oldfield M, Cooper M, Truong LD, et al. Advanced glycation end products activate Smad signaling via TGF-beta-dependent and -independent mechanisms: implications for diabetic renal and vascular disease. FASEB J 2003;18:176–178.

99. Shaw SS, Schmidt AM, Banes AK, Wang X, Stern DM, Marrero MB. S100B-RAGE-mediated augmentation of angiotensin II-induced activation of JAK2 in vascular smooth muscle cells is dependent on PLD2. Diabetes 2003;52:2381–2388.

100. Li JH, Huang XR, Zhu HJ, et al. Advanced glycation end products activate Smad signaling via TGF-beta-dependent and independent mechanisms: implications for diabetic renal and vascular disease. FASEB J 2004;18(1):176–178.

101. Xu D, Kyriakis JM. Phosphatidyl inositol-3′OH kinase-dependent activation of renal mesangial cell Ki-Ras and ERK by advanced glycation end products. J Biol Chem 2003:39,349–39,355.

102. Shanmugam N, Kim YS, Lanting L, Natarajan R. Regulation of cyclooxygenase-2 expression in monocytes by ligation of the receptor for advanced glycation end products. J Biol Chem 2003;278:34,834–34,844.

103. Li J, Schmidt AM. Characterization and functional analysis of the promoter of RAGE, the receptor for advanced glycation end products. J Biol Chem 1997;272:16,498–16,506.

104. Lee AY, Chung SS. Contributions of polyol pathway to oxidative stress in diabetic cataract. FASEB J 1999;13(1):23–30.

105. Lee AY, Chung SK, Chung SS. Demonstration that polyol accumulation is responsible for diabetic cataract by the use of transgenic mice expressing the aldose reductase gene in the lens. Proc Natl Acad Sci USA 1995;92(7):2780–2784.

106. Gopalakrishna R, Jaken S. Protein kinase C signaling and oxidative stress. Free Radic Biol Med 2000;28(9):1349–1361.

107. Gopalakrishna R, Chen ZH, Gundimeda U. Modifications of cysteine-rich regions in protein kinase C induced by oxidant tumor promoters and enzyme-specific inhibitors. Methods Enzymol 1995;252:132–146.

108. Boscoboinik D, Szewczyk A, Hensey C, Azzi A. Inhibition of cell proliferation by alpha-tocopherol. Role of protein kinase C. J Biol Chem 1991;266(10):6188–6194.

109. Gopalakrishna R, Gundimeda U, Chen Z-H. Cancer-Preventive Selenocompounds Induce a Specific Redox Modification of Cysteine-Rich Regions in Ca2+- Dependent Isoenzymes of Protein Kinase C. Arch Biochem Biophys 1997;348(1):25–36.

110. Lee T-S, Saltsman KA, Ohashi H, King GL. Activation of Protein Kinase C by Elevation of Glucose Concentration: Proposal for a Mechanism in the Development of Diabetic Vascular Complications. PNAS 1989;86(13):5141–5145.

111. Tomlinson DR. Mitogen-activated protein kinases as glucose transducers for diabetic complications. Diabetologia 1999;42(11):1271–1281.

112. Cameron NE, Cotter MA. Effects of protein kinase Cbeta inhibition on neurovascular dysfunction in diabetic rats: interaction with oxidative stress and essential fatty acid dysmetabolism. Diabetes Metab Res Rev 2002;18(4):315–323.

113. Ishii H, Koya D, King GL. Protein kinase C activation and its role in the development of vascular complications in diabetes mellitus. J Mol Med 1998;76(1):21–31.

114. Srivastava AK. High glucose-induced activation of protein kinase signaling pathways in vascular smooth muscle cells: a potential role in the pathogenesis of vascular dysfunction in diabetes (review). Int J Mol Med 2002;9(1):85–89.

115. Knebel A, Rahmsdorf HJ, Ullrich A, Herrlich P. Dephosphorylation of receptor tyrosine kinases as target of regulation by radiation, oxidants or alkylating agents. EMBO J 1996;15(19):5314–5325.

116. Sachsenmaier C, Radler-Pohl A, Zinck R, Nordheim A, Herrlich P, Rahmsdorf HJ. Involvement of growth factor receptors in the mammalian UVC response. Cell 1994;78(6): 963–972.

117. Guyton KZ, Liu Y, Gorospe M, Xu Q, Holbrook NJ. Activation of mitogen-activated protein kinase by H2O2. Role in cell survival following oxidant injury. J Biol Chem 1996;271 (8):4138–4142.

118. Iordanov MS, Magun BE. Different mechanisms of c-Jun NH(2)-terminal kinase-1 (JNK1) activation by ultraviolet-B radiation and by oxidative stressors. J Biol Chem 1999;274(36): 25,801–25,806.

119. Verheij M, Bose R, Lin XH, et al. Requirement for ceramide-initiated SAPK/JNK signalling in stress-induced apoptosis. Nature 1996;380(6569):75–79.

120. Cheng A, Chan SL, Milhavet O, Wang S, Mattson MP. p38 MAP kinase mediates nitric oxide-induced apoptosis of neural progenitor cells. J Biol Chem 2001;276(46):43,320–43,327.

121. Vincent AM GC, Brownlee M, Russell JW. Glucose induced neuronal programmed cell death is regulated by manganese superoxide dismutase and uncoupling protein-1. Endocr Soc Abstr Pt 289:210.

122. Lowell BB, Shulman GI. Mitochondrial dysfunction and type 2 diabetes. Science 2005;307 (5708):384–347.

123. Ghafourifar P, Richter C. Nitric oxide synthase activity in mitochondria. FEBS Lett 1997;418(3):291–296.

124. Coll KE, Joseph SK, Corkey BE, Williamson JR. Determination of the matrix free Ca2+ concentration and kinetics of Ca2+ efflux in liver and heart mitochondria. J Biol Chem 1982;257(15):8696–8704.

125. Carafoli E. Intracellular calcium homeostasis. Annu Rev Biochem 1987;56:395-433.

126. Fridovich I. Superoxide Radical and Superoxide Dismutases. Ann Rev Biochem 1995;64 (1):97–112.

127. Poderoso JJ, Lisdero C, Schopfer F, et al. The regulation of mitochondrial oxygen uptake by redox reactions involving nitric oxide and ubiquinol. J Biol Chem 1999;274(53): 37,709–37,716.

128. Ghafourifar P, Schenk U, Klein SD, Richter C. Mitochondrial nitric-oxide synthase stimulation causes cytochrome c release from isolated mitochondria. Evidence for intramitochondrial peroxynitrite formation. J Biol Chem 1999;274(44):31,185–31,188.

129. Boveris A, Alvarez S, Navarro A. The role of mitochondrial nitric oxide synthase in inflammation and septic shock. Free Radic Biol Med 2002;33(9):1186–1193.

130. Marcondes S, Turko IV, Murad F. Nitration of succinyl-CoA:3-oxoacid CoA-transferase in rats after endotoxin administration. Proc Natl Acad Sci USA 2001;98(13):7146–7151.

131. Aulak KS, Koeck T, Crabb JW, Stuehr DJ. Dynamics of protein nitration in cells and mitochondria. Am J Physiol Heart Circ Physiol 2004;286(1):H30–H38.

132. Clementi E, Brown GC, Feelisch M, Moncada S. Persistent inhibition of cell respiration by nitric oxide: crucial role of S-nitrosylation of mitochondrial complex I and protective action of glutathione. Proc Natl Acad Sci USA 1998;95(13):7631–7636.

133. Turko IV, Li L, Aulak KS, Stuehr DJ, Chang JY, Murad F. Protein tyrosine nitration in the mitochondria from diabetic mouse heart. Implications to dysfunctional mitochondria in diabetes. J Biol Chem 2003;278(36):33,972–33,977.

134. Ceriello A, dello Russo P, Amstad P, Cerutti P. High glucose induces antioxidant enzymes in human endothelial cells in culture. Evidence linking hyperglycemia and oxidative stress. Diabetes 1996;45(4):471–477.

135. Ceriello A, Bortolotti N, Falleti E, et al. Total radical-trapping antioxidant parameter in NIDDM patients. Diabetes Care 1997;20(2):194–197.

136. Hess DT, Patterson SI, Smith DS, Skene JHP. Neuronal growth cone collapse and inhibition of protein fatty acylation by nitric oxide. Nature 1993;366(6455):562–565.

137. Zochodne DW, Misra M, Cheng C, Sun H. Inhibition of nitric oxide synthase enhances peripheral nerve regeneration in mice. Neuroscience Lett 1997;228(2):71–74.

138. Cheng C, Zochodne DW. Sensory Neurons With Activated Caspase-3 Survive Long-Term Experimental Diabetes. Diabetes 2003;52(9):2363–2371.

139. Zochodne DW, Levy D. Nitric oxide in damage, disease and repair of the peripheral nervous system. Cell Mol Biol (Noisy-le-grand) 2005;51(3):255–267.

140. Kennedy JM, Zochodne DW. Influence of Experimental Diabetes on the Microcirculation of Injured Peripheral Nerve: Functional and Morphological Aspects. Diabetes 2002;51(7):2233–2240.

141. Sima AA. Peripheral neuropathy in the spontaneously diabetic BB-Wistar-rat. An ultrastructural study. Acta Neuropathol (Berl) 1980;51(3):223–227.

142. Powell H, Knox D, Lee S, et al. Alloxan diabetic neuropathy: electron microscopic studies. Neurology 1977;27(1):60–66.

143. Carson KA, Bossen EH, Hanker JS. Peripheral neuropathy in mouse hereditary diabetes mellitus. II. Ultrastructural correlates of degenerative and regenerative changes. Neuropathol Appl Neurobiol 1980;6(5):361–374.

144. Kniel PC, Junker U, Perrin IV, Bestetti GE, Rossi GL. Varied effects of experimental diabetes on the autonomic nervous system of the rat. Lab Invest 1986;54(5):523–530.

145. Vincent AM, Brownlee M, Russell JW. Oxidative stress and programmed cell death in diabetic neuropathy. Ann N Y Acad Sci 2002;959:368–383.

146. Matthews CC, Feldman EL. Insulin-like growth factor I rescues SH-SY5Y human neuroblastoma cells from hyperosmotic induced programmed cell death. J Cell Physiol 1996;166(2):323–331.

147. Russell JW, Golovoy D, Vincent AM, Mahendru P, Olzmann JA, Mentzer A, et al. High glucose induced oxidative stress and mitochondrial dysfunction in neurons. FASEB J 2002;16:1738–1748.

148. Layton BE, Sastry AM, Lastoskie CM, et al. In situ imaging of mitochondrial outer-membrane pores using atomic force microscopy. Biotechniques 2004;37(4):564–570, 72–73.

149. Feldman EL, Russel JW, Sullivan KA, Golovoy D. New insights into the pathogenesis of diabetic neuropathy. Curr Opin Neurol 1999;12:553–563.

150. Greene DA, Stevens MJ, Obrosova I, Feldman EL. Glucose-induced oxidative stress and programmed cell death in diabetic neuropathy. Eur J Pharmacol 1999;375:217–223.

151. Russell JW, Golovoy D, Vincent AM, et al. High glucose-induced oxidative stress and mitochondrial dysfunction in neurons. FASEB J 2002;16(13):1738–1748.

152. Leinninger GM, Russell JW, van Golen CM, Berent A, Feldman EL. Insulin-like growth factor-I regulates glucose-induced mitochondrial depolarization and apoptosis in human neuroblastoma. Cell Death Differ 2004;11(8):885–896.

153. Leinninger GM, Backus C, Uhler MD, Lentz SI, Feldman EL. Phosphatidylinositol 3-kinase and Akt effectors mediate insulin-like growth factor-I neuroprotection in dorsal root ganglia neurons. FASEB J 2004;118:1544–1546.

154. Cameron NE, Eaton SE, Cotter MA, Tesfaye S. Vascular factors and metabolic interactions in the pathogenesis of diabetic neuropathy. Diabetologia 2001;44:1973–1988.

155. Greene DA, Sima AA, Stevens MJ, Feldman EL, Lattimer SA. Complications: neuropathy, pathogenetic considerations. Diabetes Care 1992;15(12):1902–1925.

156. Chugh SN, Dhawan R, Kishore K, Sharma A, Chugh K. Glibenclamide vs gliclazide in reducing oxidative stress in patients of noninsulin dependent diabetes mellitus—a double blind randomized study. J Assoc Physicians India 2001;49:803–807.

157. Hamada Y, Nakamura J. Clinical potential of aldose reductase inhibitors in diabetic neuropathy. Treat Endocrinol 2004;3(4):245–255.

158. Chung SS, Chung SK. Aldose reductase in diabetic microvascular complications. Curr Drug Targets 2005;6(4):475–486.

159. Gonzalez AM, Sochor M, McLean P. The effect of an aldose reductase inhibitor (Sorbinil) on the level of metabolites in lenses of diabetic rats. Diabetes 1983;32(5):482–485.

160. Obrosova IG, Van Huysen C, Fathallah L, Cao XC, Greene DA, Stevens MJ. An aldose reductase inhibitor reverses early diabetes-induced changes in peripheral nerve function, metabolism, and antioxidative defense. FASEB J 2002;16(1):123–125.

161. Cameron NE, Cotter MA, Basso M, Hohman TC. Comparison of the effects of inhibitors of aldose reductase and sorbitol dehydrogenase on neurovascular function, nerve conduction and tissue polyol pathway metabolites in streptozotocin-diabetic rats. Diabetologia 1997;40(3):271–281.

162. Yamamoto T, Takakura S, Kawamura I, Seki J, Goto T. The effects of zenarestat, an aldose reductase inhibitor, on minimal F-wave latency and nerve blood flow in streptozotocin-induced diabetic rats. Life Sciences 2001;68(12):1439–1448.

163. Shimoshige Y, Ikuma K, Yamamoto T, et al. The effects of zenarestat, an aldose reductase inhibitor, on peripheral neuropathy in Zucker diabetic fatty rats. Metabolism 2000;49(11):1395–1399.

164. Greene DA, Arezzo JC, Brown MB. Effect of aldose reductase inhibition on nerve conduction and morphometry in diabetic neuropathy. Neurology 1999;53(3):580–591.

165. Yuen EC, Howe CL, Li Y, Holtzman DM, Mobley WC. Nerve growth factor and the neurotrophic factor hypothesis. Brain Dev 1996;18(5):362–368.

166. Leinninger GM, Vincent AM, Feldman EL. The role of growth factors in diabetic peripheral neuropathy. J Peripher Nerv Syst 2004;9(1):26–53.

167. Cruz-Aguado R, Turner LF, Diaz CM, Pinero J. Nerve growth factor and striatal glutathione metabolism in a rat model of Huntington's disease. Restor Neurol Neurosci 2000;17(4):217–221.

168. Apfel SC. Nerve growth factor for the treatment of diabetic neuropathy: what went wrong, what went right, and what does the future hold? Int Rev Neurobiol 2002;50:393–413.

169. Calcutt NA. Future treatments for diabetic neuropathy: clues from experimental neuropathy. Curr Diab Rep 2002;2(6):482–488.

170. Paolisso G, D'Amore A, Balbi V, et al. Plasma vitamin C affects glucose homeostasis in healthy subjects and in non-insulin-dependent diabetics. Am J Physiol 1994;266(2 Pt 1):E261–E268.

171. Jacob S, Henriksen EJ, Schiemann AL, et al. Enhancement of glucose disposal in patients with type 2 diabetes by alpha-lipoic acid. Arzneimittelforschung 1995;45(8):872–874.

172. Natarajan Sulochana K, Lakshmi S, Punitham R, Arokiasamy T, Sukumar B, Ramakrishnan S. Effect of oral supplementation of free amino acids in type 2 diabetic patients—a pilot clinical trial. Med Sci Monit 2002;8(3):CR131–CR137.

173. Reed LJ. The chemistry and function of lipoic acid. Adv Enzymol Relat Subj Biochem 1957;18:319–347.

174. Packer L, Kraemer K, Rimbach G. Molecular aspects of lipoic acid in the prevention of diabetes complications. Nutrition 2001;17(10):888–895.

175. Gregus Z, Stein AF, Varga F, Klaassen CD. Effect of lipoic acid on biliary excretion of glutathione and metals. Toxicol Appl Pharmacol 1992;114(1):88–96.

176. Kozlov AV, Gille L, Staniek K, Nohl H. Dihydrolipoic acid maintains ubiquinone in the antioxidant active form by two-electron reduction of ubiquinone and one-electron reduction of ubisemiquinone. Arch Biochem Biophys 1999;363(1):148–154.

177. Cameron NE, Cotter MA, Horrobin DH, Tritschler HJ. Effects of alpha-lipoic acid on neurovascular function in diabetic rats: interaction with essential fatty acids. Diabetologia 1998;41(4):390–399.

178. Stevens MJ, Obrosova I, Cao X, Van Huysen C, Greene DA. Effects of DL-alpha-lipoic acid on peripheral nerve conduction, blood flow, energy metabolism, and oxidative stress in experimental diabetic neuropathy. Diabetes 2000;49(6):1006–1015.

179. Coppey LJ, Gellett JS, Davidson EP, Dunlap JA, Lund DD, Yorek MA. Effect of antioxidant treatment of streptozotocin-induced diabetic rats on endoneurial blood flow, motor nerve conduction velocity, and vascular reactivity of epineurial arterioles of the sciatic nerve. Diabetes 2001;50(8):1927–1937.

180. Obrosova I, Cao X, Greene DA, Stevens MJ. Diabetes-induced changes in lens antioxidant status, glucose utilization and energy metabolism: effect of DL-alpha-lipoic acid. Diabetologia 1998;41(12):1442–1450.

181. Dincer Y, Telci A, Kayali R, Yilmaz IA, Cakatay U, Akcay T. Effect of alpha-lipoic acid on lipid peroxidation and anti-oxidant enzyme activities in diabetic rats. Clin Exp Pharmacol Physiol 2002;29(4):281–284.

182. Maritim AC, Sanders RA, Watkins JB, 3rd. Effects of alpha-lipoic acid on biomarkers of oxidative stress in streptozotocin-induced diabetic rats. J Nutr Biochem 2003;14(5):288–294.

183. Borcea V, Nourooz-Zadeh J, Wolff SP, et al. alpha-Lipoic acid decreases oxidative stress even in diabetic patients with poor glycemic control and albuminuria. Free Radic Biol Med 1999;26(11-12):1495–1500.

184. Midaoui AE, Elimadi A, Wu L, Haddad PS, de Champlain J. Lipoic acid prevents hypertension, hyperglycemia, and the increase in heart mitochondrial superoxide production. Am J Hypertens 2003;16(3):173–179.

185. Ziegler D, Hanefeld M, Ruhnau KJ, et al. Treatment of symptomatic diabetic polyneuropathy with the antioxidant alpha-lipoic acid: a 7-month multicenter randomized controlled trial (ALADIN III Study). ALADIN III Study Group. Alpha-Lipoic Acid in Diabetic Neuropathy. Diabetes Care 1999;22(8):1296–1301.

186. Ametov AS, Barinov A, Dyck PJ, et al. The Sensory Symptoms of Diabetic Polyneuropathy Are Improved With α-Lipoic Acid: The SYDNEY Trial. Diabetes Care 2003;26(3):770–776.

187. Tutuncu NB, Bayraktar M, Varli K. Reversal of defective nerve conduction with vitamin E supplementation in type 2 diabetes: a preliminary study. Diabetes Care 1998;21(11):1915–1918.

188. Celik S, Baydas G, Yilmaz O. Influence of vitamin E on the levels of fatty acids and MDA in some tissues of diabetic rats. Cell Biochem Funct 2002;20(1):67–71.

189. Li D, Devaraj S, Fuller C, Bucala R, Jialal I. Effect of alpha-tocopherol on LDL oxidation and glycation: in vitro and in vivo studies. J Lipid Res 1996;37(9):1978–1986.

190. Rosen P, Ballhausen T, Bloch W, Addicks K. Endothelial relaxation is disturbed by oxidative stress in the diabetic rat heart: influence of tocopherol as antioxidant. Diabetologia 1995;38(10):1157–1168.

191. Christen S, Woodall AA, Shigenaga MK, Southwell-Keely PT, Duncan MW, Ames BN. gamma-tocopherol traps mutagenic electrophiles such as NO(X) and complements alpha-tocopherol: physiological implications. Proc Natl Acad Sci USA 1997;94(7):3217–3222.

192. Beckman JA, Goldfine AB, Gordon MB, Garrett LA, Keaney JF, Jr., Creager MA. Oral antioxidant therapy improves endothelial function in Type 1 but not Type 2 diabetes mellitus. Am J Physiol Heart Circ Physiol 2003;285(6):H2392–H2398.

193. Peponis V, Papathanasiou M, Kapranou A, et al. Protective role of oral antioxidant supplementation in ocular surface of diabetic patients. Br J Ophthalmol 2002;86(12):1369–1373.

194. Frei B, Stocker R, England L, Ames BN. Ascorbate: the most effective antioxidant in human blood plasma. Adv Exp Med Biol 1990;264:155–163.

195. Paolisso G, Balbi V, Volpe C, et al. Metabolic benefits deriving from chronic vitamin C supplementation in aged non-insulin dependent diabetics. J Am Coll Nutr 1995;14(4):387–392.

196. Opara EC. Oxidative stress, micronutrients, diabetes mellitus and its complications. J R Soc Health 2002;122(1):28–34.

197. Meister A. Glutathione deficiency produced by inhibition of its synthesis, and its reversal; applications in research and therapy. Pharmacol Ther 1991;51(2):155–194.

INDEX